ELEMENTARY
Algebra

FIFTH ▶ EDITION

Charles P. McKeague
Cuesta College

 SAUNDERS COLLEGE PUBLISHING
Harcourt Brace College Publishers

Fort Worth   Philadelphia   San Diego   New York

Orlando   Austin   San Antonio   Toronto

Montreal   London   Sydney   Tokyo

Text Typeface: Times Roman
Compositor: Progressive Information Technologies
Acquisitions Editor: Deirdre Lynch
Developmental Editor: Marc Sherman
Managing Editor: Carol Field
Project Editor: Sarah Fitz-Hugh
Copy Editor: Martha Brown
Manager of Art and Design: Carol Bleistine
Art Director: Christine Schueler
Art Assistant: Sue Kinney
Text Designer: Gene Harris
Cover Designer: Louis Fuiano/Fuiano Art & Design
Text Artwork: Techsetters Inc./Rolin Graphics
Director of EDP: Tim Frelick
Production Manager: Carol Florence

Cover Credit: © Shinobu HIRAI/Photonica

Printed in the United States of America

ISBN: 0-03-097356-2 (Student's Edition)
       0-03-010858-6 (Instructor's Edition)

ELEMENTARY ALGEBRA, Fifth Edition

Library of Congress Catalog Card Number: 94-69194

67890123 032 10 9876543

CONTENTS

PREFACE TO THE
INSTRUCTOR

This fifth edition of *Elementary Algebra* retains the same basic format and style as the fourth edition. It is designed for use in a lecture-format class with students who have no prior experience in algebra.

Features of the Book

Chapter Openers New to this edition, each chapter opens with the following three elements:

1. *Introduction* Each chapter opens with an introduction in which a real-world application, historical example, or a link between topics is used to stimulate interest in the chapter. Whenever possible, these introductions are expanded on later in the chapter and then carried through to topics found further on in the book.
2. *Overview* A general overview of the chapter follows the chapter introduction. The overview lists the important topics that will be covered in the chapter, along with their connection to one another and to topics covered previously in the text. Most of the overviews end with a list of topics from previous chapters that students need to know in order to be successful in the current chapter.
3. *Study Skills* Found in the first six chapter openers is a list of study skills intended to help students become organized and efficient with their time. The study skills point students in the direction of success. They are intended to benefit students in this course and throughout their college careers. These skills are more detailed than the general study skills listed in the Preface to the Student.

Organization of the Problem Sets Five main ideas are incorporated into each problem set.

1. *Drill* There are enough problems in each set to ensure student proficiency in the material.

2. *Progressive Difficulty*   The problems increase in difficulty as the problem set progresses.
3. *Odd-Even Similarities*   Each pair of consecutive problems is similar. Since the answers to the odd-numbered problems are listed in the back of the book, the similarity of the odd-even pairs of problems allows your students to check their work on an odd-numbered problem and then to try the similar even-numbered problem.
4. *Application Problems*   Students are always curious about how the algebra they are learning can be applied, but at the same time many students are apprehensive about attempting application problems. I have found that they are more likely to put some time and effort into trying application problems if they do not have to work an overwhelming number of them at one time and if they work on them every day. For these reasons, I have placed a few application problems toward the end of almost every problem set in the book.
5. *Review Problems*   As was the case in the fourth edition, each problem set, beginning with Chapter 2, contains a few review problems. Where appropriate, the review problems cover material that will be needed in the next section. Otherwise they cover material from the previous chapter. That is, the review problems in Chapter 5 cover the important points covered in Chapter 4. Likewise, the review problems in Chapter 6 review the important material from Chapter 5. If you give tests on two chapters at a time, you will find this to be a time-saving feature. Your students will review one chapter as they study the next chapter.

Blueprint for Problem Solving   New to this edition, the Blueprint for Problem Solving is a detailed outline of the steps needed to successfully attempt application problems. Intended as a guide to problem solving in general, the blueprint overlays the solution process to all the application problems in the first few chapters of the book. As students become more familiar with problem solving, the steps in the blueprint are streamlined.

Research Projects   Scattered throughout this edition are problems for students to research and then report on. Although they appear at the end of a number of the problem sets, they are not intended to be part of the student's daily assignment. In my classes, these are the problems I use for extra credit. In most cases, I require students to type their reports, just as they would type an essay in their English classes. Do not be concerned if you are not familiar with the topics shown in the research projects; the idea behind these projects is to have your students do the research, and then tell you what they have learned.

Chapter Summaries   Each chapter summary lists the new properties and definitions found in the chapter. The margins in the chapter summaries contain examples that illustrate the topics being reviewed.

Chapter Reviews   Each chapter ends with a set of review problems that cover all the different types of problems found in the chapter. The chapter reviews are longer and more extensive than the chapter tests.

Chapter Tests   Each chapter test contains a representative sample of the problems covered in the chapter.

Changes in the Fifth Edition

In addition to the chapter openers, research projects, and Blueprint for Problem Solving mentioned above, the followings items are also new to this edition.

Increased Visualization of Topics   This edition contains many more diagrams, charts, and graphs than the previous edition. Their purpose is to give students additional information, in visual form, to help them understand the topics we cover.

Facts from Geometry   Many of the important facts from geometry are now listed under this heading. In most cases, an example or two accompanies each of the facts to give students a chance to see how topics from geometry are related to the algebra they are learning.

Number Sequences   An introductory coverage of number sequences is integrated throughout Chapter 1 and then expanded and continued in Chapters 2 and 3. I find that there are many interesting topics I can cover if students have some experience with number sequences. It is also the easiest way to demonstrate inductive reasoning.

Tables, Histograms, Scatter Diagrams, and Line Graphs   Beginning in Chapter 1 and then continued in Chapters 3 and 4, students are required to analyze information from tables. In addition to simply reading a table, they are given practice converting data in tabular form to data in graphical form that includes histograms and scatter diagrams. (Why not? Histograms, scatter diagrams, and line graphs are what they see if they read a newspaper or magazine.) From there they move on to graph ordered pairs and linear equations on a rectangular coordinate system. Section 3.1 has been rewritten completely to accommodate this new feature.

Unit Analysis   Chapter 6 now contains problems requiring students to convert from one unit of measure to another. The method used to accomplish the conversions is the method they will use if they take a chemistry class. Since this method is similar to the method we use to multiply rational expressions, unit analysis is covered in Section 6.2 with multiplication and division of rational expressions.

Challenging Problems and Application Problems   More challenging problems have been added to many of the problem sets. Some of the new problems are in the drill problem category, while many others are new, more realistic, application problems.

Supplements to the Textbook

This fifth edition of *Elementary Algebra* is accompanied by a number of useful supplements.

For the Instructor

- **Instructor's Edition** An instructor version of the text is available to facilitate accessible and effective teaching. All answers to the problem sets are provided.
- **Instructor's Solution Manual** The manual supplies instructor-appropriate solutions for half the odd and all the even exercises in the problem sets. Solutions for all other exercises comprise the Student's Solutions Manual.
- **Printed Test Bank and Prepared Tests** The test bank consists of multiple-choice test items organized by chapter, section, and objective. The prepared tests comprise 10 sets of ready-to-copy tests: one set for each chapter and one set for the entire book. Each set comprises two multiple-choice and four show-your-work tests. Items for half of the tests are ordered according to the sequence of topics in the book; items for the other half of the tests are in mixed-up order. Answers for every test item are provided.
- *ExaMaster +*™ **Computerized Test Bank** A flexible, powerful, computerized testing system, *ExaMaster +*™ offers teachers a wide range of integrated testing options and features. Available in IBM, Macintosh, or Windows format, it offers teachers the ability to select, edit, or create not only test items but algorithms for test items as well. Teachers can tailor tests according to a variety of criteria, scramble the order of test items, and administer tests on-line. *ExaMaster +*™ also includes full-function gradebook and graphing features.

For the Student

- **Videotape Package** Free to adopters, the videotape package consists of 9 VHS videotapes, one for each chapter of the book. Each chapter tape is an hour to an hour and a half in length and is divided into lessons that correspond to each section of the chapter.
- **Student's Solutions Manual** This manual contains complete annotated solutions to every other odd problem in problem sets and all chapter review and chapter test exercises.
- **MathCue Tutorial** This computer software package of tutorials has problems that correspond to every section in the series. The software presents problems to solve and tutors students by displaying annotated, step-by-step solutions. Students may view partial solutions to get started on a problem, see continuous record of progress, and back up to review missed problems. Student scores can also be printed. Available for IBM and Macintosh.
- **MathCue Solution Finder** This software allows students to enter their own problems into the computer and get annotated, step-by-step solutions in return. This unique program simulates working with a tutor, tracks student progress, refers students to specific sections in the text when appropriate, and prints student scores. Available for IBM and Macintosh.

- **MathCue Practice**   This algorithm-based software allows students to generate large numbers of practice problems keyed to problem types from each section of the book. Practice scores students' performance, and saves students' scores session to session. Available for IBM and Macintosh.

Acknowledgments

A project of this size cannot be completed without help from many people. In particular, Deirdre Lynch, my editor at Saunders College Publishing, contributed a number of helpful suggestions on the content of this revision. She has been encouraging throughout the process and is a pleasure to work with. Kate Pawlik and Martha Brown coordinated the copy editing, accuracy checking, and proofreading for the book. Their attention to detail and ability to get work done on time are unmatched. Marc Sherman, the developmental editor on this project, has done an exceptional job of suggesting ways in which to incorporate new material into the book. He and project editor Sarah Fitz-Hugh held the project together so that the book was published on time. My son Patrick and my daughter Amy assisted me with this revision from the beginning to the end. I am pleased with the way the book has turned out and much of what I like is due to their influence. My thanks to these seven people; this book would not have been possible without them.

Thanks also to Ann Ostberg of Central Community College for her help with problem checking, to Christine Schueler for the design of the book and cover, to Stacey Lloyd for her word-processing skills, and to my wife Diane for continuing to encourage my writing endeavors.

Finally, I am grateful to the following instructors for their suggestions and comments on this revision. Some reviewed the entire manuscript, while others were asked to evaluate the development of specific topics or the overall sequence of topics. My thanks go to the people listed below:

Betty DeGryse Truitt, Black Hawk College
Michael Hamm, Brookhaven College
Mary Henderson, Okaloosa-Walton Community College
Bonnie Hodge, Austin Peay State University
Linda Horner, Broward Community College
Carol Le Guennec, Solano Community College
Caren McClure, Rancho Santiago College
Allan Newhart, West Virginia University at Parkersburg
David Otts, Middle Tennessee State University
Larry Ozanich, Yakima Valley Community College
Jerry Shipman, Alabama A&M
Ken Wagman, Gavilan College
Matrid Whidden, Edison Community College

Charles P. McKeague

PREFACE TO THE STUDENT

I often find my students asking themselves the question "Why can't I understand this stuff the first time?" The answer is "You're not expected to." Learning a topic in algebra isn't always accomplished the first time around. There are many instances when you will find yourself reading over new material a number of times before you can begin to work problems. That's just the way things are in algebra. If you don't understand a topic the first time you see it, that doesn't mean there is something wrong with you. Understanding algebra takes time. The process of understanding requires reading the book, studying the examples, working problems, and getting your questions answered.

Here are some questions that are often asked by students starting a beginning algebra class.

How much math do I need to know before taking algebra? You should be able to do the four basic operations (addition, subtraction, multiplication, and division) with whole numbers, fractions, and decimals. Most important is your ability to work with whole numbers. If you are a bit weak at working with fractions because you haven't worked with them in a while, don't be too concerned; we will review fractions as we progress through the book. I have had students who eventually did very well in algebra, even though they were initially unsure of themselves when working with fractions.

What is the best way to study? The best way to study is to study consistently. You must work problems every day. A number of my students spend an hour or so in the morning working problems and reading over new material and then spend another hour in the evening working problems. The students of mine who are most successful in algebra are the ones who find a schedule that works for them and then stick to it. They work problems every day.

If I understand everything that goes on in class, can I take it easy on my homework? Not necessarily. There is a big difference between understanding a problem someone else is working and working the same problem yourself. There is no substitute for working problems yourself. The concepts and properties are understandable to you only if you yourself work problems involving them.

How to Be Successful in Algebra

If you have decided to be successful in algebra, then the following list will be important to you.

1. **Attend all class sessions on time**   You cannot know exactly what goes on in class unless you are there. Missing class and then expecting to find out what went on from someone else is not the same as being there yourself.

2. **Read the book**   It is best to read the section that will be covered in class beforehand. Reading in advance, even if you do not understand everything you read, is still better than going to class with no idea of what will be discussed.

3. **Work problems every day and check your answers**   The key to success in mathematics is working problems. The more problems you work, the better you will become at working them. The answers to the odd-numbered problems are given in the back of the book. When you have finished an assignment, be sure to compare your answers with those in the book. If you have made a mistake, find out what it is, and correct it.

4. **Do it on your own**   Don't be misled into thinking someone else's work is your own. Having someone else show you how to work a problem is not the same as working the same problem yourself. It is okay to get help when you are stuck. As a matter of fact, it is a good idea. Just be sure you do the work yourself.

5. **Review every day**   After you have finished the problems your instructor has assigned, take another fifteen minutes and review a section you have already completed. The more you review, the longer you will retain the material you have learned.

6. **Don't expect to understand every new topic the first time you see it**   Sometimes you will understand everything you are doing, and sometimes you won't. That's just the way things are in mathematics. Expecting to understand each new topic the first time you see it can lead to disappointment and frustration. The process of understanding algebra takes time. It requires that you read the book, work problems, and get your questions answered.

7. **Spend as much time as it takes for you to master the material**   No set formula exists for the exact amount of time you need to spend on algebra to master it. You will find out as you go along what is or isn't enough time for you. If you end up spending two or more hours on each section in order to master the material there, then that's how much time it takes; trying to get by with less will not work.

8. **Relax**   It's probably not as difficult as you think.

1

THE BASICS

INTRODUCTION

Much of what we do in mathematics is concerned with recognizing patterns and classifying together groups of numbers that share a common characteristic. Two groups of numbers that we will study in this chapter are shown below. Each of the groups forms a pattern. As you will see, the characteristic shared by the numbers in each of these sequences can be described with algebra.

$$\text{Sequence of odd numbers} = 1, 3, 5, 7, 9, \ldots$$
$$\text{Sequence of squares} = 1, 4, 9, 16, 25, \ldots$$

Once we have classified groups of numbers as to the characteristics they share, we sometimes discover that a relationship exists between the groups. Although it may not be obvious at first, there is a relationship that exists *between* the two sequences shown above. The relationship has been known for some time. In *The Book of Squares,* written by the mathematician known as Fibonacci in 1225, Fibonacci begins the introduction this way:

> I thought about the origin of all square numbers and discovered that they arise out of the increasing sequence of odd numbers.

The relationship that Fibonacci refers to (and that you will discover later in this chapter) can be described very concisely with algebra.

OVERVIEW

Chapter 1 contains some of the most important material in the book. Your goal is to master it. Your success in the following chapters depends upon how well you understand Chapter 1. Here is a list, in order of importance, of the ideas you must know after you complete this chapter.

1. You *must* know how to add, subtract, multiply, and divide positive and negative numbers.
2. You *must* understand and recognize when the commutative, associative, and distributive properties are being used.

3. You should know the major classifications of numbers. That is, you should know the difference between whole numbers, integers, rational numbers, and real numbers.

STUDY SKILLS

Some of the students enrolled in my beginning algebra classes develop difficulties early in the course. Their difficulties are not associated with their ability to learn mathematics; they all have the potential to pass the course. Students who get off to a poor start do so because they have not developed the study skills necessary to be successful in algebra; they do not put themselves on an effective homework schedule, and when they work problems, they do it their way, not my way. Here is a list of things you can do to begin to develop effective study skills.

1. **Put Yourself on a Schedule** The general rule is that you spend two hours on homework for every hour you are in class. Make a schedule for yourself in which you set aside two hours each day to work on algebra. Once you make the schedule, stick to it. Don't just complete your assignments and stop. Use all the time you have set aside. If you complete an assignment and have time left over, read the next section in the book, and then work more problems.

2. **Find Your Mistakes and Correct Them** There is more to studying algebra than just working problems. You must always check your answers with the answers in the back of the book. When you have made a mistake, find out what it is, and correct it. Making mistakes is part of the process of learning mathematics. In the prologue to *The Book of Squares,* Fibonacci has this to say about the content of his book:

> I have come to request indulgence if in any place it contains something more or less than right or necessary; for to remember everything and be mistaken in nothing is divine rather than human . . .

Fibonacci knew, as you know, that human beings make mistakes. You cannot learn algebra without making mistakes.

3. **Imitate Success** Your work should look like the work you see in this book and the work your instructor shows. The steps shown in solving problems in this book were written by someone who has been successful in mathematics. The same is true of your instructor. Your work should imitate the work of people who have been successful in mathematics.

SECTION

1.1   Notation and Symbols

Since much of what we do in algebra involves comparison of quantities, we will begin by listing some symbols used to compare mathematical quantities. The comparison symbols fall into two major groups: equality symbols and inequality symbols. Before going further, we need to say a few words about variables.

Variables: An Intuitive Look

When you filled out the application for the school you are attending, there was a space to fill in your first name. "First Name" is a variable quantity, because the value it takes on depends on who is filling out the application. For example, if your first name is Manuel, then the value of "First Name" is Manuel. On the other hand, if your first name is Christa, then the value of "First Name" is Christa.

If we denote "First Name" as *FN*, "Last Name" as *LN*, and "Whole Name" as *WN*, then we take the concept of a variable further and write the relationship between the names this way:

$$FN + LN = WN$$

(We use the + symbol loosely here to represent writing the names together with a space between them.) This relationship we have written holds for all people who have only a first name and a last name. For those people who have a middle name, the relationship between the names is

$$FN + MN + LN = WN$$

A similar situation exists in algebra when we let a letter stand for a number or a group of numbers. For instance, if we say "let *a* and *b* represent numbers," then *a* and *b* are called **variables** because the values they take on vary. We use the variables *a* and *b* in the lists below so that the relationships shown there are true for all numbers that we will encounter in this book. By using variables, the statements written below are general statements about all numbers, rather than specific statements about only a few numbers.

Comparison Symbols		
Equality:	$a = b$	a is equal to b (a and b represent the same number)
	$a \neq b$	a is not equal to b
Inequality:	$a < b$	a is less than b
	$a \not< b$	a is not less than b
	$a > b$	a is greater than b
	$a \not> b$	a is not greater than b
	$a \geq b$	a is greater than or equal to b
	$a \leq b$	a is less than or equal to b

The symbols for inequality, $<$ and $>$, always point to the smaller of the two quantities being compared. For example, $3 < x$ means 3 is smaller than x. In this case, we can say "3 is less than x" or "x is greater than 3"; both statements are correct. Similarly, the expression $5 > y$ can be read as "5 is greater than y" or as "y is less than 5" because the inequality symbol is pointing to y, meaning y is the smaller of the two quantities.

Next we consider the symbols used to represent the four basic operations—addition, subtraction, multiplication, and division.

Operation Symbols

Addition:	$a + b$	The *sum* of a and b
Subtraction:	$a - b$	The *difference* of a and b
Multiplication:	$a \cdot b, (a)(b), a(b), (a)b, ab$	The *product* of a and b
Division:	$a \div b, a/b, \dfrac{a}{b}, b\,\overline{)a}$	The *quotient* of a and b

---

When we encounter the word **sum,** the implied operation is addition. To find the sum of two numbers, we simply add them. **Difference** implies subtraction, **product** implies multiplication, and **quotient** implies division. Notice also that there is more than one way to write the product or quotient of two numbers.

Note In the past, you may have used the notation 3×5 to denote multiplication. In algebra, it is best to avoid this notation if possible, since the multiplication symbol $\times$ can be confused with the variable x when written by hand.

Grouping Symbols

Parentheses ( ) and brackets [ ] are the symbols used for grouping numbers together. (Occasionally braces { } are also used for grouping, although they are usually reserved for set notation, as we shall see.)

The following examples illustrate the relationship between the symbols for comparing, operating, and grouping and the English language.

▶ **EXAMPLES**

Mathematical Expression	*English Equivalent*
1. $4 + 1 = 5$	The sum of 4 and 1 is 5
2. $8 - 1 < 10$	The difference of 8 and 1 is less than 10
3. $2(3 + 4) = 14$	Twice the sum of 3 and 4 is 14
4. $3x \geq 15$	The product of 3 and x is greater than or equal to 15
5. $\dfrac{y}{2} = y - 2$	The quotient of y and 2 is equal to the difference of y and 2 ◀

The last type of notation we need to discuss is the notation that allows us to write repeated multiplications in a more compact form—*exponents*. In the expression 2^3, the 2 is called the **base** and the 3 is called the **exponent.** The exponent 3 tells us the number of times the base appears in the product. That is,

$$2^3 = 2 \cdot 2 \cdot 2 = 8$$

The expression 2^3 is said to be in exponential form, while $2 \cdot 2 \cdot 2$ is said to be in expanded form. Here are some additional examples of expressions involving exponents.

▶ **EXAMPLES**   Expand and multiply.

6. $5^2 = 5 \cdot 5 = 25$ Base 5, exponent 2
7. $2^5 = 2 \cdot 2 \cdot 2 \cdot 2 \cdot 2 = 32$ Base 2, exponent 5
8. $10^3 = 10 \cdot 10 \cdot 10 = 1,000$ Base 10, exponent 3   ◀

The symbols for comparing, operating, and grouping are to mathematics what punctuation symbols are to English. These symbols are the punctuation symbols for mathematics.

Consider the following sentence:

Paul said John is tall.

It can have two different meanings, depending on how it is punctuated.

1. "Paul," said John, "is tall."
2. Paul said, "John is tall."

Without the punctuation we do not know which meaning is intended. It is ambiguous without punctuation.

Let's take a look at a similar situation in mathematics. Consider the following mathematical statement:

$$5 + 2 \cdot 7$$

If we add the 5 and 2 first and then multiply by 7, we get an answer of 49. On the other hand, if we multiply the 2 and the 7 first and then add 5, we are left with 19. We have a problem that seems to have two different answers, depending on whether we add first or multiply first. We would like to avoid this type of situation. That is, every problem like $5 + 2 \cdot 7$ should have only one answer. Therefore, we have developed the following rule for the order of operations.

> **Rule (Order of Operations)**
>
> When evaluating a mathematical expression, we will perform the operations in the following order, beginning with the expression in the innermost parentheses or brackets first and working our way out.
>
> 1. Simplify all numbers with exponents, working from left to right if more than one of these expressions is present.
> 2. Then do all multiplications and divisions left to right.
> 3. Perform all additions and subtractions left to right.

▶ **EXAMPLES**　Simplify each expression using the rule for order of operations.

9.　$5 + 8 \cdot 2 = 5 + 16$　　　　　　　Multiply $8 \cdot 2$ first
　　　　　　$= 21$

10.　$12 \div 4 \cdot 2 = 3 \cdot 2$　　　　　　Work left to right
　　　　　　　$= 6$

11.　$2[5 + 2(6 + 3 \cdot 4)] = 2[5 + 2(6 + 12)]$⎫　Simplify within the
　　　　　　　　　　　$= 2[5 + 2(18)]$⎬　　innermost grouping
　　　　　　　　　　　$= 2[5 + 36]$⎭　　symbols first
　　　　　　　　　　　$= 2[41]$　　　Next, simplify inside
　　　　　　　　　　　　　　　　the brackets
　　　　　　　　　　　$= 82$　　　Multiply

12.　$10 + 12 \div 4 + 2 \cdot 3 = 10 + 3 + 6$　　Multiply and divide
　　　　　　　　　　　　　　　　　left to right
　　　　　　　　　　　$= 19$　　Add left to right

Notice in Example 12 that we divided 12 by 4 and multiplied 2 times 3 before we did any addition. The rule for order of operations indicates that we always multiply and divide before we add when simplifying expressions like the one in Example 12.

13.　$4 \cdot 2^3 - 2 \cdot 3^2 = 4 \cdot 8 - 2 \cdot 9$　　First, simplify each number
　　　　　　　　　　　　　　　　with an exponent
　　　　　　　　　$= 32 - 18$　　Then, multiply left to right
　　　　　　　　　$= 14$　　Finally, subtract

14.　$2^4 + 3^3 \div 9 - 4^2 = 16 + 27 \div 9 - 16$　Simplify numbers with
　　　　　　　　　　　　　　　　　exponents
　　　　　　　　　　　$= 16 + 3 - 16$　　Then, divide
　　　　　　　　　　　$= 19 - 16$⎫　Finally, add and subtract　◀
　　　　　　　　　　　$= 3$　⎭　　left to right

Number Sequences and Inductive Reasoning

Suppose someone asks you to give the next number in the following number sequence. (The dots mean that the sequence continues in the same pattern forever.)

$$2, 5, 8, 11, \ldots$$

If you notice that each number is 3 more than the number before it, you would say the next number in the sequence is 14 because $11 + 3 = 14$. When we reason in this way, we are using what is called **inductive reasoning.** In mathematics, we use inductive reasoning when we notice a pattern to a sequence of numbers and then use the pattern to extend the sequence.

▶ **EXAMPLE 15**   Find the next number in each sequence.
(a) 3, 8, 13, 18, . . .
(b) 2, 10, 50, 250, . . .
(c) 2, 4, 7, 11, . . .

Solution   In order to find the next number in each sequence, we need to look for a pattern or relationship.
(a) For the first sequence, each number is 5 more than the number before it; therefore, the next number will be $18 + 5 = 23$.
(b) For the sequence in part (b), each number is 5 times the number before it; therefore, the next number in the sequence will be $5 \cdot 250 = 1,250$.
(c) For the sequence in part (c), there is no number to add each time or multiply by each time. However, the pattern becomes apparent when we look at the differences between the numbers:

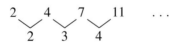

Proceeding in the same manner, we would add 5 to the next term, giving us $11 + 5 = 16$.   ◀

In the introduction to this chapter we mentioned the mathematician known as Fibonacci. There is a special sequence in mathematics named for Fibonacci. Here it is.

$$\text{Fibonacci sequence} = 1, 1, 2, 3, 5, 8, \ldots$$

Can you see the relationship among the numbers in this sequence? Start with two 1's, then add two consecutive members of the sequence to get the next number. Here is a diagram.

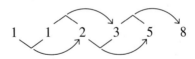

Sometimes we refer to the numbers in a sequence as **terms** of the sequence.

▶ **EXAMPLE 16** Write the first 10 terms of the Fibonacci sequence.

Solution The first six terms are given above. We extend the sequence by adding 5 and 8 to obtain the seventh term, 13. Then we add 8 and 13 to obtain 21. Continuing in this manner, the first 10 terms in the Fibonacci sequence are

$$1, 1, 2, 3, 5, 8, 13, 21, 34, 55$$ ◀

What to Do Now

Simplifying expressions using the rule for order of operations, as illustrated in examples 9–14, is a very important skill in mathematics. You need to master it, along with the other concepts presented in this section. As you work through the problems in the problem set that follows, make it your goal to use them as tools to help you understand what you have read in this section. Above all, don't just hurry through the problems in order to finish an assignment. Take enough time so that when you are finished, you are sure you understand the mathematics behind the problems.

PROBLEM SET 1.1

For each sentence below, write an equivalent expression in symbols.

1. The sum of x and 5 is 14.

2. The difference of x and 4 is 8.

3. The product of 5 and y is less than 30.

4. The product of 8 and y is greater than 16.

5. The product of 3 and y is less than or equal to the sum of y and 6.

6. The product of 5 and y is greater than or equal to the difference of y and 16.

7. The quotient of x and 3 is equal to the sum of x and 2.

8. The quotient of x and 2 is equal to the difference of x and 4.

Expand and multiply.

9. 3^2
10. 4^2
11. 7^2
12. 9^2
13. 2^3
14. 3^3
15. 4^3
16. 5^3
17. 2^4
18. 3^4
19. 10^2
20. 10^4
21. 11^2
22. 111^2

Use the rule for order of operations to simplify each expression as much as possible.

23. $2 \cdot 3 + 5$
24. $8 \cdot 7 + 1$
25. $2(3 + 5)$
26. $8(7 + 1)$
27. $5 + 2 \cdot 6$
28. $8 + 9 \cdot 4$
29. $(5 + 2) \cdot 6$
30. $(8 + 9) \cdot 4$
31. $5 \cdot 4 + 5 \cdot 2$
32. $6 \cdot 8 + 6 \cdot 3$
33. $5(4 + 2)$
34. $6(8 + 3)$
35. $8 + 2(5 + 3)$
36. $7 + 3(8 - 2)$
37. $(8 + 2)(5 + 3)$
38. $(7 + 3)(8 - 2)$
39. $20 + 2(8 - 5) + 1$
40. $10 + 3(7 + 1) + 2$
41. $5 + 2(3 \cdot 4 - 1) + 8$
42. $11 - 2(5 \cdot 3 - 10) + 2$
43. $8 + 10 \div 2$
44. $16 - 8 \div 4$
45. $4 + 8 \div 4 - 2$
46. $6 + 9 \div 3 + 2$
47. $3 + 12 \div 3 + 6 \cdot 5$
48. $18 + 6 \div 2 + 3 \cdot 4$
49. $3 \cdot 8 + 10 \div 2 + 4 \cdot 2$
50. $5 \cdot 9 + 10 \div 2 + 3 \cdot 3$
51. $(5 + 3)(5 - 3)$
52. $(7 + 2)(7 - 2)$
53. $5^2 - 3^2$
54. $7^2 - 2^2$
55. $(4 + 5)^2$
56. $(6 + 3)^2$
57. $4^2 + 5^2$
58. $6^2 + 3^2$
59. $3 \cdot 10^2 + 4 \cdot 10 + 5$

60. $6 \cdot 10^2 + 5 \cdot 10 + 4$

61. $2 \cdot 10^3 + 3 \cdot 10^2 + 4 \cdot 10 + 5$

62. $5 \cdot 10^3 + 6 \cdot 10^2 + 7 \cdot 10 + 8$

63. $10 - 2(4 \cdot 5 - 16)$ **64.** $15 - 5(3 \cdot 2 - 4)$

65. $4[7 + 3(2 \cdot 9 - 8)]$

66. $5[10 + 2(3 \cdot 6 - 10)]$

67. $5(7 - 3) + 8(6 - 4)$

68. $3(10 - 4) + 6(12 - 10)$

69. $3(4 \cdot 5 - 12) + 6(7 \cdot 6 - 40)$

70. $6(8 \cdot 3 - 4) + 5(7 \cdot 3 - 1)$

71. $3^4 + 4^2 \div 2^3 - 5^2$ **72.** $2^5 + 6^2 \div 2^2 - 3^2$

73. $5^2 + 3^4 \div 9^2 + 6^2$ **74.** $6^2 + 2^5 \div 4^2 + 7^2$

Applying the Concepts

75. On Monday Maria buys 10 shares of a certain stock. On Tuesday she buys 4 more shares of the same stock. If the stock splits Wednesday, then she has twice the number of shares she had on Tuesday. Write an expression using parentheses and the numbers 2, 4, and 10 to describe this situation.

76. Patrick has a collection of 25 baseball cards. He then buys 3 packs of gum, and each pack contains 5 baseball cards. Write an expression using the numbers 25, 3, and 5 to describe this situation.

77. A gambler begins an evening in Las Vegas with $50. After an hour she has tripled her money.

The next hour she loses $14. Write an expression using the numbers 3, 50, and 14 to describe this situation.

78. A flight from Los Angeles to New York has 128 passengers. The plane stops in Denver, where 50 of the passengers get off and 21 new passengers get on. Write an expression containing the numbers 128, 50, and 21 to describe this situation.

In 1993, the government standardized the way in which nutrition information was presented on the labels of most packaged food products. Figure 1 shows a standardized food label from a package of cookies that I ate at lunch the day I was writing the problems for this problem set. Use the information in Figure 1 to answer the following questions.

79. How many cookies are in the package?

80. If I paid $0.50 for the package of cookies, how much did each cookie cost?

81. If the ''Calories'' category stands for calories per serving, how many calories did I consume by eating the whole package of cookies?

82. Suppose that, while swimming, I burn 11 calories each minute. If I swim for 20 minutes, will I burn enough calories to cancel out the calories I added by eating 5 cookies?

SANDWICH CREMES

Nutrition Facts
Serving Size 5 Cookies (about 43g)
Servings Per Container 2
Calories 210
  Fat Calories 90
* Percent Daily Values (DV) are based on a 2,000 calorie diet.

Amount/serving	%DV*	Amount/serving	%DV*
Total Fat 9g	**15%**	**Total Carb.** 30g	**10%**
Sat. Fat 2.5g	**12%**	Fiber 1g	**2%**
Cholest. less than 5mg	**2%**	Sugars 14g	
Sodium 110mg	**5%**	**Protein** 3g	
Vitamin A 0% • Vitamin C 0% • Calcium 2% • Iron 8%			

Figure 1

The food label shown in Figure 2 was taken from a bag of corn chips. Use the information in Figure 2 to answer the following questions.

83. Approximately how many chips are in the bag?
84. If the bag of chips costs $0.99, approximately how much does one serving of chips cost?
85. The table toward the bottom of the label gives the recommended amount of total fat that should be consumed by a person eating 2,000 calories per day and by a person eating 2,500 calories per day. Use the numbers in the table to estimate the recommended fat intake for a person eating 3,000 calories per day.
86. Deirdre burns 256 calories per hour by trotting her horse at a constant rate. How long must she ride in order to burn the calories consumed by eating 4 servings of these chips?

Find the next number in each sequence.

87. 1, 2, 3, 4, . . . (The sequence of counting numbers.)
88. 0, 1, 2, 3, . . . (The sequence of whole numbers.)
89. 2, 4, 6, 8, . . . (The sequence of even numbers.)
90. 1, 3, 5, 7, . . . (The sequence of odd numbers.)
91. 1, 4, 9, 16, . . . (The sequence of squares.)
92. 1, 8, 27, 64, . . . (The sequence of cubes.)
93. 2, 2, 4, 6, . . . (A Fibonacci-like sequence.)
94. 5, 5, 10, 15, . . . (A Fibonacci-like sequence.)

Research Project 1

There are many patterns in the world around us that follow the Fibonacci sequence. Check out the book *Mathematics: A Human Endeavor*, by Harold Jacobs, from your library and use it as a reference to write a paper in which you give two applications of the Fibonacci sequence. Your paper should start with the definition of the Fibonacci sequence and then proceed to show and explain your two examples. Write in complete sentences just as you would if you were writing a paper for an English class.

Nutrition Facts

Serving Size 1 oz. (28g/About 32 chips)
Servings Per Container 7

Amount Per Serving

Calories 160   Calories from Fat 90

% Daily Value*

Total Fat 10g	**15%**
Saturated Fat 1.5g	**8%**
Cholesterol 0mg	**0%**
Sodium 160mg	**7%**
Total Carbohydrate 15g	**5%**
Dietary Fiber 1g	**4%**
Sugars 0g	
Protein 2g	

Vitamin A 0%	•	Vitamin C 0%
Calcium 2%	•	Iron 0%

* Percent Daily Values are based on a 2,000 calorie diet. Your daily values may be higher or lower depending on your calorie needs:

		Calories:	2,000	2,500
Total Fat	Less than		65g	80g
Sat Fat	Less than		20g	25g
Cholesterol	Less than		300mg	300mg
Sodium	Less than		2,400mg	2,400mg
Total Carbohydrate			300g	375g
Dietary Fiber			25g	30g

Calories per gram:
Fat 9   •   Carbohydrate 4   •   Protein 4

Figure 2

1.2   Real Numbers

In this section we will get an idea of what real numbers are. In order to do this we will draw what is called the **real number line.** We first draw a straight line and label a convenient point on the line with 0. Then we mark off equally spaced distances in both directions from 0. Label the points to the right of 0 with the numbers 1, 2, 3, . . . (the dots mean "and so on"). The points to the left of 0 we label, in order, −1, −2, −3, ... Here is what it looks like.

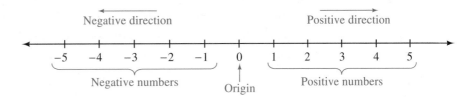

The numbers increase in value going from left to right. If we "move" to the right, we are moving in the positive direction. If we move to the left, we are moving in the negative direction. When we compare two numbers on the number line, the number on the left is always smaller than the number on the right. For instance, −3 is smaller than −1 since it is to the left of −1 on the number line.

Note   If there is no sign (+ or −) in front of a number, the number is assumed to be positive (+), except for the number 0, which is neither positive nor negative.

▶ **EXAMPLE 1**   Locate and label the points on the real number line associated with the numbers −3.5, −1$\frac{1}{4}$, $\frac{1}{2}$, $\frac{3}{4}$, 2.5.

Solution   We draw a real number line from −4 to 4 and label the points in question.

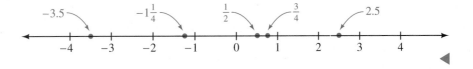

DEFINITION   The number associated with a point on the real number line is called the **coordinate** of that point.

In the preceding example, the numbers −3.5, −1$\frac{1}{4}$, $\frac{1}{2}$, $\frac{3}{4}$, and 2.5, are the coordinates of the points they represent.

DEFINITION   The numbers that can be represented with points on the real number line are called **real numbers.**

Real numbers include whole numbers, fractions, decimals, and other numbers that are not as familiar to us as these.

Fractions on the Number Line

As we proceed through Chapter 1, from time to time we will review some of the major concepts associated with fractions. To begin, here is the formal definition of a fraction.

DEFINITION   If a and b are real numbers, then the expression

$$\frac{a}{b}$$

is called a **fraction.** The top number a is called the **numerator** and the bottom number b is called the **denominator.** The restriction $b \neq 0$ keeps us from writing an expression that is undefined. (As you will see, division by zero is not allowed.)

The number line can be used to visualize fractions. Recall that for the fraction $\frac{a}{b}$, a is called the numerator and b is called the denominator. The denominator indicates the number of equal parts in the interval from 0 to 1 on the number line. The numerator indicates how many of those parts we have. If we take that part of the number line from 0 to 1 and divide it into *three equal parts,* we say that we have divided it into *thirds* (see Figure 1). Each of the three segments is $\frac{1}{3}$ (one third) of the whole segment from 0 to 1.

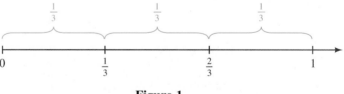

Figure 1

Two of these smaller segments together are $\frac{2}{3}$ (two thirds) of the whole segment. And three of them would be $\frac{3}{3}$ (three thirds), or the whole segment.

Let's do the same thing again with six equal divisions of the segment from 0 to 1 (see Figure 2). In this case we say each of the smaller segments has a length of $\frac{1}{6}$ (one sixth).

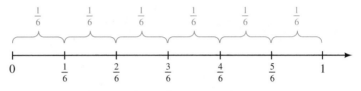

Figure 2

The same point we labeled with $\frac{1}{3}$ in Figure 1 is now labeled with $\frac{2}{6}$. Likewise, the point we labeled earlier with $\frac{2}{3}$ is now labeled $\frac{4}{6}$. It must be true then that

$$\frac{2}{6} = \frac{1}{3} \quad \text{and} \quad \frac{4}{6} = \frac{2}{3}$$

Actually, there are many fractions that name the same point as $\frac{1}{3}$. If we were to divide the segment between 0 and 1 into twelve equal parts, four of these twelve equal parts ($\frac{4}{12}$) would be the same as $\frac{2}{6}$ or $\frac{1}{3}$. That is,

$$\frac{4}{12} = \frac{2}{6} = \frac{1}{3}$$

Even though these three fractions look different, each names the same point on the number line, as shown in Figure 3. All three fractions have the same *value,* because they all represent the same number.

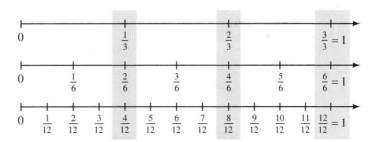

Figure 3

DEFINITION Fractions that represent the same number are said to be **equivalent.** Equivalent fractions may look different, but they must have the same value.

It is apparent that every fraction has many different representations, each of which is equivalent to the original fraction. The next two properties give us a way of changing the terms of a fraction without changing its value.

> **Property 1**
> Multiplying the numerator and denominator of a fraction by the same nonzero number never changes the value of the fraction.
>
> **Property 2**
> Dividing the numerator and denominator of a fraction by the same nonzero number never changes the value of the fraction.

▶ **EXAMPLE 2** Write $\dfrac{3}{4}$ as an equivalent fraction with denominator 20.

Solution The denominator of the original fraction is 4. The fraction we are trying to find must have a denominator of 20. We know that if we multiply 4 by 5, we get 20. Property 1 indicates that we are free to multiply the denominator by 5 as long as we do the same to the numerator.

$$\frac{3}{4} = \frac{3 \cdot 5}{4 \cdot 5} = \frac{15}{20}$$

The fraction $\frac{15}{20}$ is equivalent to the fraction $\frac{3}{4}$. ◀

Absolute Values and Opposites

Representing numbers on the number line lets us give each number two important properties: a direction from zero and a distance from zero. The direction from zero is represented by the sign in front of the number. (A number without a sign is understood to be positive.) The distance from zero is called the absolute value of the number, as the following definition indicates.

> **DEFINITION** The **absolute value** of a real number is its distance from zero on the number line. If x represents a real number, then the absolute value of x is written $|x|$.

▶ **EXAMPLES** Write each expression without absolute value symbols.

3. $|5| = 5$ The number $+5$ is 5 units from zero

4. $|-5| = 5$ The number -5 is 5 units from zero

5. $\left|-\dfrac{1}{2}\right| = \dfrac{1}{2}$ The number $-\dfrac{1}{2}$ is $\dfrac{1}{2}$ units from zero ◀

The absolute value of a number is *never* negative. It is the distance the number is from zero without regard to which direction it is from zero. When working with the absolute value of sums and differences, we must simplify the expression inside the absolute value symbols first, and then find the absolute value of the simplified expression. That is, for the rule for order of operations, absolute value symbols are treated the same as grouping symbols. We work inside the absolute value symbols first.

▶ **EXAMPLES**   Simplify each expression.

6. $|8 - 3| = |5| = 5$

7. $|3 \cdot 2^3 + 2 \cdot 3^2| = |3 \cdot 8 + 2 \cdot 9| = |24 + 18| = |42| = 42$

8. $|9 - 2| - |8 - 6| = |7| - |2| = 7 - 2 = 5$ ◀

Another important concept associated with numbers on the number line is that of opposites. Here is the definition.

DEFINITION   Numbers the same distance from zero but in opposite directions from zero are called **opposites.**

▶ **EXAMPLES**   Give the opposite of each number.

	Number	*Opposite*	
9.	5	-5	5 and -5 are opposites
10.	-3	3	-3 and 3 are opposites
11.	$\dfrac{1}{4}$	$-\dfrac{1}{4}$	$\dfrac{1}{4}$ and $-\dfrac{1}{4}$ are opposites
12.	-2.3	2.3	-2.3 and 2.3 are opposites ◀

Each negative number is the opposite of some positive number and each positive number is the opposite of some negative number. The opposite of a negative number is a positive number. In symbols, if a represents a positive number, then

$$-(-a) = a$$

Opposites always have the same absolute value. And, when you add any two opposites, the result is always zero:

$$a + (-a) = 0$$

Reciprocals and Multiplication with Fractions

The last concept we want to cover in this section is the concept of reciprocals. Understanding reciprocals requires some knowledge of multiplication with fractions. To multiply two fractions we simply multiply numerators and multiply denominators.

▶ **EXAMPLE 13** Multiply $\dfrac{3}{4} \cdot \dfrac{5}{7}$.

Solution The product of the numerators is 15 and the product of the denominators is 28:

$$\frac{3}{4} \cdot \frac{5}{7} = \frac{3 \cdot 5}{4 \cdot 7} = \frac{15}{28}$$ ◀

▶ **EXAMPLE 14** Multiply $7 \left(\dfrac{1}{3} \right)$.

Solution The number 7 can be thought of as the fraction $\frac{7}{1}$:

$$7 \left(\frac{1}{3} \right) = \frac{7}{1} \left(\frac{1}{3} \right) = \frac{7 \cdot 1}{1 \cdot 3} = \frac{7}{3}$$ ◀

Note In past math classes you may have written fractions like $\frac{7}{3}$ (improper fractions) as mixed numbers, such as $2\frac{1}{3}$. In algebra, it is usually better to write them as improper fractions rather than mixed numbers.

▶ **EXAMPLE 15** Expand and multiply $\left(\dfrac{2}{3} \right)^3$.

Solution Using the definition of exponents from the previous section, we have

$$\left(\frac{2}{3} \right)^3 = \frac{2}{3} \cdot \frac{2}{3} \cdot \frac{2}{3} = \frac{8}{27}$$ ◀

We are now ready for the definition of reciprocals.

DEFINITION Two numbers whose product is 1 are called **reciprocals**.

▶ **EXAMPLES** Give the reciprocal of each number.

	Number	Reciprocal	
16.	5	$\dfrac{1}{5}$	Because $5 \left(\dfrac{1}{5} \right) = \dfrac{5}{1} \left(\dfrac{1}{5} \right) = \dfrac{5}{5} = 1$
17.	2	$\dfrac{1}{2}$	Because $2 \left(\dfrac{1}{2} \right) = \dfrac{2}{1} \left(\dfrac{1}{2} \right) = \dfrac{2}{2} = 1$
18.	$\dfrac{1}{3}$	3	Because $\dfrac{1}{3} (3) = \dfrac{1}{3} \left(\dfrac{3}{1} \right) = \dfrac{3}{3} = 1$
19.	$\dfrac{3}{4}$	$\dfrac{4}{3}$	Because $\dfrac{3}{4} \left(\dfrac{4}{3} \right) = \dfrac{12}{12} = 1$

20. $\dfrac{2}{5}$ $\dfrac{5}{2}$ Because $\dfrac{2}{5}\left(\dfrac{5}{2}\right) = \dfrac{10}{10} = 1$

21. 0 None Because multiplying by 0 can never result in 1

◀

Although we will not develop multiplication with negative numbers until later in the chapter, you should know that the reciprocal of a negative number is also a negative number. For example, the reciprocal of -4 is $-\frac{1}{4}$. Likewise, the reciprocal of $-\frac{2}{3}$ is $-\frac{3}{2}$. Their products are 1, but we will have to wait until we develop multiplication with negative numbers to see why.

Every real number, except zero, has a reciprocal. Any time we multiply by 0, the result is 0. (It can never be 1.) This is a special property 0 has. In symbols it looks like this:

For any number a,

$$a \cdot 0 = 0 \cdot a = 0$$

Multiplying by 0 always results in 0.

▶ **EXAMPLE 22** Find the next number in each sequence.

(a) $1, \dfrac{1}{2}, \dfrac{1}{3}, \dfrac{1}{4}, \ldots$ (b) $1, 1, \dfrac{1}{2}, \dfrac{1}{3}, \dfrac{1}{5}, \ldots$

Solution As we did with the sequences in Section 1.1, we look for a pattern that defines the sequence.

(a) In the first case, the sequence is simply the reciprocals of the counting numbers. The next number in the sequence will be the reciprocal of 5, which is $\frac{1}{5}$.

(b) This sequence is formed by taking the reciprocals of the numbers in the Fibonacci sequence.

Fibonacci sequence: 1, 1, 2, 3, 5, . . .

Reciprocals of Fibonacci sequence: $1, 1, \dfrac{1}{2}, \dfrac{1}{3}, \dfrac{1}{5}, \ldots$

The next number in the Fibonacci sequence will be 8. Therefore, the next number in the sequence in question is $\frac{1}{8}$. ◀

Facts from Geometry: Formulas for Area and Perimeter

Now that we have been introduced to real numbers, variables, and multiplication with fractions, we can put these concepts to work to help classify and describe some of the geometric objects we see in the world around us: squares, rectangles, and triangles.

To discuss these objects in general, we label their dimensions with variables. For example, if we discuss the characteristics of a square in which the length of a side is x, then any properties we find can be transferred to a square of side 3 inches simply by letting $x = 3$.

A square, rectangle, and triangle are shown below. Note that we have labeled the dimensions of each with variables. The formulas for the perimeter and area of each object are given in terms of its dimensions.

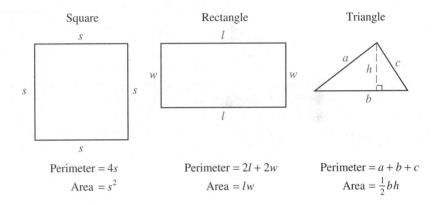

Square	Rectangle	Triangle
Perimeter $= 4s$	Perimeter $= 2l + 2w$	Perimeter $= a + b + c$
Area $= s^2$	Area $= lw$	Area $= \frac{1}{2}bh$

Note   The dotted line labeled h in the triangle is its height, or altitude. It extends from the top of the triangle down to the base, meeting the base at an angle of 90°. The altitude of a triangle is always perpendicular to the base. The small square shown where the altitude meets the base is used to indicate that the angle formed is 90°.

The formula for perimeter gives us the distance around the outside of the object along its sides, while the formula for area gives us a measure of the amount of surface the object has.

▶ **EXAMPLE 23**   Find the perimeter and area of each figure.

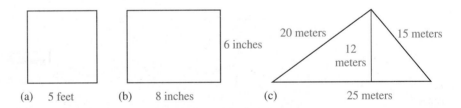

(a)   5 feet   (b)   8 inches   (c)   25 meters

Solution   We use the preceding formulas to find the perimeter and the area. In each case, the units for perimeter are linear units, while the units for area are square units.

(a) Perimeter $= 4s = 4 \cdot 5$ feet $= 20$ feet
  Area $= s^2 = (5 \text{ feet})^2 = 25 \text{ feet}^2$

(b) Perimeter $= 2l + 2w = 2(8 \text{ inches}) + 2(6 \text{ inches}) = 28$ inches
  Area $= lw = (8 \text{ inches})(6 \text{ inches}) = 48 \text{ inches}^2$

(c) Perimeter $= a + b + c = (20 \text{ meters}) + (25 \text{ meters}) + (15 \text{ meters})$
    $= 60$ meters

  Area $= \dfrac{1}{2} bh = \dfrac{1}{2}(25 \text{ meters})(12 \text{ meters}) = 150 \text{ meters}^2$ ◀

P R O B L E M   S E T   1 . 2

Draw a number line that extends from -5 to $+5$. Label the points with the following coordinates.

1. 5   **2.** -2
3. -4 **4.** -3
5. 1.5   **6.** -1.5

7. $\dfrac{9}{4}$ **8.** $\dfrac{8}{3}$

Write each of the following fractions as an equivalent fraction with denominator 24.

9. $\dfrac{3}{4}$ **10.** $\dfrac{5}{6}$

11. $\dfrac{1}{2}$ **12.** $\dfrac{1}{8}$

13. $\dfrac{5}{8}$ **14.** $\dfrac{7}{12}$

Write each fraction as an equivalent fraction with denominator 60.

15. $\dfrac{3}{5}$ **16.** $\dfrac{5}{12}$

17. $\dfrac{11}{30}$ **18.** $\dfrac{9}{10}$

For each of the following numbers, give the opposite, the reciprocal, and the absolute value. (Assume all variables are nonzero.)

19. 10   **20.** 8

21. $\dfrac{3}{4}$ **22.** $\dfrac{5}{7}$

23. $\dfrac{11}{2}$ **24.** $\dfrac{15}{3}$

25. -3 **26.** -5

27. $-\dfrac{2}{5}$ **28.** $-\dfrac{3}{8}$

29. x **30.** a

Place one of the symbols $<$ or $>$ between each of the following to make the resulting statement true.

31. -5 -3 **32.** -8 -1
33. -3 -7 **34.** -6 5
35. $|-4|$ $-|-4|$ **36.** 3 $-|-3|$
37. 7 $-|-7|$ **38.** -7 $|-7|$

39. $-\dfrac{3}{4}$ $-\dfrac{1}{4}$ **40.** $-\dfrac{2}{3}$ $-\dfrac{1}{3}$

41. $-\dfrac{3}{2}$ $-\dfrac{3}{4}$ **42.** $-\dfrac{8}{3}$ $-\dfrac{17}{3}$

Simplify each expression.

43. $|8 - 2|$ **44.** $|6 - 1|$
45. $|5 \cdot 2^3 - 2 \cdot 3^2|$ **46.** $|2 \cdot 10^2 + 3 \cdot 10|$
47. $|7 - 2| - |4 - 2|$ **48.** $|10 - 3| - |4 - 1|$
49. $10 - |7 - 2(5 - 3)|$
50. $12 - |9 - 3(7 - 5)|$
51. $15 - |8 - 2(3 \cdot 4 - 9)| - 10$
52. $25 - |9 - 3(4 \cdot 5 - 18)| - 20$

Multiply the following.

53. $\dfrac{2}{3} \cdot \dfrac{4}{5}$ **54.** $\dfrac{1}{4} \cdot \dfrac{3}{5}$

55. $\dfrac{1}{2}(3)$ **56.** $\dfrac{1}{3}(2)$

57. $\dfrac{1}{4}(5)$ **58.** $\dfrac{1}{5}(4)$

59. $\dfrac{4}{3} \cdot \dfrac{3}{4}$ **60.** $\dfrac{5}{7} \cdot \dfrac{7}{5}$

61. $6\left(\dfrac{1}{6}\right)$ **62.** $8\left(\dfrac{1}{8}\right)$

63. $3 \cdot \dfrac{1}{3}$ **64.** $4 \cdot \dfrac{1}{4}$

Expand and multiply.

65. $\left(\dfrac{3}{4}\right)^2$ **66.** $\left(\dfrac{5}{6}\right)^2$

67. $\left(\frac{2}{3}\right)^3$

68. $\left(\frac{1}{2}\right)^3$

69. $\left(\frac{1}{10}\right)^4$

70. $\left(\frac{1}{10}\right)^5$

Find the next number in each sequence.

71. $1, \frac{1}{3}, \frac{1}{5}, \frac{1}{7}, \ldots$ (Reciprocals of odd numbers.)

72. $\frac{1}{2}, \frac{1}{4}, \frac{1}{6}, \frac{1}{8}, \ldots$ (Reciprocals of even numbers.)

73. $1, \frac{1}{4}, \frac{1}{9}, \frac{1}{16}, \ldots$ (Reciprocals of squares.)

74. $1, \frac{1}{8}, \frac{1}{27}, \frac{1}{64}, \ldots$ (Reciprocals of cubes.)

Find the perimeter and area of each figure.

75.

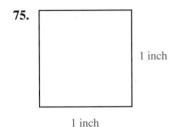

1 inch

1 inch

76.

15 millimeters

15 millimeters

77.

0.75 inch

1.5 inches

78.

1.5 centimeters

4.5 centimeters

79.

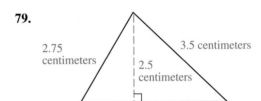

2.75 centimeters

3.5 centimeters

2.5 centimeters

4 centimeters

80.

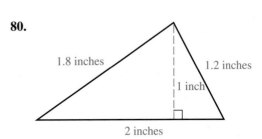

1.8 inches

1.2 inches

1 inch

2 inches

Applying the Concepts

81. A football team gains 6 yards on one play and then loses 8 yards on the next play. To what number on the number line does a loss of 8 yards correspond? The total yards gained or lost on the two plays corresponds to what negative number?

82. A woman has a balance of $20 in her checking account. If she writes a check for $30, what negative number can be used to represent the new balance in her checking account?

Temperature In the United States, temperature is measured on the Fahrenheit temperature scale. On this scale, water boils at 212 degrees and freezes at 32 degrees. To denote a temperature of 32 degrees on the Fahrenheit scale, we write

32° F which is read "32 degrees Fahrenheit"

Use this information for Problems 83 and 84.

83. Marilyn is flying from Seattle to San Francisco on a Boeing 737 jet. When the plane reaches an altitude of 35,000 feet, the temperature outside the plane is 64 degrees below zero Fahrenheit. Represent the temperature with a negative number. If the temperature outside the plane gets warmer by 10 degrees, what will the new temperature be?

84. At 10:00 in the morning in White Bear Lake, Minnesota, John notices the temperature outside is 10 degrees below zero Fahrenheit. Write the temperature as a negative number. An hour later it has warmed up by 6 degrees. What is the temperature at 11:00 that morning?

85. Steve Golden is scuba diving near his home in Maui. At one point he is 100 feet below the surface. Represent this number with a negative number. If he descends another 5 feet, what negative number will represent his new position?

86. Pete and Heidi are married and have a checking account that allows them to withdraw money from the account at an automatic teller machine. With each withdrawal from the automatic teller, the bank charges an extra $2 to their account. Suppose the balance in the checkbook is $75. If Heidi writes a check for $55 and Pete withdraws $20 from the ATM, what will their new balance be?

87. Find the area and perimeter of an $8\frac{1}{2}$-by-11-inch piece of notebook paper.

88. Find the area and perimeter of an $8\frac{1}{2}$-by-$5\frac{1}{2}$-inch piece of paper.

Table 1 gives the amount of energy expended per hour for various activities, for a person weighing 120, 150, or 180 pounds. Use Table 1 to answer questions 89–92.

T A B L E   1   Energy Expended from Exercising

Activity	Calories per Hour		
	120 lbs	**150 lbs**	**180 lbs**
Bicycling	299	374	449
Bowling	212	265	318
Handball	544	680	816
Horseback trotting	278	347	416
Jazzercise	272	340	408
Jogging	544	680	816
Skiing (downhill)	435	544	653

89. Suppose you weigh 120 pounds. How many calories will you burn if you play handball for 2 hours and then ride your bicycle for an hour?

90. How many calories are burned by a person weighing 150 pounds who jogs for $\frac{1}{2}$ hour and then goes bicycling for 2 hours?

91. Two people go skiing. One weighs 180 pounds and the other weighs 120 pounds. If they ski for 3 hours, how many more calories are burned by the person weighing 180 pounds?

92. Two people spend 3 hours bowling. If one weighs 120 pounds and the other weighs 150 pounds, how many more calories are burned during the evening by the person weighing 150 pounds?

SECTION

1.3   Addition of Real Numbers

Suppose that you are playing a friendly game of poker with some friends and you lose $3 on the first hand and $4 on the second hand. If you represent winning with positive numbers and losing with negative numbers, how can you translate this situation into symbols? Since you lost $3 and $4 for a total of $7, one way to represent this situation is with addition of negative numbers:

$$(-\$3) + (-\$4) = -\$7$$

From this equation, we see that the sum of two negative numbers is a negative number. To generalize addition with positive and negative numbers we use the number line.

Since real numbers have both a distance from zero (absolute value) and a direction from zero (sign), we can think of addition of two numbers in terms of distance and direction from zero.

Let's look at a problem for which we know the answer. Suppose we want to add the numbers 3 and 4. The problem is written $3 + 4$. To put it on the number line, we read the problem as follows:

1. The 3 tells us to "start at the origin and move 3 units in the positive direction."
2. The + sign is read "and then move."
3. The 4 means "4 units in the positive direction."

To summarize, $3 + 4$ means to start at the origin, move 3 units in the *positive* direction and then 4 units in the *positive* direction.

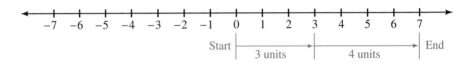

We end up at 7, which is the answer to our problem: $3 + 4 = 7$.

Let's try other combinations of positive and negative 3 and 4 on the number line.

▶ **EXAMPLE 1** Add $3 + (-4)$.

Solution Starting at the origin, move 3 units in the *positive* direction and then 4 units in the *negative* direction.

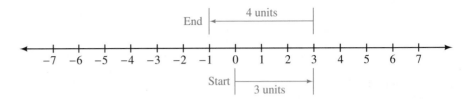

We end up at -1; therefore $3 + (-4) = -1$. ◀

▶ **EXAMPLE 2** Add $-3 + 4$.

Solution Starting at the origin, move 3 units in the *negative* direction and then 4 units in the *positive* direction.

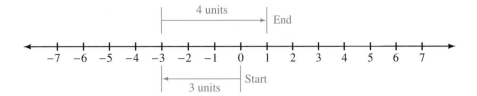

We end up at $+1$; therefore $-3 + 4 = 1$. ◀

▶ **EXAMPLE 3** Add $-3 + (-4)$.

Solution Starting at the origin, move 3 units in the *negative* direction and then 4 units in the *negative* direction.

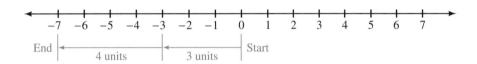

We end up at -7; therefore $-3 + (-4) = -7$. Here is a summary of what we have just completed:

$$3 + 4 = 7$$
$$3 + (-4) = -1$$
$$-3 + 4 = 1$$
$$-3 + (-4) = -7$$ ◀

Let's do four more problems on the number line and then summarize our results into a rule we can use to add any two real numbers.

▶ **EXAMPLE 4** Add $5 + 7 = 12$.

Solution

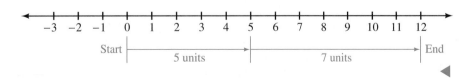

▶ **EXAMPLE 5** Add $5 + (-7) = -2$.

Solution

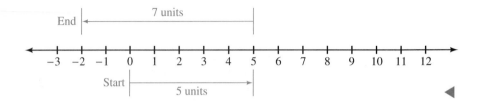

▶ **EXAMPLE 6** Add $-5 + 7 = 2$.

Solution

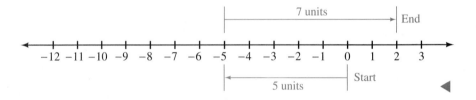

▶ **EXAMPLE 7** Add $-5 + (-7) = -12$.

Solution

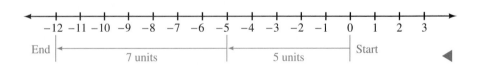

If we look closely at the results of the preceding addition problems, we can see that they support (or justify) the following rule.

Rule

To add two real numbers with

1. the *same* sign: Simply add their absolute values and use the common sign. (Both numbers are positive, the answer is positive. Both numbers are negative, the answer is negative.)
2. *different* signs: Subtract the smaller absolute value from the larger. The answer will have the sign of the number with the larger absolute value.

This rule covers all possible combinations of addition with real numbers. You must memorize it. After you have worked a number of problems, it will seem almost automatic.

▶ **EXAMPLE 8**   Add all combinations of positive and negative 10 and 13.

Solution   Rather than work these problems on the number line, we use the rule for adding positive and negative numbers to obtain our answers:

$$10 + 13 = 23$$
$$10 + (-13) = -3$$
$$-10 + 13 = 3$$
$$-10 + (-13) = -23 \qquad ◀$$

▶ **EXAMPLE 9**   Add all possible combinations of positive and negative 12 and 17.

Solution   Applying the rule for adding positive and negative numbers, we have

$$12 + 17 = 29$$
$$12 + (-17) = -5$$
$$-12 + 17 = 5$$
$$-12 + (-17) = -29 \qquad ◀$$

▶ **EXAMPLE 10**   Add $-3 + 2 + (-4)$.

Solution   Applying the rule for order of operations, we add left to right:

$$-3 + 2 + (-4) = -1 + (-4)$$
$$= -5 \qquad ◀$$

▶ **EXAMPLE 11**   Add $-8 + [2 + (-5)] + (-1)$.

Solution   Adding inside the brackets first, and then left to right, we have

$$-8 + [2 + (-5)] + (-1) = -8 + (-3) + (-1)$$
$$= -11 + (-1)$$
$$= -12 \qquad ◀$$

▶ **EXAMPLE 12**   Simplify $(-8 + 4) + [3 + (-5)] + (-3)$.

Solution   We begin by simplifying inside the parentheses and brackets. Then, we add, left to right:

$$(-8 + 4) + [3 + (-5)] + (-3) = -4 + (-2) + (-3)$$
$$= -6 + (-3)$$
$$= -9 \qquad ◀$$

▶ **EXAMPLE 13**   Simplify $-10 + 2(-8 + 11) + (-4)$.

Solution   First, we simplify inside the parentheses. Then, we multiply. Finally, we add, left to right:

$$-10 + 2(-8 + 11) + (-4) = -10 + 2(3) + (-4)$$
$$= -10 + 6 + (-4)$$
$$= -4 + (-4)$$
$$= -8 \qquad \blacktriangleleft$$

Arithmetic Sequences

The pattern in a sequence of numbers is easy to identify when each number in the sequence comes from the preceding number by adding the same amount each time. This leads us to our next level of classification, in which we classify together groups of sequences with a common characteristic.

DEFINITION   An **arithmetic sequence** is a sequence of numbers in which each number (after the first number) comes from adding the same amount to the number before it.

Here is an example of an arithmetic sequence:

$$2, 5, 8, 11, \ldots$$

Each number is obtained by adding 3 to the number before it.

▶ **EXAMPLE 14**   Each sequence below is an arithmetic sequence. Find the next two numbers in each sequence.
(a) 7, 10, 13, . . .
(b) 9.5, 10, 10.5, . . .
(c) 5, 0, −5, . . .

Solution   Since we know that each sequence is arithmetic, we know to look for the number that is added to each term to produce the next consecutive term.
(a) 7, 10, 13, . . . : Each term is found by adding 3 to the term before it. Therefore, the next two terms will be 16 and 19.
(b) 9.5, 10, 10.5, . . . : Each term comes from adding 0.5 to the term before it. Therefore, the next two terms will be 11 and 11.5.
(c) 5, 0, −5, . . . : Each term comes from adding −5 to the term before it. Therefore, the next two terms will be $-5 + (-5) = -10$ and $-10 + (-5) = -15$. ◀

PROBLEM SET 1.3

1. Add all positive and negative combinations of 3 and 5. (Look back to Examples 8 and 9.)
2. Add all positive and negative combinations of 6 and 4.
3. Add all positive and negative combinations of 15 and 20.
4. Add all positive and negative combinations of 18 and 12.

Work the following problems. You may want to begin by doing a few on the number line.

5. $6 + (-3)$
6. $7 + (-8)$
7. $13 + (-20)$
8. $15 + (-25)$
9. $18 + (-32)$
10. $6 + (-9)$
11. $-6 + 3$
12. $-8 + 7$
13. $-30 + 5$
14. $-18 + 6$
15. $-6 + (-6)$
16. $-5 + (-5)$
17. $-9 + (-10)$
18. $-8 + (-6)$
19. $-10 + (-15)$
20. $-18 + (-30)$

Work the following problems using the rule for addition of real numbers. You may want to refer back to the rule for order of operations.

21. $5 + (-6) + (-7)$
22. $6 + (-8) + (-10)$
23. $-7 + 8 + (-5)$
24. $-6 + 9 + (-3)$
25. $5 + [6 + (-2)] + (-3)$
26. $10 + [8 + (-5)] + (-20)$
27. $[6 + (-2)] + [3 + (-1)]$
28. $[18 + (-5)] + [9 + (-10)]$
29. $20 + (-6) + [3 + (-9)]$
30. $18 + (-2) + [9 + (-13)]$
31. $-3 + (-2) + [5 + (-4)]$
32. $-6 + (-5) + [-4 + (-1)]$
33. $(-9 + 2) + [5 + (-8)] + (-4)$
34. $(-7 + 3) + [9 + (-6)] + (-5)$
35. $[-6 + (-4)] + [7 + (-5)] + (-9)$
36. $[-8 + (-1)] + [8 + (-6)] + (-6)$
37. $(-6 + 9) + (-5) + (-4 + 3) + 7$
38. $(-10 + 4) + (-3) + (-3 + 8) + 6$

The problems that follow involve some multiplication. Be sure that you work inside the parentheses first, then multiply, and, finally, add left to right.

39. $-5 + 2(-3 + 7)$
40. $-3 + 4(-2 + 7)$
41. $9 + 3(-8 + 10)$
42. $4 + 5(-2 + 6)$
43. $-10 + 2(-6 + 8) + (-2)$
44. $-20 + 3(-7 + 10) + (-4)$
45. $2(-4 + 7) + 3(-6 + 8)$
46. $5(-2 + 5) + 7(-1 + 6)$

Each sequence below is an arithmetic sequence. In each case, find the next two numbers in the sequence.

47. $3, 8, 13, 18, \ldots$
48. $1, 5, 9, 13, \ldots$
49. $10, 15, 20, 25, \ldots$
50. $10, 16, 22, 28, \ldots$
51. $20, 15, 10, 5, \ldots$
52. $24, 20, 16, 12, \ldots$
53. $6, 0, -6, \ldots$
54. $1, 0, -1, \ldots$
55. $8, 4, 0, \ldots$
56. $5, 2, -1, \ldots$
57. Is the sequence of odd numbers an arithmetic sequence?
58. Is the sequence of squares an arithmetic sequence?

Recall that the word *sum* indicates addition. Write the numerical expression that is equivalent to each of the following phrases and then simplify.

59. The sum of 5 and 9
60. The sum of 6 and -3
61. Four added to the sum of -7 and -5
62. Six added to the sum of -9 and 1
63. The sum of -2 and -3 increased by 10
64. The sum of -4 and -12 increased by 2

Answer the following questions.

65. What number do you add to -8 to get -5?
66. What number do you add to 10 to get 4?
67. The sum of what number and -6 is -9?
68. The sum of what number and -12 is 8?

Applying the Concepts

69. The temperature at noon is 12 degrees below zero Fahrenheit. By 1:00 it has risen 4 degrees. Write an expression using the numbers -12 and 4 to describe this situation.

70. On Monday a certain stock gains 2 points. On Tuesday it loses 3 points. Write an expression using positive and negative numbers with addition to describe this situation and then simplify.

71. On three consecutive hands of draw poker a gambler wins $10, loses $6, and then loses another $8. Write an expression using positive and negative numbers and addition to describe this situation and then simplify.

72. You know from your past experience with numbers that subtracting 5 from 8 results in 3: $(8 - 5 = 3)$. What addition problem that starts with the number 8 gives the same result?

73. Suppose that you balance your checkbook and find that you are overdrawn by $30. That is, your balance is $-$30. Then you go to the bank and deposit $40. Translate this situation into an addition problem, the answer to which gives the new balance in your checkbook.

74. The balance in your checkbook is $-$25. If you make a deposit of $75, and then write a check for $18, what is the new balance?

75. Colleen has a job for which her starting wage is $6.50 per hour. If she gets a raise of $0.25 every six months, write a sequence of numbers that gives her wage every six months for the first three years. Is the sequence an arithmetic sequence?

76. Benja has a coin collection that she purchased for $650. Each year the collection increases $50 in value. Write a sequence of numbers that gives the value of the coin collection every year for the first five years. Is this sequence an arithmetic sequence?

Research Project 2

As we mentioned in the introduction to this chapter, there is a relationship between the sequence of odd numbers and the sequence of squares.

Sequence of odd numbers = 1, 3, 5, 7, . . .

Sequence of squares = 1, 4, 9, 16, . . .

The relationship begins to appear when we add successive terms in the sequence of odd numbers. Give a written description of this relationship, along with a diagram that illustrates the relationship. Write neatly in complete sentences.

SECTION

1.4 Subtraction of Real Numbers

Suppose that the temperature at noon is 20° Fahrenheit and 12 hours later, at midnight, it has dropped to $-15°$ Fahrenheit. What is the difference between the temperature at noon and the temperature at midnight? Intuitively, we know the difference in the two temperatures is 35°. We also know that the word difference indicates subtraction. The difference between 20 and -15 is written

$$20 - (-15)$$

It must be true that $20 - (-15) = 35$. In this section we will see how our definition for subtraction confirms that this last statement is in fact correct.

In the last section we spent some time developing the rule for addition of real numbers. Since we want to make as few rules as possible, we can define subtraction in terms of addition. By doing so, we can then use the rule for addition to solve our subtraction problems.

> **Rule**
> To subtract one real number from another, simply add its opposite.
>   Algebraically, the rule is written like this: If a and b represent two real numbers, then it is always true that
>
> $$\underbrace{a - b}_{} \quad = \quad \underbrace{a + (-b)}_{}$$
>
> To subtract b add the opposite of b

This is how subtraction is defined in algebra. This definition of subtraction will not conflict with what you already know about subtraction, but it will allow you to do subtraction using negative numbers.

▶ **EXAMPLE 1**   Subtract all possible combinations of positive and negative 7 and 2.

Solution

$$\left.\begin{array}{l} 7 - 2 = \quad 7 + (-2) = 5 \\ -7 - 2 = -7 + (-2) = -9 \end{array}\right\} \quad \begin{array}{l} \text{Subtracting 2 is the same} \\ \quad \text{as adding } -2 \end{array}$$

$$\left.\begin{array}{l} 7 - (-2) = \quad 7 + 2 = 9 \\ -7 - (-2) = -7 + 2 = -5 \end{array}\right\} \quad \begin{array}{l} \text{Subtracting } -2 \text{ is the same} \\ \quad \text{as adding 2} \end{array} \quad ◀$$

Notice that each subtraction problem is first changed to an addition problem. The rule for addition is then used to arrive at the answer.

We have defined subtraction in terms of addition, and we still obtain answers consistent with the answers we are used to getting with subtraction. Moreover, we can now do subtraction problems involving both positive and negative numbers.

As you proceed through the following examples and the problem set, you will begin to notice shortcuts you can use in working the problems. You will not always have to change subtraction to addition of the opposite to be able to get answers quickly. Use all the shortcuts you wish as long as you consistently get the correct answers.

▶ **EXAMPLE 2**   Subtract all combinations of positive and negative 8 and 13.

Solution

$$\left.\begin{array}{l} 8 - 13 = \quad 8 + (-13) = -5 \\ -8 - 13 = -8 + (-13) = -21 \end{array}\right\} \quad \begin{array}{l} \text{Subtracting } +13 \text{ is the} \\ \quad \text{same as adding } -13 \end{array}$$

$$\left.\begin{array}{l} 8 - (-13) = \quad 8 + 13 = 21 \\ -8 - (-13) = -8 + 13 = 5 \end{array}\right\} \quad \begin{array}{l} \text{Subtracting } -13 \text{ is the} \\ \quad \text{same as adding } +13 \end{array} \quad ◀$$

▶ **EXAMPLES** Simplify each expression as much as possible.

3. $7 + (-3) - 5 = 7 + (-3) + (-5)$ Begin by changing all
$\qquad\qquad\quad\; = 4 + (-5)$ subtractions to additions
$\qquad\qquad\quad\; = -1$ Then add left to right

4. $8 - (-2) - 6 = 8 + 2 + (-6)$ Begin by changing all
$\qquad\qquad\quad\;\; = 10 + (-6)$ subtractions to additions
$\qquad\qquad\quad\;\; = 4$ Then add left to right

5. $-2 - (-3 + 1) - 5 = -2 - (-2) - 5$ Do what is in the parentheses
$\qquad\qquad\qquad\;\; = -2 + 2 + (-5)$ first
$\qquad\qquad\qquad\;\; = -5$ ◀

The next two examples involve multiplication and exponents as well as subtraction. Remember, according to the rule for order of operations, we evaluate the numbers containing exponents and multiply before we subtract.

▶ **EXAMPLE 6** Simplify $2 \cdot 5 - 3 \cdot 8 - 4 \cdot 9$.

Solution First, we multiply left to right, and then we subtract.

$$2 \cdot 5 - 3 \cdot 8 - 4 \cdot 9 = 10 - 24 - 36$$
$$= -14 - 36$$
$$= -50$$

▶ **EXAMPLE 7** Simplify $3 \cdot 2^3 - 2 \cdot 4^2$.

Solution We begin by evaluating each number that contains an exponent. Then, we multiply before we subtract:

$$3 \cdot 2^3 - 2 \cdot 4^2 = 3 \cdot 8 - 2 \cdot 16$$
$$= 24 - 32$$
$$= -8$$

▶ **EXAMPLE 8** Subtract 7 from -3.

Solution First, we write the problem in terms of subtraction. We then change to addition of the opposite:

$$-3 - 7 = -3 + (-7)$$
$$= -10$$

▶ **EXAMPLE 9**   Subtract −5 from 2.

Solution   Subtracting −5 is the same as adding +5:

$$2 - (-5) = 2 + 5$$
$$= 7 \quad ◀$$

▶ **EXAMPLE 10**   Find the difference of 9 and 2.

Solution   Written in symbols, the problem looks like this:

$$9 - 2 = 7$$

The difference of 9 and 2 is 7.   ◀

▶ **EXAMPLE 11**   Find the difference of 3 and −5.

Solution   Subtracting −5 from 3 we have

$$3 - (-5) = 3 + 5$$
$$= 8 \quad ◀$$

In the sport of drag racing, two cars at the starting line race to the finish line $\frac{1}{4}$ mile away. The car that crosses the finish line first wins the race.

Jim Rizzoli owns and races an alcohol-fueled dragster. On board the dragster is a computer that records data during each of Jim's races. Table 1 gives some of the data from a race Jim was in during the 1993 Winternationals.

TABLE 1   Speed of Race Car

Time in Seconds	Speed in Miles/Hour
0	0
1	72.7
2	129.9
3	162.8
4	192.2
5	212.4
6	228.1

▶ **EXAMPLE 12**   Use Table 1 to find the difference in speed after 5 seconds and after 2 seconds have elapsed during the race.

Solution   We know the word difference implies subtraction. The speed at 2 seconds is 129.9 miles per hour, while the speed at 5 seconds is 212.4 miles per hour. Therefore, the expression that represents the solution to our problem looks like this:

$$212.4 - 129.9 = 82.5 \text{ miles per hour}$$ ◀

Facts from Geometry: Complementary and Supplementary Angles

If you have studied geometry at all, you know that there are 360° in a full rotation; the number of degrees swept out by the radius of a circle as it rotates once around the circle.

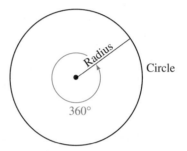

The number 360 was decided upon by early civilizations because it was believed that the earth was at the center of the universe and the sun would rotate once around the earth every 360 days.

We can apply our knowledge of algebra to help solve some simple geometry problems. Before we do, however, we need to review some of the vocabulary associated with angles.

DEFINITION   In geometry, two angles that add to 90° are called **complementary angles.** In a similar manner, two angles that add to 180° are called **supplementary angles.** The diagrams below illustrate the relationships between angles that are complementary and between angles that are supplementary.

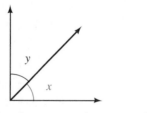

Complementary angles: $x + y = 90°$

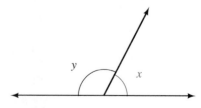

Supplementary angles: $x + y = 180°$

▶ **EXAMPLE 13**   Find x in each of the following diagrams.

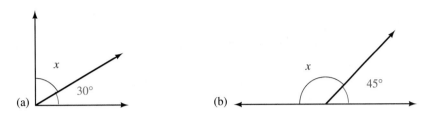

Solution   We use subtraction to find each angle.

(a) Since the two angles are complementary, we can find x by subtracting 30° from 90°:

$$x = 90° - 30° = 60°$$

We say 30° and 60° are complementary angles. The complement of 30° is 60°.

(b) The two angles in the diagram are supplementary. To find x we subtract 45° from 180°:

$$x = 180° - 45° = 135°$$

We say 45° and 135° are supplementary angles. The supplement of 45° is 135°.   ◀

Subtraction and Taking Away

For some people taking algebra for the first time, subtraction of positive and negative numbers can be a problem. These people may believe that $-5 - 9$ should be -4 or 4, not -14. If this is happening to you, you are probably thinking of subtraction in terms of taking one number away from another. Thinking of subtraction in this way works well with positive numbers if you always subtract the smaller number from the larger. In algebra, however, we encounter many situations other than this. The definition of subtraction, that $a - b = a + (-b)$, clearly indicates the correct way to use subtraction. That is, when working subtraction problems, you should think "addition of the opposite," not "take one number away from another." To be successful in algebra, you need to apply properties and definitions exactly as they are presented here.

P R O B L E M   S E T   1 . 4

The following problems are intended to give you practice with subtraction of positive and negative numbers. Remember, in algebra, subtraction is not taking one number away from another. Instead, subtracting a number is equivalent to adding its opposite.

Subtract.

1. $5 - 8$ **2.** $6 - 7$
3. $3 - 9$ **4.** $2 - 7$
5. $5 - 5$ **6.** $8 - 8$
7. $-8 - 2$ **8.** $-6 - 3$
9. $-4 - 12$ **10.** $-3 - 15$
11. $-6 - 6$ **12.** $-3 - 3$
13. $-8 - (-1)$ **14.** $-6 - (-2)$
15. $15 - (-20)$ **16.** $20 - (-5)$
17. $-4 - (-4)$ **18.** $-5 - (-5)$

Simplify each expression by applying the rule for order of operations.

19. $3 - 2 - 5$ **20.** $4 - 8 - 6$
21. $9 - 2 - 3$ **22.** $8 - 7 - 12$
23. $-6 - 8 - 10$ **24.** $-5 - 7 - 9$
25. $-22 + 4 - 10$ **26.** $-13 + 6 - 5$
27. $10 - (-20) - 5$ **28.** $15 - (-3) - 20$
29. $8 - (2 - 3) - 5$ **30.** $10 - (4 - 6) - 8$
31. $7 - (3 - 9) - 6$ **32.** $4 - (3 - 7) - 8$
33. $5 - (-8 - 6) - 2$ **34.** $4 - (-3 - 2) - 1$
35. $-(5 - 7) - (2 - 8)$
36. $-(4 - 8) - (2 - 5)$
37. $-(3 - 10) - (6 - 3)$
38. $-(3 - 7) - (1 - 2)$
39. $16 - [(4 - 5) - 1]$
40. $15 - [(4 - 2) - 3]$
41. $5 - [(2 - 3) - 4]$ **42.** $6 - [(4 - 1) - 9]$
43. $21 - [-(3 - 4) - 2] - 5$
44. $30 - [-(10 - 5) - 15] - 25$

The following problems involve multiplication and exponents. Use the rule for order of operations to simplify each expression as much as possible.

45. $2 \cdot 8 - 3 \cdot 5$ **46.** $3 \cdot 4 - 6 \cdot 7$
47. $3 \cdot 5 - 2 \cdot 7$ **48.** $6 \cdot 10 - 5 \cdot 20$

49. $5 \cdot 9 - 2 \cdot 3 - 6 \cdot 2$
50. $4 \cdot 3 - 7 \cdot 1 - 9 \cdot 4$
51. $3 \cdot 8 - 2 \cdot 4 - 6 \cdot 7$
52. $5 \cdot 9 - 3 \cdot 8 - 4 \cdot 5$
53. $2 \cdot 3^2 - 5 \cdot 2^2$ **54.** $3 \cdot 7^2 - 2 \cdot 8^2$
55. $4 \cdot 3^3 - 5 \cdot 2^3$
56. $3 \cdot 6^2 - 2 \cdot 3^2 - 8 \cdot 6^2$

Rewrite each of the following phrases as an equivalent expression in symbols and then simplify.

57. Subtract 4 from -7.
58. Subtract 5 from -19.
59. Subtract -8 from 12.
60. Subtract -2 from 10.
61. Subtract -7 from -5.
62. Subtract -9 from -3.
63. Subtract 17 from the sum of 4 and -5.
64. Subtract -6 from the sum of 6 and -3.

Recall that the word *difference* indicates subtraction. The difference of a and b is $a - b$, in that order. Write a numerical expression that is equivalent to each of the following phrases and then simplify.

65. The difference of 8 and 5.
66. The difference of 5 and 8.
67. The difference of -8 and 5.
68. The difference of -5 and 8.
69. The difference of 8 and -5.
70. The difference of 5 and -8.

Answer the following questions.

71. What number do you subtract from 8 to get -2?
72. What number do you subtract from 1 to get -5?
73. What number do you subtract from 8 to get 10?
74. What number do you subtract from 1 to get 5?

Applying the Concepts

75. A man with $1,500 in a savings account makes a withdrawal of $730. Write an expression using subtraction that describes this situation.

76. The temperature inside a space shuttle is 73 degrees F before reentry. During reentry the temperature inside the craft increases 10 degrees. Upon landing it drops 8 degrees. Write an expression using the numbers 73, 10, and 8 to describe this situation. What is the temperature inside the shuttle upon landing?

77. A man who has lost $35 playing roulette in Las Vegas wins $15 playing blackjack. He then loses $20 playing the wheel of fortune. Write an expression using the numbers −35, 15, and 20 to describe this situation and then simplify it.

78. An airplane flying at 10,000 feet lowers its altitude by 1,500 feet to avoid other air traffic. Then, it increases its altitude by 3,000 feet to clear a mountain range. Write an expression that describes this situation and then simplify it.

79. Bob has $98 in his checking account when he writes a check for $65 and then another check for $53. Write a subtraction problem that gives the new balance in Bob's checkbook. What is his new balance?

80. The temperature at noon is 23° F. Six hours later it has dropped 19° F and by midnight it has dropped another 10° F. Write a subtraction problem that gives the temperature at midnight. What is the temperature at midnight?

81. Stacey buys a used car for $4,500. With each year that passes, the car drops $550 in value. Write a sequence of numbers that gives the value of the car at the beginning of each of the first 5 years she owns it. Can this sequence be considered an arithmetic sequence?

82. Wade buys a computer system for $6,575. Each year after that he finds that the system is worth $1,250 less than it was the year before. Write a sequence of numbers that gives the value of the computer system at the beginning of each of the first four years he owns it. Can this sequence be considered an arithmetic sequence?

Table 2 extends the information given in Table 1 of this section. In addition to showing the time and speed of Jim Rizzoli's dragster during a race, it also shows the distance past the starting line that his dragster has traveled. Use the information in Table 2 to answer the following questions.

TABLE 2   Speed and Distance for a Race Car

Time in Seconds	Speed in Miles/Hour	Distance Traveled in Feet
0	0	0
1	72.7	69
2	129.9	231
3	162.8	439
4	192.2	728
5	212.4	1,000
6	228.1	1,373

83. Find the difference in the distance traveled by the dragster after 5 seconds and after 2 seconds.

84. How much faster is he traveling after 4 seconds than he is after 2 seconds?

85. How far from the starting line is he after 3 seconds?

86. How far from the starting line is he when his speed is 192.2 miles per hour?

87. How many seconds have gone by between the time his speed is 162.8 miles per hour and the time at which he has traveled 1,000 feet?

88. How many seconds have gone by between the time at which he has traveled 231 feet and the time at which his speed is 228.1 miles per hour?

Table 3 gives the weight of a lamb at the end of each month of the first six months of its life. (Note that when time is 0, the lamb weighs 12 pounds. This is its birth weight.) Use the information in the table to answer questions 89–92.

TABLE 3   Weight of a Lamb

Time in Months	Weight in Pounds
0	12
1	46
2	68
3	98
4	122
5	149
6	173

89. How much more does the lamb weigh when it is 5 months old than when it is 2 months old?

90. Find the difference in weight at 6 months and at 3 months.

91. During which month does the lamb gain the most weight?

92. During which month does the lamb gain the least weight?

Find x in each of the following diagrams.

93.

94.

95.

96.

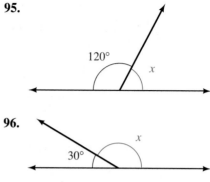

SECTION

1.5

Properties of Real Numbers

In this section we will list all the facts (properties) that you know from past experience are true about numbers in general. We will give each property a name so we can refer to it later in the book. Mathematics is very much like a game. The game involves numbers. The rules of the game are the properties and rules we are developing in this chapter. The goal of the game is to extend the basic rules to as many new situations as possible.

You know from past experience with numbers that it makes no difference in which order you add two numbers. That is, $3 + 5$ is the same as $5 + 3$. This fact about numbers is called the **commutative property of addition.** We say addition is a commutative operation. Changing the order of the numbers does not change the answer.

There is one other basic operation that is commutative. Since $3(5)$ is the same as $5(3)$, we say multiplication is a commutative operation. Changing the order of the two numbers you are multiplying does not change the answer.

For all properties listed in this section, a, b, and c represent real numbers.

Commutative Property of Addition

In symbols: $a + b = b + a$

In words: Changing the *order* of the numbers in a sum will not change the result.

Commutative Property of Multiplication

In symbols: $a \cdot b = b \cdot a$

In words: Changing the *order* of the numbers in a product will not change the result.

▶ **EXAMPLES**

1. The statement $5 + 8 = 8 + 5$ is an example of the commutative property of addition.
2. The statement $2 \cdot y = y \cdot 2$ is an example of the commutative property of multiplication.
3. The expression $5 + x + 3$ can be simplified using the commutative property of addition:

$$5 + x + 3 = x + 5 + 3 \qquad \text{Commutative property of addition}$$
$$= x + 8 \qquad \text{Addition} \qquad ◀$$

Note At this point, some students are confused by the expression $x + 8$; they feel that there is more to do, but they don't know what. At this point, there isn't any more that can be done with $x + 8$, unless we know what x is. So $x + 8$ is as far as we can go with this problem.

The other two basic operations, subtraction and division, are not commutative. The order in which we subtract or divide two numbers makes a difference in the answer.

Another property of numbers that you have used many times has to do with grouping. You know that when we add three numbers it makes no difference which two we add first. When adding $3 + 5 + 7$, we can add the 3 and 5 first and then the 7, or we can add the 5 and 7 first and then the 3. Mathematically, it looks like this: $(3 + 5) + 7 = 3 + (5 + 7)$. This property is true of multiplication as well. Operations that behave in this manner are called **associative** operations. The answer will not change when we change the association (or grouping) of the numbers.

Associative Property of Addition

In symbols: $a + (b + c) = (a + b) + c$

In words: Changing the *grouping* of the numbers in a sum will not change the result.

Associative Property of Multiplication

In symbols: $a(bc) = (ab)c$

In words: Changing the *grouping* of the numbers in a product will not change the result.

The following examples illustrate how the associative properties can be used to simplify expressions that involve both numbers and variables.

▶ **EXAMPLES** Simplify.

4. $4 + (5 + x) = (4 + 5) + x$ Associative property of addition
$\qquad\qquad\quad = 9 + x$ Addition

5. $5(2x) = (5 \cdot 2)x$ Associative property of multiplication
$\qquad\ = 10x$ Multiplication

6. $\dfrac{1}{5}(5x) = \left(\dfrac{1}{5} \cdot 5\right) x$ Associative property of multiplication

$\qquad\ = 1x$ Multiplication
$\qquad\ = x$ ◀

The associative and commutative properties apply to problems that are either all multiplication or all addition. There is a third basic property that involves both addition and multiplication. It is called the **distributive property** and looks like this:

Distributive Property

In symbols: $a(b + c) = ab + ac$
In words: Multiplication *distributes* over addition.

You will see as we progress through the book that the distributive property is used very frequently in algebra. We can give a visual justification to the distributive property by finding the areas of rectangles. Figure 1 shows a large rectangle that is made up of two smaller rectangles.

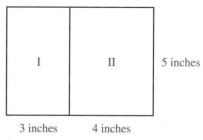

Figure 1

We can find the area of the large rectangle two different ways.

Method 1   We can calculate the area of the large rectangle directly by finding its length and width. The width is 5 inches, and the length is $(3 + 4)$ inches.

$$\text{Area of large rectangle} = 5(3 + 4)$$
$$= 5(7)$$
$$= 35 \text{ square inches}$$

Method 2 Since the area of the large rectangle is the sum of the areas of the two smaller rectangles, we find the area of each small rectangle and then add to find the area of the large rectangle.

Area of large rectangle	=	Area of rectangle I	+	Area of rectangle II
	=	5(3)	+	5(4)
	=	15	+	20
	=	35 square inches		

In both cases the result is 35 square inches. Since the results are the same, the two original expressions must be equal. Stated mathematically, $5(3 + 4) = 5(3) + 5(4)$. We can either add the 3 and the 4 first and then multiply that sum by 5, or we can multiply the 3 and the 4 separately by 5 and then add the results. In either case, we get the same answer.

Note Since subtraction is defined in terms of addition, it is also true that the distributive property applies to subtraction as well as addition. That is, $a(b - c) = ab - ac$ for any three real numbers a, b, and c.

Here are some examples that illustrate how we use the distributive property:

▶ **EXAMPLES** Apply the distributive property to each expression and then simplify the result.

7. $2(x + 3) = 2(x) + 2(3)$ Distributive property
        $= 2x + 6$ Multiplication

8. $5(2x - 8) = 5(2x) - 5(8)$ Distributive property
        $= 10x - 40$ Multiplication

Notice in this example that multiplication distributes over subtraction as well as addition.

9. $4(x + y) = 4x + 4y$ Distributive property

10. $5(2x + 4y) = 5(2x) + 5(4y)$ Distributive property
         $= 10x + 20y$ Multiplication

11. $\dfrac{1}{2}(3x + 6) = \dfrac{1}{2}(3x) + \dfrac{1}{2}(6)$ Distributive property

         $= \dfrac{3}{2}x + 3$ Multiplication

12. $4(2a + 3) + 8 = 4(2a) + 4(3) + 8$ Distributive property
         $= 8a + 12 + 8$ Multiplication
         $= 8a + 20$ Addition

Special Numbers

In addition to the three properties mentioned so far, we want to include in our list two special numbers that have unique properties. They are the numbers zero and one.

Additive Identity Property

There exists a unique number 0 such that

In symbols: $a + 0 = a$ and $0 + a = a$

In words: Zero preserves identities under addition. (The identity of the number is unchanged after addition with 0.)

Multiplicative Identity Property

There exists a unique number 1 such that

In symbols: $a(1) = a$ and $1(a) = a$

In words: The number 1 preserves identities under multiplication. (The identity of the number is unchanged after multiplication by 1.)

Additive Inverse Property

For each real number a, there exists a unique number $-a$ such that

In symbols: $a + (-a) = 0$

In words: Opposites add to 0.

Multiplicative Inverse Property

For every real number a, except 0, there exists a unique real number $\dfrac{1}{a}$ such that

In symbols: $a\left(\dfrac{1}{a}\right) = 1$

In words: Reciprocals multiply to 1.

Note that these last two properties indicate that the number 0 has an additive inverse but not a multiplicative inverse. We indicated this previously when we said 0 was the only number without a reciprocal.

Of all the basic properties listed, the commutative, associative, and distributive properties are the ones we will use most often. They are important because they will be used as justifications or reasons for many of the things we will do in the future.

The following examples illustrate how we use the preceding properties. Each one contains an algebraic expression that has been changed in some way. The property that justifies the change is written to the right.

▶ **EXAMPLES**   State the property that justifies the given statement.

13. $x + 5 = 5 + x$ Commutative property of addition
14. $(2 + x) + y = 2 + (x + y)$ Associative property of addition
15. $6(x + 3) = 6x + 18$ Distributive property
16. $2 + (-2) = 0$ Additive inverse property

17. $3\left(\dfrac{1}{3}\right) = 1$ Multiplicative inverse property

18. $(2 + 0) + 3 = 2 + 3$ 0 is the identity element for addition
19. $(2 + 3) + 4 = 3 + (2 + 4)$ Commutative and associative
                                                          properties of addition
20. $(x + 2) + y = (x + y) + 2$ Commutative and associative
                                                          properties of addition   ◀

As a final note on the properties of real numbers, we should mention that although some of the properties are stated for only two or three real numbers, they hold for as many numbers as needed. For example, the distributive property holds for expressions like $3(x + y + z + 5 + 2)$. That is,

$$3(x + y + z + 5 + 2) = 3x + 3y + 3z + 15 + 6$$

It is not important how many numbers are contained in the sum, only that it is a sum. Multiplication, you see, distributes over addition, whether there are two numbers in the sum or two hundred.

P R O B L E M   S E T   1 . 5

State the property or properties that justify the following.

1. $3 + 2 = 2 + 3$ **2.** $5 + 0 = 5$

3. $4\left(\dfrac{1}{4}\right) = 1$ **4.** $10(0.1) = 1$

5. $4 + x = x + 4$
6. $3(x - 10) = 3x - 30$
7. $2(y + 8) = 2y + 16$
8. $3 + (4 + 5) = (3 + 4) + 5$

9. $(3 + 1) + 2 = 1 + (3 + 2)$
10. $(5 + 2) + 9 = (2 + 5) + 9$
11. $(8 + 9) + 10 = (8 + 10) + 9$
12. $(7 + 6) + 5 = (5 + 6) + 7$
13. $3(x + 2) = 3(2 + x)$
14. $2(7y) = (7 \cdot 2)y$ **15.** $x(3y) = 3(xy)$
16. $a(5b) = 5(ab)$ **17.** $4(xy) = 4(yx)$
18. $3[2 + (-2)] = 3(0)$
19. $8[7 + (-7)] = 8(0)$ **20.** $7(1) = 7$

Each of the following problems has a mistake in it. Correct the right-hand side.

21. $3(x + 2) = 3x + 2$
22. $5(4 + x) = 4 + 5x$
23. $9(a + b) = 9a + b$
24. $2(y + 1) = 2y + 1$

25. $3(0) = 3$ **26.** $5\left(\dfrac{1}{5}\right) = 5$

27. $3 + (-3) = 1$ **28.** $8(0) = 8$

29. $10(1) = 0$ **30.** $3 \cdot \dfrac{1}{3} = 0$

Use the associative property to rewrite each of the following expressions and then simplify the result. (See Examples 4, 5, and 6.)

31. $4 + (2 + x)$ **32.** $5 + (6 + x)$
33. $(x + 2) + 7$ **34.** $(x + 8) + 2$
35. $3(5x)$ **36.** $5(3x)$
37. $9(6y)$ **38.** $6(9y)$

39. $\dfrac{1}{2}(3a)$ **40.** $\dfrac{1}{3}(2a)$

41. $\dfrac{1}{3}(3x)$ **42.** $\dfrac{1}{4}(4x)$

43. $\dfrac{1}{2}(2y)$ **44.** $\dfrac{1}{7}(7y)$

45. $\dfrac{3}{4}\left(\dfrac{4}{3}x\right)$ **46.** $\dfrac{3}{2}\left(\dfrac{2}{3}x\right)$

47. $\dfrac{6}{5}\left(\dfrac{5}{6}a\right)$ **48.** $\dfrac{2}{5}\left(\dfrac{5}{2}a\right)$

Apply the distributive property to each of the following expressions. Simplify when possible.

49. $8(x + 2)$ **50.** $5(x + 3)$
51. $8(x - 2)$ **52.** $5(x - 3)$
53. $4(y + 1)$ **54.** $4(y - 1)$
55. $3(6x + 5)$ **56.** $3(5x + 6)$
57. $2(3a + 7)$ **58.** $5(3a + 2)$
59. $9(6y - 8)$ **60.** $2(7y - 4)$

61. $\dfrac{1}{2}(3x - 6)$ **62.** $\dfrac{1}{3}(2x - 6)$

63. $\dfrac{1}{3}(3x + 6)$ **64.** $\dfrac{1}{2}(2x + 4)$

65. $3(x + y)$ **66.** $2(x - y)$
67. $8(a - b)$ **68.** $7(a + b)$
69. $6(2x + 3y)$ **70.** $8(3x + 2y)$
71. $4(3a - 2b)$ **72.** $5(4a - 8b)$

73. $\dfrac{1}{2}(6x + 4y)$ **74.** $\dfrac{1}{3}(6x + 9y)$

75. $4(a + 4) + 9$ **76.** $6(a + 2) + 8$
77. $2(3x + 5) + 2$ **78.** $7(2x + 1) + 3$
79. $7(2x + 4) + 10$ **80.** $3(5x + 6) + 20$

Applying the Concepts

81. While getting dressed for work, a man puts on his socks and puts on his shoes. Are the two statements "put on your socks" and "put on your shoes" commutative?

82. Are the statements "put on your left shoe" and "put on your right shoe" commutative?

83. A skydiver flying over the jump area is about to do two things: jump out of the plane and pull the rip cord. Are the two events "jump out of the plane" and "pull the rip cord" commutative? That is, will changing the order of the events always produce the same result?

84. Give an example of two events in your daily life that are commutative.

85. Give an example that shows that division is not a commutative operation. That is, find two numbers for which changing the order of division gives two different answers.

86. Simplify the expression $10 - (5 - 2)$ and the expression $(10 - 5) - 2$ to show that subtraction is not an associative operation.

87. Jose works at a winery. His monthly salary is $2,400. To cover his taxes and retirement, the winery withholds $480 from each check. Calculate his yearly "take-home" pay using the numbers 2,400, 480, and 12. Do the calculation two different ways so that the results give further justification for the distributive property.

88. Carlo works as a waiter. He works double shifts 4 days a week. The lunch shift is 2 hours and the dinner shift is 3 hours. Find the total number of hours he works per week using the numbers 2, 3, and 4. Do the calculation two different ways so that the results give further justification for the distributive property.

89. Use the distributive property to rewrite the formula for the perimeter of a rectangle: $P = 2l + 2w$.

90. Maria is estimating her expenses for attending college for a year. Tuition is $650 per academic quarter. She estimates she will spend $225 on books each quarter. If she plans on attending 3 academic quarters during the year, how much can she expect to spend? Do the calculation two different ways so that the results give further justification for the distributive property.

SECTION

1.6   ## Multiplication of Real Numbers

Suppose that you own 5 shares of a stock and the price per share drops $3. How much money have you lost? Intuitively we know the loss is $15. Since it is a loss, we can express it as −$15. To describe this situation with numbers we would write

5 shares   each lose $3   for a total loss of $15

$$5(-3) = -15$$

Reasoning in this manner, we conclude that the product of a positive number with a negative number is a negative number. Let's look at multiplication in more detail.

From our experience with counting numbers, we know that multiplication is simply repeated addition. That is, $3(5) = 5 + 5 + 5$. We will use this fact, along with our knowledge of negative numbers, to develop the rule for multiplication of any two real numbers. The following examples illustrate multiplication with three of the possible combinations of positive and negative numbers.

▶ **EXAMPLES**   Multiply.

1. Two positives: $3(5) = 5 + 5 + 5$
                                     $= 15$ Positive answer

2. One positive: $3(-5) = -5 + (-5) + (-5)$
                                     $= -15$ Negative answer

3. One negative: $-3(5) = 5(-3)$ Commutative property
                                     $= -3 + (-3) + (-3) + (-3) + (-3)$
                                     $= -15$ Negative answer

4. Two negatives: $-3(-5) = ?$

With two negatives, $-3(-5)$, it is not possible to work the problem in terms of repeated addition. (It doesn't ''make sense'' to write −5 down a −3 number of times.) The answer is probably +15 (that's just a guess), but we need some justification for saying so. We will solve a different problem and in so doing get the answer to the problem $(-3)(-5)$.

Here is a problem to which we know the answer. We will work it two different ways.

$$-3[5 + (-5)] = -3(0) = 0$$

The answer is zero. We can also work the problem using the distributive property.

$$-3[5 + (-5)] = -3(5) + (-3)(-5) \qquad \text{Distributive property}$$
$$= -15 + \quad ?$$

Since the answer to the problem is 0, our ? must be $+15$. (What else could we add to -15 to get 0? Only $+15$.) ◀

Note You may have to read the explanation for Example 4 several times before you understand it completely. The purpose of the explanation in Example 4 is simply to justify the fact that the product of two negative numbers is a positive number. If you have no trouble believing that, then it is not so important that you understand everything in the explanation.

Here is a summary of the results we have obtained from the first four examples:

Original numbers have		The answer is
the same sign	$3(5) = 15$	positive
different signs	$3(-5) = -15$	negative
different signs	$-3(5) = -15$	negative
the same sign	$-3(-5) = 15$	positive

By examining Examples 1 through 4 and the preceding table, we can use the information there to write the following rule. This rule tells us how to multiply any two real numbers.

Rule

To multiply any two real numbers, simply multiply their absolute values. The sign of the answer is

1. *positive* if both numbers have the same sign (both + or both −).
2. *negative* if the numbers have opposite signs (one +, the other −).

The following examples illustrate how we use the preceding rule to multiply real numbers.

▶ **EXAMPLES** Multiply.

5. $-8(-3) = 24$
6. $-10(-5) = 50$
7. $-4(-7) = 28$
$\left.\right\}$ If the two numbers in the product have the same sign, the answer is positive.

8. $5(-7) = -35$
9. $-4(8) = -32$
10. $-6(10) = -60$
$\left.\right\}$ If the two numbers in the product have different signs, the answer is negative. ◀

Note Some students have trouble with the expression $-8(-3)$ because they want to subtract rather than multiply. Because we are very precise with the notation we use in algebra, the expression $-8(-3)$ has only one meaning—multiplication. A subtraction problem that uses the same numbers is $-8 - 3$. Compare the two lists below

All Multiplication	*No Multiplication*
$5(4)$	$5 + 4$
$-5(4)$	$-5 + 4$
$5(-4)$	$5 - 4$
$-5(-4)$	$-5 - 4$

In the following examples, we combine the rule for order of operations with the rule for multiplication to simplify expressions. Remember, the rule for order of operations specifies that we are to work inside the parentheses first and then simplify numbers containing exponents. After this, we multiply and divide, left to right. The last step is to add and subtract, left to right.

▶ **EXAMPLES** Simplify as much as possible.

11. $\begin{aligned} -5(-3)(-4) &= 15(-4) \\ &= -60 \end{aligned}$

12. $\begin{aligned} 4(-3) + 6(-5) - 10 &= -12 + (-30) - 10 \qquad \text{Multiply} \\ &= -42 - 10 \qquad \text{Add} \\ &= -52 \qquad \text{Subtract} \end{aligned}$

13. $\begin{aligned} (-2)^3 &= (-2)(-2)(-2) \qquad \text{Definition of} \\ & \qquad\qquad\qquad\qquad\quad \text{exponents} \\ &= -8 \qquad\qquad\qquad\quad \text{Multiply, left to right} \end{aligned}$

14. $\begin{aligned} -3(-2)^3 - 5(-4)^2 &= -3(-8) - 5(16) \qquad \text{Exponents first} \\ &= 24 - 80 \qquad\qquad\quad \text{Multiply} \\ &= -56 \qquad\qquad\qquad \text{Subtract} \end{aligned}$

15. $\begin{aligned} 6 - 4(7 - 2) &= 6 - 4(5) \qquad \text{Inside parentheses} \\ & \qquad\qquad\qquad\quad \text{first} \\ &= 6 - 20 \qquad\quad \text{Multiply} \\ &= -14 \qquad\quad\;\; \text{Subtract} \end{aligned}$ ◀

Multiplying Fractions

Previously we mentioned that to multiply two fractions we multiply numerators and multiply denominators. We can apply the rule for multiplication of positive and negative numbers to fractions in the same way we apply it to other numbers. We multiply absolute values: the product is positive if both fractions have the same sign and negative if they have different signs. Here are some examples.

▶ **EXAMPLES** Multiply.

16. $-\dfrac{3}{4}\left(\dfrac{5}{7}\right) = -\dfrac{3 \cdot 5}{4 \cdot 7}$ Different signs give a negative answer

$\quad\quad\quad\quad\;\; = -\dfrac{15}{28}$

17. $-6\left(\dfrac{1}{2}\right) = -\dfrac{6}{1}\left(\dfrac{1}{2}\right)$ Different signs give a negative answer

$\quad\quad\quad\quad\;\; = -\dfrac{6}{2}$

$\quad\quad\quad\quad\;\; = -3$

18. $-\dfrac{2}{3}\left(-\dfrac{3}{2}\right) = \dfrac{2 \cdot 3}{3 \cdot 2}$ Same signs give a positive answer

$\quad\quad\quad\quad\;\; = \dfrac{6}{6}$

$\quad\quad\quad\quad\;\; = 1$

19. $\left(-\dfrac{5}{6}\right)^2 = \left(-\dfrac{5}{6}\right)\left(-\dfrac{5}{6}\right)$ Definition of exponents

$\quad\quad\quad\quad\;\; = \dfrac{25}{36}$ Multiply (same signs, positive answer)   ◀

We can use the rule for multiplication of real numbers, along with the associative property, to multiply expressions that contain numbers and variables.

▶ **EXAMPLES** Apply the associative property and then multiply.

20. $-3(2x) = (-3 \cdot 2)x$ Associative property
$\quad\quad\quad\;\; = -6x$ Multiplication

21. $6(-5y) = [6(-5)]y$ Associative property
$\quad\quad\quad\;\; = -30y$ Multiplication

22. $-2\left(-\dfrac{1}{2}x\right) = \left[(-2)\left(-\dfrac{1}{2}\right)\right]x$ Associative property

$\quad\quad\quad\quad\quad\; = 1x$ Multiplication
$\quad\quad\quad\quad\quad\; = x$ Multiplication   ◀

The following examples show how we can use both the distributive property and multiplication with real numbers.

▶ **EXAMPLES**   Apply the distributive property to each expression.

23. $-2(a + 3) = -2a + (-2)(3)$ Distributive property
$= -2a + (-6)$ Multiplication
$= -2a - 6$

24. $-3(2x + 1) = -3(2x) + (-3)(1)$ Distributive property
$= -6x + (-3)$ Multiplication
$= -6x - 3$

25. $-\dfrac{1}{3}(2x - 6) = -\dfrac{1}{3}(2x) - \left(-\dfrac{1}{3}\right)(6)$ Distributive property

$= -\dfrac{2}{3}x - (-2)$ Multiplication

$= -\dfrac{2}{3}x + 2$

26. $-4(3x - 5) - 8 = -4(3x) - (-4)(5) - 8$ Distributive property
$= -12x - (-20) - 8$ Multiplication
$= -12x + 20 - 8$ Definition of
                                    subtraction
$= -12x + 12$ Subtraction   ◀

Geometric Sequences

A **geometric sequence** is a sequence of numbers in which each number (after the first number) comes from the number before it by multiplying by the same amount each time. For example, the sequence

$$2, 6, 18, 54, \ldots$$

is a geometric sequence because each number is obtained by multiplying the number before it by 3.

▶ **EXAMPLE 27**   Each sequence below is a geometric sequence. Find the next number in each sequence.
(a) 5, 10, 20, . . .
(b) 3, −15, 75, . . .
(c) $\dfrac{1}{8}, \dfrac{1}{4}, \dfrac{1}{2}, \ldots$

Solution   Since each sequence is a geometric sequence, we know that each term is obtained from the previous term by multiplying by the same number each time.

(a) 5, 10, 20, . . . : Starting with 5, each number is obtained from the previous number by multiplying by 2 each time. The next number will be $20 \cdot 2 = 40$.

(b) 3, $-15, 75$, . . . : The sequence starts with 3. After that, each number is obtained by multiplying by -5 each time. The next number will be $75(-5) = -375$.

(c) $\frac{1}{8}, \frac{1}{4}, \frac{1}{2}$, . . . : This sequence starts with $\frac{1}{8}$. Multiplying each number in the sequence by 2 produces the next number in the sequence. To extend the sequence we multiply $\frac{1}{2}$ by 2:

$$\frac{1}{2} \cdot 2 = 1$$

The next number in the sequence is 1.   ◀

PROBLEM SET 1.6

Use the rule for multiplying two real numbers to find each of the following products.

1. $7(-6)$ **2.** $8(-4)$

3. $-7(3)$ **4.** $-5(4)$

5. $-8(2)$ **6.** $-16(3)$

7. $-3(-1)$ **8.** $-7(-1)$

9. $-11(-11)$ **10.** $-12(-12)$

Use the rule for order of operations to simplify each expression as much as possible.

11. $-3(2)(-1)$ **12.** $-2(3)(-4)$

13. $-3(-4)(-5)$ **14.** $-5(-6)(-7)$

15. $-2(-4)(-3)(-1)$ **16.** $-1(-3)(-2)(-1)$

17. $(-7)^2$ **18.** $(-8)^2$

19. $(-3)^3$ **20.** $(-2)^4$

21. $-2(2 - 5)$ **22.** $-3(3 - 7)$

23. $-5(8 - 10)$ **24.** $-4(6 - 12)$

25. $(4 - 7)(6 - 9)$ **26.** $(3 - 10)(2 - 6)$

27. $(-3 - 2)(-5 - 4)$ **28.** $(-3 - 6)(-2 - 8)$

29. $-3(-6) + 4(-1)$ **30.** $-4(-5) + 8(-2)$

31. $2(3) - 3(-4) + 4(-5)$

32. $5(4) - 2(-1) + 5(6)$

33. $4(-3)^2 + 5(-6)^2$ **34.** $2(-5)^2 + 4(-3)^2$

35. $7(-2)^3 - 2(-3)^3$ **36.** $10(-2)^3 - 5(-2)^4$

37. $6 - 4(8 - 2)$ **38.** $7 - 2(6 - 3)$

39. $9 - 4(3 - 8)$ **40.** $8 - 5(2 - 7)$

41. $-4(3 - 8) - 6(2 - 5)$

42. $-8(2 - 7) - 9(3 - 5)$

43. $7 - 2[-6 - 4(-3)]$ **44.** $6 - 3[-5 - 3(-1)]$

45. $7 - 3[2(-4 - 4) - 3(-1 - 1)]$

46. $5 - 3[7(-2 - 2) - 3(-3 + 1)]$

47. $8 - 6[-2(-3 - 1) + 4(-2 - 3)]$

48. $4 - 2[-3(-1 + 8) + 5(-5 + 7)]$

Multiply the following fractions. (See Examples 16–19.)

49. $-\dfrac{2}{3} \cdot \dfrac{5}{7}$ **50.** $-\dfrac{6}{5} \cdot \dfrac{2}{7}$

51. $-8\left(\dfrac{1}{2}\right)$ **52.** $-12\left(\dfrac{1}{3}\right)$

53. $-\dfrac{3}{4}\left(-\dfrac{4}{3}\right)$ **54.** $-\dfrac{5}{8}\left(-\dfrac{8}{5}\right)$

55. $\left(-\dfrac{3}{4}\right)^2$ **56.** $\left(-\dfrac{2}{5}\right)^2$

57. $\left(-\dfrac{2}{3}\right)^3$ **58.** $\left(-\dfrac{1}{2}\right)^3$

Find the following products. (See Examples 20, 21, and 22.)

59. $-2(4x)$ **60.** $-8(7x)$

61. $-7(-6x)$ **62.** $-8(-9x)$

63. $-\dfrac{1}{3}(-3x)$ **64.** $-\dfrac{1}{5}(-5x)$

65. $-4\left(-\dfrac{1}{4}x\right)$ **66.** $-2\left(-\dfrac{1}{2}x\right)$

Apply the distributive property to each expression and then simplify the result. (See Examples 23–26.)

67. $-4(a + 2)$ **68.** $-7(a + 6)$

69. $-\dfrac{1}{2}(3x - 6)$ **70.** $-\dfrac{1}{4}(2x - 4)$

71. $-3(2x - 5) - 7$ **72.** $-4(3x - 1) - 8$
73. $-5(3x + 4) - 10$ **74.** $-3(4x + 5) - 20$

75. Five added to the product of 3 and -10 is what number?

76. If the product of -8 and -2 is decreased by 4, what number results?

77. Write an expression for twice the product of -4 and x and then simplify it.

78. Write an expression for twice the product of -2 and $3x$ and then simplify it.

79. What number results if 8 is subtracted from the product of -9 and 2?

80. What number results if -8 is subtracted from the product of -9 and 2?

Each sequence below is a geometric sequence. In each case, find the next number in the sequence.

81. $1, 2, 4, \ldots$ **82.** $1, 5, 25, \ldots$
83. $10, -20, 40, \ldots$ **84.** $10, -30, 90, \ldots$

85. $1, \dfrac{1}{2}, \dfrac{1}{4}, \ldots$ **86.** $1, \dfrac{1}{3}, \dfrac{1}{9}, \ldots$

87. $8, 4, 2, \ldots$ **88.** $20, 10, 5, \ldots$
89. $3, -6, 12, \ldots$ **90.** $-3, 6, -12, \ldots$

Applying the Concepts

91. Suppose you own 20 shares of a stock. If the price per share drops $3, how much money have you lost?

92. Imagine that you purchase 50 shares of a stock at a price of $18 per share. If the stock is selling for $11 a share a week after you purchased it, how much money have you lost?

93. The temperature is 25° F at 5:00 in the afternoon. If the temperature drops 6° every hour after that, what is the temperature at 9:00 in the evening?

94. The temperature is -5° F at 6:00 in the evening. If the temperature drops 3° F every hour after that, what is the temperature at midnight?

95. Suppose you purchase $500 worth of a mutual fund and find that the value of your purchase doubles every 2 years. Write a sequence of numbers that gives the value of your purchase every 2 years for the first 10 years you own it. Is this sequence a geometric sequence?

96. A colony of 150 bacteria doubles in size every 8 hours. Starting with the original 150 bacteria, write a sequence of numbers that gives the number of bacteria in the colony every 8 hours during a 40-hour time span. Is this sequence a geometric sequence?

Research Project 3

As we mentioned when we began working with sequences, the type of reasoning used to extend a sequence is called inductive reasoning. Write an essay in which you start with a definition for inductive reasoning, then show an example of how it is used to extend a sequence of numbers. Follow this with some examples of everyday situations, not involving mathematics, in which inductive reasoning is used.

SECTION
....
1.7   # Division of Real Numbers

Suppose that you and four friends bought equal shares of an investment for a total of $15,000 and then sold it later for only $13,000. How much did each person lose? Since the total amount of money that was lost can be represented by

−$2,000, and there are five people with equal shares, we can represent each person's loss with division:

$$\frac{-\$2,000}{5} = -\$400$$

From this discussion it seems reasonable to say that a negative number divided by a positive number is a negative number. Here is a more detailed discussion of division with positive and negative numbers.

We will use the same approach to define division as we used for subtraction. That is, we will define division in terms of rules we already know.

Recall that we developed the rule for subtraction of real numbers by defining subtraction in terms of addition. We changed our subtraction problems to addition problems and then added to get our answers. Since we already have a rule for multiplication of real numbers, and division is the inverse operation of multiplication, we will simply define division in terms of multiplication.

We know that division by the number 2 is the same as multiplication by $\frac{1}{2}$. That is, 6 divided by 2 is 3, which is the same as 6 times $\frac{1}{2}$. Similarly, dividing a number by 5 gives the same result as multiplying it by $\frac{1}{5}$. We can extend this idea to all real numbers with the following rule.

Rule

If a and b represent any two real numbers (b cannot be 0), then it is always true that

$$a \div b = \frac{a}{b} = a\left(\frac{1}{b}\right)$$

Division by a number is the same as multiplication by its reciprocal. Since every division problem can be written as a multiplication problem and since we already know the rule for multiplication of two real numbers, we do not have to write a new rule for division of real numbers. We will simply replace our division problem with multiplication and use the rule we already have.

▶ **EXAMPLES** Write each division problem as an equivalent multiplication problem and then multiply.

1. $\dfrac{6}{2} = 6\left(\dfrac{1}{2}\right) = 3$ The product of two positives is positive

2. $\dfrac{6}{-2} = 6\left(-\dfrac{1}{2}\right) = -3$

3. $\dfrac{-6}{2} = -6\left(\dfrac{1}{2}\right) = -3$ The product of a positive and a negative is a negative

4. $\dfrac{-6}{-2} = -6\left(-\dfrac{1}{2}\right) = 3$ The product of two negatives is positive ◀

The second step in these examples is used only to show that we *can* write division in terms of multiplication. (In actual practice we wouldn't write $\frac{6}{2}$ as $6(\frac{1}{2})$.) The answers, therefore, follow from the rule for multiplication. That is, like signs produce a positive answer, and unlike signs produce a negative answer.

Here are some examples. This time we will not show division as multiplication by the reciprocal. We will simply divide. If the original numbers have the same signs, the answer will be positive. If the original numbers have different signs, the answer will be negative.

▶ **EXAMPLES**   Divide.

5. $\dfrac{12}{6} = 2$ Like signs give a positive answer

6. $\dfrac{12}{-6} = -2$ Unlike signs give a negative answer

7. $\dfrac{-12}{6} = -2$ Unlike signs give a negative answer

8. $\dfrac{-12}{-6} = 2$ Like signs give a positive answer

9. $\dfrac{15}{-3} = -5$ Unlike signs give a negative answer

10. $\dfrac{-40}{-5} = 8$ Like signs give a positive answer

11. $\dfrac{-14}{2} = -7$ Unlike signs give a negative answer   ◀

Division with Fractions

We can apply the definition of division to fractions. Since dividing by a fraction is equivalent to multiplying by its reciprocal, we can divide a number by the fraction $\frac{3}{4}$ by multiplying it by the reciprocal of $\frac{3}{4}$, which is $\frac{4}{3}$. For example,

$$\frac{2}{5} \div \frac{3}{4} = \frac{2}{5} \cdot \frac{4}{3} = \frac{8}{15}$$

You may have learned this rule in previous math classes. In some math classes, multiplication by the reciprocal is referred to as "inverting the divisor and multiplying." No matter how you say it, division by any number (except 0) is always equivalent to multiplication by its reciprocal. Here are additional examples that involve division by fractions.

▶ **EXAMPLES** Divide.

12. $\dfrac{2}{3} \div \dfrac{5}{7} = \dfrac{2}{3} \cdot \dfrac{7}{5}$ Rewrite as multiplication by the reciprocal

$= \dfrac{14}{15}$ Multiply

13. $-\dfrac{3}{4} \div \dfrac{7}{9} = -\dfrac{3}{4} \cdot \dfrac{9}{7}$ Rewrite as multiplication by the reciprocal

$= -\dfrac{27}{28}$ Multiply

14. $8 \div \left(-\dfrac{4}{5}\right) = \dfrac{8}{1}\left(-\dfrac{5}{4}\right)$ Rewrite as multiplication by the reciprocal

$= -\dfrac{40}{4}$ Multiply

$= -10$ Divide 40 by 4   ◀

The last step in each of the following examples involves reducing a fraction to lowest terms. To reduce a fraction to lowest terms, we divide the numerator and denominator by the largest number that divides each of them exactly. For example, to reduce $\frac{15}{20}$ to lowest terms, we divide 15 and 20 by 5 to get $\frac{3}{4}$.

▶ **EXAMPLES** Simplify as much as possible.

15. $\dfrac{-4(5)}{6} = \dfrac{-20}{6}$ Simplify numerator

$= -\dfrac{10}{3}$ Reduce to lowest terms by dividing numerator and denominator by 2

16. $\dfrac{30}{-4-5} = \dfrac{30}{-9}$ Simplify denominator

$= -\dfrac{10}{3}$ Reduce to lowest terms by dividing numerator and denominator by 3   ◀

In the examples that follow, the numerators and denominators contain expressions that are somewhat more complicated than those we have seen thus far. To apply the rule for order of operations to these examples, we treat fraction bars the same way we treat grouping symbols. That is, fraction bars separate numerators and denominators so that each will be simplified separately.

▶ **EXAMPLES** Simplify.

17. $\dfrac{2(-3)+4}{12} = \dfrac{-6+4}{12}$ In the numerator, we multiply before we add

$$= \frac{-2}{12} \qquad \text{Addition}$$

$$= -\frac{1}{6} \qquad \text{Reduce to lowest terms by dividing numerator and denominator by 2}$$

18. $\dfrac{5(-4) + 6(-1)}{2(3) - 4(1)} = \dfrac{-20 + (-6)}{6 - 4}$ Multiplication before addition

$$= \frac{-26}{2} \qquad \text{Simplify numerator and denominator}$$

$$= -13 \qquad \text{Divide } -26 \text{ by } 2 \qquad \blacktriangleleft$$

We must be careful when we are working with expressions such as $(-5)^2$ and -5^2, that we include the negative sign with the base only when parentheses indicate we are to do so.

Unless there are parentheses to indicate otherwise, we consider the base to be only the number directly below and to the left of the exponent. If we want to include a negative sign with the base, we must use parentheses.

To simplify a more complicated expression, we follow the same rule. For example,

$$7^2 - 3^2 = 49 - 9 \qquad \text{The bases are 7 and 3; the sign between the two terms is a subtraction sign}$$

For another example,

$$5^3 - 3^4 = 125 - 81 \qquad \text{We simplify exponents first, then subtract}$$

▶ **EXAMPLES**   Simplify.

19. $\dfrac{5^2 - 3^2}{-5 + 3} = \dfrac{25 - 9}{-2}$ Simplify numerator and denominator separately

$$= \frac{16}{-2}$$

$$= -8$$

20. $\dfrac{(3 + 2)^2}{-3^2 - 2^2} = \dfrac{5^2}{-9 - 4}$ Simplify numerator and denominator separately

$$= \frac{25}{-13}$$

$$= -\frac{25}{13}$$

21. $\dfrac{(2-8)-(3-6)}{-9+12} = \dfrac{(-6)-(-3)}{3}$ Simplify inside parentheses

$$= \dfrac{-6+3}{3}$$

$$= \dfrac{-3}{3}$$

$$= -1$$ ◄

Division with the Number 0

For every division problem there is an associated multiplication problem involving the same numbers. For example, the following two problems say the same thing about the numbers 2, 3, and 6:

Division   *Multiplication*

$$\dfrac{6}{3} = 2 \qquad 6 = 2(3)$$

We can use this relationship between division and multiplication to clarify division involving the number 0.

First, dividing 0 by a number other than 0 is allowed and always results in 0. To see this, consider dividing 0 by 5. We know the answer is 0 because of the relationship between multiplication and division. This is how we write it:

$$\dfrac{0}{5} = 0 \quad \text{because} \quad 0 = 0(5)$$

On the other hand, dividing a nonzero number by 0 is not allowed in the real numbers. Suppose we were attempting to divide 5 by 0. We don't know if there is an answer to this problem, but if there is, let's say the answer is a number that we can represent with the letter n. If 5 divided by 0 is a number n, then

$$\dfrac{5}{0} = n \quad \text{and} \quad 5 = n(0)$$

This is impossible, however, because no matter what number n is, when we multiply it by 0 the answer must be 0. It can never be 5. In algebra, we say expressions like $\frac{5}{0}$ are undefined, because there is no answer to them. That is, division by 0 is not allowed in the real numbers.

The only other possibility for division involving the number 0 is 0 divided by 0. We will treat problems like $\frac{0}{0}$ as if they were undefined also.

P R O B L E M   S E T   1 . 7

Find the following quotients (divide).

1. $\dfrac{8}{-4}$

2. $\dfrac{10}{-5}$

3. $\dfrac{-48}{16}$

4. $\dfrac{-32}{4}$

5. $\dfrac{-7}{21}$

6. $\dfrac{-25}{100}$

7. $\dfrac{-39}{-13}$

8. $\dfrac{-18}{-6}$

9. $\dfrac{-6}{-42}$

10. $\dfrac{-4}{-28}$

11. $\dfrac{0}{-32}$

12. $\dfrac{0}{17}$

The following problems review all four operations with positive and negative numbers. Perform the indicated operations.

13. $-3 + 12$

14. $5 + (-10)$

15. $-3 - 12$

16. $5 - (-10)$

17. $-3(12)$

18. $5(-10)$

19. $-3 \div 12$

20. $5 \div (-10)$

Divide and reduce all answers to lowest terms.

21. $\dfrac{4}{5} \div \dfrac{3}{4}$

22. $\dfrac{6}{8} \div \dfrac{3}{4}$

23. $-\dfrac{5}{6} \div \left(-\dfrac{5}{8}\right)$

24. $-\dfrac{7}{9} \div \left(-\dfrac{1}{6}\right)$

25. $\dfrac{10}{13} \div \left(-\dfrac{5}{4}\right)$

26. $\dfrac{5}{12} \div \left(-\dfrac{10}{3}\right)$

27. $-\dfrac{5}{6} \div \dfrac{5}{6}$

28. $-\dfrac{8}{9} \div \dfrac{8}{9}$

29. $-\dfrac{3}{4} \div \left(-\dfrac{3}{4}\right)$

30. $-\dfrac{6}{7} \div \left(-\dfrac{6}{7}\right)$

The following problems involve more than one operation. Simplify as much as possible.

31. $\dfrac{3(-2)}{-10}$

32. $\dfrac{4(-3)}{24}$

33. $\dfrac{-5(-5)}{-15}$

34. $\dfrac{-7(-3)}{-35}$

35. $\dfrac{-8(-7)}{-28}$

36. $\dfrac{-3(-9)}{-6}$

37. $\dfrac{27}{4 - 13}$

38. $\dfrac{27}{13 - 4}$

39. $\dfrac{20 - 6}{5 - 5}$

40. $\dfrac{10 - 12}{3 - 3}$

41. $\dfrac{-3 + 9}{2 \cdot 5 - 10}$

42. $\dfrac{-4 + 8}{2 \cdot 4 - 8}$

43. $\dfrac{15(-5) - 25}{2(-10)}$

44. $\dfrac{10(-3) - 20}{5(-2)}$

45. $\dfrac{27 - 2(-4)}{-3(5)}$

46. $\dfrac{20 - 5(-3)}{10(-3)}$

47. $\dfrac{12 - 6(-2)}{12(-2)}$

48. $\dfrac{3(-4) + 5(-6)}{10 - 6}$

49. $\dfrac{5^2 - 2^2}{-5 + 2}$

50. $\dfrac{7^2 - 4^2}{-7 + 4}$

51. $\dfrac{8^2 - 2^2}{8^2 + 2^2}$

52. $\dfrac{4^2 - 6^2}{4^2 + 6^2}$

53. $\dfrac{(5 + 3)^2}{-5^2 - 3^2}$

54. $\dfrac{(7 + 2)^2}{-7^2 - 2^2}$

55. $\dfrac{(8 - 4)^2}{8^2 - 4^2}$

56. $\dfrac{(6 - 2)^2}{6^2 - 2^2}$

57. $\dfrac{-4 \cdot 3^2 - 5 \cdot 2^2}{-8(7)}$

58. $\dfrac{-2 \cdot 5^2 + 3 \cdot 2^3}{-3(13)}$

59. $\dfrac{3 \cdot 10^2 + 4 \cdot 10 + 5}{345}$

60. $\dfrac{5 \cdot 10^2 + 6 \cdot 10 + 7}{567}$

61. $\dfrac{7 - [(2 - 3) - 4]}{-1 - 2 - 3}$

62. $\dfrac{2 - [(3 - 5) - 8]}{-3 - 4 - 5}$

63. $\dfrac{6(-4) - 2(5 - 8)}{-6 - 3 - 5}$

64. $\dfrac{3(-4) - 5(9 - 11)}{-9 - 2 - 3}$

65. $\dfrac{3(-5-3)+4(7-9)}{5(-2)+3(-4)}$

66. $\dfrac{-2(6-10)-3(8-5)}{6(-3)-6(-2)}$

67. $\dfrac{|3-9|}{3-9}$

68. $\dfrac{|4-7|}{4-7}$

69. $\dfrac{(2-9)-(5-7)}{10-15}$

70. $\dfrac{(3-10)-(1-5)}{9-12}$

71. $\dfrac{(5-8)+(2-6)}{(5-8)-(2-6)}$

72. $\dfrac{(4-9)+(2-8)}{(4-9)-(2-8)}$

73. $\dfrac{(-3+7)-(-4+2)}{-1-5}$

74. $\dfrac{(-2+6)-(3-5)}{-2-4}$

Answer the following questions.

75. What is the quotient of -12 and -4?

76. The quotient of -4 and -12 is what number?

77. What number do we divide by -5 to get 2?

78. What number do we divide by -3 to get 4?

79. Twenty-seven divided by what number is -9?

80. Fifteen divided by what number is -3?

81. If the quotient of -20 and 4 is decreased by 3, what number results?

82. If -4 is added to the quotient of 24 and -8, what number results?

Applying the Concepts

83. Suppose that you and three friends bought equal shares of an investment for a total of $15,000 and then sold it later for only $13,600. How much did each person lose?

84. If eight people invest $500 each in a stamp collection and after a year the collection is worth $3,800, how much did each person lose?

85. Suppose that the temperature outside is dropping at a constant rate. If the temperature is 75° F at noon and drops to 61° F by 4:00 in the afternoon, by how much did the temperature change each hour?

86. In a chemistry class a thermometer is placed in a beaker of hot water. The initial temperature of the water is 165° F. After 10 minutes, the water has cooled to 72° F. If the water temperature drops at a constant rate, by how much does the water temperature change each minute?

SECTION

1.8   **Subsets of the Real Numbers**

In Section 1.2 we introduced the real numbers and defined them as the numbers associated with points on the real number line. At that time we said the real numbers include whole numbers, fractions, and decimals, as well as other numbers that are not as familiar to us as these numbers. In this section we take a more detailed look at the kinds of numbers that make up the set of real numbers.

The numbers that make up the set of real numbers can be classified as **counting numbers, whole numbers, integers, rational numbers,** and **irrational numbers;** each is said to be a **subset** of the real numbers.

> **DEFINITION**   Set A is called a **subset** of set B if set A is contained in set B, that is, if each and every element in set A is also a member of set B.

Here is a detailed description of the major subsets of the real numbers.

The counting numbers are the numbers with which we count. They are the numbers 1, 2, 3, and so on. The notation we use to specify a group of numbers like this is **set notation.** We use the symbols { and } to enclose the members of the set.

$$\text{Counting numbers} = \{1, 2, 3, \ldots\}$$

▶ **EXAMPLE 1** Which of the numbers in the following set are not counting numbers?

$$\left\{-3, 0, \frac{1}{2}, 1, 1.5, 3\right\}$$

Solution The numbers -3, 0, $\frac{1}{2}$, and 1.5 are not counting numbers. ◀

The whole numbers include the counting numbers and the number 0.

$$\text{Whole numbers} = \{0, 1, 2, \ldots\}$$

The set of integers includes the whole numbers and the opposites of all the counting numbers.

$$\text{Integers} = \{\ldots, -3, -2, -1, 0, 1, 2, 3, \ldots\}$$

When we refer to positive integers we are referring to the numbers 1, 2, 3, ... Likewise, the negative integers are $-1, -2, -3, \ldots$. The number 0 is neither positive nor negative.

▶ **EXAMPLE 2** Which of the numbers in the following set are not integers?

$$\left\{-5, -1.75, 0, \frac{2}{3}, 1, \pi, 3\right\}$$

Solution The only numbers in the set that are not integers are -1.75, $\frac{2}{3}$, and π. ◀

The set of rational numbers is the set of numbers commonly called "fractions" together with the integers. The set of rational numbers is difficult to list in the same way we have listed the other sets, so we will use a different kind of notation:

$$\text{Rational numbers} = \left\{\frac{a}{b} \,\middle|\, a \text{ and } b \text{ are integers } (b \neq 0)\right\}$$

This notation is read "The set of elements $\frac{a}{b}$ such that a and b are integers (and b is not 0)." If a number can be put in the form $\frac{a}{b}$, where a and b are both from the set of integers, then it is called a rational number.

Rational numbers include any number that can be written as the ratio of two integers. That is, rational numbers are numbers that can be put in the form

$$\frac{\text{integer}}{\text{integer}}$$

▶ **EXAMPLE 3**   Show why each of the numbers in the following set is a rational number.

$$\left\{ -3, -\frac{2}{3}, 0, 0.333 \ldots, 0.75 \right\}$$

Solution   The number -3 is a rational number because it can be written as the ratio of -3 to 1. That is,

$$-3 = \frac{-3}{1}$$

Similarly, the number $-\frac{2}{3}$ can be thought of as the ratio of -2 to 3, while the number 0 can be thought of as the ratio of 0 to 1.

Any repeating decimal, such as $0.333 \ldots$ (the dots indicate that the 3's repeat forever), can be written as the ratio of two integers. In this case, $0.333 \ldots$ is the same as the fraction $\frac{1}{3}$.

Finally, any decimal that terminates after a certain number of digits can be written as the ratio of two integers. The number 0.75 is equal to the fraction $\frac{3}{4}$ and is therefore a rational number.   ◀

Still other numbers exist, each of which is associated with a point on the real number line, that cannot be written as the ratio of two integers. In decimal form they never terminate and never repeat a sequence of digits indefinitely. They are called irrational numbers (because they are not rational numbers):

Irrational numbers = {nonrational numbers; nonrepeating, nonterminating decimals}

We cannot write any irrational number in a form that is familiar to us, because they are all nonterminating, nonrepeating decimals. Since they are not rational, they cannot be written as the ratio of two integers. They have to be represented in other ways. One irrational number you have probably seen before is π. It is not 3.14. Rather, 3.14 is an approximation to π. It cannot be written as a decimal number. Other representations for irrational numbers are $\sqrt{2}, \sqrt{3}, \sqrt{5}, \sqrt{6}$, and, in general, the square root of any number that is not itself a perfect square. (If you are not familiar with square roots, you will be after Chapter 7.) Right now it is enough to know that some numbers on the number line cannot be written as the ratio of two integers or in decimal form. We call them irrational numbers.

The set of real numbers is the set of numbers that are either rational or irrational. That is, a real number is either rational or irrational.

Real numbers = {all rational numbers and all irrational numbers}

Prime Numbers and Factoring

The following diagram shows the relationship between multiplication and factoring:

Multiplication

$$\text{Factors} \rightarrow 3 \cdot 4 = 12 \leftarrow \text{Product}$$

Factoring

When we read the problem from left to right, we say the product of 3 and 4 is 12. Or we multiply 3 and 4 to get 12. When we read the problem in the other direction, from right to left, we say we have *factored* 12 into 3 times 4, or 3 and 4 are **factors** of 12.

The number 12 can be factored still further:

$$12 = 4 \cdot 3$$
$$= 2 \cdot 2 \cdot 3$$
$$= 2^2 \cdot 3$$

The numbers 2 and 3 are called **prime factors** of 12, because neither of them can be factored any further.

DEFINITION   If a and b represent integers, then a is said to be a **factor** (or divisor) of b if a divides b evenly—that is, if a divides b with no remainder.

DEFINITION   A **prime number** is any positive integer larger than 1 whose only positive factors (divisors) are itself and 1.

Note   The number 15 is not a prime number since it has factors of 3 and 5. That is, $15 = 3 \cdot 5$. When a whole number larger than 1 is not prime, it is said to be **composite.**

Here is a list of the first few prime numbers.

Prime numbers $= \{2, 3, 5, 7, 11, 13, 17, 19, 23, 29, 31, 37, 41, \ldots\}$

When a number is not prime, we can factor it into the product of prime numbers. To factor a number into the product of primes, we simply factor it until it cannot be factored further.

▶ **EXAMPLE 4**   Factor the number 60 into the product of prime numbers.

Solution   We begin by writing 60 as the product of any two positive integers whose product is 60, like 6 and 10:

$$60 = 6 \cdot 10$$

We then factor these numbers:

$$
\begin{aligned}
60 &= 6 \cdot 10 \\
&= (2 \cdot 3) \cdot (2 \cdot 5) \\
&= 2 \cdot 2 \cdot 3 \cdot 5 \\
&= 2^2 \cdot 3 \cdot 5
\end{aligned}
$$ ◀

Note   It is customary to write the prime factors in order from smallest to largest.

▶ **EXAMPLE 5**   Factor the number 630 into the product of primes.

Solution   Let's begin by writing 630 as the product of 63 and 10:

$$
\begin{aligned}
630 &= 63 \cdot 10 \\
&= (7 \cdot 9) \cdot (2 \cdot 5) \\
&= 7 \cdot 3 \cdot 3 \cdot 2 \cdot 5 \\
&= 2 \cdot 3^2 \cdot 5 \cdot 7
\end{aligned}
$$

It makes no difference which two numbers we start with, as long as their product is 630. We will always get the same result because a number has only one set of prime factors.

$$
\begin{aligned}
630 &= 18 \cdot 35 \\
&= 3 \cdot 6 \cdot 5 \cdot 7 \\
&= 3 \cdot 2 \cdot 3 \cdot 5 \cdot 7 \\
&= 2 \cdot 3^2 \cdot 5 \cdot 7
\end{aligned}
$$ ◀

Note   There are some "tricks" to finding the divisors of a number. For instance, if a number ends in 0 or 5, then it is divisible by 5. If a number ends in an even number (0, 2, 4, 6, or 8), then it is divisible by 2. A number is divisible by 3 if the sum of its digits is divisible by 3. For example, 921 is divisible by 3 because the sum of its digits is $9 + 2 + 1 = 12$, which is divisible by 3.

When we have factored a number into the product of its prime factors, we not only know what prime numbers divide the original number, but we also know all of the other numbers that divide it as well. For instance, if we were to factor 210 into its prime factors, we would have $210 = 2 \cdot 3 \cdot 5 \cdot 7$, which means that 2, 3, 5, and 7 divide 210, as well as any combination of products of 2, 3, 5, and 7. That is, since 3 and 7 divide 210, then so does their product 21. Since 3, 5, and 7 each divide 210, then so does their product 105:

$$21 \text{ divides } 210$$

$$210 = 2 \cdot 3 \cdot 5 \cdot 7$$

$$105 \text{ divides } 210$$

Although there are many ways in which factoring is used in arithmetic and algebra, one simple application is in reducing fractions to lowest terms.

Recall that we reduce fractions to lowest terms by dividing the numerator and denominator by the same number. We can use the prime factorization of numbers to help us reduce fractions with large numerators and denominators.

▶ **EXAMPLE 6** Reduce $\dfrac{210}{231}$ to lowest terms.

Solution First we factor 210 and 231 into the product of prime factors. Then we reduce to lowest terms by dividing the numerator and denominator by any factors they have in common.

$$\frac{210}{231} = \frac{2 \cdot 3 \cdot 5 \cdot 7}{3 \cdot 7 \cdot 11} \qquad \text{Factor the numerator and} \\ \text{denominator completely}$$

$$= \frac{2 \cdot \cancel{3} \cdot 5 \cdot \cancel{7}}{\cancel{3} \cdot \cancel{7} \cdot 11} \qquad \text{Divide the numerator and} \\ \text{denominator by } 3 \cdot 7$$

$$= \frac{2 \cdot 5}{11}$$

$$= \frac{10}{11} \qquad \blacktriangleleft$$

Note The small lines we have drawn through the factors that are common to the numerator and denominator are used to indicate that we have divided the numerator and denominator by those factors.

In summary, the diagram on the next page illustrates the relationships that exist among the subsets of the real numbers.

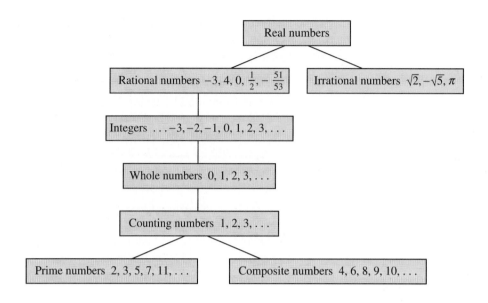

PROBLEM SET 1.8

Given the numbers in the set $\{-3, -2.5, 0, 1, \frac{3}{2}, \sqrt{15}\}$:

1. List all the whole numbers.
2. List all the integers.
3. List all the rational numbers.
4. List all the irrational numbers.
5. List all the real numbers.

Given the numbers in the set $\{-10, -8, -0.333\ldots, -2, 9, \frac{25}{3}, \pi\}$:

6. List all the whole numbers.
7. List all the integers.
8. List all the rational numbers.
9. List all the irrational numbers.
10. List all the real numbers.

Identify the following statements as either true or false.

11. Every whole number is also an integer.
12. The set of whole numbers is a subset of the set of integers.
13. A number can be both rational and irrational.

14. The set of rational numbers and the set of irrational numbers have some elements in common.
15. Some whole numbers are also negative integers.
16. Every rational number is also a real number.
17. All integers are also rational numbers.
18. The set of integers is a subset of the set of rational numbers.

Label each of the following numbers as *prime* or *composite*. If a number is composite, then factor it completely.

19. 48   **20.** 72
21. 37   **22.** 23
23. 1,023   **24.** 543

Factor the following into the product of primes. When the number has been factored completely, write its prime factors from smallest to largest.

25. 144   **26.** 288
27. 38   **28.** 63

29. 105	**30.** 210
31. 180	**32.** 900
33. 385	**34.** 1,925
35. 121	**36.** 546
37. 420	**38.** 598
39. 620	**40.** 2,310

Reduce each fraction to lowest terms by first factoring the numerator and denominator into the product of prime factors and then dividing out any factors they have in common.

41. $\dfrac{105}{165}$ **42.** $\dfrac{165}{385}$

43. $\dfrac{525}{735}$ **44.** $\dfrac{550}{735}$

45. $\dfrac{385}{455}$ **46.** $\dfrac{385}{735}$

47. $\dfrac{322}{345}$ **48.** $\dfrac{266}{285}$

49. $\dfrac{205}{369}$ **50.** $\dfrac{111}{185}$

51. $\dfrac{215}{344}$ **52.** $\dfrac{279}{310}$

53. Factor 6^3 into the product of prime factors by first factoring 6 and then raising each of its factors to the third power.

54. Factor 12^2 into the product of prime factors by first factoring 12 and then raising each of its factors to the second power.

55. Factor $9^4 \cdot 16^2$ into the product of prime factors by first factoring 9 and 16 completely.

56. Factor $10^2 \cdot 12^3$ into the product of prime factors by first factoring 10 and 12 completely.

57. Simplify the expression $3 \cdot 8 + 3 \cdot 7 + 3 \cdot 5$ and then factor the result into the product of primes. (Notice one of the factors of the answer is 3.)

58. Simplify the expression $5 \cdot 4 + 5 \cdot 9 + 5 \cdot 3$ and then factor the result into the product of primes.

Recall the Fibonacci sequence we introduced earlier in this chapter.

Fibonacci sequence = 1, 1, 2, 3, 5, 8, . . .

Any number in the Fibonacci sequence is a *Fibonacci number*.

59. The Fibonacci numbers are not a subset of which of the following sets?

Real numbers, rational numbers,
  irrational numbers, whole numbers

60. Name three Fibonacci numbers that are prime numbers.

61. Name three Fibonacci numbers that are composite numbers.

62. Is the sequence of odd numbers a subset of the Fibonacci numbers?

SECTION

1.9   **Addition and Subtraction with Fractions**

You may recall from previous math classes that to add two fractions with the same denominator, you simply add their numerators and put the result over the common denominator:

$$\frac{3}{4} + \frac{2}{4} = \frac{3 + 2}{4} = \frac{5}{4}$$

The reason we add numerators but do not add denominators is that we must follow the distributive property. To see this, you first have to recall that $\frac{3}{4}$ can be written as $3 \cdot \frac{1}{4}$, and $\frac{2}{4}$ can be written as $2 \cdot \frac{1}{4}$ (dividing by 4 is equivalent to multiplying by $\frac{1}{4}$). Here is the addition problem again, this time showing the use of the distributive property:

$$\frac{3}{4} + \frac{2}{4} = 3 \cdot \frac{1}{4} + 2 \cdot \frac{1}{4}$$

$$= (3 + 2) \cdot \frac{1}{4} \qquad \text{Distributive property}$$

$$= 5 \cdot \frac{1}{4}$$

$$= \frac{5}{4}$$

What we have here is the sum of the numerators placed over the **common denominator.** In symbols we have the following:

Addition and Subtraction of Fractions

If a, b, and c are integers and c is not equal to 0, then

$$\frac{a}{c} + \frac{b}{c} = \frac{a + b}{c}$$

This rule holds for subtraction as well. That is,

$$\frac{a}{c} - \frac{b}{c} = \frac{a - b}{c}$$

In Examples 1–4, find the sum or difference. (Add or subtract as indicated.) Reduce all answers to lowest terms. (Assume all variables represent nonzero numbers.)

▶ **EXAMPLES**

1. $\dfrac{3}{8} + \dfrac{1}{8} = \dfrac{3 + 1}{8}$ Add numerators; keep the same denominator

$\qquad = \dfrac{4}{8}$ The sum of 3 and 1 is 4

$\qquad = \dfrac{1}{2}$ Reduce to lowest terms

2. $\dfrac{a+5}{8} - \dfrac{3}{8} = \dfrac{a+5-3}{8}$ Combine numerators; keep the same denominator

$$= \dfrac{a+2}{8}$$

3. $\dfrac{9}{x} - \dfrac{3}{x} = \dfrac{9-3}{x}$ Subtract numerators; keep the same denominator

$$= \dfrac{6}{x}$$ The difference of 9 and 3 is 6

4. $\dfrac{3}{7} + \dfrac{2}{7} - \dfrac{9}{7} = \dfrac{3+2-9}{7}$

$$= \dfrac{-4}{7}$$

$$= -\dfrac{4}{7}$$ Unlike signs give a negative answer   ◄

As Examples 1–4 indicate, addition and subtraction are simple, straightforward processes when all the fractions have the same denominator. We will now turn our attention to the process of adding fractions that have different denominators. In order to get started, we need the following definition.

DEFINITION The **least common denominator (LCD)** for a set of denominators is the smallest number that is exactly divisible by each denominator. (Note that in some books the least common denominator is also called the *least common multiple.*)

In other words, all the denominators of the fractions involved in a problem must divide into the least common denominator exactly. That is, they divide it without giving a remainder.

▶ **EXAMPLE 5** Find the LCD for the fractions $\dfrac{5}{12}$ and $\dfrac{7}{18}$.

Solution The least common denominator for the denominators 12 and 18 must be the smallest number divisible by both 12 and 18. We can factor 12 and 18 completely and then build the LCD from these factors. Factoring 12 and 18 completely gives us

$$12 = 2 \cdot 2 \cdot 3 \qquad 18 = 2 \cdot 3 \cdot 3$$

Now, if 12 is going to divide the LCD exactly, then the LCD must have factors of $2 \cdot 2 \cdot 3$. If 18 is to divide it exactly, it must have factors of $2 \cdot 3 \cdot 3$. We don't need to repeat the factors that 12 and 18 have in common:

$$12 \text{ divides the LCD}$$

$$\left.\begin{array}{l} 12 = 2 \cdot 2 \cdot 3 \\ 18 = 2 \cdot 3 \cdot 3 \end{array}\right\} \quad \text{LCD} = 2 \cdot 2 \cdot 3 \cdot 3 = 36$$

$$18 \text{ divides the LCD}$$

In other words, first we write down the factors of 12, then we attach the factors of 18 that do not already appear as factors of 12. We start with $2 \cdot 2 \cdot 3$ because those are the factors of 12. Then we look at the first factor of 18. It is 2. Since 2 already appears in the expression $2 \cdot 2 \cdot 3$ we don't need to attach another one. Next we look at the factors $3 \cdot 3$. The expression $2 \cdot 2 \cdot 3$ has one 3. In order for it to contain the expression $3 \cdot 3$, we attach another 3. The final expression, our LCD, is $2 \cdot 2 \cdot 3 \cdot 3$.

The LCD for 12 and 18 is 36. It is the smallest number that is divisible by both 12 and 18; 12 divides it exactly three times, and 18 divides it exactly two times. ◀

We can use the results of Example 5 to find the sum of the fractions $\frac{5}{12}$ and $\frac{7}{18}$.

▶ **EXAMPLE 6** Add $\dfrac{5}{12} + \dfrac{7}{18}$.

Solution We can add fractions only when they have the same denominators. In Example 5 we found the LCD for $\frac{5}{12}$ and $\frac{7}{18}$ to be 36. We change $\frac{5}{12}$ and $\frac{7}{18}$ to equivalent fractions that each have 36 for a denominator by applying Property 1 (see p. 14) for fractions:

$$\frac{5}{12} = \frac{5 \cdot \mathbf{3}}{12 \cdot \mathbf{3}} = \frac{15}{36}$$

$$\frac{7}{18} = \frac{7 \cdot \mathbf{2}}{18 \cdot \mathbf{2}} = \frac{14}{36}$$

The fraction $\frac{15}{36}$ is equivalent to $\frac{5}{12}$, since it was obtained by multiplying both the numerator and denominator by 3. Likewise, $\frac{14}{36}$ is equivalent to $\frac{7}{18}$, since it was obtained by multiplying the numerator and denominator by 2. All we have left to do is to add numerators:

$$\frac{15}{36} + \frac{14}{36} = \frac{29}{36}$$

The sum of $\frac{5}{12}$ and $\frac{7}{18}$ is the fraction $\frac{29}{36}$. Let's write the complete problem again step-by-step.

$$\frac{5}{12} + \frac{7}{18} = \frac{5 \cdot \mathbf{3}}{12 \cdot \mathbf{3}} + \frac{7 \cdot \mathbf{2}}{18 \cdot \mathbf{2}}$$

Rewrite each fraction as an equivalent fraction with denominator 36

$$= \frac{15}{36} + \frac{14}{36}$$

$$= \frac{29}{36}$$

Add numerators; keep the common denominator ◀

▶ **EXAMPLE 7** Find the LCD for $\frac{3}{4}$ and $\frac{1}{6}$.

Solution We factor 4 and 6 into products of prime factors and build the LCD from these factors:

$$\left. \begin{array}{l} 4 = 2 \cdot 2 \\ 6 = 2 \cdot 3 \end{array} \right\} \quad \text{LCD} = 2 \cdot 2 \cdot 3 = 12$$

The LCD is 12. Both denominators divide it exactly; 4 divides 12 exactly 3 times, and 6 divides 12 exactly 2 times. ◀

▶ **EXAMPLE 8** Add $\frac{3}{4} + \frac{1}{6}$.

Solution In Example 7 we found that the LCD for these two fractions is 12. We begin by changing $\frac{3}{4}$ and $\frac{1}{6}$ to equivalent fractions with denominator 12:

$$\frac{3}{4} = \frac{3 \cdot \mathbf{3}}{4 \cdot \mathbf{3}} = \frac{9}{12}$$

$$\frac{1}{6} = \frac{1 \cdot \mathbf{2}}{6 \cdot \mathbf{2}} = \frac{2}{12}$$

The fraction $\frac{9}{12}$ is equal to the fraction $\frac{3}{4}$, since it was obtained by multiplying the numerator and denominator of $\frac{3}{4}$ by 3. Likewise, $\frac{2}{12}$ is equivalent to $\frac{1}{6}$, since it was obtained by multiplying the numerator and denominator of $\frac{1}{6}$ by 2. To complete the problem we add numerators:

$$\frac{9}{12} + \frac{2}{12} = \frac{11}{12}$$

The sum of $\frac{3}{4}$ and $\frac{1}{6}$ is $\frac{11}{12}$. Here is how the complete problem looks:

$$\frac{3}{4} + \frac{1}{6} = \frac{3 \cdot \mathbf{3}}{4 \cdot \mathbf{3}} + \frac{1 \cdot \mathbf{2}}{6 \cdot \mathbf{2}}$$

Rewrite each fraction as an equivalent fraction with denominator 12

$$= \frac{9}{12} + \frac{2}{12}$$

$$= \frac{11}{12}$$

Add numerators; keep the same denominator ◀

▶ **EXAMPLE 9** Subtract $\dfrac{7}{15} - \dfrac{3}{10}$.

Solution Let's factor 15 and 10 completely and use these factors to build the LCD:

$$\left.\begin{array}{l} 15 = 3 \cdot 5 \\ 10 = 2 \cdot 5 \end{array}\right\} \quad \text{LCD} = 2 \cdot 3 \cdot 5 = 30$$

15 divides the LCD

10 divides the LCD

Changing to equivalent fractions and subtracting, we have

$$\frac{7}{15} - \frac{3}{10} = \frac{7 \cdot \mathbf{2}}{15 \cdot \mathbf{2}} - \frac{3 \cdot \mathbf{3}}{10 \cdot \mathbf{3}} \qquad \text{Rewrite as equivalent fractions with the LCD for denominator}$$

$$= \frac{14}{30} - \frac{9}{30}$$

$$= \frac{5}{30} \qquad \text{Subtract numerators; keep the LCD}$$

$$= \frac{1}{6} \qquad \text{Reduce to lowest terms} \qquad ◀$$

As a summary of what we have done so far, and as a guide to working other problems, we will now list the steps involved in adding and subtracting fractions with different denominators.

Strategy for Adding or Subtracting Two Fractions

Step 1: Factor each denominator completely and use the factors to build the LCD. (Remember, the LCD is the smallest number divisible by each of the denominators in the problem.)

Step 2: Rewrite each fraction as an equivalent fraction that has the LCD for its denominator. This is done by multiplying both the numerator and denominator of the fraction in question by the appropriate whole number.

Step 3: Add or subtract the numerators of the fractions produced in step 2. This is the numerator of the sum or difference. The denominator of the sum or difference is the LCD.

Step 4: Reduce the fraction produced in step 3 to lowest terms if it is not already in lowest terms.

The idea behind adding or subtracting fractions is really very simple. We can add or subtract only fractions that have the same denominators. If the fractions we are trying to add or subtract do not have the same denominators, we rewrite each of them as an equivalent fraction with the LCD for a denominator.

Here are some further examples of sums and differences of fractions.

▶ **EXAMPLE 10**   Add $\dfrac{1}{6} + \dfrac{1}{8} + \dfrac{1}{4}$.

Solution   We begin by factoring the denominators completely and building the LCD from the factors that result:

$$6 = 2 \cdot 3$$

$$8 = 2 \cdot 2 \cdot 2$$

$$4 = 2 \cdot 2$$

8 divides the LCD

$$\text{LCD} = 2 \cdot 2 \cdot 2 \cdot 3 = 24$$

4 divides the LCD   6 divides the LCD

We then change to equivalent fractions and add as usual:

$$\frac{1}{6} + \frac{1}{8} + \frac{1}{4} = \frac{1 \cdot \mathbf{4}}{6 \cdot \mathbf{4}} + \frac{1 \cdot \mathbf{3}}{8 \cdot \mathbf{3}} + \frac{1 \cdot \mathbf{6}}{4 \cdot \mathbf{6}}$$

$$= \frac{4}{24} + \frac{3}{24} + \frac{6}{24}$$

$$= \frac{13}{24}$$

◀

▶ **EXAMPLE 11**   Subtract $3 - \dfrac{5}{6}$.

Solution   The denominators are 1 (because $3 = \frac{3}{1}$) and 6. The smallest number divisible by both 1 and 6 is 6.

$$3 - \frac{5}{6} = \frac{3}{1} - \frac{5}{6}$$

$$= \frac{3 \cdot \mathbf{6}}{1 \cdot \mathbf{6}} - \frac{5}{6}$$

$$= \frac{18}{6} - \frac{5}{6}$$

$$= \frac{13}{6}$$

◀

▶ **EXAMPLE 12**   Find the next number in each sequence.

(a) $\dfrac{1}{2}, 0, -\dfrac{1}{2}, \cdots$

(b) $\dfrac{1}{2}, 1, \dfrac{3}{2}, \ldots$

(c) $\dfrac{1}{2}, \dfrac{1}{4}, \dfrac{1}{8}, \ldots$

Solution

(a) $\frac{1}{2}, 0, -\frac{1}{2}, \ldots$: Adding $-\frac{1}{2}$ to each term produces the next term. The fourth term will be $-\frac{1}{2} + (-\frac{1}{2}) = -1$. This is an arithmetic sequence.

(b) $\frac{1}{2}, 1, \frac{3}{2}, \ldots$: Each term comes from the term before it by adding $\frac{1}{2}$. The fourth term will be $\frac{3}{2} + \frac{1}{2} = 2$. This sequence is also an arithmetic sequence.

(c) $\frac{1}{2}, \frac{1}{4}, \frac{1}{8}, \ldots$: This is a geometric sequence in which each term comes from the term before it by multiplying by $\frac{1}{2}$ each time. The next term will be $\frac{1}{8} \cdot \frac{1}{2} = \frac{1}{16}$. ◀

PROBLEM SET 1.9

Find the following sums and differences, and reduce to lowest terms. Add and subtract as indicated. Assume all variables represent nonzero numbers.

1. $\dfrac{3}{6} + \dfrac{1}{6}$

2. $\dfrac{2}{5} + \dfrac{3}{5}$

3. $\dfrac{3}{8} - \dfrac{5}{8}$

4. $\dfrac{1}{7} - \dfrac{6}{7}$

5. $-\dfrac{1}{4} + \dfrac{3}{4}$

6. $-\dfrac{4}{9} + \dfrac{7}{9}$

7. $\dfrac{x}{3} - \dfrac{1}{3}$

8. $\dfrac{x}{8} - \dfrac{1}{8}$

9. $\dfrac{1}{4} + \dfrac{2}{4} + \dfrac{3}{4}$

10. $\dfrac{2}{5} + \dfrac{3}{5} + \dfrac{4}{5}$

11. $\dfrac{x+7}{2} - \dfrac{1}{2}$

12. $\dfrac{x+5}{4} - \dfrac{3}{4}$

13. $\dfrac{1}{10} - \dfrac{3}{10} - \dfrac{4}{10}$

14. $\dfrac{3}{20} - \dfrac{1}{20} - \dfrac{4}{20}$

15. $\dfrac{1}{a} + \dfrac{4}{a} + \dfrac{5}{a}$

16. $\dfrac{5}{a} + \dfrac{4}{a} + \dfrac{3}{a}$

Find the LCD for each of the following; then use the methods developed in this section to add and subtract as indicated. Assume all variables represent nonzero numbers.

17. $\dfrac{1}{8} + \dfrac{3}{4}$

18. $\dfrac{1}{6} + \dfrac{2}{3}$

19. $\dfrac{3}{10} - \dfrac{1}{5}$

20. $\dfrac{5}{6} - \dfrac{1}{12}$

21. $\dfrac{4}{9} + \dfrac{1}{3}$

22. $\dfrac{1}{2} + \dfrac{1}{4}$

23. $2 + \dfrac{1}{3}$

24. $3 + \dfrac{1}{2}$

25. $-\dfrac{3}{4} + 1$

26. $-\dfrac{3}{4} + 2$

27. $\dfrac{1}{2} + \dfrac{2}{3}$

28. $\dfrac{2}{3} + \dfrac{1}{4}$

29. $\dfrac{5}{12} - \left(-\dfrac{3}{8}\right)$

30. $\dfrac{9}{16} - \left(-\dfrac{7}{12}\right)$

31. $-\dfrac{1}{20} + \dfrac{8}{30}$

32. $-\dfrac{1}{30} + \dfrac{9}{40}$

33. $\dfrac{17}{30} + \dfrac{11}{42}$

34. $\dfrac{19}{42} + \dfrac{13}{70}$

35. $\dfrac{25}{84} + \dfrac{41}{90}$

36. $\dfrac{23}{70} + \dfrac{29}{84}$

37. $\dfrac{13}{126} - \dfrac{13}{180}$ **38.** $\dfrac{17}{84} - \dfrac{17}{90}$

Find the fourth term in each sequence.

39. $\dfrac{3}{4} + \dfrac{1}{8} + \dfrac{5}{6}$ **40.** $\dfrac{3}{8} + \dfrac{2}{5} + \dfrac{1}{4}$

47. $\dfrac{1}{3}, 0, -\dfrac{1}{3}, \ldots$ **48.** $\dfrac{2}{3}, 0, -\dfrac{2}{3}, \ldots$

41. $\dfrac{1}{2} + \dfrac{1}{3} + \dfrac{1}{4} + \dfrac{1}{6}$ **42.** $\dfrac{1}{8} + \dfrac{1}{4} + \dfrac{1}{5} + \dfrac{1}{10}$

49. $\dfrac{1}{3}, 1, \dfrac{5}{3}, \ldots$ **50.** $1, \dfrac{3}{2}, 2, \ldots$

43. Find the sum of $\frac{3}{7}$, 2, and $\frac{1}{9}$.

44. Find the sum of 6, $\frac{6}{11}$, and 11.

51. $1, \dfrac{1}{5}, \dfrac{1}{25}, \ldots$ **52.** $1, -\dfrac{1}{2}, \dfrac{1}{4}, \ldots$

45. Give the difference of $\frac{7}{8}$ and $\frac{1}{4}$.

46. Give the difference of $\frac{9}{10}$ and $\frac{1}{100}$.

CHAPTER 1 SUMMARY

Note: We will use the margins in the chapter summaries to give examples that correspond to the topic being reviewed whenever it is appropriate.

The number(s) in brackets next to each heading indicates the section(s) in which that topic is discussed.

Symbols [1.1]

$a = b$ a is equal to b
$a \neq b$ a is not equal to b
$a < b$ a is less than b
$a \not< b$ a is not less than b
$a > b$ a is greater than b
$a \not> b$ a is not greater than b

Examples

1. $2^5 = 2 \cdot 2 \cdot 2 \cdot 2 \cdot 2 = 32$

$5^2 = 5 \cdot 5 = 25$

$10^3 = 10 \cdot 10 \cdot 10 = 1{,}000$

$1^4 = 1 \cdot 1 \cdot 1 \cdot 1 = 1$

Exponents [1.1]

Exponents are notation used to indicate repeated multiplication. In the expression 3^4, 3 is the *base* and 4 is the *exponent*.

$$3^4 = 3 \cdot 3 \cdot 3 \cdot 3 = 81$$

2. $10 + (2 \cdot 3^2 - 4 \cdot 2)$

$= 10 + (2 \cdot 9 - 4 \cdot 2)$

$= 10 + (18 - 8)$

$= 10 + 10$

$= 20$

Order of Operations [1.1]

When evaluating a mathematical expression, we will perform the operations in the following order, beginning with the expression in the innermost parentheses or brackets and working our way out.

1. Simplify all numbers with exponents, working from left to right if more than one of these numbers is present.
2. Then do all multiplications and divisions left to right.
3. Finally, perform all additions and subtractions left to right.

3. Add all combinations of positive and negative 10 and 13:

$$10 + 13 = 23$$
$$10 + (-13) = -3$$
$$-10 + 13 = 3$$
$$-10 + (-13) = -23$$

Addition of Real Numbers [1.3]

To add two real numbers with

1. the same sign: Simply add their absolute values and use the common sign.
2. different signs: Subtract the smaller absolute value from the larger absolute value. The answer has the same sign as the number with the larger absolute value.

4. Subtracting 2 is the same as adding -2:

$$7 - 2 = 7 + (-2) = 5$$

Subtraction of Real Numbers [1.4]
To subtract one number from another, simply add the opposite of the number you are subtracting. That is, if a and b represent real numbers, then

$$a - b = a + (-b)$$

5.
$$3(5) = 15$$
$$3(-5) = -15$$
$$-3(5) = -15$$
$$-3(-5) = 15$$

Multiplication of Real Numbers [1.6]
To multiply two real numbers, simply multiply their absolute values. Like signs give a positive answer. Unlike signs give a negative answer.

6. $\dfrac{-6}{2} = -6\left(\dfrac{1}{2}\right) = -3$

$\dfrac{-6}{-2} = -6\left(-\dfrac{1}{2}\right) = 3$

Division of Real Numbers [1.7]
Division by a number is the same as multiplication by its reciprocal. Like signs give a positive answer. Unlike signs give a negative answer.

7. $|5| = 5$

$|-5| = 5$

Absolute Value [1.2]
The absolute value of a real number is its distance from zero on the real number line. Absolute value is never negative.

8. The numbers 3 and -3 are opposites; their sum is 0:

$$3 + (-3) = 0$$

Opposites [1.2, 1.5]
Any two real numbers the same distance from zero on the number line but in opposite directions from zero are called opposites. Opposites always add to zero.

9. The numbers 2 and $\frac{1}{2}$ are reciprocals; their product is 1:

$$2\left(\dfrac{1}{2}\right) = 1$$

Reciprocals [1.2, 1.5]
Any two real numbers whose product is one are called reciprocals. Every real number has a reciprocal except zero.

Properties of Real Numbers [1.5]

	For Addition	For Multiplication
Commutative:	$a + b = b + a$	$a \cdot b = b \cdot a$
Associative:	$a + (b + c) = (a + b) + c$	$a \cdot (b \cdot c) = (a \cdot b) \cdot c$
Identity:	$a + 0 = a$	$a \cdot 1 = a$
Inverse:	$a + (-a) = 0$	$a \left(\dfrac{1}{a} \right) = 1$
Distributive:		$a(b + c) = ab + ac$

10. (a) 7 and 100 are counting numbers, but 0 and -2 are not.

(b) 0 and 241 are whole numbers, but -4 and $\frac{1}{2}$ are not.

(c) -15, 0, and 20 are integers.

(d) -4, $-\frac{1}{2}$, 0.75, and 0.666 . . . are rational numbers.

(e) $-\pi$, $\sqrt{3}$, and π are irrational numbers.

(f) All the numbers listed above are real numbers.

Subsets of the Real Numbers [1.8]

Counting numbers:	$\{1, 2, 3, \ldots\}$
Whole numbers:	$\{0, 1, 2, 3, \ldots\}$
Integers:	$\{\ldots, -3, -2, -1, 0, 1, 2, 3, \ldots\}$
Rational numbers:	{all numbers that can be expressed as the ratio of two integers}
Irrational numbers:	{all numbers on the number line that cannot be expressed as the ratio of two integers}
Real numbers:	{all numbers that are either rational or irrational}

11. The number 150 can be factored into the product of prime numbers:

$$150 = 15 \cdot 10$$
$$= (3 \cdot 5)(2 \cdot 5)$$
$$= 2 \cdot 3 \cdot 5^2$$

Factoring [1.8]

Factoring is the reverse of multiplication.

Multiplication

Factors $\rightarrow 3 \cdot 5 = 15 \leftarrow$ Product

Factoring

12. The LCD for $\frac{5}{12}$ and $\frac{7}{18}$ is 36.

Least Common Denominator (LCD) [1.9]

The *least common denominator* (LCD) for a set of denominators is the smallest number that is exactly divisible by each denominator.

13. $\dfrac{5}{12} + \dfrac{7}{18} = \dfrac{5}{12} \cdot \dfrac{3}{3} + \dfrac{7}{18} \cdot \dfrac{2}{2}$

$= \dfrac{15}{36} + \dfrac{14}{36}$

$= \dfrac{29}{36}$

Addition and Subtraction of Fractions [1.9]

To add (or subtract) two fractions with a common denominator, add (or subtract) numerators and use the common denominator.

$$\frac{a}{c} + \frac{b}{c} = \frac{a + b}{c} \quad \text{and} \quad \frac{a}{c} - \frac{b}{c} = \frac{a - b}{c}$$

COMMON MISTAKES
1. Interpreting absolute value as changing the sign of the number inside the absolute value symbols. $|-5| = +5, |+5| = -5$. (The first expression is correct, the second one is not.) To avoid this mistake, remember: Absolute value is a distance and distance is always measured in positive units.
2. Using the phrase "two negatives make a positive." This works only with multiplication and division. With addition, two negative numbers produce a negative answer. It is best not to use the phrase "two negatives make a positive" at all.

CHAPTER 1 REVIEW

The numbers in brackets refer to the sections of the text in which similar problems can be found.

Write the numerical expression that is equivalent to each phrase and then simplify. [1.3, 1.4, 1.6, 1.7]

1. The sum of -7 and -10
2. Five added to the sum of -7 and 4
3. The sum of -3 and 12 increased by 5
4. The difference of 4 and 9
5. The difference of 9 and -3
6. The difference of -7 and -9
7. The product of -3 and -7 decreased by 6
8. Ten added to the product of 5 and -6
9. Twice the product of -8 and $3x$
10. The quotient of -25 and -5
11. The quotient of -40 and 8 decreased by 7
12. The quotient of -45 and 15 increased by 9

Locate and label the points on the number line with the following coordinates. [1.2]

13. 5

14. -4

15. $\dfrac{24}{8}$

16. $-\dfrac{11}{2}$

17. 1.25

18. -1.5

For each number give the absolute value. [1.2]

19. 12

20. -3

21. $-\dfrac{4}{5}$

22. $-\dfrac{7}{10}$

Simplify. [1.2]

23. $|-1.8|$

24. $-|-10|$

For each number, give the opposite and the reciprocal. [1.2]

25. 6

26. $\dfrac{3}{10}$

27. -9

28. $-\dfrac{12}{5}$

Multiply. [1.2, 1.6]

29. $\left(\dfrac{2}{5}\right)\left(\dfrac{3}{7}\right)$

30. $\left(\dfrac{4}{5}\right)\left(\dfrac{5}{4}\right)$

31. $\dfrac{1}{2}(-10)$

32. $\dfrac{1}{6}(-18)$

33. $\left(-\dfrac{4}{5}\right)\left(\dfrac{25}{16}\right)$

34. $-7\left(-\dfrac{1}{7}\right)$

Add. [1.3]

35. $-9 + 12$ **36.** $-40 + (-5)$
37. $-18 + (-20)$ **38.** $-20 + 14$
39. $-3 + 8 + (-5)$ **40.** $9 + (-7) + (-8)$
41. $(-5) + (-10) + (-7)$
42. $(-3) + (-12) + (-9)$
43. $-2 + (-8) + [-9 + (-6)]$
44. $-5 + (-7) + [-3 + (-1)]$
45. $(-21) + 40 + (-23) + 5$
46. $(-31) + 52 + (-39) + 18$

Subtract. [1.4]

47. $6 - 9$ **48.** $8 - 11$
49. $14 - (-8)$ **50.** $20 - (-9)$
51. $-12 - (-8)$ **52.** $-15 - (-3)$
53. $4 - 9 - 15$ **54.** $10 - 8 - 11$

Simplify. [1.4]

55. $-14 + 7 - 8$ **56.** $-18 + 10 - 3$
57. $5 - (-10 - 2) - 3$
58. $5 - (-2 - 8) - 1$
59. $6 - [(3 - 4) - 5]$
60. $8 - [(7 - 8) - 9]$
61. $20 - [-(10 - 3) - 8] - 7$
62. $20 - [-(8 - 6) - 10] - 12$

Find the products. [1.6]

63. $(-5)(6)$ **64.** $(-9)(-7)$
65. $4(-3)$ **66.** $(-9)(-5)$
67. $-2(3)(4)$ **68.** $-3(4)(-2)$
69. $(-1)(-3)(-1)(-4)$
70. $(-2)(-3)(-1)(-5)$

Find the following quotients. [1.7]

71. $\frac{12}{-3}$ **72.** $\frac{-48}{12}$
73. $\frac{-9}{36}$ **74.** $\frac{-63}{-7}$
75. $-\frac{8}{9} \div \frac{4}{3}$ **76.** $-\frac{4}{5} \div \frac{8}{15}$

Simplify. [1.1, 1.6, 1.7]

77. $4 \cdot 5 + 3$ **78.** $2 \cdot 7 + 10$
79. $9 \cdot 3 + 4 \cdot 5$ **80.** $6 \cdot 7 + 7 \cdot 9$
81. $2^3 - 4 \cdot 3^2 + 5^2$

82. $1^3 - 2 \cdot 3^2 + 10^2$
83. $12 - 3(2 \cdot 5 - 7) + 4$
84. $16 - 2(3 \cdot 3 - 4) + 6$
85. $20 + 8 \div 4 + 2 \cdot 5$
86. $30 + 6 \div 2 + 4 \cdot 5$
87. $2(3 - 5) - (2 - 8)$
88. $4(5 - 7) - (3 - 4)$
89. $-4(-5) + 10$
90. $-3(-7) + 15$
91. $(-2)(3) - (4)(-3) - 9$
92. $(-4)(2) - 5(2) - 10$
93. $3(4 - 7)^2 - 5(3 - 8)^2$
94. $6(6 - 7)^2 - 9(3 - 6)^2$
95. $(-5 - 2)(-3 - 7)$
96. $(-3 - 7)(-2 - 8)$

97. $\dfrac{4(-3)}{-6}$ **98.** $\dfrac{9(-2)}{-2}$

99. $\dfrac{3^2 + 5^2}{(3 - 5)^2}$ **100.** $\dfrac{4^2 - 8^2}{(4 - 8)^2}$

101. $\dfrac{15 - 10}{6 - 6}$ **102.** $\dfrac{12 - 3}{8 - 8}$

103. $\dfrac{8(-5) - 24}{4(-2)}$ **104.** $\dfrac{9(-6) - 10}{2(-8)}$

105. $\dfrac{2(-7) + (-11)(-4)}{7 - (-3)}$

106. $\dfrac{-3(4 - 7) - 5(7 - 2)}{-5 - 2 - 1}$

State the property or properties that justify the following. [1.5]

107. $9(3y) = (9 \cdot 3)y$ **108.** $8(1) = 8$
109. $(4 + y) + 2 = (y + 4) + 2$

110. $5 + (-5) = 0$ **111.** $6\left(\dfrac{1}{6}\right) = 1$

112. $8 + 0 = 8$
113. $(4 + 2) + y = (4 + y) + 2$
114. $5(w - 6) = 5w - 30$

Use the associative property to rewrite each expression and then simplify the result. [1.5]

115. $7 + (5 + x)$ **116.** $4(7a)$

117. $\dfrac{1}{9}(9x)$

118. $\dfrac{4}{5}\left(\dfrac{5}{4}y\right)$

Apply the distributive property to each of the following expressions. Simplify when possible. [1.5, 1.6]

119. $7(2x + 3)$

120. $3(2a - 4)$

121. $\dfrac{1}{2}(5x - 6)$

122. $-\dfrac{1}{2}(3x - 6)$

For the set $\{\sqrt{7}, -\frac{1}{3}, 0, 5, -4.5, \frac{2}{5}, \pi, -3\}$ list all the [1.8]

123. rational numbers   **124.** whole numbers
125. irrational numbers   **126.** integers

Factor into the product of primes. [1.8]

127. 90

128. 120

129. 840

130. 1,024

Combine. [1.9]

131. $\dfrac{18}{35} + \dfrac{13}{42}$

132. $\dfrac{9}{70} + \dfrac{11}{84}$

Find the next number in each sequence. [1.1, 1.2, 1.3, 1.6, 1.9]

133. 10, 7, 4, 1, . . .
134. 10, −30, 90, −270, . . .
135. 1, 1, 2, 3, 5, . . .
136. 4, 6, 8, 10, . . .

137. $1, \dfrac{1}{2}, 0, -\dfrac{1}{2}, \ldots$

138. $1, -\dfrac{1}{2}, \dfrac{1}{4}, -\dfrac{1}{8}, \ldots$

CHAPTER 1 TEST

Translate into symbols. [1.1]

1. The sum of x and 3 is 8.
2. The product of 5 and y is 15.

Simplify according to the rule for order of operations. [1.1]

3. $5^2 + 3(9 - 7) + 3^2$
4. $10 - 6 \div 3 + 2^3$

For each number, name the opposite, reciprocal, and absolute value. [1.2]

5. -4

6. $\dfrac{3}{4}$

Add. [1.3]

7. $3 + (-7)$
8. $(-9 + (-6)) + (-3 + 5)$

Subtract. [1.4]

9. $-4 - 8$

10. $9 - (7 - 2) - 4$

Match each expression on the left with the letter of the property that justifies it. [1.5]

11. $(x + y) + z$
$= x + (y + z)$

12. $3(x + 5) = 3x + 15$

13. $5(3x) = (5 \cdot 3)x$

14. $(x + 5) + 7$
$= 7 + (x + 5)$

a. Commutative property of addition
b. Commutative property of multiplication
c. Associative property of addition
d. Associative property of multiplication
e. Distributive property

Multiply. [1.6]

15. $-3(7)$

16. $-4(8)(-2)$

17. $8\left(-\dfrac{1}{4}\right)$ **18.** $\left(-\dfrac{2}{3}\right)^3$

Simplify using the rule for order of operations. [1.6]

19. $-3(-4) - 8$ **20.** $5(-6)^2 - 3(-2)^3$
21. $7 - 3(2 - 8)$
22. $4 - 2[-3(-1 + 5) + 4(-3)]$

Simplify as much as possible. [1.7]

23. $\dfrac{4(-5) - 2(7)}{-10 - 7}$

24. $\dfrac{2(-3 - 1) + 4(-5 + 2)}{-3(2) - 4}$

Apply the associative property and then simplify. [1.5, 1.6]

25. $3 + (5 + 2x)$ **26.** $-2(-5x)$

Multiply by applying the distributive property. [1.5, 1.6]

27. $2(3x + 5)$ **28.** $-\dfrac{1}{2}(4x - 2)$

From the set of numbers $\{1, 1.5, \sqrt{2}, \frac{3}{4}, -8\}$ list [1.8]

29. all the integers
30. all the rational numbers
31. all the irrational numbers
32. all the real numbers

Factor into the product of primes. [1.8]

33. 592   **34.** 1,340

Combine. [1.9]

35. $\dfrac{5}{15} + \dfrac{11}{42}$ **36.** $\dfrac{7}{12} + \dfrac{23}{30} + \dfrac{11}{45}$

Write an expression in symbols that is equivalent to each English phrase and then simplify it.

37. The sum of 8 and -3 [1.1, 1.3]
38. The difference of -24 and 2 [1.1, 1.4]
39. The product of -5 and -4 [1.1, 1.6]
40. The quotient of -24 and -2 [1.1, 1.7]

Find the next number in each sequence. [1.1, 1.2, 1.3, 1.6, 1.9]

41. $-8, -3, 2, 7, \ldots$
42. $8, -4, 2, -1, \ldots$

2 LINEAR EQUATIONS AND INEQUALITIES

INTRODUCTION

When I was preparing to produce the videotapes that accompany this book, I obtained bids from a number of stores for the equipment I needed. The lowest bid was $3,200, for one of the items I needed. At that time, the sales tax rate in California was 7.5%. I expected to pay the $3,200, plus another 7.5% in sales tax. When I went to pick up the equipment, the sales person gave me a bill for exactly $3,200. I asked her how she computed the sales tax to pay the state when the total for the equipment plus the tax was $3,200. She said, ''I just back tax it, by dividing by 1.075.'' The reason why she divides by 1.075 can be explained very clearly with algebra. As you will see as you progress through this chapter, if we let x represent the retail price of the video equipment, then the situation in question can be described by the equation below:

$$x + 0.075x = 3,200$$

Setting up and solving equations like the one above is one of the main topics in this chapter.

Also in this chapter are the techniques we use to simplify expressions and formulas. Formulas give us a way to describe the pattern shown by a sequence of numbers. For example, if n is a counting number, then

$2n$ is always an even number, and

$2n - 1$ is always an odd number

Likewise, the complementary and supplementary angles we studied in Chapter 1 we describe using algebraic expressions this way:

The complement of x is $90° - x$

The supplement of x is $180° - x$

OVERVIEW

In this chapter we will develop most of the properties necessary to solve many different types of equations. It may seem surprising, but most equations can be solved by applying just two main properties—the addition property of equality and the multiplication property of equality.

To be successful in this chapter you must know the following concepts from Chapter 1.

1. Addition, subtraction, multiplication, and division of positive and negative real numbers.
2. The commutative, associative, and distributive properties.
3. That reciprocals multiply to one and opposites add to zero.
4. That the least common denominator for a set of fractions is divisible by each of the denominators used to find it.

STUDY SKILLS

If you have successfully completed Chapter 1, then you have made a good start at developing the study skills necessary to succeed in all math classes. Here is the list of study skills for this chapter. Some are a continuation of the skills from Chapter 1, while others are new to this chapter.

1. **Continue to Set and Keep a Schedule**   Sometimes I find students do well in Chapter 1 and then become overconfident. They will begin to put in less time with their homework. Don't do it. Keep to the same schedule.
2. **Increase Effectiveness**   You want to become more and more effective with the time you spend on your homework. You want to increase the amount of learning you obtain in the time you have set aside. Increase those activities that you feel are the most beneficial and decrease those that have not given you the results you want.
3. **List Difficult Problems**   Begin to make lists of problems that give you the most difficulty. These are the problems in which you are repeatedly making mistakes.
4. **Begin to Develop Confidence with Word Problems**   It seems that the major difference between those people who are good at working word problems and those who are not is confidence. The people with confidence know that no matter how long it takes them, they will eventually be able to solve the problem. Those without confidence begin by saying to themselves, ''I'll never be able to work this problem.'' Are you like that? If you are, what you need to do is put your old ideas about you and word problems aside for a while and make a decision to be successful. Sometimes that's all it takes. Instead of telling yourself that you can't do word problems, that you don't like them, or that they're not good for anything anyway, decide to do whatever it takes to master them.

Many of my students keep a notebook that contains everything that they need for the course: class notes, homework, quizzes, tests, and research projects.

A three-ring binder with tabs is ideal. Organize your notebook so that you can easily get to any item you wish to look at.

Simplifying Expressions

As you will see in the next few sections, the first step in solving an equation is to simplify both sides as much as possible. In the first part of this section, we will practice simplifying expressions by combining what are called **similar** (or **like**) **terms.**

For our immediate purposes, a term is a number or a number and one or more variables multiplied together. For example, the number 5 is a term, as are the expressions $3x$, $-7y$, and $15xy$.

DEFINITION Two or more terms with the same variable part are called **similar** (or **like**) **terms.**

The terms $3x$ and $4x$ are similar, since their variable parts are identical. Likewise, the terms $18y$, $-10y$, and $6y$ are similar terms.

To simplify an algebraic expression, we simply reduce the number of terms in the expression. We accomplish this by applying the distributive property along with our knowledge of addition and subtraction of positive and negative real numbers. The following examples illustrate the procedure.

▶ **EXAMPLES** Simplify by combining similar terms.

1. $3x + 4x = (3 + 4)x$ Distributive property
   $= 7x$ Addition of 3 and 4

2. $7a - 10a = (7 - 10)a$ Distributive property
   $= -3a$ Addition of 7 and -10

3. $18y - 10y + 6y = (18 - 10 + 6)y$ Distributive property
   $= 14y$ Addition of 18, -10, and 6 ◀

When the expression we intend to simplify is more complicated, we use the commutative and associative properties first.

▶ **EXAMPLES** Simplify each expression.

4. $3x + 5 + 2x - 3 = 3x + 2x + 5 - 3$ Commutative property
   $= (3x + 2x) + (5 - 3)$ Associative property
   $= (3 + 2)x + (5 - 3)$ Distributive property
   $= 5x + 2$ Addition

5. $4a - 7 - 2a + 3 = (4a - 2a) + (-7 + 3)$ Commutative and
    associative properties
$\qquad\qquad\qquad = (4 - 2)a + (-7 + 3)$ Distributive property
$\qquad\qquad\qquad = 2a - 4$ Addition

6. $5x + 8 - x - 6 = (5x - x) + (8 - 6)$ Commutative and
    associative properties
$\qquad\qquad\qquad = (5 - 1)x + (8 - 6)$ Distributive property
$\qquad\qquad\qquad = 4x + 2$ Addition   ◀

Notice that in each case the result has fewer terms than the original expression. Since there are fewer terms, the resulting expression is said to be simpler than the original expression.

Simplifying Expressions Containing Parentheses

If an expression contains parentheses, it is often necessary to apply the distributive property to remove the parentheses before combining similar terms.

▶ **EXAMPLE 7**   Simplify the expression $5(2x - 8) - 3$.

Solution   We begin by distributing the 5 across $2x - 8$. We then combine similar terms:

$$5(2x - 8) - 3 = 10x - 40 - 3 \qquad \text{Distributive property}$$
$$= 10x - 43 \qquad\qquad\qquad ◀$$

▶ **EXAMPLE 8**   Simplify $7 - 3(2y + 1)$.

Solution   By the rule for order of operations we must multiply before we add or subtract. For that reason, it would be incorrect to subtract 3 from 7 first. Instead we multiply -3 and $2y + 1$ to remove the parentheses and then combine similar terms:

$$7 - 3(2y + 1) = 7 - 6y - 3 \qquad \text{Distributive property}$$
$$= -6y + 4 \qquad\qquad\qquad ◀$$

▶ **EXAMPLE 9**   Simplify $5(x - 2) - (3x + 4)$.

Solution   We begin by applying the distributive property to remove the parentheses. The expression $-(3x + 4)$ can be thought of as $-1(3x + 4)$. Thinking of it in this way allows us to apply the distributive property:

$$-1(3x + 4) = -1(3x) + (-1)(4)$$
$$= -3x - 4$$

The complete solution looks like this:

$$5(x - 2) - (3x + 4) = 5x - 10 - 3x - 4 \qquad \text{Distributive property}$$
$$= 2x - 14 \qquad \text{Combine similar terms} \quad \blacktriangleleft$$

As you can see from the explanation in Example 9, we use the distributive property to simplify expressions in which parentheses are preceded by a negative sign. In general we can write

$$-(a + b) = -1(a + b)$$
$$= -a + (-b)$$
$$= -a - b$$

The negative sign outside the parentheses ends up changing the sign of each term within the parentheses. In words, we say "the opposite of a sum is the sum of the opposites."

The Value of an Expression

An expression like $3x + 2$ will have a certain value depending on what number we assign to x. For instance, when x is 4, $3x + 2$ becomes $3(4) + 2$, or 14. When x is -8, $3x + 2$ becomes $3(-8) + 2$, or -22. The value of an expression is found by replacing the variable with a given number.

▶ **EXAMPLES** Find the value of the following expressions by replacing the variable with the given number.

	Expression	Value of the Variable	Value of the Expression
10.	$3x - 1$	$x = 2$	$3(2) - 1 = 6 - 1 = 5$
11.	$7a + 4$	$a = -3$	$7(-3) + 4 = -21 + 4 = -17$
12.	$2x - 3 + 4x$	$x = -1$	$2(-1) - 3 + 4(-1)$ $= -2 - 3 + (-4) = -9$
13.	$2x - 5 - 8x$	$x = 5$	$2(5) - 5 - 8(5)$ $= 10 - 5 - 40 = -35$
14.	$y^2 - 6y + 9$	$y = 4$	$4^2 - 6(4) + 9$ $= 16 - 24 + 9 = 1$ $\blacktriangleleft$

Simplifying an expression should not change its value. That is, if an expression has a certain value when x is 5, then it will always have that value no matter how much it has been simplified as long as x is 5. If we were to simplify the expression in Example 13 first, it would look like

$$2x - 5 - 8x = -6x - 5$$

When x is 5, the simplified expression $-6x - 5$ is

$$-6(5) - 5 = -30 - 5 = -35$$

It has the same value as the original expression when x is 5.

We can also find the value of an expression that contains two variables if we know the values for both variables.

▶ **EXAMPLE 15**   Find the value of the expression $2x - 3y + 4$ when x is -5 and y is 6.

Solution   Substituting -5 for x and 6 for y, the expression becomes

$$2(-5) - 3(6) + 4 = -10 - 18 + 4$$
$$= -28 + 4$$
$$= -24 \qquad ◀$$

▶ **EXAMPLE 16**   Find the value of the expression $x^2 - 2xy + y^2$ when x is 3 and y is -4.

Solution   Replacing each x in the expression with the number 3 and each y in the expression with the number -4 gives us

$$3^2 - 2(3)(-4) + (-4)^2 = 9 - 2(3)(-4) + 16$$
$$= 9 - (-24) + 16$$
$$= 33 + 16$$
$$= 49 \qquad ◀$$

More about Sequences

As the next example indicates, when we substitute the counting numbers, in order, into algebraic expressions, we form some of the sequences of numbers that we studied in Chapter 1. To review, recall that the sequence of counting numbers (also called the sequence of positive integers) is

$$\text{Counting numbers} = 1, 2, 3, \ldots$$

▶ **EXAMPLE 17**   Substitute 1, 2, 3, and 4 for n in the expression $2n - 1$.

Solution   Substituting as indicated, we have

$$\text{When } n = 1, 2n - 1 = 2 \cdot 1 - 1 = 1$$
$$\text{When } n = 2, 2n - 1 = 2 \cdot 2 - 1 = 3$$
$$\text{When } n = 3, 2n - 1 = 2 \cdot 3 - 1 = 5$$
$$\text{When } n = 4, 2n - 1 = 2 \cdot 4 - 1 = 7$$

As you can see, substituting the first four counting numbers into the formula $2n - 1$ produces the first four numbers in the sequence of odd numbers. ◀

▶ **EXAMPLE 18** What sequences are formed by substituting the counting numbers, in order, into the expressions below?

(a) $2n$ (b) n^2

Solution Proceeding as we did in the previous example we substitute the numbers 1, 2, 3, and so on into the given expressions.

(a) When $n = 1$, $2n = 2 \cdot 1 = 2$
    When $n = 2$, $2n = 2 \cdot 2 = 4$
    When $n = 3$, $2n = 2 \cdot 3 = 6$
    When $n = 4$, $2n = 2 \cdot 4 = 8$

As you can see, the expression $2n$ produces the sequence of even numbers when n is replaced by the counting numbers.

(b) When $n = 1$, $n^2 = 1^2 = 1$
    When $n = 2$, $n^2 = 2^2 = 4$
    When $n = 3$, $n^2 = 3^2 = 9$
    When $n = 4$, $n^2 = 4^2 = 16$

The expression n^2 produces the sequence of squares when n is replaced by 1, 2, 3, and so on. ◀

PROBLEM SET 2.1

Simplify the following expressions.

1. $3x - 6x$
2. $7x - 5x$
3. $-2a + a$
4. $3a - a$
5. $7x + 3x + 2x$
6. $8x - 2x - x$
7. $3a - 2a + 5a$
8. $7a - a + 2a$
9. $4x - 3 + 2x$
10. $5x + 6 - 3x$
11. $3a + 4a + 5$
12. $6a + 7a + 8$
13. $2x - 3 + 3x - 2$
14. $6x + 5 - 2x + 3$
15. $3a - 1 + a + 3$
16. $-a + 2 + 8a - 7$
17. $-4x + 8 - 5x - 10$
18. $-9x - 1 + x - 4$
19. $7a + 3 + 2a + 3a$
20. $8a - 2 + a + 5a$
21. $5(2x - 1) + 4$
22. $2(4x - 3) + 2$
23. $7(3y + 2) - 8$
24. $6(4y + 2) - 7$
25. $-3(2x - 1) + 5$
26. $-4(3x - 2) - 6$
27. $5 - 2(a + 1)$
28. $7 - 8(2a + 3)$

Simplify the following expressions.

29. $6 - 4(x - 5)$
30. $12 - 3(4x - 2)$
31. $-9 - 4(2 - y) + 1$
32. $-10 - 3(2 - y) + 3$
33. $-6 + 2(2 - 3x) + 1$

34. $-7 - 4(3 - x) + 1$
35. $(4x - 7) - (2x + 5)$
36. $(7x - 3) - (4x + 2)$
37. $8(2a + 4) - (6a - 1)$
38. $9(3a + 5) - (8a - 7)$
39. $3(x - 2) - (x - 3)$
40. $2(2x + 1) - (x + 4)$
41. $4(2y - 8) - (y + 7)$
42. $5(y - 3) - (y - 4)$
43. $-9(2x + 1) - (x + 5)$
44. $-3(3x - 2) - (2x + 3)$

Evaluate the following expressions when x is 2. (Find the value of the expressions if x is 2.)

45. $3x - 1$
46. $4x + 3$
47. $-2x - 5$
48. $-3x + 6$
49. $x^2 - 8x + 16$
50. $x^2 - 10x + 25$
51. $(x - 4)^2$
52. $(x - 5)^2$

Find the value of the following expressions when x is -5. Then, simplify the expression and check to see that it has the same value for $x = -5$.

53. $7x - 4 - x - 3$
54. $3x + 4 + 7x - 6$

55. $5(2x + 1) + 4$

56. $2(3x - 10) + 5$

Find the value of each expression when x is -3 and y is 5.

57. $x^2 - 2xy + y^2$

58. $x^2 + 2xy + y^2$

59. $(x - y)^2$ **60.** $(x + y)^2$

61. $x^2 + 6xy + 9y^2$

62. $x^2 + 10xy + 25y^2$

63. $(x + 3y)^2$

64. $(x + 5y)^2$

Find the value of $12x - 3$ for each of the following values of x.

65. $\dfrac{1}{2}$ **66.** $\dfrac{1}{3}$

67. $\dfrac{1}{4}$ **68.** $\dfrac{1}{6}$

69. $\dfrac{3}{2}$ **70.** $\dfrac{2}{3}$

71. $\dfrac{3}{4}$ **72.** $\dfrac{5}{6}$

Substitute 1, 2, 3, and 4 for n in each expression below.

73. $2n + 3$ **74.** $5n + 2$

75. n^3 **76.** $3n$

77. $n^2 + 1$ **78.** $(n + 1)^2$

Find the sequences formed by substituting the counting numbers, in order, into the expressions below.

79. $3n - 2$ **80.** $2n - 3$

81. $n^2 - 2n + 1$ **82.** $(n - 1)^2$

Translate each phrase into an algebraic expression. Then find the value of the expression when x is -2.

83. The sum of x and 5

84. The product of x and 5

85. The difference of x and 5

86. The difference of 5 and x

87. Twice the sum of x and 10

88. The sum of twice x and 10

89. The quotient of 10 and x

90. The quotient of x and 10

91. The sum of $3x$ and -2, decreased by 5

92. The product of $3x$ and -2, decreased by 5

Review Problems

From here on, each problem set will end with some review problems. In mathematics, it is very important to review work you have done previously. The more you review, the better you will understand the topics we cover and the longer you will remember them. Also, there are times when material that seemed confusing earlier becomes less confusing the second time around.

The following problems review material we covered previously in Section 1.9. Reviewing these problems will help you with the material in the next section.

Add and subtract as indicated.

93. $-3 - \dfrac{1}{2}$ **94.** $-5 - \dfrac{1}{3}$

95. $\dfrac{4}{5} + \dfrac{1}{10} + \dfrac{3}{8}$

96. $\dfrac{3}{10} + \dfrac{7}{25} + \dfrac{3}{4}$

Research Project 4

We have been using the equal sign, $=$, for some time now. It is interesting to note that the first published use of the symbol was in 1557 with the publication of *The Whetstone of Witte* by the English mathematician and physician Robert Recorde. Research the first use of the symbols we use for addition, subtraction, multiplication, and division and then write an essay on the subject from your results.

Addition Property of Equality

In this section we will solve some simple equations. To solve an equation, we must find all replacements for the variable that make the equation a true statement.

DEFINITION   The **solution set** for an equation is the set of all numbers that when used in place of the variable make the equation a true statement.

For example, the equation $x + 2 = 5$ has solution set $\{3\}$ because when x is 3 the equation becomes the true statement $3 + 2 = 5$, or $5 = 5$.

▶ **EXAMPLE 1**   Is 5 a solution to $2x - 3 = 7$?

Solution   We substitute 5 for x in the equation, and then simplify to see if a true statement results. A true statement means we have a solution; a false statement indicates the number we are using is not a solution.

$$\text{When} \qquad\qquad x = 5$$

$$\text{the equation} \ \ 2x - 3 = 7$$

$$\text{becomes} \qquad 2(5) - 3 \overset{?}{=} 7$$

$$10 - 3 \overset{?}{=} 7$$

$$7 = 7 \qquad \text{A true statement}$$

Since $x = 5$ turns the equation into the true statement $7 = 7$, we know 5 is a solution to the equation.   ◀

Note   We can use a question mark over the equal signs above to show that we don't know yet whether the two sides of the equation are equal.

▶ **EXAMPLE 2**   Is -2 a solution to $8 = 3x + 4$?

Solution   Substituting -2 for x in the equation, we have

$$8 \overset{?}{=} 3(-2) + 4$$

$$8 \overset{?}{=} -6 + 4$$

$$8 = -2 \qquad \text{A false statement}$$

Substituting -2 for x in the equation produces a false statement. Therefore, $x = -2$ is not a solution to the equation.   ◀

The important thing about an equation is its solution set. Because of this, we make the following definition to classify together all equations with the same solution set.

DEFINITION Two or more equations with the same solution set are said to be **equivalent equations.**

Equivalent equations may look different but must have the same solution set.

▶ **EXAMPLES**

3. $x + 2 = 5$ and $x = 3$ are equivalent equations, since both have solution set $\{3\}$.

4. $a - 4 = 3$, $a - 2 = 5$, and $a = 7$ are equivalent equations, since they all have solution set $\{7\}$.

5. $y + 3 = 4$, $y - 8 = -7$, and $y = 1$ are equivalent equations, since they all have solution set $\{1\}$. ◀

If two numbers are equal and we increase (or decrease) both of them by the same amount, the resulting quantities are also equal. We can apply this concept to equations. Adding the same amount to both sides of an equation always produces an equivalent equation—one with the same solution set. This fact about equations is called the **addition property of equality** and can be stated more formally as follows.

Addition Property of Equality

For any three algebraic expressions A, B, and C,

$$\text{if} \qquad A = B$$

$$\text{then} \quad A + C = B + C$$

In words: Adding the same quantity to both sides of an equation will not change the solution set.

This property is just as simple as it seems. We can add any amount to both sides of an equation and always be sure we have not changed the solution set.

Consider the equation $x + 6 = 5$. We want to solve this equation for the value of x that makes it a true statement. We want to end up with x on one side of the equal sign and a number on the other side. Since we want x by itself, we will add -6 to both sides:

$$x + 6 + (\mathbf{-6}) = 5 + (\mathbf{-6}) \qquad \text{Addition property of equality}$$

$$x + 0 = -1 \qquad \text{Addition}$$

$$x = -1$$

All three equations say the same thing about x. They all say that x is -1. All three equations are equivalent. The last one is just easier to read.

Here are some further examples of how the addition property of equality can be used to solve equations.

▶ **EXAMPLE 6** Solve the equation $x - 5 = 12$ for x.

Solution Since we want x alone on the left side, we choose to add $+5$ to both sides:

$$x - 5 + \mathbf{5} = 12 + \mathbf{5} \qquad \text{Addition property of equality}$$

$$x + 0 = 17$$

$$x = 17$$

To check our solution, we substitute 17 for x in the original equation:

$$\text{When} \qquad x = 17$$

$$\text{the equation } x - 5 = 12$$

$$\text{becomes} \quad 17 - 5 \overset{?}{=} 12$$

$$12 = 12 \qquad \text{A true statement}$$

As you can see, our solution checks. The purpose for checking a solution to an equation is to catch any mistakes we may have made in the process of solving the equation. ◀

▶ **EXAMPLE 7** Solve for a: $a + \dfrac{3}{4} = -\dfrac{1}{2}$.

Solution Since we want a by itself on the left side of the equal sign, we add the opposite of $\frac{3}{4}$ to each side of the equation.

$$a + \frac{3}{4} + \left(-\frac{\mathbf{3}}{\mathbf{4}}\right) = -\frac{1}{2} + \left(-\frac{\mathbf{3}}{\mathbf{4}}\right) \qquad \begin{array}{l}\text{Addition property of}\\ \text{equality}\end{array}$$

$$a + 0 = -\frac{1}{2} \cdot \frac{\mathbf{2}}{\mathbf{2}} + \left(-\frac{3}{4}\right) \qquad \text{LCD on the right side is 4}$$

$$a = -\frac{2}{4} + \left(-\frac{3}{4}\right) \qquad \frac{2}{4} \text{ is equivalent to } \frac{1}{2}$$

$$a = -\frac{5}{4} \qquad \text{Add fractions}$$

The solution is $a = -\frac{5}{4}$. To check our result we replace a with $-\frac{5}{4}$ in the original equation. The left side then becomes $-\frac{5}{4} + \frac{3}{4}$, which reduces to $-\frac{1}{2}$, so our solution checks. ◄

▶ **EXAMPLE 8** Solve for x: $7.3 + x = -2.4$

Solution Again, we want to isolate x so we add the opposite of 7.3 to both sides:

$$7.3 + (\mathbf{-7.3}) + x = -2.4 + (\mathbf{-7.3}) \qquad \text{Addition property of equality}$$
$$0 + x = -9.7$$
$$x = -9.7 \qquad\qquad ◄$$

Sometimes it is necessary to simplify each side of an equation before using the addition property of equality. The reason we simplify both sides first is that we want as few terms as possible on each side of the equation before we use the addition property of equality. The following examples illustrate this procedure.

▶ **EXAMPLE 9** Solve for x: $-x + 2 + 2x = 7 + 5$.

Solution We begin by combining similar terms on each side of the equation. Then we use the addition property to solve the simplified equation.

$$x + 2 = 12 \qquad\qquad \text{Simplify each side separately}$$
$$x + 2 + (\mathbf{-2}) = 12 + (\mathbf{-2}) \qquad \text{Addition property of equality}$$
$$x + 0 = 10$$
$$x = 10 \qquad\qquad ◄$$

▶ **EXAMPLE 10** Solve $3x - 4 - 2x = 11 - 4$.

Solution The first step is to simplify each side separately.

$$x - 4 = 7 \qquad\qquad \text{Simplify each side}$$
$$x - 4 + \mathbf{4} = 7 + \mathbf{4} \qquad \text{Addition property of equality}$$
$$x + 0 = 11$$
$$x = 11 \qquad\qquad ◄$$

▶ **EXAMPLE 11** Solve $4(2a - 3) - 7a = 2 - 5$.

Solution We must begin by applying the distributive property to separate terms on the left side of the equation. Following that, we combine similar terms and then apply the addition property of equality.

$$4(2a - 3) - 7a = 2 - 5 \qquad \text{Original equation}$$
$$8a - 12 - 7a = 2 - 5 \qquad \text{Distributive property}$$

$$a - 12 = -3 \qquad \text{Simplify each side}$$
$$a - 12 + \mathbf{12} = -3 + \mathbf{12} \qquad \text{Add } \mathbf{12} \text{ to each side}$$
$$a = 9 \qquad \text{Addition}$$

To check our solution, we replace a with 9 in the original equation.

$$4(2 \cdot 9 - 3) - 7 \cdot 9 \stackrel{?}{=} 2 - 5$$
$$4(18 - 3) - 63 \stackrel{?}{=} -3$$
$$4(15) - 63 \stackrel{?}{=} -3$$
$$60 - 63 \stackrel{?}{=} -3$$
$$-3 = -3 \qquad \text{A true statement}$$

Our solution checks. ◀

So far in this section we have used the addition property of equality to add only numbers to both sides of an equation. It is often necessary to add a term involving a variable to both sides of an equation, as the following examples indicate.

▶ **EXAMPLE 12** Solve for x: $5x = 4x + 2$.

Solution To isolate x on the left side of the equation, we add $-4x$ to both sides:

$$5x + (\mathbf{-4x}) = 4x + (\mathbf{-4x}) + 2 \qquad \text{Add } \mathbf{-4x} \text{ to both sides}$$
$$x = 2 \qquad \text{Simplify} \qquad ◀$$

▶ **EXAMPLE 13** Solve $3x - 5 = 2x + 7$.

Solution We can solve this equation in two steps. First, we add $-2x$ to both sides of the equation. When this has been done, x will appear on the left side only. Second, we add 5 to both sides:

$$3x + (\mathbf{-2x}) - 5 = 2x + (\mathbf{-2x}) + 7 \qquad \text{Add } \mathbf{-2x} \text{ to both sides}$$
$$x - 5 = 7 \qquad \text{Simplify each side}$$
$$x - 5 + \mathbf{5} = 7 + \mathbf{5} \qquad \text{Add } \mathbf{5} \text{ to both sides}$$
$$x = 12 \qquad \text{Simplify each side} \qquad ◀$$

Many of the equations in the problem set that follows will seem very simple. You will be able to recognize solutions to some of them without doing any work. It won't always be this way. The idea here is to develop a method of solving equations. To do so, you must show all your work until you can consistently solve equations correctly using the method shown in the examples in this section.

A Note on Subtraction

Although the addition property of equality is stated for addition only, we can subtract the same number from both sides of an equation as well. Because subtraction is defined as addition of the opposite, subtracting the same quantity from both sides of an equation will not change the solution. If we were to solve the equation in Example 7 using subtraction instead of addition, the steps would look like this:

$$x + 2 = 12 \qquad \text{Original equation}$$

$$x + 2 - 2 = 12 - 2 \qquad \text{Subtract } \mathbf{2} \text{ from each side}$$

$$x = 10 \qquad \text{Subtraction}$$

In my experience teaching algebra, I find that students make fewer mistakes if they think in terms of addition rather than subtraction. So, you are probably better off if you continue to use the addition property just the way we have used it in the examples in this section. But, if you are curious as to whether you can subtract the same number from both sides of an equation, the answer is yes.

P R O B L E M   S E T   2 . 2

Find the solution set for the following equations. Be sure to show when you have used the addition property of equality.

1. $x - 3 = 8$

2. $x - 2 = 7$

3. $x + 2 = 6$

4. $x + 5 = 4$

5. $a + \dfrac{1}{2} = -\dfrac{1}{4}$

6. $a + \dfrac{1}{3} = -\dfrac{5}{6}$

7. $x + 2.3 = -3.5$

8. $x + 7.9 = -3.4$

9. $y + 11 = -6$

10. $y - 3 = -1$

11. $x - \dfrac{5}{8} = -\dfrac{3}{4}$

12. $x - \dfrac{2}{5} = -\dfrac{1}{10}$

13. $m - 6 = -10$

14. $m - 10 = -6$

15. $6.9 + x = 3.3$

16. $7.5 + x = 2.2$

17. $5 = a + 4$

18. $12 = a - 3$

19. $-\dfrac{5}{9} = x - \dfrac{2}{5}$

20. $-\dfrac{7}{8} = x - \dfrac{4}{5}$

Simplify both sides of the following equations as much as possible and then solve. Check each solution.

21. $4x + 2 - 3x = 4 + 1$

22. $5x + 2 - 4x = 7 - 3$

23. $8a - \dfrac{1}{2} - 7a = \dfrac{3}{4} + \dfrac{1}{8}$

24. $9a - \dfrac{4}{5} - 8a = \dfrac{3}{10} - \dfrac{1}{5}$

25. $-3 - 4x + 5x = 18$

26. $10 - 3x + 4x = 20$

27. $-11x + 2 + 10x + 2x = 9$

28. $-10x + 5 - 4x + 15x = 0$

29. $-2.5 + 4.8 = 8x - 1.2 - 7x$

30. $-4.8 + 6.3 = 7x - 2.7 - 6x$

31. $2y - 10 + 3y - 4y = 18 - 6$

32. $3y - 20 + 6y - 8y = 21$

33. $15 - 21 = 8x + 3x - 10x$

34. $23 - 17 = -7x - x + 9x$

35. $24 - 3 + 8a - 5a - 2a = 21$

36. $30 - 4 + 7a - 2 - 6a = 30$

The following equations contain parentheses. Apply the distributive property to remove the parentheses; then simplify each side before using the addition property of equality.

37. $2(x + 3) - x = 4$

38. $5(x + 1) - 4x = 2$

39. $-3(x - 4) + 4x = 3 - 7$
40. $-2(x - 5) + 3x = 4 - 9$
41. $5(2a + 1) - 9a = 8 - 6$
42. $4(2a - 1) - 7a = 9 - 5$
43. $-(x + 3) + 2x - 1 = 6$
44. $-(x - 7) + 2x - 8 = 4$
45. $4y - 3(y - 6) + 2 = 8$
46. $7y - 6(y - 1) + 3 = 9$
47. $2(3x + 1) - 5(x + 2) = 1 - 10$
48. $4(2x + 1) - 7(x - 1) = 2 - 6$
49. $-3(2m - 9) + 7(m - 4) = 12 - 9$
50. $-5(m - 3) + 2(3m + 1) = 15 - 8$

Solve the following equations by the method used in Examples 12 and 13 in this section. Check each solution in the original equation.

51. $4x = 3x + 2$
52. $6x = 5x - 4$
53. $8a = 7a - 5$
54. $9a = 8a - 3$
55. $2x = 3x + 1$
56. $4x = 3x + 5$
57. $3y + 4 = 2y + 1$
58. $5y + 6 = 4y + 2$
59. $2m - 3 = m + 5$
60. $8m - 1 = 7m - 3$
61. $4x - 7 = 5x + 1$

62. $3x - 7 = 4x - 6$

63. $5x - \dfrac{2}{3} = 4x + \dfrac{4}{3}$

64. $3x - \dfrac{5}{4} = 2x + \dfrac{1}{4}$

65. $8a - 7.1 = 7a + 3.9$
66. $10a - 4.3 = 9a + 4.7$

Review Problems

The problems below review material we covered in Section 1.5. Reviewing this material will help you in the next section.

Apply the associative property to each expression and then simplify the result.

67. $3(6x)$ **68.** $5(4x)$

69. $\dfrac{1}{5}(5x)$ **70.** $\dfrac{1}{3}(3x)$

71. $8\left(\dfrac{1}{8}y\right)$ **72.** $6\left(\dfrac{1}{6}y\right)$

73. $-2\left(-\dfrac{1}{2}x\right)$ **74.** $-4\left(-\dfrac{1}{4}x\right)$

75. $-\dfrac{4}{3}\left(-\dfrac{3}{4}a\right)$ **76.** $-\dfrac{5}{2}\left(-\dfrac{2}{5}a\right)$

SECTION

2.3   Multiplication Property of Equality

In the previous section we found that adding the same number to both sides of an equation never changed the solution set. The same idea holds for multiplication by numbers other than zero. We can multiply both sides of an equation by the same nonzero number and always be sure we have not changed the solution set. (The reason we cannot multiply both sides by zero will become apparent later.) This fact about equations is called the **multiplication property of equality** and can be stated formally as follows.

Multiplication Property of Equality

For any three algebraic expressions A, B, and C, where $C \neq 0$,

$$\text{if} \quad A = B$$

$$\text{then } AC = BC$$

In words: Multiplying both sides of an equation by the same nonzero number will not change the solution set.

Suppose we want to solve the equation $5x = 30$. We have $5x$ on the left side but would like to have just x. We choose to multiply both sides by $\frac{1}{5}$ since $(\frac{1}{5})(5) = 1$. Here is the solution:

$$5x = 30$$

$$\frac{1}{5}(5x) = \frac{1}{5}(30) \qquad \text{Multiplication property of equality}$$

$$\left(\frac{1}{5} \cdot 5\right)x = \frac{1}{5}(30) \qquad \text{Associative property of multiplication}$$

$$1x = 6$$

$$x = 6$$

We chose to multiply by $\frac{1}{5}$ because it is the reciprocal of 5. We can see that multiplication by any number except zero will not change the solution set. If, however, we were to multiply both sides by zero, the result would always be $0 = 0$, since multiplication by zero always results in zero. Although the statement $0 = 0$ is true, we have lost our variable and cannot solve the equation. This is the only restriction on the multiplication property of equality. We are free to multiply both sides of an equation by any number except zero.

Here are some more examples that use the multiplication property of equality.

▶ **EXAMPLE 1** Solve for a: $-4a = 24$.

Solution Since we want a alone on the left side, we choose to multiply both sides by $-\frac{1}{4}$:

$$-\frac{1}{4}(-4a) = -\frac{1}{4}(24) \qquad \text{Multiplication property of equality}$$

$$\left[-\frac{1}{4}(-4)\right]a = -\frac{1}{4}(24) \qquad \text{Associative property}$$

$$a = -6 \qquad \blacktriangleleft$$

▶ **EXAMPLE 2**   Solve for t: $-\dfrac{t}{3} = 5$.

Solution   Since division by 3 is the same as multiplication by $\frac{1}{3}$, we can write $-\frac{t}{3}$ as $-\frac{1}{3}t$. To solve the equation, we multiply each side by the reciprocal of $-\frac{1}{3}$, which is -3.

$$-\frac{t}{3} = 5 \qquad \text{Original equation}$$

$$-\frac{1}{3}t = 5 \qquad \begin{array}{l}\text{Dividing by 3 is equivalent to}\\ \text{multiplying by } \frac{1}{3}\end{array}$$

$$-3\left(-\frac{1}{3}t\right) = -3(5) \qquad \text{Multiply each side by } -3$$

$$\left[-3\left(-\frac{1}{3}\right)\right]t = -3(5) \qquad \text{Associative property}$$

$$t = -15 \qquad \text{Multiplication} \qquad \blacktriangleleft$$

▶ **EXAMPLE 3**   Solve $\dfrac{2}{3}y = 4$.

Solution   We can multiply both sides by $\frac{3}{2}$ and have $1y$ on the left side:

$$\frac{3}{2}\left(\frac{2}{3}y\right) = \frac{3}{2}(4) \qquad \text{Multiplication property of equality}$$

$$\left(\frac{3}{2}\cdot\frac{2}{3}\right)y = \frac{3}{2}(4) \qquad \text{Associative property}$$

$$y = 6 \qquad \text{Simplify } \tfrac{3}{2}(4) = \tfrac{3}{2}(\tfrac{4}{1}) = \tfrac{12}{2} = 6 \qquad \blacktriangleleft$$

Notice in Examples 1 through 3 that if the variable is being multiplied by a number like -4 or $\frac{2}{3}$, we always multiply by the number's reciprocal, $-\frac{1}{4}$ or $\frac{3}{2}$, to end up with just x on one side of the equation.

▶ **EXAMPLE 4**   Solve $5 + 8 = 10x + 20x - 4x$.

Solution   Our first step will be to simplify each side of the equation:

$$13 = 26x \qquad \text{Simplify both sides first}$$

$$\frac{1}{26}(13) = \frac{1}{26}(26x) \qquad \text{Multiplication property of equality}$$

$$\frac{13}{26} = x \qquad \text{Multiplication}$$

$$\frac{1}{2} = x \qquad \text{Reduce to lowest terms} \qquad \blacktriangleleft$$

In the next three examples we will use both the addition property of equality and the multiplication property of equality.

▶ **EXAMPLE 5** Solve for x: $6x + 5 = -13$.

Solution We begin by adding -5 to both sides of the equation:

$$6x + 5 + (-5) = -13 + (-5) \qquad \text{Add } -5 \text{ to both sides}$$

$$6x = -18 \qquad \text{Simplify}$$

$$\frac{1}{6}(6x) = \frac{1}{6}(-18) \qquad \text{Multiply both sides by } \frac{1}{6}$$

$$x = -3 \qquad \blacktriangleleft$$

▶ **EXAMPLE 6** Solve for x: $5x = 2x + 12$.

Solution We begin by adding $-2x$ to both sides of the equation:

$$5x + (-2x) = 2x + (-2x) + 12 \qquad \text{Add } -2x \text{ to both sides}$$

$$3x = 12 \qquad \text{Simplify}$$

$$\frac{1}{3}(3x) = \frac{1}{3}(12) \qquad \text{Multiply both sides by } \frac{1}{3}$$

$$x = 4 \qquad \text{Simplify} \qquad \blacktriangleleft$$

Notice that in Example 6 we used the addition property of equality first in order to combine all the terms containing x on the left side of the equation. Once this had been done, we used the multiplication property to isolate x on the left side.

▶ **EXAMPLE 7** Solve for x: $3x - 4 = -2x + 6$.

Solution We begin by adding $2x$ to both sides:

$$3x + 2x - 4 = -2x + 2x + 6 \qquad \text{Add } 2x \text{ to both sides}$$

$$5x - 4 = 6 \qquad \text{Simplify}$$

Now we add 4 to both sides:

$$5x - 4 + 4 = 6 + 4 \qquad \text{Add } 4 \text{ to both sides}$$

$$5x = 10 \qquad \text{Simplify}$$

$$\frac{1}{5}(5x) = \frac{1}{5}(10) \qquad \text{Multiply by } \frac{1}{5}$$

$$x = 2 \qquad \text{Simplify} \qquad \blacktriangleleft$$

The next example involves fractions. You will see that the properties we use to solve equations containing fractions are the same as the properties we used to solve the previous equations. Also, the LCD that we used previously to add fractions can be used with the multiplication property of equality to simplify equations containing fractions.

▶ **EXAMPLE 8** Solve $\dfrac{2}{3}x + \dfrac{1}{2} = -\dfrac{3}{4}$.

Solution We can solve this equation by applying our properties and working with the fractions, or we can begin by eliminating the fractions. Let's use both methods.

Method 1 Working with the fractions.

$$\frac{2}{3}x + \frac{1}{2} + \left(-\frac{1}{2}\right) = -\frac{3}{4} + \left(-\frac{1}{2}\right) \qquad \text{Add } -\frac{1}{2} \text{ to each side}$$

$$\frac{2}{3}x = -\frac{5}{4} \qquad\qquad -\frac{3}{4} + \left(-\frac{1}{2}\right) = -\frac{3}{4} + \left(-\frac{2}{4}\right)$$

$$\frac{3}{2}\left(\frac{2}{3}x\right) = \frac{3}{2}\left(-\frac{5}{4}\right) \qquad\qquad \text{Multiply each side by } \frac{3}{2}$$

$$x = -\frac{15}{8}$$

Method 2 Eliminating the fractions in the beginning.

Our original equation has denominators of 3, 2, and 4. The LCD for these three denominators is 12, and it has the property that all three denominators will divide it evenly. Therefore, if we multiply both sides of our equation by 12, each denominator will divide into 12 and we will be left with an equation that does not contain any denominators other than 1.

$$12\left(\frac{2}{3}x + \frac{1}{2}\right) = 12\left(-\frac{3}{4}\right) \qquad \begin{array}{l}\text{Multiply each side by the}\\ \text{LCD } 12\end{array}$$

$$12\left(\frac{2}{3}x\right) + 12\left(\frac{1}{2}\right) = 12\left(-\frac{3}{4}\right) \qquad \begin{array}{l}\text{Distributive property on the}\\ \text{left side}\end{array}$$

$$8x + 6 = -9 \qquad\qquad \text{Multiply}$$

$$8x = -15 \qquad\qquad \text{Add } -6 \text{ to each side}$$

$$x = -\frac{15}{8} \qquad\qquad \text{Multiply each side by } \frac{1}{8}$$

As the third line above indicates, multiplying each side of the equation by the LCD eliminates all the fractions from the equation.

As you can see, both methods yield the same solution. To check our solution, we substitute $x = -\frac{15}{8}$ back into our original equation to obtain

$$\frac{2}{3}\left(-\frac{15}{8}\right) + \frac{1}{2} \overset{?}{=} -\frac{3}{4} \qquad \text{Substitute } -\frac{15}{8} \text{ for } x$$

$$-\frac{5}{4} + \frac{1}{2} \overset{?}{=} -\frac{3}{4} \qquad \text{Multiply } \frac{2}{3}\left(-\frac{15}{8}\right)$$

$$-\frac{5}{4} + \frac{2}{4} \overset{?}{=} -\frac{3}{4} \qquad \text{Change } \frac{1}{2} \text{ to } \frac{2}{4}$$

$$-\frac{3}{4} = -\frac{3}{4} \qquad \text{A true statement} \qquad \blacktriangleleft$$

A Note on Division

Since division is defined as multiplication by the reciprocal, multiplying both sides of an equation by the same number is equivalent to dividing both sides of the equation by the reciprocal of that number. That is, multiplying each side of an equation by $\frac{1}{3}$ and dividing each side of the equation by 3 are equivalent operations. If we were to solve the equation $3x = 18$ using division instead of multiplication, the steps would look like this:

$$3x = 18 \qquad \text{Original equation}$$

$$\frac{3x}{\mathbf{3}} = \frac{18}{\mathbf{3}} \qquad \text{Divide each side by } \mathbf{3}$$

$$x = 6 \qquad \text{Division}$$

Using division instead of multiplication on a problem like this may save you some writing. On the other hand, with multiplication, it is easier to explain "why" we end up with just one x on the left side of the equation. (The "why" has to do with the associative property of multiplication.) My suggestion is that you continue to use multiplication to solve equations like the one above until you understand the process completely. Then, if you find it more convenient, you can use division instead of multiplication.

P R O B L E M   S E T   2 . 3

Solve the following equations. Be sure to show your work.

1. $5x = 10$

2. $6x = 12$

3. $7a = 28$

4. $4a = 36$

5. $-8x = 4$

6. $-6x = 2$

7. $8m = -16$

8. $5m = -25$

9. $-3x = -9$

10. $-9x = -36$

11. $-7y = -28$

12. $-15y = -30$

13. $2x = 0$

14. $7x = 0$

15. $-5x = 0$

16. $-3x = 0$

17. $\dfrac{x}{3} = 2$

18. $\dfrac{x}{4} = 3$

19. $-\dfrac{m}{5} = 10$

20. $-\dfrac{m}{7} = 1$

21. $-\dfrac{x}{2} = -\dfrac{3}{4}$

22. $-\dfrac{x}{3} = \dfrac{5}{6}$

23. $\dfrac{2}{3}a = 8$

24. $\dfrac{3}{4}a = 6$

25. $-\dfrac{3}{5}x = \dfrac{9}{5}$ **26.** $-\dfrac{2}{5}x = \dfrac{6}{15}$

27. $-\dfrac{5}{8}y = -20$ **28.** $-\dfrac{7}{2}y = -14$

Simplify both sides as much as possible and then solve. Check each solution.

29. $-4x - 2x + 3x = 24$
30. $7x - 5x + 8x = 20$
31. $4x + 8x - 2x = 15 - 10$
32. $5x + 4x + 3x = 4 + 8$
33. $-3 - 5 = 3x + 5x - 10x$
34. $10 - 16 = 12x - 6x - 3x$
35. $18 - 13 = \dfrac{1}{2}a + \dfrac{3}{4}a - \dfrac{5}{8}a$
36. $20 - 14 = \dfrac{1}{3}a + \dfrac{5}{6}a - \dfrac{2}{3}a$

Solve each of the following equations by multiplying both sides by -1.

37. $-x = 4$ **38.** $-x = -3$
39. $-x = -4$ **40.** $-x = 3$
41. $15 = -a$ **42.** $-15 = -a$
43. $-y = \dfrac{1}{2}$ **44.** $-y = -\dfrac{3}{4}$

Solve each of the following equations using the method shown in Examples 5–8 in this section.

45. $3x - 2 = 7$ **46.** $2x - 3 = 9$
47. $2a + 1 = 3$ **48.** $5a - 3 = 7$
49. $\dfrac{1}{8} + \dfrac{1}{2}x = \dfrac{1}{4}$ **50.** $\dfrac{1}{3} + \dfrac{1}{7}x = -\dfrac{8}{21}$

51. $6x = 2x - 12$ **52.** $8x = 3x - 10$
53. $2y = -4y + 18$ **54.** $3y = -2y - 15$
55. $-7x = -3x - 8$ **56.** $-5x = -2x - 12$
57. $8x + 4 = 2x - 5$ **58.** $5x + 6 = 3x - 6$

59. $x + \dfrac{1}{2} = \dfrac{1}{4}x - \dfrac{5}{8}$

60. $\dfrac{1}{3}x + \dfrac{2}{5} = \dfrac{1}{5}x - \dfrac{2}{5}$

Solve.

61. $6m - 3 = m + 2$ **62.** $6m - 5 = m + 5$

63. $\dfrac{1}{2}m - \dfrac{1}{4} = \dfrac{1}{12}m + \dfrac{1}{6}$

64. $\dfrac{1}{2}m - \dfrac{5}{12} = \dfrac{1}{12}m + \dfrac{5}{12}$

65. $9y + 2 = 6y - 4$ **66.** $6y + 14 = 2y - 2$

67. $\dfrac{3}{2}y + \dfrac{1}{3} = y - \dfrac{2}{3}$ **68.** $\dfrac{3}{2}y + \dfrac{7}{2} = \dfrac{1}{2}y - \dfrac{1}{2}$

Review Problems

The problems below review material we covered in Section 2.1. Reviewing this material will help you in the next section.

Simplify each expression.

69. $5(2x - 8) - 3$ **70.** $4(3x - 1) + 7$
71. $-2(3x + 5) + 3(x - 1)$
72. $6(x + 3) - 2(2x + 4)$
73. $7 - 3(2y + 1)$ **74.** $8 - 5(3y - 4)$
75. $4x - (9x - 3) + 4$ **76.** $x - (5x + 2) - 3$

SECTION

2.4   **Solving Linear Equations**

We will now use the material we have developed in the first three sections of this chapter to build a method for solving any linear equation.

> **DEFINITION**   A **linear equation** in one variable is any equation that can be put in the form $ax + b = 0$, where a and b are real numbers and a is not zero.

Each of the equations we will solve in this section is a linear equation in one variable. The steps we use to solve a linear equation in one variable are listed here.

Strategy for Solving Linear Equations in One Variable

Step 1a: Use the distributive property to separate terms, if necessary.

1b: If fractions are present, consider multiplying both sides by the LCD to eliminate the fractions. If decimals are present, consider multiplying both sides by a power of 10 to clear the equation of decimals.

1c: Combine similar terms on each side of the equation.

Step 2: Use the addition property of equality to get all variable terms on one side of the equation and all constant terms on the other side. A variable term is a term that contains the variable (for example, $5x$). A constant term is a term that does not contain the variable (the number 3, for example).

Step 3: Use the multiplication property of equality to get x (that is, $1x$) by itself on one side of the equation.

Step 4: Check your solution in the original equation to be sure that you have not made a mistake in the solution process.

As you will see as you work through the examples in this section, it is not always necessary to use all four steps when solving equations. The number of steps used depends upon the equation. In Example 1 below, there are no fractions or decimals in the original equation, so step 1b will not be used. Likewise, after applying the distributive property to the left side of the equation in Example 1, there are no similar terms to combine on either side of the equation, making step 1c also unnecessary.

▶ **EXAMPLE 1**   Solve $2(x + 3) = 10$.

Solution   To begin, we apply the distributive property to the left side of the equation to separate terms:

Step 1a: $\qquad 2x + 6 = 10 \qquad\qquad$ Distributive property

Step 2: $\begin{cases} 2x + 6 + (-6) = 10 + (-6) & \text{Addition property of equality} \\ \qquad\quad 2x = 4 \end{cases}$

Step 3: $\begin{cases} \dfrac{1}{2}\,(2x) = \dfrac{1}{2}\,(4) & \text{Multiply each side by } \dfrac{1}{2} \\[2mm] \qquad\quad x = 2 & \text{The solution is 2} \end{cases}$

The solution to our equation is 2. We check our work (to be sure we have not made either a mistake in applying the properties or an arithmetic mistake) by substituting 2 into our original equation and simplifying each side of the result separately.

Check: *Step 4:* $\begin{cases} \text{When} \qquad\qquad x = 2 \\ \text{the equation } 2(x + 3) = 10 \\ \text{becomes} \qquad 2(2 + 3) \overset{?}{=} 10 \\ \qquad\qquad\qquad 2(5) \overset{?}{=} 10 \\ \qquad\qquad\qquad 10 = 10 \qquad \text{A true statement} \end{cases}$

Our solution checks. ◀

The general method of solving linear equations is actually very simple. It is based on the properties we developed in Chapter 1 and on two very simple new properties. We can add any number to both sides of the equation and multiply both sides by any nonzero number. The equation may change in form, but the solution set will not. If we look back to Example 1, each equation looks a little different from each preceding equation. What is interesting and useful is that each equation says the same thing about x. They all say x is 2. The last equation, of course, is the easiest to read, and that is why our goal is to end up with x by itself.

▶ **EXAMPLE 2** Solve for x: $3(x - 5) + 4 = 13$.

Solution Our first step will be to apply the distributive property to the left side of the equation:

Step 1a: $\qquad 3x - 15 + 4 = 13 \qquad\qquad$ Distributive property

Step 1c: $\qquad\quad 3x - 11 = 13 \qquad\qquad$ Simplify the left side

Step 2: $\begin{cases} 3x - 11 + 11 = 13 + 11 & \text{Add } 11 \text{ to both sides} \\ \qquad\quad 3x = 24 \end{cases}$

Step 3: $\begin{cases} \dfrac{1}{3}\,(3x) = \dfrac{1}{3}\,(24) & \text{Multiply both sides by } \dfrac{1}{3} \\[2mm] \qquad\quad x = 8 & \text{The solution is 8} \end{cases}$

Check:

$$\text{When} \qquad\qquad\qquad x = 8$$

Step 4: the equation $3(x - 5) + 4 = 13$

becomes $\qquad 3(8 - 5) + 4 \stackrel{?}{=} 13$

$$3(3) + 4 \stackrel{?}{=} 13$$

$$9 + 4 \stackrel{?}{=} 13$$

$$13 = 13 \qquad \text{A true statement} \qquad \blacktriangleleft$$

▶ **EXAMPLE 3** Solve $5(x - 3) + 2 = 5(2x - 8) - 3$.

Solution In this case, we apply the distributive property on each side of the equation:

Step 1a: $\qquad 5x - 15 + 2 = 10x - 40 - 3 \qquad$ Distributive property

Step 1b: $\qquad\quad 5x - 13 = 10x - 43 \qquad$ Simplify each side

Step 2:

$$5x + (-5x) - 13 = 10x + (-5x) - 43 \quad \text{Add } -5x \text{ to both sides}$$

$$-13 = 5x - 43$$

$$-13 + 43 = 5x - 43 + 43 \qquad \text{Add } 43 \text{ to both sides}$$

$$30 = 5x$$

Step 3:

$$\frac{1}{5}(30) = \frac{1}{5}(5x) \qquad\qquad \text{Multiply both sides by } \frac{1}{5}$$

$$6 = x \qquad\qquad \text{The solution is 6}$$

Check:

Step 4: Replacing x with 6 in the original equation, we have

$$5(6 - 3) + 2 \stackrel{?}{=} 5(2 \cdot 6 - 8) - 3$$

$$5(3) + 2 \stackrel{?}{=} 5(12 - 8) - 3$$

$$5(3) + 2 \stackrel{?}{=} 5(4) - 3$$

$$15 + 2 \stackrel{?}{=} 20 - 3$$

$$17 = 17 \qquad \text{A true statement} \qquad \blacktriangleleft$$

It makes no difference on which side of the equal sign x ends up. Most people prefer to have x on the left side because we read from left to right and it seems to sound better to say x is 6 rather than 6 is x. Both expressions, however, have exactly the same meaning.

▶ **EXAMPLE 4** Solve the equation $0.08x + 0.09(x + 2{,}000) = 690$.

Solution We can solve the equation in its original form by working with the decimals, or we can eliminate the decimals first by using the multiplication property of equality and solving the resulting equation. Both methods follow.

Method 1   Working with the decimals.

$$0.08x + 0.09(x + 2,000) = 690$$ Original equation

Step 1a: $0.08x + 0.09x + 0.09(2,000) = 690$ Distributive property

Step 1c: $0.17x + 180 = 690$ Simplify the left side

Step 2: $\begin{cases} 0.17x + 180 + (-180) = 690 + (-180) & \text{Add } -180 \text{ to each side} \\ 0.17x = 510 \end{cases}$

Step 3: $\begin{cases} \dfrac{0.17x}{0.17} = \dfrac{510}{0.17} & \text{Divide each side by } \textbf{0.17} \\ x = 3,000 \end{cases}$

Note that we divided each side of the equation by 0.17 to obtain our solution. This is still an application of the multiplication property of equality, since dividing by 0.17 is equivalent to multiplying by $\frac{1}{0.17}$.

Method 2   Eliminating the decimals in the beginning.

$$0.08x + 0.09(x + 2,000) = 690$$ Original equation

Step 1a: $0.08x + 0.09x + 180 = 690$ Distributive property

Step 1b: $\begin{cases} \textbf{100}(0.08x + 0.09x + 180) = \textbf{100}(690) & \text{Multiply both sides} \\ & \text{by } \textbf{100} \\ 8x + 9x + 18,000 = 69,000 \end{cases}$

Step 1c: $17x + 18,000 = 69,000$ Simplify the left side

Step 2: $17x = 51,000$ Add $-18,000$ to each side

Step 3: $\begin{cases} \dfrac{17x}{17} = \dfrac{51,000}{17} & \text{Divide each side by} \\ & \textbf{17} \\ x = 3,000 \end{cases}$

Check: Substituting 3,000 for x in the original equation, we have

Step 4: $\begin{cases} 0.08(3,000) + 0.09(3,000 + 2,000) \overset{?}{=} 690 \\ 0.08(3,000) + 0.09(5,000) \overset{?}{=} 690 \\ 240 + 450 \overset{?}{=} 690 \\ 690 = 690 \qquad \text{A true statement} \end{cases}$

◀

▶ **EXAMPLE 5**   Solve $7 - 3(2y + 1) = 16$.

Solution   We begin by multiplying -3 times the sum of $2y$ and 1:

Step 1a: $\qquad\qquad 7 - 6y - 3 = 16 \qquad\qquad$ Distributive property

Step 1c: $\qquad\qquad\quad -6y + 4 = 16 \qquad\qquad$ Simplify the left side

Step 2: $\left\{\begin{array}{l} -6y + 4 + (\mathbf{-4}) = 16 + (\mathbf{-4}) \quad \text{Add } \mathbf{-4} \text{ to both sides} \\ \qquad\qquad\quad -6y = 12 \end{array}\right.$

Step 3: $\left\{\begin{array}{l} -\dfrac{1}{6}(-6y) = -\dfrac{1}{6}(12) \qquad \text{Multiply both sides by } -\dfrac{1}{6} \\[2mm] \qquad\qquad y = -2 \end{array}\right.$

Check: $\left\{\begin{array}{l} \text{When} \qquad\qquad\qquad\qquad y = -2 \\[1mm] \text{the equation } 7 - 3(2y + 1) = 16 \\[1mm] \text{becomes } 7 - 3[2(-2) + 1] \stackrel{?}{=} 16 \\[1mm] \qquad\qquad 7 - 3(-4 + 1) \stackrel{?}{=} 16 \\[1mm] \qquad\qquad\qquad 7 - 3(-3) \stackrel{?}{=} 16 \\[1mm] \qquad\qquad\qquad\qquad 7 + 9 \stackrel{?}{=} 16 \\[1mm] \qquad\qquad\qquad\qquad 16 = 16 \qquad \text{A true statement} \quad \blacktriangleleft \end{array}\right.$

Step 4:

There are two things to notice about the example that follows: first, the distributive property is used to remove parentheses that are preceded by a negative sign, and, second, the addition property and the multiplication property are not shown in as much detail as in the previous examples.

▶ **EXAMPLE 6** Solve $3(2x - 5) - (2x - 4) = 6 - (4x + 5)$.

Solution When we apply the distributive property to remove the grouping symbols and separate terms, we have to be careful with the signs. Remember, we can think of $-(2x - 4)$ as $-1(2x - 4)$, so that

$$-(2x - 4) = -1(2x - 4) = -2x + 4$$

It is not uncommon for students to make a mistake with this type of simplification and write the result as $-2x - 4$, which is incorrect. Here is the complete solution to our equation:

$3(2x - 5) - (2x - 4) = 6 - (4x + 5)$	Original equation
$6x - 15 - 2x + 4 = 6 - 4x - 5$	Distributive property
$4x - 11 = -4x + 1$	Simplify each side
$8x - 11 = 1$	Add $4x$ to each side
$8x = 12$	Add 11 to each side

$$x = \frac{12}{8} \qquad\qquad \text{Multiply each side by } \frac{1}{8}$$

$$x = \frac{3}{2} \qquad\qquad \text{Reduce to lowest terms}$$

The solution, $\frac{3}{2}$, checks when replacing x in the original equation.   ◄

▶ **EXAMPLE 7**   Solve for x: $\frac{1}{2}(2x - 6) + 4 = \frac{2}{3}(3x + 9) - 8$.

Solution   We could multiply each side of the equation by the LCD 6 and clear the equation of fractions. In this case, however, it is easier to simply apply the distributive property.

$$\frac{1}{2}(2x - 6) + 4 = \frac{2}{3}(3x + 9) - 8$$

$$\frac{1}{2}(2x) - \frac{1}{2}(6) + 4 = \frac{2}{3}(3x) + \frac{2}{3}(9) - 8$$

$$x - 3 + 4 = 2x + 6 - 8$$

$$x + 1 = 2x - 2$$

$$-x + 1 = -2$$

$$-x = -3$$

$$x = 3 \qquad\qquad ◄$$

P R O B L E M   S E T   2 . 4

Solve each of the following equations using the four steps shown in this section.

Solve each equation.

1. $2(x + 3) = 12$ **2.** $3(x - 2) = 6$

3. $6(x - 1) = -18$ **4.** $4(x + 5) = 16$

5. $2(4a + 1) = -6$ **6.** $3(2a - 4) = 12$

7. $14 = 2(5x - 3)$ **8.** $-25 = 5(3x + 4)$

9. $-2(3y + 5) = 14$ **10.** $-3(2y - 4) = -6$

11. $-5(2a + 4) = 0$ **12.** $-3(3a - 6) = 0$

13. $1 = \frac{1}{2}(4x + 2)$ **14.** $1 = \frac{1}{3}(6x + 3)$

15. $3(t - 4) + 5 = -4$

16. $5(t - 1) + 6 = -9$

17. $4(2y + 1) - 7 = 1$

18. $6(3y + 2) - 8 = -2$

19. $\frac{1}{2}(x - 3) = \frac{1}{4}(x + 1)$

20. $\frac{1}{3}(x - 4) = \frac{1}{2}(x - 6)$

21. $-0.7(2x - 7) = 0.3(11 - 4x)$

22. $-0.3(2x - 5) = 0.7(3 - x)$

23. $-2(3y + 1) = 3(1 - 6y) - 9$

24. $-5(4y - 3) = 2(1 - 8y) + 11$

25. $\dfrac{3}{4}(8x - 4) + 3 = \dfrac{2}{5}(5x + 10) - 1$

26. $\dfrac{5}{6}(6x + 12) + 1 = \dfrac{2}{3}(9x - 3) + 5$

27. $0.06x + 0.08(100 - x) = 6.5$
28. $0.05x + 0.07(100 - x) = 6.2$
29. $6 - 5(2a - 3) = 1$
30. $-8 - 2(3 - a) = 0$
31. $0.2x - 0.5 = 0.5 - 0.2(2x - 13)$
32. $0.4x - 0.1 = 0.7 - 0.3(6 - 2x)$
33. $2(t - 3) + 3(t - 2) = 28$
34. $-3(t - 5) - 2(2t + 1) = -8$
35. $5(x - 2) - (3x + 4) = 3(6x - 8) + 10$
36. $3(x - 1) - (4x - 5) = 2(5x - 1) - 7$
37. $2(5x - 3) - (2x - 4) = 5 - (6x + 1)$
38. $3(4x - 2) - (5x - 8) = 8 - (2x + 3)$
39. $-(3x + 1) - (4x - 7) = 4 - (3x + 2)$
40. $-(6x + 2) - (8x - 3) = 8 - (5x + 1)$

Review Problems

The problems that follow review material we covered in Sections 1.2, 1.5, and 1.6. Reviewing these problems will help you understand the next section.

Multiply.

41. $\dfrac{1}{2}(3)$

42. $\dfrac{1}{3}(2)$

43. $\dfrac{2}{3}(6)$

44. $\dfrac{3}{2}(4)$

45. $\dfrac{5}{9} \cdot \dfrac{9}{5}$

46. $\dfrac{3}{7} \cdot \dfrac{7}{3}$

Apply the distributive property and then simplify each expression as much as possible.

47. $2(3x - 5)$

48. $4(2x - 6)$

49. $\dfrac{1}{2}(3x + 6)$

50. $\dfrac{1}{4}(2x + 8)$

51. $\dfrac{1}{3}(-3x + 6)$

52. $\dfrac{1}{2}(-2x + 6)$

SECTION

2.5

Formulas

In this section, we continue solving equations by working with formulas. To begin, here is the definition for a formula.

> **DEFINITION** In mathematics, a **formula** is an equation that contains more than one variable.

The equation $P = 2l + 2w$, which tells us how to find the perimeter of a rectangle, is an example of a formula.

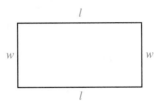

To begin our work with formulas, we will consider some examples in which we are given numerical replacements for all but one of the variables.

▶ **EXAMPLE 1** The perimeter P of a rectangular livestock pen is 40 feet. If the width w is 6 feet, find the length.

Solution First we substitute 40 for P and 6 for w in the formula $P = 2l + 2w$. Then we solve for l:

When $\quad\quad P = 40$ and $w = 6$

the formula $\quad P = 2l + 2w$

becomes $\quad 40 = 2l + 2(6)$

or $\quad\quad\quad 40 = 2l + 12$ $\quad\quad$ Multiply 2 and 6

$\quad\quad\quad\quad\quad 28 = 2l$ $\quad\quad\quad\quad$ Add -12 to each side

$\quad\quad\quad\quad\quad 14 = l$ $\quad\quad\quad\quad$ Multiply each side by $\dfrac{1}{2}$

To summarize our results, if a rectangular pen has a perimeter of 40 feet and a width of 6 feet, then the length must be 14 feet. ◀

▶ **EXAMPLE 2** Find y when $x = 4$ in the formula $3x + 2y = 6$.

Solution We substitute 4 for x in the formula and then solve for y:

When $\quad\quad\quad\quad\quad x = 4$

the formula $\quad 3x + 2y = 6$

becomes $\quad 3(4) + 2y = 6$

or $\quad\quad\quad\quad 12 + 2y = 6$ $\quad\quad$ Multiply 3 and 4

$\quad\quad\quad\quad\quad\quad 2y = -6$ $\quad\quad$ Add -12 to each side

$\quad\quad\quad\quad\quad\quad y = -3$ $\quad\quad$ Multiply each side by $\dfrac{1}{2}$ ◀

In the next examples, we will solve a formula for one of its variables without being given numerical replacements for the other variables.

Consider the formula for the area of a triangle:

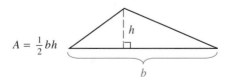

$$A = \tfrac{1}{2}bh$$

where A = area, b = length of the base, and h = the height of the triangle.

Suppose we want to solve this formula for h. What we must do is isolate the variable h on one side of the equal sign. We begin by multiplying both sides by 2:

$$2 \cdot A = 2 \cdot \frac{1}{2} bh$$

$$2A = bh$$

Then we divide both sides by b:

$$\frac{2A}{b} = \frac{bh}{b}$$

$$h = \frac{2A}{b}$$

The original formula $A = \frac{1}{2}bh$ and the final formula $h = \dfrac{2A}{b}$ both give the same relationship among A, b, and h. The first one has been solved for A and the second one has been solved for h.

Rule

To solve a formula for one of its variables, we must isolate that variable on either side of the equal sign. All other variables and constants will appear on the other side.

▶ **EXAMPLE 3**   Solve $3x + 2y = 6$ for y.

Solution   To solve for y, we must isolate y on the left side of the equation. To begin, we use the addition property of equality to add $-3x$ to each side:

$3x + 2y = 6$	Original formula
$3x + (-3x) + 2y = (-3x) + 6$	Add $-3x$ to each side
$2y = -3x + 6$	Simplify the left side
$\dfrac{1}{2}(2y) = \dfrac{1}{2}(-3x + 6)$	Multiply each side by $\dfrac{1}{2}$
$y = -\dfrac{3}{2}x + 3$	Multiplication

◀

▶ **EXAMPLE 4**   Solve $h = vt - 16t^2$ for v.

Solution   Let's begin by interchanging the left and right sides of the equation. That way, the variable we are solving for, v, will be on the left side.

$vt - 16t^2 = h$	Exchange sides
$vt - 16t^2 + \mathbf{16t^2} = h + \mathbf{16t^2}$	Add $\mathbf{16t^2}$ to each side

$$vt = h + 16t^2$$

$$\frac{vt}{t} = \frac{h + 16t^2}{t} \qquad \text{Divide each side by } t$$

$$v = \frac{h + 16t^2}{t}$$

We know we are finished because we have isolated the variable we are solving for on the left side of the equation and it does not appear on the other side. ◀

In Chapter 1 we defined complementary angles as angles that add to 90°. That is, if x and y are complementary angles, then

$$x + y = 90°$$

If we solve this formula for y, we obtain a formula equivalent to our original formula:

$$y = 90° - x$$

Since y is the complement of x, we can generalize by saying that the complement of angle x is the angle $90° - x$. By a similar reasoning process, we can say that the supplement of angle x is the angle $180° - x$. To summarize, if x is an angle, then

The complement of x is $90° - x$, and

The supplement of x is $180° - x$

If you go on to take a trigonometry class, you will see this formula again.

▶ **EXAMPLE 5**   Find the complement and the supplement of 25°.

Solution   We can use the formulas above with $x = 25°$.

The complement of 25° is $90° - 25° = 65°$

The supplement of 25° is $180° - 25° = 155°$ ◀

Basic Percent Problems

The last examples in this section show how basic percent problems can be translated directly into equations. To understand these examples, you must recall that percent means "per hundred." That is, 75% is the same as $\frac{75}{100}$, 0.75, and, in reduced fraction form, $\frac{3}{4}$. Likewise, the decimal 0.25 is equivalent to 25%. To change a decimal to a percent, we move the decimal point two places to the right and write the % symbol. To change from a percent to a decimal, we drop the % symbol and move the decimal point two places to the left. The table that follows gives some of the most commonly used fractions and decimals and their equivalent percents.

Fraction	Decimal	Percent
$\frac{1}{2}$	0.5	50%
$\frac{1}{4}$	0.25	25%
$\frac{3}{4}$	0.75	75%
$\frac{1}{3}$	$0.33\frac{1}{3}$	$33\frac{1}{3}\%$
$\frac{2}{3}$	$0.66\frac{2}{3}$	$66\frac{2}{3}\%$
$\frac{1}{5}$	0.2	20%
$\frac{2}{5}$	0.4	40%
$\frac{3}{5}$	0.6	60%
$\frac{4}{5}$	0.8	80%

▶ **EXAMPLE 6**   What number is 25% of 60?

Solution   To solve a problem like this, we let x = the number in question (that is, the number we are looking for). Then, we translate the sentence directly into an equation by using an equal sign for the word "is" and multiplication for the word "of." Here is how it is done:

$$\underbrace{\text{What number}}_{x} \quad \overset{\downarrow}{\text{is}} \; \overset{\downarrow}{25\%} \; \overset{\downarrow}{\text{of}} \; \overset{\downarrow}{60?}$$
$$x = 0.25 \;\cdot\; 60$$
$$x = 15$$

Notice that we must write 25% as a decimal in order to do the arithmetic in the problem.

The number 15 is 25% of 60.   ◀

▶ **EXAMPLE 7**   What percent of 24 is 6?

Solution   Translating this sentence into an equation, as we did in Example 6, we have

$$\underbrace{\text{What percent}}_{x} \quad \overset{\downarrow}{\text{of}} \; \overset{\downarrow}{24} \; \overset{\downarrow}{\text{is}} \; \overset{\downarrow}{6?}$$
$$x \;\cdot\; 24 = 6$$
$$\text{or} \qquad 24x = 6$$

Next we multiply each side by $\frac{1}{24}$. (This is the same as dividing each side by 24):

$$\frac{1}{24}(24x) = \frac{1}{24}(6)$$

$$x = \frac{6}{24}$$

$$= \frac{1}{4}$$

$$= 0.25 \text{ or } 25\%$$

The number 6 is 25% of 24.   ◀

▶ **EXAMPLE 8**   45 is 75% of what number?

Solution   Again, we translate the sentence directly:

45 is 75% of what number?
$\downarrow$ $\downarrow$ $\downarrow$ $\downarrow$
$45 = 0.75 \cdot \qquad x$

Next, we multiply each side by $\frac{1}{0.75}$ (which is the same as dividing each side by 0.75):

$$\frac{1}{0.75}(45) = \frac{1}{0.75}(0.75x)$$

$$\frac{45}{0.75} = x$$

$$60 = x$$

The number 45 is 75% of 60.   ◀

We can solve application problems involving percent by translating each problem into one of the three basic percent problems shown in Examples 6, 7, and 8 above.

▶ **EXAMPLE 9**   The American Dietetics Association (ADA) recommends eating foods in which the calories from fat are less than 30% of the total calories. The nutrition labels from two kinds of granola bars are shown on the next page. For each bar, what percent of the total calories come from fat?

Solution   The information needed to solve this problem is located toward the top of each label. Each serving of Bar I contains 210 calories of which 70

Bar I

Bar II

Nutrition Facts

Serving Size 2 bars (47g)
Servings Per Container 6

Amount Per Serving

Calories	210
Calories from Fat	70

% Daily Value*

Total Fat 8g	**12%**
Saturated Fat 1g	**5%**
Cholesterol 0mg	**0%**
Sodium 150mg	**6%**
Total Carbohydrate 32g	**11%**
Dietary Fiber 2g	**10%**
Sugars 12g	
Protein 4g	

* Percent Daily Values are based on a 2,000 calorie diet.
Your daily values may be higher or lower depending on
your calorie needs.

Nutrition Facts

Serving Size 1 Bar (21g)
Servings Per Container 8

Amount Per Serving

Calories	80
Calories from Fat	15

% Daily Value*

Total Fat 1.5g	**2%**
Saturated Fat 0g	**0%**
Cholesterol 0mg	**0%**
Sodium 60mg	**3%**
Total Carbohydrate 16g	**5%**
Dietary Fiber 1g	**4%**
Sugars 5g	
Protein 2g	

* Percent Daily Values are based on a 2,000 calorie diet.
Your daily values may be higher or lower depending on
your calorie needs.

calories come from fat. To find the percent of total calories that come from fat, we must answer this question:

70 is what percent of 210?

For Bar II, one serving contains 80 calories of which 15 calories come from fat. To find the percent of total calories that come from fat, we must answer this question:

15 is what percent of 80?

Translating each question into symbols, we have

70 is what percent of 210

$$70 = x \cdot 210$$

$$x = \frac{70}{210}$$

$x = 0.33$ to the nearest hundredth

$x = 33\%$

15 is what percent of 80

$$15 = x \cdot 80$$

$$x = \frac{15}{80}$$

$x = 0.19$ to the nearest hundredth

$x = 19\%$

Comparing the two bars, 33% of the calories in Bar I are fat calories while 19% of the calories in Bar II are fat calories. According to the ADA, Bar II is the healthier choice. ◀

P R O B L E M   S E T   2 . 5

Use the formula $P = 2l + 2w$ to find the length l of a rectangular lot if

1. the width w is 50 feet and the perimeter P is 300 feet

2. the width w is 75 feet and the perimeter P is 300 feet

Use the formula $2x + 3y = 6$ to find y if

3. x is 3   **4.** x is -2
5. x is 0   **6.** x is -3

Use the formula $2x - 5y = 20$ to find x if

7. y is 2   **8.** y is -4
9. y is 0   **10.** y is -6

Use the equation $y = 2x - 1$ to find x when

11. y is 7   **12.** y is 9
13. y is 3   **14.** y is -1

Solve each of the following for the indicated variable.

15. $A = lw$ for l **16.** $A = lw$ for w
17. $d = rt$ for r **18.** $d = rt$ for t
19. $V = lwh$ for h **20.** $V = lwh$ for l
21. $PV = nRT$ for P **22.** $PV = nRT$ for T
23. $P = a + b + c$ for a
24. $P = a + b + c$ for b
25. $x - 3y = -1$ for x **26.** $x + 3y = 2$ for x
27. $-3x + y = 6$ for y
28. $2x + y = -17$ for y
29. $2x + 3y = 6$ for y **30.** $4x + 5y = 20$ for y
31. $6x + 3y = 12$ for y **32.** $3x + 6y = 12$ for y
33. $5x - 2y = 3$ for y **34.** $7x - 3y = 5$ for y

35. $P = 2l + 2w$ for w **36.** $P = 2l + 2w$ for l
37. $h = vt + 16t^2$ for v **38.** $h = vt - 16t^2$ for v
39. $A = \pi r^2 + 2\pi rh$ for h
40. $A = 2\pi r^2 + 2\pi rh$ for h

Solve each formula for y.

41. $\dfrac{x}{2} + \dfrac{y}{3} = 1$ **42.** $\dfrac{x}{5} + \dfrac{y}{4} = 1$

43. $\dfrac{x}{7} - \dfrac{y}{3} = 1$ **44.** $\dfrac{x}{4} - \dfrac{y}{9} = 1$

45. $-\dfrac{1}{4}x + \dfrac{1}{8}y = 1$ **46.** $-\dfrac{1}{9}x + \dfrac{1}{3}y = 1$

Find the complement and the supplement of each angle.

47. $30°$ **48.** $60°$
49. $45°$ **50.** $15°$

Translate each of the following into an equation and then solve that equation.

51. What number is 25% of 40?
52. What number is 75% of 40?
53. What number is 12% of 2,000?
54. What number is 9% of 3,000?
55. What percent of 28 is 7?
56. What percent of 28 is 21?
57. What percent of 40 is 14?
58. What percent of 20 is 14?
59. 32 is 50% of what number?
60. 16 is 50% of what number?
61. 240 is 12% of what number?
62. 360 is 12% of what number?

Applying the Concepts

More about Temperature   As we mentioned in Chapter 1, in the U.S. system, temperature is measured on the Fahrenheit scale. In the metric system, temperature is measured on the Celsius scale. On the Celsius scale, water boils at 100 degrees and freezes at 0 degrees. To denote a temperature of 100 degrees on the Celsius scale, we write

100° C, which is read ''100 degrees Celsius''

Table 1 is intended to give you an intuitive idea of the relationship between the two temperature scales.

TABLE 1

Situation	Temperature	
	Fahrenheit	**Celsius**
Water freezes	32° F	0° C
Room temperature	68° F	20° C
Normal body temperature	98.6° F	37° C
Water boils	212° F	100° C
Bake cookies	365° F	185° C

Table 2 gives the formulas, in both symbols and words, that are used to convert between the two scales.

63. Let $F = 212$ in the formula $C = \frac{5}{9}(F - 32)$, and solve for C. Does the value of C agree with the information in Table 1?

64. Let $C = 100$ in the formula $F = \frac{9}{5}C + 32$, and solve for F. Does the value of F agree with the information in Table 1?

65. Let $F = 68$ in the formula $C = \frac{5}{9}(F - 32)$, and solve for C. Does the value of C agree with the information in Table 1?

66. Let $C = 37$ in the formula $F = \frac{9}{5}C + 32$, and solve for F. Does the value of F agree with the information in Table 1?

67. Solve the formula $F = \frac{9}{5}C + 32$ for C.

68. Solve the formula $C = \frac{5}{9}(F - 32)$ for F.

The nutrition label below is from a quart of vanilla ice cream. The label on the next page is from a pint of vanilla frozen yogurt. Use the information on these labels for problems 69–72. Round your answers to the nearest tenth of a percent.

Vanilla Ice Cream

Nutrition Facts

Serving Size 1/2 cup (65g)
Servings 8

Amount/Serving	
Calories 150	Calories from Fat 90

	% Daily Value*
Total Fat 10g	**16%**
Saturated Fat 6g	**32%**
Cholesterol 35mg	**12%**
Sodium 30mg	**1%**
Total Carbohydrate 14g	**5%**
Dietary Fiber 0g	**0%**
Sugars 11g	
Protein 2g	

Vitamin A 6%	•	Vitamin C 0%
Calcium 6%	•	Iron 0%

* Percent Daily Values are based on a 2,000 calorie diet.

TABLE 2

To Convert from	Formula in Symbols	Formula in Words
Fahrenheit to Celsius	$C = \dfrac{5}{9}(F - 32)$	Subtract 32, multiply by 5, then divide by 9.
Celsius to Fahrenheit	$F = \dfrac{9}{5}C + 32$	Multiply by $\dfrac{9}{5}$, then add 32.

Vanilla Frozen Yogurt

Nutrition Facts

Serving Size 1/2 cup (98g)
Servings Per Container 4

Amount Per Serving

Calories 160   Calories from Fat 25

% Daily Value*

Total Fat 2.5g   **4%**

    Saturated Fat 1.5g   **7%**

Cholesterol 45mg   **15%**

Sodium 55mg   **2%**

Total Carbohydrate 26g   **9%**

    Dietary Fiber 0g   **0%**

    Sugars 19g

Protein 8g

Vitamin A 0%   •   Vitamin C 0%

Calcium 25%   •   Iron 0%

* Percent Daily Values are based on a 2,000 calorie diet.

69. What percent of the calories in one serving of the vanilla ice cream are fat calories?

70. What percent of the calories in one serving of the frozen yogurt are fat calories?

71. One serving of frozen yogurt is 98 grams of which 26 grams are carbohydrates. What percent of one serving are carbohydrates?

72. One serving of vanilla ice cream is 65 grams. Find the percent of one serving that is sugar.

The circumference of a circle is given by the formula $C = 2\pi r$. Find r if

73. the circumference C is 44 meters and π is $\frac{22}{7}$

74. the circumference C is 176 meters and π is $\frac{22}{7}$

75. the circumference is 9.42 inches and π is 3.14

76. the circumference is 12.56 inches and π is 3.14

The volume of a cylinder is given by the formula $V = \pi r^2 h$. Find the height h if

77. the volume V is 42 cubic feet, the radius is $\frac{7}{22}$ feet, and π is $\frac{22}{7}$

78. the volume V is 84 cubic inches, the radius is $\frac{7}{11}$ inches, and π is $\frac{22}{7}$

79. the volume is 6.28 cubic centimeters, the radius is 3 centimeters, and π is 3.14

80. the volume is 12.56 cubic centimeters, the radius is 2 centimeters, and π is 3.14

Review Problems

The following problems review material we covered in Chapter 1. Reviewing these problems will help you in the next section.

Write an equivalent expression in English. Include the words *sum* and *difference* when possible.

81. $4 + 1 = 5$ **82.** $7 + 3 = 10$

83. $6 - 2 = 4$ **84.** $8 - 1 = 7$

For each of the following expressions, write an equivalent expression with numbers.

85. Twice the sum of 6 and 3

86. Four added to the product of 5 and -1

87. The sum of twice 5 and 3 is 13

88. Twice the difference of 8 and 2 is 12

SECTION

....

2.6   Applications

As you begin reading through the examples in this section, you may find yourself asking why some of these problems seem so contrived. The title of the section is "Applications," but many of the problems here don't seem to have much to do with "real life." Example 3 is what we refer to as an "age problem." Realistically, it is not the kind of problem you would expect to find if you choose a career in which you use algebra. However, solving "age problems" is good practice for someone with little experience with application problems, since the solution process has a form that can be applied to all similar age problems. Although many of the other "application" problems in this section are contrived, they too are good for practicing the strategy we will use to solve all application problems.

To begin this section, we list the steps used in solving application problems. We call this strategy the *Blueprint for Problem Solving.* It is an outline that will overlay the solution process we use on all application problems.

Blueprint for Problem Solving

Step 1: **Read** the problem and then mentally **list** the items that are known and the items that are unknown.

Step 2: **Assign a variable** to one of the unknown items. (In most cases this will amount to letting $x =$ the item that is asked for in the problem.) Then **translate** the other **information** in the problem to expressions involving the variable.

Step 3: **Reread** the problem, and then **write an equation,** using the items and variables listed in steps 1 and 2, that describes the situation.

Step 4: **Solve the equation** found in step 3.

Step 5: **Write** your **answer** using a complete sentence.

Step 6: **Reread** the problem and **check** your solution with the original words in the problem.

There are a number of substeps within each of the steps in our blueprint. For instance, with steps 1 and 2 it is always a good idea to draw a diagram or picture, if it helps visualize the relationship between the items in the problem. In other cases, a table helps organize the information. As you gain more experience using the blueprint to solve application problems, you will find additional techniques that expand the blueprint.

To help with problems of the type shown next in Example 1, here are some common English words and phrases and their mathematical translations.

English	Algebra
The sum of a and b	$a + b$
The difference of a and b	$a - b$
The product of a and b	$a \cdot b$
The quotient of a and b	$\dfrac{a}{b}$
of	$\cdot$ (multiply)
is	$=$ (equals)
A number	x
4 more than x	$x + 4$
4 times x	$4x$
4 less than x	$x - 4$

Number Problems

▶ **EXAMPLE 1**   The sum of twice a number and three is seven. Find the number.

Solution   Using the Blueprint for Problem Solving as an outline, we solve the problem as follows:

Step 1: ***Read*** the problem and then mentally ***list*** the items that are known and the items that are unknown.

> *Known items:* The numbers 3 and 7
>
> *Unknown item:* The number in question

Step 2: ***Assign a variable*** to one of the unknown items. Then ***translate*** the other ***information*** in the problem to expressions involving the variable.

> Let x = the number asked for in the problem,
>
> then ''The sum of twice a number and three'' translates to $2x + 3$.

Step 3: ***Reread*** the problem and then ***write an equation,*** using the items and variables listed in steps 1 and 2, that describes the situation.

> With all word problems, the word ''is'' translates to $=$.

$$\underbrace{\text{The sum of twice } x \text{ and } 3}_{2x + 3} \text{ is } \begin{matrix} \downarrow \\ = \end{matrix} \begin{matrix} 7 \\ \downarrow \\ 7 \end{matrix}$$

Step 4: **Solve the equation** found in step 3.

$$2x + 3 = 7$$

$$2x + 3 + (-3) = 7 + (-3)$$

$$2x = 4$$

$$\frac{1}{2}(2x) = \frac{1}{2}(4)$$

$$x = 2$$

Step 5: **Write** your **answer** using a complete sentence.

The number is 2.

Step 6: **Reread** the problem and **check** your solution with the original words in the problem.

The sum of twice 2 and 3 is 7; a true statement.   ◄

You may find some examples and problems in this section, and the problem set that follows, that you can solve without using algebra or our blueprint. It is very important that you solve those problems using the methods we are showing here. The purpose behind these problems is to give you experience using the blueprint as a guide to solving problems written in words. Your answers are much less important than the work that you show to obtain your answer. You will be able to condense the steps in the blueprint later in the course. For now, though, you need to show your work in the same detail that we are showing in the examples in this section.

In our next example, we abbreviate the steps in the blueprint.

▶ **EXAMPLE 2**   One number is three more than twice another; their sum is eighteen. Find the numbers.

Solution

Step 1: **Read and list**
Known items: Two numbers that add to 18. One is 3 more than twice the other.
Unknown items: The numbers in question.

Step 2: **Assign a variable and translate information**
Let x = the first number. The other is $2x + 3$.

Step 3: **Reread and write an equation**

$$\underbrace{\text{Their sum}}\quad \text{is 18}$$

$$x + (2x + 3) = 18$$

Step 4: **Solve the equation**

$$x + (2x + 3) = 18$$

$$3x + 3 = 18$$

$$3x + 3 + (-3) = 18 + (-3)$$

$$3x = 15$$

$$x = 5$$

Step 5: **Write answer**
The first number is 5. The other is $2 \cdot 5 + 3 = 13$.

Step 6: **Reread and check**
The sum of 5 and 13 is 18 and 13 is 3 more than twice 5. ◄

Age Problem

Remember as you read through the steps in the solutions to the examples in this section that step 1 is done mentally. Read the problem and then mentally list the items that you know and the items that you don't know. The purpose of step 1 is to give you direction as you begin to work application problems. Finding the solution to an application problem is a process; it doesn't happen all at once. The first step is to read the problem with a purpose in mind. That purpose is to mentally note the items that are known and the items that are unknown.

▶ **EXAMPLE 3** Bill is 6 years older than Tom. Three years ago Bill's age was four times Tom's age. Find the age of each boy now.

Solution Applying the Blueprint for Problem Solving we have

Step 1: **Read and list**
Known items: Bill is 6 years older than Tom. Three years ago Bill's age was four times Tom's age.
Unknown items: Bill's age and Tom's age

Step 2: **Assign a variable and translate information**
Let x = Tom's age now. That makes Bill $x + 6$ years old now. A table like the one shown below can help organize the information in an age problem. Notice how we placed the x in the box that corresponds to Tom's age now.

	Three Years Ago	Now
Bill		$x + 6$
Tom		x

If Tom is x years old now, three years ago he was $x - 3$ years old. If Bill is $x + 6$ years old now, three years ago he was $x + 6 - 3 = x + 3$ years old. We use this information to fill in the remaining squares in the table.

	Three Years Ago	Now
Bill	$x + 3$	$x + 6$
Tom	$x - 3$	x

Step 3: *Reread and write an equation*

Reading the problem again, we see that three years ago Bill's age was four times Tom's age. Writing this as an equation, we have Bill's age three years ago = 4 (Tom's age three years ago):

$$x + 3 = 4(x - 3)$$

Step 4: *Solve the equation*

$$x + 3 = 4(x - 3)$$
$$x + 3 = 4x - 12$$
$$x + (-x) + 3 = 4x + (-x) - 12$$
$$3 = 3x - 12$$
$$3 + 12 = 3x - 12 + 12$$
$$15 = 3x$$
$$x = 5$$

Step 5: *Write answer*

Tom is 5 years old. Bill is 11 years old.

Step 6: *Reread and check*

If Tom is 5 and Bill is 11, then Bill is 6 years older than Tom. Three years ago Tom was 2 and Bill was 8. At that time, Bill's age was four times Tom's age. As you can see, the answers check with the original problem.   ◀

Geometry Problem

To understand Example 4 completely, you need to recall from Chapter 1 that the perimeter of a rectangle is the sum of the lengths of the sides. The formula for the perimeter is $P = 2l + 2w$.

▶ **EXAMPLE 4**   The length of a rectangle is 5 inches more than twice the width. The perimeter is 34 inches. Find the length and width.

Solution   When working problems that involve geometric figures, a sketch of the figure helps organize and visualize the problem.

Step 1: **Read and list**
Known items: The figure is a rectangle. The length is 5 inches more than twice the width. The perimeter is 34 inches.
Unknown items: The length and the width

Step 2: **Assign a variable and translate information**
Since the length is given in terms of the width (the length is 5 more than twice the width), we let x = the width of the rectangle. The length is 5 more than twice the width, so it must be $2x + 5$. The diagram below is a visual description of the relationships we have listed so far.

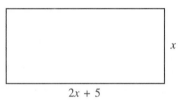

Step 3: **Reread and write an equation**
The equation that describes the situation is

$$\text{twice the length} + \text{twice the width is the perimeter}$$
$$2(2x + 5) \quad + \quad 2x \quad = \quad 34$$

Step 4: **Solve the equation**

$$2(2x + 5) + 2x = 34$$

$$4x + 10 + 2x = 34 \qquad \text{Distributive property}$$

$$6x + 10 = 34 \qquad \text{Add } 4x \text{ and } 2x$$

$$6x = 24 \qquad \text{Add } -10 \text{ to each side}$$

$$x = 4 \qquad \text{Divide each side by 6}$$

Step 5: **Write answer**
The width, x, is 4 inches. The length is $2x + 5 = 2(4) + 5 = 13$ inches.

Step 6: **Reread and check**
If the length is 13 and the width is 4, then the perimeter must be $2(13) + 2(4) = 26 + 8 = 34$, which checks with the original problem. ◀

Coin Problem

▶ **EXAMPLE 5**   Jennifer has $2.45 in dimes and nickels. If she has 8 more dimes than nickels, how many of each coin does she have?

Solution

Step 1: ***Read and list***
Known items: The type of coins, the total value of the coins, and that there are 8 more dimes than nickels
Unknown items: The number of nickels and the number of dimes

Step 2: ***Assign a variable and translate information***
If we let x = the number of nickels, then $x + 8$ = the number of dimes. Since the value of each nickel is 5 cents, the amount of money in nickels is $5x$. Similarly, since each dime is worth 10 cents, the amount of money in dimes is $10(x + 8)$. Here is a table that summarizes the information we have so far:

	Nickels	Dimes
Number	x	$x + 8$
Value (in cents)	$5x$	$10(x + 8)$

Step 3: ***Reread and write an equation***
Since the total value of all the coins is 245 cents, the equation that describes this situation is

Amount of money in nickels	+	Amount of money in dimes	=	Total amount of money
$5x$	+	$10(x + 8)$	=	245

Step 4: ***Solve the equation***
To solve the equation, we apply the distributive property first.

$$5x + 10x + 80 = 245 \qquad \text{Distributive property}$$
$$15x + 80 = 245 \qquad \text{Add } 5x \text{ and } 10x$$
$$15x = 165 \qquad \text{Add } -80 \text{ to each side}$$
$$x = 11 \qquad \text{Divide each side by 15}$$

Step 5: ***Write answer***
The number of nickels is $x = 11$.
The number of dimes is $x + 8 = 11 + 8 = 19$.

Step 6: Reread and check

To check our results

11 nickels are worth 5(11) =	55 cents
19 dimes are worth 10(19) =	190 cents
The total value is	245 cents = $2.45	◀

When you begin working the problems in the problem set that follows, there are a couple of things to remember. The first is that you may have to read the problems over a number of times before you begin to see how to solve them. The second thing to remember is that word problems are not always solved correctly the first time you try them. Sometimes it takes a couple of attempts and some wrong answers before you can set up and solve these problems correctly.

PROBLEM SET 2.6

Solve the following applications. Follow the steps given in the Blueprint for Problem Solving.

Number Problems

1. The sum of a number and five is thirteen. Find the number.
2. The difference of ten and a number is negative eight. Find the number.
3. The sum of twice a number and four is fourteen. Find the number.
4. The difference of four times a number and eight is sixteen. Find the number.
5. Five times the sum of a number and seven is thirty. Find the number.
6. Five times the difference of twice a number and six is negative twenty. Find the number.
7. One number is two more than another. Their sum is eight. Find both numbers.
8. One number is three less than another. Their sum is fifteen. Find the numbers.
9. One number is four less than three times another. If their sum is increased by five, the result is twenty-five. Find the numbers.
10. One number is five more than twice another. If their sum is decreased by ten, the result is twenty-two. Find the numbers.

Age Problems

11. Fred is 4 years older than Barney. Five years ago the sum of their ages was 48. How old are they now? (Begin by filling in the table below.)

	Five Years Ago	Now
Fred		
Barney		x

12. Tim is 5 years older than JoAnn. Six years from now the sum of their ages will be 79. How old are they now?

	Now	Six Years from Now
Tim		
JoAnn	x	

13. Jack is twice as old as Lacy. In three years the sum of their ages will be 54. How old are they now?

14. John is four times as old as Martha. Five years ago the sum of their ages was 50. How old are they now?

15. Nancy was 28 years old when her son Erik was born. In 14 years the sum of their ages will be 100. How old is Nancy now?

16. Diane was 23 years old when her daughter Amy was born. In 6 years Diane will be twice as old as Amy. How old are they now?

Geometry Problems

17. The length of a rectangle is 5 inches more than the width. The perimeter is 34 inches. Find the length and width.

18. The width of a rectangle is 3 feet less than the length. The perimeter is 10 feet. Find the length and width.

19. The perimeter of a square is 48 meters. Find the length of one side.

20. One side of a triangle is twice the shortest side. The third side is 3 feet more than the shortest side. The perimeter is 19 feet. Find all 3 sides.

21. The length of a rectangle is 3 inches less than twice the width. The perimeter is 54 inches. Find the length and width.

22. The length of a rectangle is 4 feet less than 3 times the width. The sum of the length and width is 14 more than the width. Find the width.

Coin Problems

23. Sue has $2.10 in dimes and nickels. If she has 9 more dimes than nickels, how many of each coin does she have? (Completing the table below may help you get started.)

	Nickels	Dimes
Number	x	
Value (in cents)		

24. Mike has $1.55 in dimes and nickels. If he has 7 more nickels than dimes, how many of each coin does he have?

	Nickels	Dimes
Number		x
Value (in cents)		

25. Suppose you have $9.00 in dimes and quarters. How many of each coin do you have if you have twice as many quarters as dimes?

26. A collection of dimes and quarters has a total value of $2.20. If there are 3 times as many dimes as quarters, how many of each coin is in the collection?

27. Katie has a collection of nickels, dimes, and quarters with a total value of $4.35. There are 3 more dimes than nickels and 5 more quarters than nickels. How many of each coin is in her collection? (*Hint:* Let x = the number of nickels.)

28. Mary Jo has $3.90 worth of nickels, dimes, and quarters. The number of nickels is 3 more than the number of dimes. The number of quarters is 7 more than the number of dimes. How many of each coin does she have? (*Hint:* Let x = the number of dimes.)

Review Problems

The following problems review material we covered in Sections 1.1 and 1.2.

Write an equivalent statement in English. [1.1]

29. $4 < 10$

30. $4 \leq 10$

31. $9 \geq -5$

32. $x - 2 > 4$

Place the symbol $<$ or the symbol $>$ between the quantities in each expression. [1.2]

33. 12   20

34. -12 20

35. -8 -6

36. -10 -20

Simplify. [1.2]

37. $|8 - 3| - |5 - 2|$

38. $|9 - 2| - |10 - 8|$

39. $15 - |9 - 3(7 - 5)|$

40. $10 - |7 - 2(5 - 3)|$

SECTION

2.7   **More Applications**

Now that you have worked through a number of application problems using our blueprint, you have probably noticed that step 3, in which we write an equation that describes the situation, is the key step. Anyone with experience solving application problems will tell you that there will be times when your first attempt at step 3 results in the wrong equation. Remember, mistakes are part of the process of learning to do things correctly. Many times the correct equation will become obvious after you have written an equation that is partially wrong. In any case, it is better to write an equation that is partially wrong and be actively involved with the problem than to write nothing at all. Application problems, like other problems in algebra, are not always solved correctly the first time.

In this section we continue our work with application problems. There are two main differences between the problems in this section and the problems you worked with in the previous section. First of all, not all of the problems in the problem set that follows this section fall into specific categories. Second, many of the problems in this section and problem set are more realistic in nature.

▶ **EXAMPLE 1**   If you read the introduction to this chapter, you know that finding the amount of sales tax if the total bill is $3,200 and the sales tax rate is 7.5% is a matter of dividing by 1.075. Show why division is the key to solving this type of problem.

Solution   We solve this problem using our six-step Blueprint for Problem Solving:

Step 1: Read and list
Known items: The total bill is $3,200. The sales tax rate is 7.5%, which is 0.075 in decimal form.
Unknown item: The retail price of the video equipment

Step 2: Assign a variable and translate information
If we let x = the retail price of the video equipment, then to calculate the sales tax, we multiply the retail price x by the sales tax rate:

$$\text{Sales tax} = (\text{sales tax rate})(\text{retail price})$$
$$= 0.075x$$

Step 3: Reread and write an equation

$$\text{Retail price} + \text{sales tax} = \text{total price}$$
$$x \quad + \quad 0.075x \quad = \quad 3{,}200$$

Step 4: **Solve the equation**

$$x + 0.075x = 3,200$$

$$1.075x = 3,200$$

$$x = \frac{3,200}{1.075}$$

$$= 2,976.74 \quad \text{to the nearest hundredth}$$

Step 5: **Write answer**

The retail price of the equipment is $2,976.74. The sales tax is $3,200 − $2,976.74 = $223.26.

Step 6: **Reread and check**

The retail price of the equipment is $2,976.74. The tax on this is 0.075(2,976.74) = $223.26. Adding the retail price and the sales tax we have the total bill, $3,200.00. ◀

▶ **EXAMPLE 2** Suppose you invest a certain amount of money in an account that earns 8% in annual interest. At the same time, you invest $2,000 more than that in an account that pays 9% in annual interest. If the total interest from both accounts at the end of the year is $690, how much is invested in each account?

Solution

Step 1: **Read and list**

Known items: The interest rates, the total interest earned, and how much more is invested at 9%

Unknown items: The amounts invested in each account

Step 2: **Assign a variable and translate information**

Let x = the amount of money invested at 8%. From this, $x + 2,000$ = the amount of money invested at 9%. The interest earned on x dollars invested at 8% is $0.08x$. The interest earned on $x + 2,000$ dollars invested at 9% is $0.09(x + 2,000)$.

Here is a table that summarizes this information:

	Dollars Invested at 8%	Dollars Invested at 9%
Number of	x	$x + 2,000$
Interest on	$0.08x$	$0.09(x + 2,000)$

Step 3: **Reread and write an equation**

Since the total amount of interest earned from both accounts is $690, the equation that describes the situation is

| Interest earned at 8% | + | interest earned at 9% | = | total interest earned |

$$0.08x + 0.09(x + 2,000) = 690$$

Step 4: Solve the equation

$$
\begin{aligned}
0.08x + 0.09(x + 2,000) &= 690 \\
0.08x + 0.09x + 180 &= 690 && \text{Distributive property} \\
0.17x + 180 &= 690 && \text{Add } 0.08x \text{ and } 0.09x \\
0.17x &= 510 && \text{Add } -180 \text{ to each side} \\
x &= 3,000 && \text{Divide each side by } 0.17
\end{aligned}
$$

Step 5: Write answer

The amount of money invested at 8% is $3,000, while the amount of money invested at 9% is $x + 2,000 = 3,000 + 2,000 = \$5,000$.

Step 6: Reread and check

The interest at 8% is 8% of 3,000 = 0.08(3,000) = \$240

The interest at 9% is 9% of 5,000 = 0.09(5,000) = \$450

The total interest is   \$690 ◀

Facts from Geometry: Labeling Triangles and the Sum of the Angles in a Triangle

One way to label the important parts of a triangle is to label the vertices with capital letters and the sides with small letters, as shown in Figure 1.

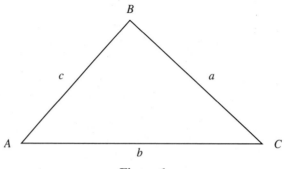

Figure 1

In Figure 1, notice that side a is opposite vertex A, side b is opposite vertex B, and side c is opposite vertex C. Also, since each vertex is the vertex of one of the angles of the triangle, we refer to the three interior angles as A, B, and C.

In any triangle, the sum of the interior angles is 180°. For the triangle shown in Figure 1, the relationship is written

$$A + B + C = 180°$$

▶ **EXAMPLE 3** The angles in a triangle are such that one angle is twice the smallest angle, while the third angle is three times as large as the smallest angle. Find the measure of all three angles.

Solution

Step 1: ***Read and list***
Known items: The sum of all three angles is 180°, one angle is twice the smallest angle, the largest angle is three times the smallest angle
Unknown items: The measure of each angle

Step 2: ***Assign a variable and translate information***
Let x be the smallest angle, then $2x$ will be the measure of another angle and $3x$ will be the measure of the largest angle.

Step 3: ***Reread and write an equation***
When working with geometric objects, drawing a generic diagram will sometimes help us visualize what it is that we are asked to find. In Figure 2, we draw a triangle with angles A, B, and C.

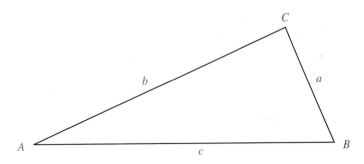

Figure 2

We can let the value of $A = x$, the value of $B = 2x$ and the value of $C = 3x$. We know that the sum of angles A, B, and C will be 180°, so our equation becomes

$$x + 2x + 3x = 180°$$

Step 4: ***Solve the equation***

$$x + 2x + 3x = 180°$$
$$6x = 180°$$
$$x = 30°$$

Step 5: **Write answer**

The smallest angle, *A*, measures 30°

Angle *B* measures 2*x* or 2(30°) = 60°

Angle *C* measures 3*x* or 3(30°) = 90°

Step 6: **Reread and check**

The angles must add to 180°:

$$A + B + C = 180°$$

$$30° + 60° + 90° = 180°$$

$$180° = 180°$$

Our answers check.   ◀

P R O B L E M   S E T   2 . 7
· ·

Interest Problems

1. Suppose you invest money in two accounts. One of the accounts pays 8% annual interest, while the other pays 9% annual interest. If you have $2,000 more invested at 9% than you have invested at 8%, how much do you have invested in each account if the total amount of interest you earn in a year is $860? (Begin by completing the following table.)

	Dollars Invested at 8%	Dollars Invested at 9%
Number of	*x*	
Interest on		

2. Suppose you invest a certain amount of money in an account that pays 11% interest annually, and $4,000 more than that in an account that pays 12% annually. How much money do you

have in each account if the total interest for a year is $940?

	Dollars Invested at 11%	Dollars Invested at 12%
Number of	*x*	
Interest on		

3. Tyler has two savings accounts that his grandparents opened for him. The two accounts pay 10% and 12% in annual interest; there is $500 more in the account that pays 12% than there is in the other account. If the total interest for a year is $214, how much money does he have in each account?

4. Travis has a savings account that his parents opened for him. It pays 6% annual interest. His uncle also opened an account for him, but it pays 8% annual interest. If there is $800 more in the account that pays 6%, and the total interest from both accounts is $104, how much money is in each of the accounts?

5. A stockbroker has money in three accounts. The interest rates on the three accounts are 8%, 9%, and 10%. If she has twice as much money invested at 9% as she has invested at 8%, and three times as much at 10% as she has at 8%, and the total interest for the year is $280, how much is invested at each rate? (*Hint:* Let $x =$ the amount invested at 8%.)

6. An accountant has money in three accounts that pay 9%, 10%, and 11% in annual interest. He has twice as much invested at 9% as he does at 10%, and three times as much invested at 11% as he does at 10%. If the total interest from the three accounts is $610 for the year, how much is invested at each rate? (*Hint:* Let $x =$ the amount invested at 10%.)

Geometry Problems

7. Two angles in a triangle are equal and their sum is equal to the third angle in the triangle. What are the measures of each of the three interior angles?

8. One angle in a triangle measures twice the smallest angle, while the largest angle is six times the smallest angle. Find the measures of all three angles.

9. The smallest angle in a triangle is $\frac{1}{5}$ as large as the largest angle. The third angle is twice the smallest angle. Find the three angles.

10. One angle in a triangle is half the largest angle, but three times the smallest. Find all three angles.

11. A right triangle has one $37°$ angle. Find the other two angles.

12. In a right triangle, one of the acute angles is twice as large as the other acute angle. Find the measure of the two acute angles.

Miscellaneous Problems

13. The cost of a long-distance phone call is $0.41 for the first minute and $0.32 for each additional minute. If the total charge for a long-distance call is $5.21, how many minutes was the call?

14. Danny, who is 1 year old, is playing with the telephone when he accidentally presses one of the buttons his mother has programmed to dial her friend Sue's number. Sue answers the phone and realizes Danny is on the other end. She talks to Danny, trying to get him to hang up. The cost for a call is $0.23 for the first minute and $0.14 for every minute after that. If the total charge for the call is $3.73, how long did it take Sue to convince Danny to hang up the phone?

15. JoAnn works in the publicity office at the state university. She is paid $12 an hour for the first 35 hours she works each week and $18 an hour for every hour after that. If she makes $492 one week, how many hours did she work?

16. Diane has a part-time job that pays her $6.50 an hour. During one week she works 26 hours and is paid $178.10. She realizes when she sees her check that she has been given a raise. How much per hour is that raise?

17. Stacey is selling tickets to the school play. The tickets are $6.00 for adults and $4.50 for children. She sells twice as many adult tickets as children's tickets and brings in a total of $115.50. How many of each kind of ticket did she sell?

18. Tyler is taking piano lessons. Since he doesn't practice as often as his parents would like him to, he has to pay for part of the lessons himself. His parents pay him $0.50 to do the laundry and $1.25 to mow the lawn. In one month, he does the laundry 6 more times than he mows the lawn. If his parents pay him $13.50 that month, how many times did he mow the lawn?

19. A 4-H club has collected $215.00 from the sales of raffle tickets. They are going to use the money to buy fencing material for a pasture for their sheep. If fencing costs $1.25 per foot, what is the largest area they can fence? Assume the fenced area is rectangular.

20. If the fencing material in the previous problem costs $1.75 per foot, what is the largest area the 4-H club can fence?

21. Jeff and Carla Cole are driving separately from San Luis Obispo, California, to the north shore of Lake Tahoe, a distance of 425 miles. Jeff leaves San Luis Obispo at 11:00 AM and averages 55 miles/hour on the drive. Carla leaves later, at 1:00 PM, but averages 65 miles/hour. Which person arrives in Lake Tahoe first?

22. Referring to the information in Problem 21, how many minutes after the first person arrives does the second person arrive?

In 1993, the Texas Junior College Teachers Association annual conference was held in Austin, in February. At that time a taxi ride in Austin was $1.25 for the first $\frac{1}{5}$ of a mile and $0.25 for each additional $\frac{1}{5}$ of a mile. The charge for a taxi to wait is $12.00 per hour. Use this information for Problems 23–26.

23. If the distance from a convention hotel to the airport is 7.5 miles, how much will it cost to take a taxi from that hotel to the airport?

24. If you were to tip the driver of the taxi in Problem 23 15%, how much would it cost to take a taxi from the hotel to the airport?

25. Suppose the distance from one of the hotels to one of the western dance clubs in Austin is 12.4 miles. If the fare meter in the taxi gives the charge for that trip as $16.50, is the meter working correctly?

26. Suppose that the distance from a hotel to the airport is 8.2 miles and the ride takes 20 minutes. Is it more expensive to take a taxi to the airport or to just sit in the taxi?

Ike and Nancy Lara give western dance lessons at the Elk's Lodge on Sunday nights. The lessons cost $3.00 for members of the lodge and $5.00 for nonmembers. Half of the money collected for the lessons is paid to Ike and Nancy. The Elk's Lodge keeps the other half. On Sunday night, March 13, 1993, Ike counts 36 people in the dance lesson. Use this information to work Problems 27–30.

27. What is the least amount of money Ike and Nancy will make?

28. What is the largest amount of money Ike and Nancy will make?

29. At the end of the evening, the Elk's Lodge gives Ike and Nancy a check for $80 to cover half of the receipts. Can this amount be correct?

30. Besides the number of people in the dance lesson, what additional information does Ike need to know in order to always be sure he is being paid the correct amount?

Review Problems

The problems that follow review material we covered in Chapter 1 on number sequences.

Find the next number in each sequence.

31. $8, 4, 0, -4, \ldots$
32. $8, -16, 32, -64, \ldots$
33. $12, -6, 3, -\frac{3}{2}, \ldots$
34. $-12, -6, 0, 6, \ldots$
35. $1, -4, 9, -16, \ldots$
36. $-1, 8, -27, 64, \ldots$
37. $2, \frac{3}{2}, 1, \frac{1}{2}, \ldots$
38. $1, \frac{1}{2}, 0, -\frac{1}{2}, \ldots$

SECTION

2.8 Linear Inequalities

Linear inequalities are solved by a method similar to the one used in solving linear equations. The only real differences between the methods are in the multiplication property for inequalities and in graphing the solution set.

An inequality differs from an equation only with respect to the comparison symbol between the two quantities being compared. In place of the equal sign, we use $<$ (less than), $\leq$ (less than or equal to), $>$ (greater than), or $\geq$ (greater than or equal to). The addition property for inequalities is almost identical to the addition property for equality.

Addition Property for Inequalities

For any three algebraic expressions A, B, and C,

$$\text{if} \qquad A < B$$

$$\text{then } A + C < B + C$$

In words: Adding the same quantity to both sides of an inequality will not change the solution set.

It makes no difference which inequality symbol we use to state the property. Adding the same amount to both sides always produces an inequality equivalent to the original inequality. Also, since subtraction can be thought of as addition of the opposite, this property holds for subtraction as well as addition.

▶ **EXAMPLE 1**   Solve the inequality $x + 5 < 7$.

Solution   To isolate x we add -5 to both sides of the inequality:

$$x + 5 < 7$$

$$x + 5 + (-5) < 7 + (-5) \qquad \text{Addition property for inequalities}$$

$$x < 2 \qquad\qquad\qquad\qquad\qquad\qquad\qquad ◀$$

We can go one step further here and graph the solution set. The solution set is all real numbers less than 2. To graph this set, we simply draw a straight line and label the center 0 (zero) for reference. Then we label the 2 on the right side of zero and extend an arrow beginning at 2 and pointing to the left. We use an open circle at 2, since it is not included in the solution set. Here is the graph.

▶ **EXAMPLE 2**   Solve $x - 6 \leq -3$.

Solution   Adding 6 to each side will isolate x on the left side:

$$x - 6 \leq -3$$

$$x - 6 + 6 \leq -3 + 6 \qquad \text{Add **6** to both sides}$$

$$x \leq 3$$

The graph of the solution set is

Notice that the dot at the 3 is darkened because 3 is included in the solution set. We will always use open circles on the graphs of solution sets with $<$ or $>$ and closed (darkened) circles on the graphs of solution sets with $\leq$ or $\geq$.

To see the idea behind the multiplication property for inequalities, we will consider three true inequality statements and explore what happens when we multiply both sides by a positive number and then what happens when we multiply by a negative number.

Consider the following three true statements:

$$3 < 5 \qquad -3 < 5 \qquad -5 < -3$$

Now multiply both sides by the positive number 4:

$$4(3) < 4(5) \qquad 4(-3) < 4(5) \qquad 4(-5) < 4(-3)$$
$$12 < 20 \qquad\quad -12 < 20 \qquad\quad -20 < -12$$

In each case, the inequality symbol in the result points in the same direction it did in the original inequality. We say the ''sense'' of the inequality doesn't change when we multiply both sides by a positive quantity.

Notice what happens when we go through the same process but multiply both sides by -4 instead of 4:

$$3 < 5 \qquad\qquad -3 < 5 \qquad\qquad -5 < -3$$

$$-4(3) > -4(5) \qquad -4(-3) > -4(5) \qquad -4(-5) > -4(-3)$$
$$-12 > -20 \qquad\quad 12 > -20 \qquad\qquad 20 > 12$$

In each case, we have to change the direction in which the inequality symbol points to keep each statement true. Multiplying both sides of an inequality by a negative quantity *always* reverses the sense of the inequality. Our results are summarized in the multiplication property for inequalities.

Multiplication Property for Inequalities
For any three algebraic expressions A, B, and C,

$$\text{if} \quad A < B$$
$$\text{then } AC < BC \quad \text{when } C \text{ is positive}$$
$$\text{and } \quad AC > BC \quad \text{when } C \text{ is negative}$$

In words: Multiplying both sides of an inequality by a positive number does not change the solution set. When multiplying both sides of an inequality by a negative number, it is necessary to reverse the inequality symbol in order to produce an equivalent inequality.

We can multiply both sides of an inequality by any nonzero number we choose. If that number happens to be negative, we must also reverse the sense of the inequality.

Note   Since division is defined in terms of multiplication, this property is also true for division. We can divide both sides of an inequality by any number we choose. If that number happens to be negative, we must also reverse the direction of the inequality symbol.

▶ **EXAMPLE 3**   Solve $3a < 15$ and graph the solution.

Solution   We begin by multiplying each side by $\frac{1}{3}$. Since $\frac{1}{3}$ is a positive number, we do not reverse the direction of the inequality symbol:

$$3a < 15$$

$$\frac{1}{3}(3a) < \frac{1}{3}(15) \qquad \text{Multiply each side by } \frac{1}{3}$$

$$a < 5$$

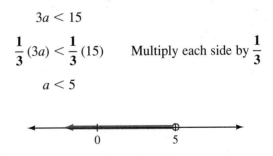

▶ **EXAMPLE 4**   Solve $-3a \leq 18$ and graph the solution.

Solution   We begin by multiplying both sides by $-\frac{1}{3}$. Since $-\frac{1}{3}$ is a negative number, we must reverse the direction of the inequality symbol at the same time that we multiply by $-\frac{1}{3}$:

$$-3a \leq 18$$

$$-\frac{1}{3}(-3a) \geq -\frac{1}{3}(18)$$

Multiply both sides by $-\frac{1}{3}$ and reverse the direction of the inequality symbol

$$a \geq -6$$

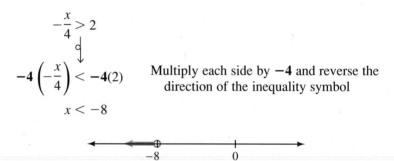

▶ **EXAMPLE 5** Solve $-\dfrac{x}{4} > 2$ and graph the solution.

Solution To isolate x, we multiply each side by -4. Since -4 is a negative number, we must also reverse the direction of the inequality symbol:

$$-\frac{x}{4} > 2$$

$$-4\left(-\frac{x}{4}\right) < -4(2)$$

Multiply each side by **-4** and reverse the direction of the inequality symbol

$$x < -8$$

To solve more complicated inequalities we use the following steps.

Strategy for Solving Linear Inequalities in One Variable

Step 1a: Use the distributive property to separate terms, if necessary.

1b: If fractions are present, consider multiplying both sides by the LCD to eliminate the fractions. If decimals are present, consider multiplying both sides by a power of 10 to clear the inequality of decimals.

1c: Combine similar terms on each side of the inequality.

Step 2: Use the addition property for inequalities to get all variable terms on one side of the inequality and all constant terms on the other side.

Step 3: Use the multiplication property for inequalities to get x by itself on one side of the inequality.

Step 4: Graph the solution set.

▶ **EXAMPLE 6** Solve $2.5x - 3.48 < 4.9x + 2.07$.

Solution We have two methods we can use to solve this inequality. We can simply apply our properties to the inequality the way it is currently written and work with the decimal numbers, or we can eliminate the decimals to begin with and solve the resulting inequality.

Method 1 Working with the decimals.

$$2.5x - 3.48 < -4.9x + 2.07 \qquad \text{Original inequality}$$

$$2.5x + \mathbf{4.9x} - 3.48 < -4.9x + \mathbf{4.9x} + 2.07 \qquad \text{Add } \mathbf{4.9x} \text{ to each side}$$

$$7.4x - 3.48 < 2.07$$

$$7.4x - 3.48 + \mathbf{3.48} < 2.07 + \mathbf{3.48} \qquad \text{Add } \mathbf{3.48} \text{ to each side}$$

$$7.4x < 5.55$$

$$\frac{7.4x}{\mathbf{7.4}} < \frac{5.55}{\mathbf{7.4}} \qquad \text{Divide each side by } \mathbf{7.4}$$

$$x < 0.75$$

Method 2 Eliminating the decimals in the beginning.

Since the greatest number of places to the right of the decimal point in any of the numbers is 2, we can multiply each side of the inequality by 100 and we will be left with an equivalent inequality that contains only whole numbers.

$$2.5x - 3.48 < -4.9x + 2.07 \qquad \text{Original inequality}$$

$$\mathbf{100}(2.5x - 3.48) < \mathbf{100}(-4.9x + 2.07) \qquad \text{Multiply each side by } \mathbf{100}$$

$$\mathbf{100}(2.5x) - \mathbf{100}(3.48) < \mathbf{100}(-4.9x) + \mathbf{100}(2.07) \qquad \text{Distributive property}$$

$$250x - 348 < -490x + 207 \qquad \text{Multiplication}$$

$$740x - 348 < 207 \qquad \text{Add } 490x \text{ to each side}$$

$$740x < 555 \qquad \text{Add } 348 \text{ to each side}$$

$$\frac{740x}{\mathbf{740}} < \frac{555}{\mathbf{740}} \qquad \text{Divide each side by } \mathbf{740}$$

$$x < 0.75$$

The solution by either method is $x < 0.75$. Here is the graph:

▶ **EXAMPLE 7**   Solve $3(x - 4) \geq -2$.

Solution

$$3x - 12 \geq -2 \qquad \text{Distributive property}$$

$$3x - 12 + \mathbf{12} \geq -2 + \mathbf{12} \qquad \text{Add } \mathbf{12} \text{ to both sides}$$

$$3x \geq 10$$

$$\frac{1}{3}(3x) \geq \frac{1}{3}(10) \qquad \text{Multiply both sides by } \frac{1}{3}$$

$$x \geq \frac{10}{3}$$

▶ **EXAMPLE 8**   Solve and graph $2(1 - 3x) + 4 < 4x - 14$.

Solution

$$2 - 6x + 4 < 4x - 14 \qquad \text{Distributive property}$$

$$-6x + 6 < 4x - 14 \qquad \text{Simplify}$$

$$-6x + 6 + (\mathbf{-6}) < 4x - 14 + (\mathbf{-6}) \qquad \text{Add } \mathbf{-6} \text{ to both sides}$$

$$-6x < 4x - 20$$

$$-6x + (\mathbf{-4x}) < 4x + (\mathbf{-4x}) - 20 \qquad \text{Add } \mathbf{-4x} \text{ to both sides}$$

$$-10x < -20$$

$$\left(-\frac{1}{10}\right)(-10x) > \left(-\frac{1}{10}\right)(-20) \qquad \text{Multiply by } -\frac{1}{10}, \text{ reverse}$$
$$\text{the sense of the inequality}$$

$$x > 2$$

▶ **EXAMPLE 9**   Solve $2x - 3y < 6$ for y.

Solution   We can solve this formula for y by first adding $-2x$ to each side and then multiplying each side by $-\frac{1}{3}$. When we multiply by $-\frac{1}{3}$ we must reverse the direction of the inequality symbol. Since this is a formula, we will not graph the solution.

$$2x - 3y < 6 \qquad \text{Original formula}$$

$$2x + (\mathbf{-2x}) - 3y < (\mathbf{-2x}) + 6 \qquad \text{Add } \mathbf{-2x} \text{ to each side}$$

$$-3y < -2x + 6$$

$$-\frac{1}{3}(-3y) > -\frac{1}{3}(-2x + 6) \qquad \text{Multiply each side by } -\frac{1}{3}$$

$$y > \frac{2}{3}x - 2 \qquad \text{Distributive property} \qquad \blacktriangleleft$$

When working application problems that involve inequalities, the phrases "at least" and "at most" translate as follows:

In Words	*In Symbols*
x is at least 30	$x \geq 30$
x is at most 20	$x \leq 20$

Our next example involves consecutive integers. When we ask for consecutive integers, we mean integers that are next to each other on the number line, like 5 and 6, or 13 and 14, or -4 and -3. In the dictionary, consecutive is defined as following one another in uninterrupted order. If we ask for consecutive *odd* integers, then we mean odd integers that follow one another on the number line. For example, 3 and 5, 11 and 13, and -9 and -7 are consecutive odd integers. As you can see, to get from one odd integer to the next consecutive odd integer we add 2.

We can modify our Blueprint for Problem Solving to solve application problems whose solutions depend on writing and then solving inequalities.

▶ **EXAMPLE 10** The sum of two consecutive odd integers is at most 28. What are the possibilities for the first of the two integers?

Solution When we use the phrase "their sum is at most 28," we mean that their sum is less than or equal to 28.

Step 1: Read and list
Known items: Two consecutive odd integers. Their sum is less than or equal to 28.
Unknown items: The numbers in question.

Step 2: Assign a variable and translate information
If we let $x =$ the first of the two consecutive odd integers, then $x + 2$ is the next consecutive one.

Step 3: Reread and write an inequality
Their sum is at most 28.

$$x + (x + 2) \leq 28$$

Step 4: **Solve the inequality**

$$2x + 2 \le 28 \qquad \text{Simplify the left side}$$

$$2x \le 26 \qquad \text{Add } -2 \text{ to each side}$$

$$x \le 13 \qquad \text{Multiply each side by } \frac{1}{2}$$

Step 5: **Write answer**

The first of the two integers must be an odd integer that is less than or equal to 13. The second of the two integers will be two more than whatever the first one is.

Step 6: **Reread and check**

Suppose the first integer is 13. The next consecutive odd integer is 15. The sum of 15 and 13 is 28. If the first odd integer is less than 13, the sum of it and the next consecutive odd integer will be less than 28.   ◀

P R O B L E M   S E T   2 . 8

Solve the following inequalities using the addition property of inequalities. Graph each solution set.

1. $x - 5 < 7$ **2.** $x + 3 < -5$
3. $a - 4 \le 8$ **4.** $a + 3 \le 10$
5. $x - 4.3 > 8.7$ **6.** $x - 2.6 > 10.4$
7. $y + 6 \ge 10$ **8.** $y + 3 \ge 12$
9. $2 < x - 7$ **10.** $3 < x + 8$

Solve the following inequalities using the multiplication property of inequalities. If you multiply both sides by a negative number, be sure to reverse the direction of the inequality symbol. Graph the solution set.

11. $3x < 6$ **12.** $2x < 14$
13. $5a \le 25$ **14.** $4a \le 16$
15. $\dfrac{x}{3} > 5$ **16.** $\dfrac{x}{7} > 1$
17. $-2x > 6$ **18.** $-3x \ge 9$
19. $-3x \ge -18$ **20.** $-8x \ge -24$
21. $-\dfrac{x}{5} \le 10$ **22.** $-\dfrac{x}{9} \ge -1$
23. $-\dfrac{2}{3}y > 4$ **24.** $-\dfrac{3}{4}y > 6$

Solve the following inequalities. Graph the solution set in each case.

25. $2x - 3 < 9$ **26.** $3x - 4 < 17$
27. $-\dfrac{1}{5}y - \dfrac{1}{3} \le \dfrac{2}{3}$ **28.** $-\dfrac{1}{6}y - \dfrac{1}{2} \le \dfrac{2}{3}$
29. $-4x + 1 > -11$ **30.** $-6x - 1 > 17$
31. $\dfrac{2}{3}x - 5 \le 7$ **32.** $\dfrac{3}{4}x - 8 \le 1$
33. $-\dfrac{2}{5}a - 3 > 5$ **34.** $-\dfrac{4}{5}a - 2 > 10$
35. $5 - \dfrac{3}{5}y > -10$ **36.** $4 - \dfrac{5}{6}y > -11$
37. $0.3(a + 1) \le 1.2$ **38.** $0.4(a - 2) \le 0.4$
39. $2(5 - 2x) \le -20$ **40.** $7(8 - 2x) > 28$
41. $3x - 5 > 8x$ **42.** $8x - 4 > 6x$
43. $\dfrac{1}{3}y - \dfrac{1}{2} \le \dfrac{5}{6}y + \dfrac{1}{2}$
44. $\dfrac{7}{6}y + \dfrac{4}{3} \le \dfrac{11}{6}y - \dfrac{7}{6}$
45. $-0.4x + 1.2 < -2x - 0.4$
46. $-1.2x - 0.4 < -0.4x + 2$

47. $3(m - 2) - 4 \geq 7m + 14$
48. $2(3m - 1) + 5 \geq 8m - 7$
49. $3 - 4(x - 2) \leq -5x + 6$
50. $8 - 6(x - 3) \leq -4x + 12$

Solve each of the following formulas for y.

51. $3x + 2y < 6$ **52.** $-3x + 2y < 6$
53. $2x - 5y > 10$ **54.** $-2x - 5y > 5$
55. $-3x + 7y \leq 21$ **56.** $-7x + 3y \leq 21$
57. $2x - 4y \geq -4$ **58.** $4x - 2y \geq -8$

Number Problems

59. The sum of twice a number and six is less than ten. Find all solutions.
60. Twice the difference of a number and three is greater than or equal to the number increased by five. Find all solutions.
61. The product of a number and four is greater than the number minus eight. Find the solution set.
62. The quotient of a number and five is less than the sum of seven and two. Find the solution set.
63. Twice the sum of a number and five is less than or equal to twelve. Find all solutions.
64. Three times the difference of a number and four is greater than twice the number. Find all solutions.
65. The difference of three times a number and five is less than the sum of the number and seven. Find all solutions.
66. If twice a number is added to three, the result is less than the sum of three times the number and two. Find all solutions.

Geometry Problems

67. The length of a rectangle is three times the width. If the perimeter is to be at least 48 meters, what are the possible values for the width? (If the perimeter is at least 48 meters, then it is greater than or equal to 48 meters.)

68. The length of a rectangle is three more than twice the width. If the perimeter is to be at least 51 meters, what are the possible values for the width? (If the perimeter is at least 51 meters, then it is greater than or equal to 51 meters.)
69. The numerical values of the three sides of a triangle are given by three consecutive even integers. If the perimeter is greater than 24 inches, what are the possibilities for the shortest side?
70. The numerical values of the three sides of a triangle are given by three consecutive odd integers. If the perimeter is greater than 27 inches, what are the possibilities for the shortest side?

Review Problems

The problems below review material we covered in Sections 1.5 and 1.8.

Match each expression in Problems 71–76 with one or more of the properties labeled a–e. [1.5]
71. $x + 4 = 4 + x$
72. $2(3x) = (2 \cdot 3)x$
73. $5(x - 3) = 5x - 15$
74. $x + (y + 4) = (x + y) + 4$
75. $x + (y + 4) = (x + 4) + y$
76. $7 \cdot 5 = 5 \cdot 7$

   a. Distributive property
   b. Commutative property of addition
   c. Associative property of addition
   d. Commutative property of multiplication
   e. Associative property of multiplication

For the set $\{-3, -\frac{5}{3}, 0, 2, \pi, \frac{15}{2}\}$ list all the elements that are in the following sets. [1.8]

77. Whole numbers   **78.** Integers
79. Rational numbers   **80.** Irrational numbers

Reduce to lowest terms.

81. $\dfrac{130}{858}$ **82.** $\dfrac{455}{546}$

CHAPTER 2 SUMMARY

Examples

1. The terms $2x$, $5x$, and $-7x$ are all similar since their variable parts are the same.

Similar Terms [2.1]

A *term* is a number or a number and one or more variables multiplied together. *Similar terms* are terms with the same variable part.

2. Simplify $3x + 4x$.

$$3x + 4x = (3 + 4)x$$
$$= 7x$$

Simplifying Expressions [2.1]

In this chapter we simplified expressions that contained variables by using the distributive property to combine similar terms.

3. The solution set for the equation $x + 2 = 5$ is $\{3\}$ because when x is 3 the equation is $3 + 2 = 5$ or $5 = 5$.

Solution Set [2.2]

The *solution set* for an equation (or inequality) is all the numbers that, when used in place of the variable, make the equation (or inequality) a true statement.

4. The equations $a - 4 = 3$ and $a - 2 = 5$ are equivalent since both have solution set $\{7\}$.

Equivalent Equations [2.2]

Two equations are called *equivalent* if they have the same solution set.

5. Solve $x - 5 = 12$.

$$x - 5 + \mathbf{5} = 12 + \mathbf{5}$$
$$x + 0 = 17$$
$$x = 17$$

Addition Property of Equality [2.2]

When the same quantity is added to both sides of an equation, the solution set for the equation is unchanged. Adding the same amount to both sides of an equation produces an equivalent equation.

6. Solve $3x = 18$.

$$\frac{1}{3}(3x) = \frac{1}{3}(18)$$
$$x = 6$$

Multiplication Property of Equality [2.3]

If both sides of an equation are multiplied by the same nonzero number, the solution set is unchanged. Multiplying both sides of an equation by a nonzero quantity produces an equivalent equation.

7. Solve $2(x + 3) = 10$.

$$2x + 6 = 10$$
$$2x + 6 + (\mathbf{-6}) = 10 + (\mathbf{-6})$$
$$2x = 4$$
$$\frac{1}{2}(2x) = \frac{1}{2}(4)$$
$$x = 2$$

Strategy for Solving Linear Equations in One Variable [2.4]

Step 1a: Use the distributive property to separate terms, if necessary.

1b: If fractions are present, consider multiplying both sides by the LCD to eliminate the fractions. If decimals are present, consider multiplying both sides by a power of 10 to clear the equation of decimals.

1c: Combine similar terms on each side of the equation.

Step 2: Use the addition property of equality to get all variable terms on one side of the equation and all constant terms on the other side. A variable term is a term that contains the variable (for example, $5x$). A constant term is a term that does not contain the variable (the number 3, for example).

Step 3: Use the multiplication property of equality to get x (that is, $1x$) by itself on one side of the equation.

Step 4: Check your solution in the original equation to be sure that you have not made a mistake in the solution process.

8. Solving $P = 2l + 2w$ for l, we have

$$P - 2w = 2l$$

$$\frac{P - 2w}{2} = l$$

Formulas [2.5]

A formula is an equation with more than one variable. To solve a formula for one of its variables, we use the addition and multiplication properties of equality to move everything except the variable in question to one side of the equal sign so the variable in question is alone on the other side.

9. The sum of twice a number and three is seven. Find the number.

Step 1: **Read and list**
Known items: If we add 3 to twice a number, the result is 7
Unknown item: The number in question

Step 2: **Assign a variable and translate information**
Let $x =$ the number asked for.
"The sum of twice x and 3" translates to $2x + 3$.

Step 3: **Reread and write an equation**

$$2x + 3 = 7$$

Step 4: **Solve the equation**

$$2x = 4$$

$$x = 2$$

Step 5: **Write answer**
The number is 2.

Step 6: **Reread and check**
The sum of twice 2 and 3 is $4 + 3$, which is 7.

Blueprint for Problem Solving [2.6, 2.7]

Step 1: **Read** the problem and then mentally **list** the items that are known and the items that are unknown.

Step 2: **Assign a variable** to one of the unknown items. (In most cases this will amount to letting $x =$ the item that is asked for in the problem.) Then **translate** the other **information** in the problem to expressions involving the variable.

Step 3: **Reread** the problem, and then **write an equation,** using the items and variables listed in steps 1 and 2, that describes the situation.

Step 4: **Solve the equation** found in step 3.

Step 5: **Write** your **answer** using a complete sentence.

Step 6: **Reread** the problem and **check** your solution with the original words in the problem.

10. Solve $x + 5 < 7$.

$$x + 5 + (-5) < 7 + (-5)$$

$$x < 2$$

Addition Property for Inequalities [2.8]

Adding the same quantity to both sides of an inequality produces an equivalent inequality, one with the same solution set.

11. Solve $-3a \leq 18$.

$$\Downarrow$$

$$-\frac{1}{3}(-3a) \geq -\frac{1}{3}(18)$$

$$a \geq -6$$

Multiplication Property for Inequalities [2.8]
Multiplying both sides of an inequality by a positive number never changes the solution set. If both sides are multiplied by a negative number, the sense of the inequality must be reversed to produce an equivalent inequality.

12. Solve $3(x - 4) \geq -2$.

$$3x - 12 \geq -2$$

$$3x - 12 + 12 \geq -2 + 12$$

$$3x \geq 10$$

$$\frac{1}{3}(3x) \geq \frac{1}{3}(10)$$

$$x \geq \frac{10}{3}$$

Strategy for Solving Linear Inequalities in One Variable [2.8]

Step 1a: Use the distributive property to separate terms, if necessary.

1b: If fractions are present, consider multiplying both sides by the LCD to eliminate the fractions. If decimals are present, consider multiplying both sides by a power of 10 to clear the inequality of decimals.

1c: Combine similar terms on each side of the inequality.

Step 2: Use the addition property for inequalities to get all variable terms on one side of the inequality and all constant terms on the other side.

Step 3: Use the multiplication property for inequalities to get x by itself on one side of the inequality.

Step 4: Graph the solution set.

COMMON MISTAKES

1. Trying to subtract away coefficients (the number in front of variables) when solving equations. For example

$$4x = 12$$

$$4x - 4 = 12 - 4$$

$$x = 8 \leftarrow \text{Mistake}$$

It is not incorrect to add (-4) to both sides, it's just that $4x - 4$ is not equal to x. Both sides should be multiplied by $\frac{1}{4}$ to solve for x.

2. Forgetting to reverse the direction of the inequality symbol when multiplying both sides of an inequality by a negative number. For instance:

$$-3x < 12$$

$$-\frac{1}{3}(-3x) < -\frac{1}{3}(12) \leftarrow \text{Mistake}$$

$$x < -4$$

It is not incorrect to multiply both sides by $-\frac{1}{3}$. But if we do, we must also reverse the sense of the inequality.

CHAPTER 2 REVIEW

The numbers in brackets refer to the sections of the text in which similar problems can be found.

Simplify each expression as much as possible. [2.1]

1. $5x - 8x$

2. $4x - 7x$

3. $6x - 3 - 8x$

4. $5x - 4 - 9x$

5. $-a + 2 + 5a - 9$

6. $-a + 3 + 6a - 8$

7. $5(2a - 1) - 4(3a - 2)$

8. $6(4a + 2) - 3(5a - 1)$

9. $6 - 2(3y + 1) - 4$

10. $7 - 5(2a - 3) + 7$

11. $4 - 2(3x - 1) - 5$

12. $6 - 2(4a + 2) - 5$

Find the value of each expression when x is 3. [2.1]

13. $7x - 2$

14. $8x - 3$

15. $-4x - 5 + 2x$

16. $-3x + 7 + 5x$

17. $-x - 2x - 3x$

18. $-x - 4x - 2x$

Find the value of each expression when x is -2. [2.1]

19. $5x - 3$

20. $2x + 9$

21. $-3x + 2$

22. $-4x - 5$

23. $7 - x - 3$

24. $8 - x - 4$

Solve each equation. [2.2, 2.3]

25. $x + 2 = -6$

26. $x + 3 = -4$

27. $x - \dfrac{1}{2} = \dfrac{4}{7}$

28. $x - \dfrac{3}{4} = \dfrac{5}{6}$

29. $10 - 3y + 4y = 12$

30. $8 - 2y + 3y = 12$

31. $-3 - 4 = -y - 2 + 2y$

32. $-5 - 6 = -y - 3 + 2y$

33. $2x = -10$

34. $3x = -18$

35. $3x = 0$

36. $-2x = 0$

37. $\dfrac{x}{3} = 4$

38. $\dfrac{x}{2} = 5$

39. $-\dfrac{x}{4} = 2$

40. $-\dfrac{x}{3} = 7$

41. $3a - 2 = 5a$

42. $6a - 5 = 4a$

43. $\dfrac{7}{10}a = \dfrac{1}{5}a + \dfrac{1}{2}$

44. $\dfrac{2}{3}a = \dfrac{1}{6}a + 1$

45. $3x + 2 = 5x - 8$

46. $4x - 3 = 8x + 5$

47. $6x - 3 = 2x + 7$

48. $5x - 1 = 9x + 9$

49. $0.7x - 0.1 = 0.5x - 0.1$

50. $0.2x - 0.3 = 0.8x - 0.3$

Solve each equation. Be sure to simplify each side first. [2.4]

51. $2(x - 5) = 10$

52. $3(x - 4) = 9$

53. $12 = 2(5x - 4)$

54. $18 = 3(2x - 2)$

55. $\dfrac{1}{2}(3t - 2) + \dfrac{1}{2} = \dfrac{5}{2}$

56. $\dfrac{1}{2}(4t - 1) + \dfrac{1}{3} = -\dfrac{25}{6}$

57. $\dfrac{3}{5}(5x - 10) = \dfrac{2}{3}(9x + 3)$

58. $\dfrac{3}{4}(8x - 12) = \dfrac{1}{2}(4x + 4)$

59. $2(3x + 7) = 4(5x - 1) + 18$

60. $3(5x - 1) = 6(2x + 3) - 21$

61. $7 - 3(y + 4) = 10$

62. $8 - 2(y + 4) = 12$

63. $10 - 2(2x + 3) = -5x$

64. $7 - 4(3x + 4) = -9x$

Use the formula $4x - 5y = 20$ to find y if [2.5]

65. x is 5

66. x is 0

67. x is -5

68. x is 10

Solve each of the following formulas for the indicated variable. [2.5]

69. $2x - 5y = 10$ for y

70. $5x - 2y = 10$ for y

71. $V = \pi r^2 h$ for h

72. $P = 2l + 2w$ for w

73. What number is 86% of 240? [2.5]

74. What percent of 2,000 is 180? [2.5]

Solve each of the following application problems. In each case, be sure to show the equation that describes the situation. [2.6, 2.7]

75. The sum of twice a number and 6 is 28. Find the number.

76. The sum of three times a number and 5 is 23. Find the number.

77. Bob is 4 years older than Tom. In 3 years the sum of their ages will be 40. How old is each boy now?

78. Debra is 6 years older than her sister Tracey. In 5 years the sum of their ages will be 36. How old is each girl now?

79. The length of a rectangle is 5 times as long as the width. If the perimeter is 60 meters, find the length and the width.

80. The length of a rectangle is 4 times as long as the width. If the perimeter is 60 meters, find the length and the width.

81. A collection of 15 coins is worth $1.00. If the coins are dimes and nickels, how many of each coin are there?

82. A collection of dimes and quarters is worth $2.55. If there are a total of 15 coins in the collection, how many of each type are there?

83. A man invests a certain amount of money in an account that pays 9% annual interest. He in-

vests $300 more than that in an account that pays 10% annual interest. If his total interest after a year is $125, how much does he have invested in each account?

84. Mary has money invested in two accounts. One account pays 8% annual interest and the other pays 9% annual interest. She has $400 more in the account that pays 9% than she does in the other account. If the total interest after a year is $155, how much is invested in each account?

Solve each inequality. [2.8]

85. $-2x < 4$ **86.** $-3x < 9$

87. $-5x > -10$ **88.** $-2x > -8$

89. $-\dfrac{a}{2} \le -3$ **90.** $-\dfrac{a}{3} \le -2$

91. $-\dfrac{a}{3} > 5$ **92.** $-\dfrac{a}{6} > 4$

Solve each inequality and graph the solution. [2.8]

93. $-4x + 5 > 37$

94. $-0.3x + 0.7 \le -2$

95. $\dfrac{1}{3} - \dfrac{1}{4}x < \dfrac{19}{12}$

96. $8 - 2x \ge 0$

97. $2x + 10 < 5x - 11$

98. $5x + 10 \le 7x - 14$

99. $2(3t + 1) + 6 \ge 5(2t + 4)$

100. $3(2t - 5) - 7 \le 5(3t + 1)$

CHAPTER 2 TEST

Simplify each of the following expressions. [2.1]

1. $3x + 2 - 7x + 3$ **2.** $4a - 5 - a + 1$

3. $7 - 3(y + 5) - 4$

4. $8(2x + 1) - 5(x - 4)$

5. Find the value of $2x - 3 - 7x$ when $x = -5$. [2.1]

6. Find the value of $x^2 + 2xy + y^2$ when $x = 2$ and $y = 3$. [2.1]

Solve the following equations. [2.2, 2.3, 2.4]

7. $2x - 5 = 7$ **8.** $2y + 4 = 5y$

9. $\dfrac{1}{2}x - \dfrac{1}{10} = \dfrac{1}{5}x + \dfrac{1}{2}$

10. $\dfrac{2}{5}(5x - 10) = -5$

11. $-5(2x + 1) - 6 = 19$

12. $0.04x + 0.06(100 - x) = 4.6$

13. $2(t - 4) + 3(t + 5) = 2t - 2$

14. $2x - 4(5x + 1) = 3x + 17$

15. What number is 15% of 38? [2.5]

16. 240 is 12% of what number? [2.5]

17. If $2x - 3y = 12$, find x when $y = -2$. [2.5]

18. The formula for the volume of a cone is $V = \frac{1}{3}\pi r^2 h$. Find h if $V = 88$ cubic inches, $\pi = \frac{22}{7}$, and $r = 3$ inches. [2.5]

19. Solve $2x + 5y = 20$ for y. [2.5]

20. Solve $h = x + vt + 16t^2$ for v. [2.5]

Solve each application problem. [2.6, 2.7]

21. Dave is twice as old as Rick. Ten years ago the sum of their ages was 40. How old are they now?

22. A rectangle is twice as long as it is wide. The perimeter is 60 inches. What are the length and width?

23. A man has a collection of dimes and quarters with a total value of $3.50. If he has 7 more dimes than quarters, how many of each coin does he have?

24. A woman has money in two accounts. One account pays 7% annual interest, while the other pays 9% annual interest. If she has $600 more invested at 9% than she does at 7%, and her total interest for a year is $182, how much does she have in each account?

Solve each inequality and graph the solution. [2.8]

25. $2x + 3 < 5$

26. $-5a > 20$

27. $0.4 - 0.2x \geq 1$

28. $4 - 5(m + 1) \leq 9$

3

GRAPHING AND LINEAR SYSTEMS

INTRODUCTION

In this chapter we will take a second look at some of the topics we have covered previously, this time from a visual standpoint. For example, the information in the table that gave the speed of a dragster at each second during a quarter-mile race is shown visually in Figure 1. The diagram in Figure 1 is called a **bar chart** or **histogram.** Histograms are used frequently to summarize information from a table.

TABLE 1   Speed of a Race Car

Time in Seconds	Speed in Miles/Hour
0	0
1	72.7
2	129.9
3	162.8
4	192.2
5	212.4
6	228.1

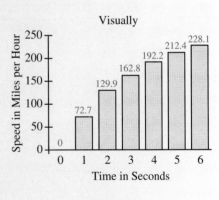

Figure 1

The relationship between the odd numbers and squares that we mentioned in the introduction to Chapter 1 can be visualized with the squares shown in Figure 2 below.

Odds and Squares

$$1 = 1$$
$$1 + 3 = 4$$
$$1 + 3 + 5 = 9$$
$$1 + 3 + 5 + 7 = 16$$

Figure 2

When we drew the diagram showing the relationship between two angles that are complementary, we were giving a visual description of that concept (Figure 3). Another way to visualize the same concept is shown in Figure 4.

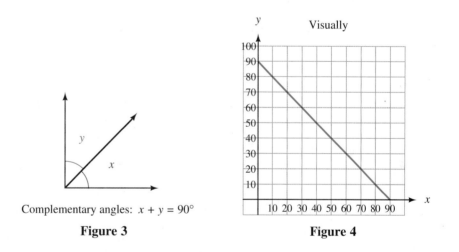

Complementary angles: $x + y = 90°$

Figure 3

Figure 4

The diagram in Figure 4, on which the diagonal line is drawn, is called a **rectangular coordinate system.** Most of the visualization we do in this chapter will be in connection with the rectangular coordinate system.

Overview

In the previous chapter we spent most of our time solving linear equations in one variable. The equations we worked with had the form $ax + b = c$. (If they didn't have this form, they could be put into this form.) In this chapter we will expand our work with equations to include linear equations in two variables. We are going to include another variable so that most of the equations in this chapter will have the form $ax + by = c$, where a, b, and c are constants and x and y are variables. We will also extend the technique of graphing to include points associated with two number lines instead of just one. Your first goal for this chapter is to be able to graph linear equations in two variables on a rectangular coordinate system. Your second goal is to solve systems of linear equations using any one of three methods. Then you will put your knowledge of systems of equations to work by solving some application problems. The background material needed for this chapter is in Chapter 2. You need to know how to solve a linear equation in one variable.

STUDY SKILLS

1. **Getting Ready to Take an Exam**   Try to arrange your daily study habits so that you have very little studying to do the night before your next exam. The next two goals will help you achieve goal number 1.
2. **Review with the Exam in Mind**   Each day, you should review material that will be covered on the next exam. Your review should consist of working problems. Preferably, the problems you work should be problems from your list of difficult problems.
3. **Continue to List Difficult Problems**   This study skill was started in the previous chapter. You should continue to list and rework the problems that give you the most difficulty. It is this list that you will use to study for the next exam. Your goal is to go into the next exam knowing that you can successfully work any problem from your list of hard problems.
4. **Pay Attention to Instructions**   Taking a test is different from doing homework. When you take a test, the problems will be mixed up. When you do your homework, you usually work a number of similar problems. I sometimes have students do very well on their homework, but become confused when they see the same problems on a test because they have not paid attention to the instructions on their homework. For example, suppose you see the equation $y = 3x - 2$ on your next test. By itself, the equation is simply a statement. There isn't anything to do unless the equation is accompanied by instructions. Each of the following is a valid instruction with respect to the equation $y = 3x - 2$, and the result of applying the instructions will be different in each case:

Find x when y is 10	(Section 2.5)
Solve for x	(Section 2.5)
Graph the equation	(Section 3.3)
Find the intercepts	(Section 3.4)
Find the slope	(Section 3.5)

There are many things to do with the equation $y = 3x - 2$. If you train yourself to pay attention to the instructions that accompany a problem as you work through the assigned problems, you will not find yourself confused about what to do with a problem when you see it on a test.

SECTION

3.1   Paired Data and Graphing Ordered Pairs

In the introduction to this chapter we showed the relationship between the table of values for the speed of a race car and the corresponding histogram. Table 1 and Figure 1 from the introduction are reproduced here for reference.

TABLE 1   Speed of a Race Car

Time in Seconds	Speed in Miles/Hour
0	0
1	72.7
2	129.9
3	162.8
4	192.2
5	212.4
6	228.1

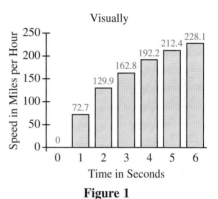

Figure 1

For the histogram in Figure 1, the horizontal line that shows the elapsed time in seconds is called the **horizontal axis,** while the vertical line that shows the speed in miles per hour is called the **vertical axis.** The key to constructing a readable histogram is in labeling the axes in a simple and straightforward manner.

The data in Table 1 are called **paired data** because the information is organized so that each number in the first column is paired with a specific number in the second column. Each pair of numbers is associated with one of the solid bars in the histogram in Figure 1. For example, the third bar in the histogram is associated with the pair of numbers 3 seconds and 162.8 miles per hour. The first number, 3 seconds, is associated with the horizontal axis, while the second number, 162.8 miles per hour, is associated with the vertical axis.

Scatter Diagrams and Line Graphs

The information in Table 1 can be visualized with a **scatter diagram** and **line graph** as well. Figure 2 is a scatter diagram of the information in Table 1. We use dots instead of the bars shown in Figure 1 to show the speed of the race car at each second during the race. Figure 3 is called a *line graph.* It is constructed by taking the dots in Figure 2 and connecting each one to the next with a straight line. Notice that we have labeled the axes in these two figures a little differently than we did with the histogram, by making the axes intersect at the number 0.

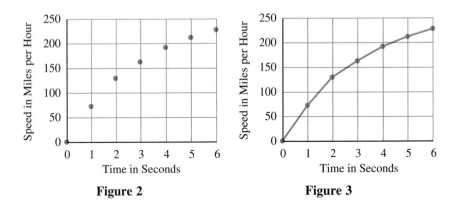

Figure 2   **Figure 3**

▶ **EXAMPLE 1**   Table 2 below gives the weight of a lamb at the beginning of each of the first six months of its life. Use the information in the table to construct a histogram and a line graph.

TABLE 2 Weight of a Lamb

Time in Months	Weight in Pounds
0	12
1	46
2	68
3	98
4	122
5	149
6	173

Solution   The histogram and line graph are shown in Figures 4 and 5, respectively.

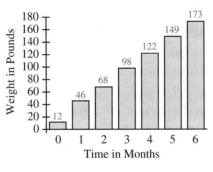

Figure 4   **Figure 5**

The number sequences we have worked with in the past can also be written as paired data by associating each number in the sequence with its position in the sequence. For instance, in the sequence of odd numbers

$$1, 3, 5, 7, 9, \ldots$$

the number 7 is the fourth number in the sequence. Its position is 4 and its value is 7. Here is the sequence of odd numbers written so that the position of each term is noted:

Position   1, 2, 3, 4, 5, . . .

Value   1, 3, 5, 7, 9, . . .

▶ **EXAMPLE 2**   Below are two tables that give the first five terms of the sequence of odd numbers and the sequence of squares as paired data. In each case, construct a scatter diagram.

TABLE 3	**Odd Numbers**
Position	**Value**
1	1
2	3
3	5
4	7
5	9

TABLE 4	**Squares**
Position	**Value**
1	1
2	4
3	9
4	16
5	25

Solution   The two scatter diagrams are shown on the following page. Notice how the dots in Figure 6 seem to line up in a straight line, while the dots in Figure 7 give the impression of a curve. We say the points in Figure 6 suggest a **linear** relationship between the two sets of data, while the points in Figure 7 suggest a **nonlinear** relationship.

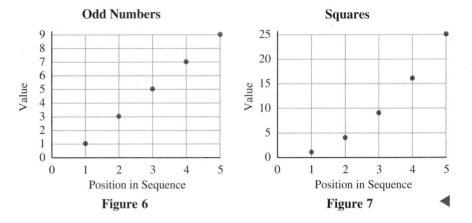

Figure 6

Figure 7

As you know, each dot in Figures 6 and 7 corresponds to a pair of numbers, one of which is associated with the horizontal axis and the other with the vertical axis. Paired data play a very important role in the equations we will solve in the next section. To prepare ourselves for those equations, we need to expand the concept of paired data to include negative numbers. At the same time, we want to standardize the position of the axes in the diagrams that we use to visualize paired data.

> **DEFINITION**   A pair of numbers enclosed in parentheses and separated by a comma, such as $(-2, 1)$, is called an **ordered pair** of numbers. The first number in the pair is called the **x-coordinate** of the ordered pair, while the second number is called the **y-coordinate.** For the ordered pair $(-2, 1)$, the x-coordinate is -2 and the y-coordinate is 1.

Ordered pairs of numbers are important in the study of mathematics because they give us a way to visualize solutions to equations. In order to see the visual component of ordered pairs we need the diagram shown in Figure 8. It is called the **rectangular coordinate system.**

The rectangular coordinate system is built from two number lines oriented perpendicular to each other. The horizontal number line is exactly the same as our real number line and is called the x-axis. The vertical number line is also the same as our real number line with the positive direction up and the negative direction down. It is called the y-axis. The point where the two axes intersect is called the **origin.** As you can see from Figure 8, the axes divide the plane into four **quadrants,** which are numbered I through IV in a counterclockwise direction.

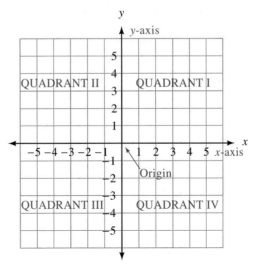

Figure 8

Graphing Ordered Pairs

To graph the ordered pair (a, b), we start at the origin and move a units forward or back (forward if a is positive and back if a is negative). Then we move b units up or down (up if b is positive, down if b is negative). The point where we end up is the graph of the ordered pair (a, b).

▶ **EXAMPLE 3**   Graph the ordered pairs $(3, 4)$, $(3, -4)$, $(-3, 4)$, and $(-3, -4)$.

Solution

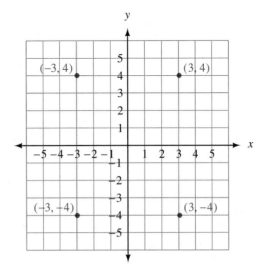

Figure 9

As indicated in Figure 9 that when we graph ordered pairs the x-coordinate corresponds to movement parallel to the x-axis (horizontal) and the y-coordinate corresponds to movement parallel to the y-axis (vertical). ◄

Note   It is very important that you graph ordered pairs quickly and accurately. Remember, the first coordinate goes with the horizontal axis and the second coordinate with the vertical axis.

▶ **EXAMPLE 4**   Graph the ordered pairs $(-1, 3)$, $(2, 5)$, $(0, 0)$, $(0, -3)$, and $(4, 0)$.

Solution

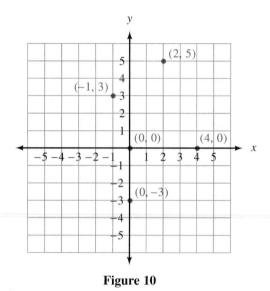

Figure 10   ◄

Note   If we do not label the axes of a coordinate system, we assume that each square is one unit long and one unit wide.

P R O B L E M   S E T   3 . 1

Table 5 extends the information given in Table 1 of this section. In addition to showing the time and speed of Jim Rizzoli's dragster during a race, it also shows the distance past the starting line that his dragster has traveled. Use Table 5 to work Problems 1 and 2.

1. Construct a histogram that shows the relationship between time and distance.

2. Construct a line graph that shows the relationship between time and distance.

Table 6 gives more information about the race car. In addition to giving the speed at one-second intervals, it gives the revolutions per minute (RPM) at which the front axle is turning. Use Table 6 to work Problems 3 and 4.

3. Construct a histogram that shows the relationship between time and front axle RPM.

4. Construct a line graph that shows the relationship between time and front axle RPM.

TABLE 5   Speed and Distance for a Race Car

Time in Seconds	Speed in Miles/Hour	Distance Traveled in Feet
0	0	0
1	72.7	69
2	129.9	231
3	162.8	439
4	192.2	728
5	212.4	1,000
6	228.1	1,373

TABLE 6   Speed and Front Axle Revolutions per Minute

Time in Seconds	Speed in Miles/Hour	Front Axle RPM
0	0	0
1	72.7	1,107
2	129.9	1,978
3	162.8	2,486
4	192.2	2,919
5	212.4	3,233
6	228.1	3,473

Table 7 gives corresponding temperatures on the Celsius and Fahrenheit temperature scales. By reading the table you will see that 100 degrees Celsius is the same temperature as 212 degrees Fahrenheit. Use Table 7 to work Problems 5–8.

TABLE 7   Corresponding Temperatures

In Degrees Celsius	In Degrees Fahrenheit
0	32
25	77
50	122
75	167
100	212

5. Construct a scatter diagram from the information in the table.

6. Construct a histogram from the information in the table.

7. Does your scatter diagram suggest that the relationship between the two temperature scales is a linear one?

8. Use your scatter diagram to predict the Fahrenheit temperature that corresponds to a Celsius temperature of 125 degrees.

Table 8 gives the interest earned on $500 that is invested for one year at a variety of simple interest rates. Use Table 8 to work Problems 9–12.

TABLE 8   Interest Earned in One Year

Interest Rate in Percent	Interest Earned in Dollars
2.0	10.00
2.5	12.50
3.0	15.00
3.5	17.50
4.0	20.00

9. Construct a scatter diagram from the information in the table.

10. Construct a histogram from the information in the table.

11. Does your scatter diagram suggest that the relationship between interest rate and interest earned is a linear one?

12. Use your scatter diagram to predict the amount of interest that is earned on $500 invested at 4.5% for 1 year.

Graph the following ordered pairs.

13. $(3, 2)$

14. $(3, -2)$

15. $(-3, 2)$

16. $(-3, -2)$

17. $(5, 1)$

18. $(5, -1)$

19. $(1, 5)$

20. $(1, -5)$

21. $(-1, 5)$

22. $(-1, -5)$

23. $(2, \frac{1}{2})$

24. $(3, \frac{3}{2})$

25. $(-4, -\frac{5}{2})$

26. $(-5, -\frac{3}{2})$

27. $(3, 0)$

28. $(-2, 0)$

29. $(0, 5)$

30. $(0, 0)$

Give the coordinates of each numbered point in the figure below.

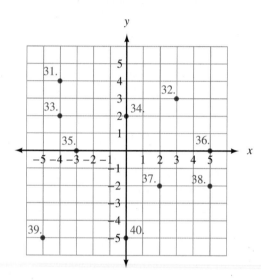

Graph the points $(4, 3)$ and $(-4, -1)$ and draw a straight line that passes through both of them. Then answer the following questions.

41. Does the graph of $(2, 2)$ lie on the line?

42. Does the graph of $(-2, 0)$ lie on the line?

43. Does the graph of $(0, -2)$ lie on the line?

44. Does the graph of $(-6, 2)$ lie on the line?

Graph the points $(-2, 4)$ and $(2, -4)$ and draw a straight line that passes through both of them. Then answer the following questions.

45. Does the graph of $(0, 0)$ lie on the line?

46. Does the graph of $(-1, 2)$ lie on the line?

47. Does the graph of $(2, -1)$ lie on the line?

48. Does the graph of $(1, -2)$ lie on the line?

Draw a straight line that passes through the points $(3, 4)$ and $(3, -4)$. Then answer the following questions.

49. Is the graph of $(3, 0)$ on this line?

50. Is the graph of $(0, 3)$ on this line?

51. Is there any point on this line with an x-coordinate other than 3?

52. If you extended the line, would it pass through a point with a y-coordinate of 10?

Draw a straight line that passes through the points $(3, 4)$ and $(-3, 4)$. Then answer the following questions.

53. Is the graph of $(4, 0)$ on this line?

54. Is the graph of $(0, 4)$ on this line?

55. Is there any point on this line with a y-coordinate other than 4?

56. If you extended the line, would it pass through a point with an x-coordinate of 10?

57. Draw a straight line that passes through the points $(-2, -3)$ and $(4, -3)$. What do all the points on this line have in common?

58. Draw a straight line that passes through the points $(-2, -3)$ and $(-2, 4)$. What do all the points on this line have in common?

59. For the points on a rectangular coordinate system, where will you find all the ordered pairs of the form $(0, y)$?

60. For the points on a rectangular coordinate system, where will you find all the ordered pairs of the form $(x, 0)$?

Review Problems

The problems that follow review the material we covered on number sequences in Chapter 1.

Find the next number in each sequence.

61. 3, 10, 17, 24, . . .

62. 3, -4, -11, -18, . . .

63. 3, 1, $\frac{1}{3}$, $\frac{1}{9}$, . . .

64. 3, 21, 147, 1,029, . . .

65. 7, 4, 1, -2, . . .

66. 7, 10, 13, 16, . . .

67. 7, 21, 63, 189, . . .

68. 7, 1, $\frac{1}{7}$, $\frac{1}{49}$, . . .

69. 5, 6, 8, 11, . . .

70. 10, 11, 13, 16, . . .

Solutions to Linear Equations in Two Variables

In this section we will begin to investigate equations in two variables. As you will see, equations in two variables have pairs of numbers for solutions. Since we know how to use paired data to construct tables, histograms, and other charts, we can take our work with paired data further by using equations in two variables to construct tables of paired data. Let's begin this section by reviewing the relationship between equations in one variable and their solutions.

If we solve the equation $3x - 2 = 10$, the solution is $x = 4$. If we graph this solution, we simply draw the real number line and place a dot at the point whose coordinate is 4. The relationship between linear equations in one variable, their solutions, and the graphs of those solutions looks like this:

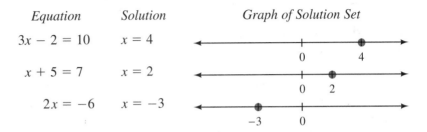

Equation	*Solution*	*Graph of Solution Set*
$3x - 2 = 10$	$x = 4$	
$x + 5 = 7$	$x = 2$	
$2x = -6$	$x = -3$	

When the equation has one variable, the solution is a single number whose graph is a point on a line.

Now consider the equation $2x + y = 3$. The first thing we notice is that there are two variables instead of one. Therefore a solution to the equation $2x + y = 3$ will be not a single number but a pair of numbers, one for x and one for y, that makes the equation a true statement. One pair of numbers that works is $x = 2, y = -1$, because when we substitute them for x and y in the equation, we get a true statement. That is,

$$2(2) + (-1) \overset{?}{=} 3$$

$$4 - 1 \overset{?}{=} 3$$

$$3 = 3 \qquad \text{A true statement}$$

The pair of numbers $x = 2, y = -1$ is written as $(2, -1)$. As you know from Section 3.1, $(2, -1)$ is called an **ordered pair** because it is a pair of numbers written in a specific order. The first number is always associated with the variable x, and the second number is always associated with the variable y. We call the first number in the ordered pair the **x-coordinate** (or x component) and the second number the **y-coordinate** (or y component) of the ordered pair.

Let's look back to the equation $2x + y = 3$. The ordered pair $(2, -1)$ is not the only solution. Another solution is $(0, 3)$, because when we substitute 0 for x and 3 for y we get

$$2(0) + 3 \stackrel{?}{=} 3$$

$$0 + 3 \stackrel{?}{=} 3$$

$$3 = 3 \qquad \text{A true statement}$$

Still another solution is the ordered pair $(5, -7)$, because

$$2(5) + (-7) \stackrel{?}{=} 3$$

$$10 - 7 \stackrel{?}{=} 3$$

$$3 = 3 \qquad \text{A true statement}$$

As a matter of fact, for any number we want to use for x there is another number we can use for y that will make the equation a true statement. There is an infinite number of ordered pairs that satisfy (are solutions to) the equation $2x + y = 3$; we have listed just a few of them.

▶ **EXAMPLE 1** Given the equation $2x + 3y = 6$, complete the following ordered pairs so they will be solutions to the equation: $(0, \ )$, $(\ , 1)$, $(3, \ )$.

Solution To complete the ordered pair $(0, \ )$, we substitute 0 for x in the equation and then solve for y:

$$2(0) + 3y = 6$$

$$3y = 6$$

$$y = 2$$

The ordered pair is $(0, 2)$.

To complete the ordered pair $(\ , 1)$, we substitute 1 for y in the equation and solve for x:

$$2x + 3(1) = 6$$

$$2x + 3 = 6$$

$$2x = 3$$

$$x = \frac{3}{2}$$

The ordered pair is $(\frac{3}{2}, 1)$.

To complete the ordered pair $(3, \ )$, we substitute 3 for x in the equation and solve for y:

$$2(3) + 3y = 6$$

$$6 + 3y = 6$$

$$3y = 0$$

$$y = 0$$

The ordered pair is $(3, 0)$. ◀

Notice in each case that once we have used a number in place of one of the variables, the equation becomes a linear equation in one variable. We then use the method explained in Chapter 2 to solve for that variable.

▶ **EXAMPLE 2**   Complete the following table for the equation $2x - 5y = 20$.

x	y
0	
	2
	0
-5	

Solution   Filling in the table is equivalent to completing the following ordered pairs: $(0, \ ), (\ , 2), (\ , 0), (-5, \ )$. So we proceed as in Example 1.

When $x = 0$, we have

$$2(0) - 5y = 20$$
$$0 - 5y = 20$$
$$-5y = 20$$
$$y = -4$$

When $y = 0$, we have

$$2x - 5(0) = 20$$
$$2x - 0 = 20$$
$$2x = 20$$
$$x = 10$$

When $y = 2$, we have

$$2x - 5(2) = 20$$
$$2x - 10 = 20$$
$$2x = 30$$
$$x = 15$$

When $x = -5$, we have

$$2(-5) - 5y = 20$$
$$-10 - 5y = 20$$
$$-5y = 30$$
$$y = -6$$

The completed table looks like this:

x	y
0	-4
15	2
10	0
-5	-6

which is equivalent to the ordered pairs $(0, -4)$, $(15, 2)$, $(10, 0)$, and $(-5, -6)$.   ◀

▶ **EXAMPLE 3**   Complete the following table for the equation $y = 2x - 1$.

x	y
0	
5	
	7
	3

Solution   When $x = 0$, we have   When $x = 5$, we have

$$y = 2(0) - 1 \qquad\qquad y = 2(5) - 1$$
$$y = 0 - 1 \qquad\qquad\quad y = 10 - 1$$
$$y = -1 \qquad\qquad\qquad y = 9$$

When $y = 7$, we have   When $y = 3$, we have

$$7 = 2x - 1 \qquad\qquad\quad 3 = 2x - 1$$
$$8 = 2x \qquad\qquad\qquad 4 = 2x$$
$$4 = x \qquad\qquad\qquad\; 2 = x$$

The completed table is

x	y
0	-1
5	9
4	7
2	3

which means the ordered pairs $(0, -1)$, $(5, 9)$, $(4, 7)$, and $(2, 3)$ are among the solutions to the equation $y = 2x - 1$.   ◀

▶ **EXAMPLE 4**   Which of the ordered pairs (2, 3), (1, 5), and (−2, −4) are solutions to the equation $y = 3x + 2$?

Solution   If an ordered pair is a solution to the equation, then it must satisfy the equation. That is, when the coordinates are used in place of the variables in the equation, the equation becomes a true statement.

Try (2, 3) in $y = 3x + 2$:

$$3 \overset{?}{=} 3(2) + 2$$

$$3 \overset{?}{=} 6 + 2$$

$$3 = 8 \qquad \text{A false statement}$$

Try (1, 5) in $y = 3x + 2$:

$$5 \overset{?}{=} 3(1) + 2$$

$$5 \overset{?}{=} 3 + 2$$

$$5 = 5 \qquad \text{A true statement}$$

Try (−2, −4) in $y = 3x + 2$:

$$-4 \overset{?}{=} 3(-2) + 2$$

$$-4 \overset{?}{=} -6 + 2$$

$$-4 = -4 \qquad \text{A true statement}$$

The ordered pairs (1, 5) and (−2, −4) are solutions to the equation $y = 3x + 2$, and (2, 3) is not.   ◀

▶ **EXAMPLE 5**   One of the rates listed in the GTE Mobilnet rate card in 1994 called for a flat rate of $15 per month plus $1.50 for each minute you use the phone. Write an equation in two variables that will let you calculate the monthly charge for talking x minutes. Then make a table that shows the cost for talking 10, 15, 20, 25, or 30 minutes in one month. Use the data in the table to construct a histogram.

Solution   If we let x = the number of minutes of phone use then the charge for using the phone x minutes in one month will be $1.5x + 15$. If y is the charge for talking x minutes, then

$$y = 1.5x + 15$$

Now, GTE Mobilnet charges $1.50 for each minute or fraction of a minute, so the equation above will work only when x is a nonnegative integer. If we substitute $x = 10, 15, 20, 25,$ and 30 into the equation for x, then the equation gives us the corresponding costs as shown in Table 1. A histogram of the data in Table 1 is shown in Figure 1.

TABLE 1	Monthly Phone Costs
Time in Minutes	**Cost in Dollars**
10	$30.00
15	$37.50
20	$45.00
25	$52.50
30	$60.00

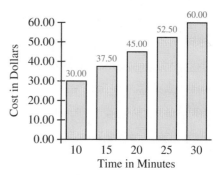

Figure 1

PROBLEM SET 3.2

For each equation, complete the given ordered pairs.

1. $2x + y = 6$ (0,), (3,), (, −6)
2. $3x − y = 5$ (0,), (1,), (, 5)
3. $3x + 4y = 12$ (0,), (, 0), (−4,)
4. $5x − 5y = 20$ (0,), (, −2), (1,)
5. $y = 4x − 3$ (1,), (, 0), (5,)
6. $y = 3x − 5$ (, 13), (0,), (−2,)
7. $y = 7x − 1$ (2,), (, 6), (0,)
8. $y = 8x + 2$ (3,), (, 0), (, −6)
9. $x = −5$ (, 4), (, −3), (, 0)

10. $y = 2$ (5,), (−8,), $\left(\dfrac{1}{2}, \quad \right)$

For each of the following equations complete the given table.

11. $y = 3x$

x	y
1	
−3	
	12
	18

12. $y = −2x$

x	y
−4	
0	
	10
	12

13. $y = 4x$

x	y
0	
	−2
−3	
	12

14. $y = −5x$

x	y
3	
	0
−2	
	−20

15. $x + y = 5$

x	y
2	
3	
	0
	−4

16. $x − y = 8$

x	y
0	
4	
	−3
	−2

17. $2x − y = 4$

x	y
	0
	2
1	
−3	

18. $3x − y = 9$

x	y
	0
	−9
5	
−4	

19. $y = 6x - 1$

x	y
0	
−1	
−3	
	8

20. $y = 5x + 7$

x	y
0	
−2	
−4	
	−8

For the following equations, tell which of the given ordered pairs are solutions.

21. $2x - 5y = 10$ $(2, 3), (0, -2), \left(\dfrac{5}{2}, 1\right)$

22. $3x + 7y = 21$ $(0, 3), (7, 0), (1, 2)$
23. $y = 7x - 2$ $(1, 5), (0, -2), (-2, -16)$
24. $y = 8x - 3$ $(0, 3), (5, 16), (1, 5)$
25. $y = 6x$ $(1, 6), (-2, -12), (0, 0)$
26. $y = -4x$ $(0, 0), (2, 4), (-3, 12)$
27. $x + y = 0$ $(1, 1), (2, -2), (3, 3)$
28. $x - y = 1$ $(0, 1), (0, -1), (1, 2)$
29. $x = 3$ $(3, 0), (3, -3), (5, 3)$
30. $y = -4$ $(3, -4), (-4, 4), (0, -4)$
31. If the perimeter of a rectangle is 30 inches, then the relationship between the length l and the width w is given by the equation

$$2l + 2w = 30$$

What is the length when the width is 3 inches?
32. The relationship between the perimeter P of a square and the length of its side s is given by the formula $P = 4s$. If each side of a square is 5 inches, what is the perimeter? If the perimeter of a square is 28 inches, how long is a side?
33. If every ordered pair that satisfies an equation has a y-coordinate that is twice as large as its x-coordinate, then the equation is $y = 2x$, and every ordered pair has the form $(x, 2x)$. Write the ordered pair that has 5 as its x-coordinate.
34. If every ordered pair that satisfies an equation has a y-coordinate that is three times as large as its x-coordinate, then the equation is $y = 3x$. What is the form of each ordered pair? (*Hint:*

Fill in the second coordinate in $(x, \ )$; see Problem 33.) If x is 4, what is y?
35. For the equation $y = |x|$, y is equal to the absolute value of x. Complete the following ordered pairs so that they are solutions to $y = |x|$.
$$(3, \ ), (-3, \ ), (5, \ ), (-5, \ )$$
36. If $y = |x|$ and y is 4, there are two possible x values: either 4 or −4, because both the ordered pairs (4, 4) and (−4, 4) satisfy the equation. What two ordered pairs that are solutions to $y = |x|$ have a y-coordinate of 6?
37. San Luis Garbage charges $13.00 per month for their basic service, plus $1.50 per bag for each trash bag left by the curb on the day the trash is picked up. Write an equation in two variables that will let you calculate the total monthly charge y for picking up x trash bags. Then use this equation to make a table that gives the cost for picking up 5, 7, 9, 11, and 13 trash bags in one month. Use the information in the table to construct a histogram.
38. Five Cities Garbage charges $21.85 per month for their basic service, plus $1.00 per bag for each trash bag left by the curb on the day the trash is picked up. Write an equation in two variables that will let you calculate the total monthly charge y for picking up x trash bags. Then use this equation to make a table that gives the cost for picking up 5, 7, 9, 11, and 13 trash bags in one month. Use the information in the table to construct a histogram.
39. A bottled water company charges $7.00 per month for their water dispenser and $1.10 for each gallon of water delivered. Write an equation in two variables that will let you calculate the total monthly charge y for drinking x gallons of water in a month. Then use this equation to make a table that gives the cost for drinking 15, 20, 25, 30, and 35 gallons of water in one month. Use the information in the table to construct a histogram.
40. A car rental company charges $21.99 per day and $0.20 per mile to rent one of their cars. Write an equation in two variables that will let you calculate the total charge to rent one of these cars for a week and drive it x miles.

Review Problems

The following problems review material we covered in Section 2.5.

41. Find y when x is 4 in the formula $3x + 2y = 6$.
42. Find y when x is 0 in the formula $3x + 2y = 6$.

43. Find y when x is 0 in $y = -\dfrac{1}{3}x + 2$.

44. Find y when x is 3 in $y = -\dfrac{1}{3}x + 2$.

45. Find y when x is 2 in $y = \dfrac{3}{2}x - 3$.

46. Find y when x is 4 in $y = \dfrac{3}{2}x - 3$.

47. Solve $5x + y = 4$ for y.
48. Solve $-3x + y = 5$ for y.
49. Solve $3x - 2y = 6$ for y.
50. Solve $2x - 3y = 6$ for y.

SECTION
3.3
Graphing Linear Equations in Two Variables

At the end of the previous section we used a histogram to obtain a visual picture of *some* of the solutions to the equation $y = 1.5x + 15$. In this section we will use the rectangular coordinate system introduced in Section 3.1 to obtain a visual picture of *all* solutions to a linear equation in two variables. The process we use to obtain a visual picture of all solutions to an equation is called **graphing.** The picture itself is called the **graph** of the equation.

▶ **EXAMPLE 1**   Graph the solution set for $x + y = 5$.

Solution   We know from the last section that there is an infinite number of ordered pairs that are solutions to the equation $x + y = 5$. We can't possibly list them all. What we can do is list a few of them and see if there is any pattern to their graphs.

Some ordered pairs that are solutions to $x + y = 5$ are $(0, 5)$, $(2, 3)$, $(3, 2)$, $(5, 0)$. The graph of each is shown in Figure 1.

Now, by passing a straight line through these points we can graph the solution set for the equation $x + y = 5$. Linear equations in two variables always have graphs that are straight lines. The graph of the solution set for $x + y = 5$ is shown in Figure 2.

Every ordered pair that satisfies $x + y = 5$ has its graph on the line, and any point on the line has coordinates that satisfy the equation. That is, there is a one-to-one correspondence between points on the line and solutions to the equation.

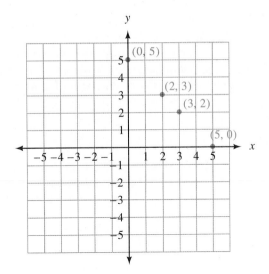

Figure 1

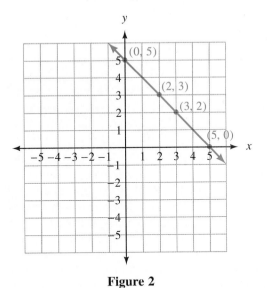

Figure 2

Our ability to graph an equation as we have done in Example 1 is due to the invention of the rectangular coordinate system. The French philosopher René Descartes (1595–1650) is the person usually credited with the invention of the rectangular coordinate system. As a philosopher, Descartes is responsible for the statement "I think, therefore I am." Until Descartes invented his coordinate system in 1637, algebra and geometry were treated as separate subjects. The rectangular coordinate system allows us to connect algebra and geometry by associating geometric shapes with algebraic equations.

Here is the precise definition for a linear equation in two variables:

DEFINITION   Any equation that can be put in the form $ax + by = c$, where a, b, and c are real numbers and a and b are not both 0, is called a **linear equation in two variables.** The graph of any equation of this form is a straight line (that is why these equations are called "linear"). The form $ax + by = c$ is called **standard form.**

To graph a linear equation in two variables, we simply graph its solution set. That is, we draw a line through all the points whose coordinates satisfy the equation. Here are the steps to follow:

Strategy for Graphing a Straight Line

Step 1:   Find any three ordered pairs that satisfy the equation. This can be done by using a convenient number for one variable and solving for the other variable.

Step 2:   Graph the three ordered pairs found in step 1. Actually, we need only two points to graph a straight line. The third point serves as a check. If all three points do not line up, there is a mistake in our work.

Step 3:   Draw a straight line through the three points graphed in step 2.

▶ **EXAMPLE 2**   Graph the equation $y = 3x - 1$.

Solution   Since $y = 3x - 1$ can be put in the form $ax + by = c$, it is a linear equation in two variables. Hence, the graph of its solution set is a straight line. We can find some specific solutions by substituting numbers for x and then solving for the corresponding values of y. We are free to choose any numbers for x, so let's use 0, 2, and -1.

Let $x = 0$: $y = 3(0) - 1$
$y = 0 - 1$
$y = -1$
The ordered pair $(0, -1)$ is one solution.

Let $x = 2$: $y = 3(2) - 1$
$y = 6 - 1$
$y = 5$
The ordered pair $(2, 5)$ is a second solution.

Let $x = -1$: $y = 3(-1) - 1$
$y = -3 - 1$
$y = -4$
The ordered pair $(-1, -4)$ is a third solution.

In table form

x	y
0	-1
2	5
-1	-4

Next we graph the ordered pairs $(0, -1)$, $(2, 5)$, $(-1, -4)$ and draw a straight line through them.

The line we have drawn in Figure 3 is the graph of $y = 3x - 1$.

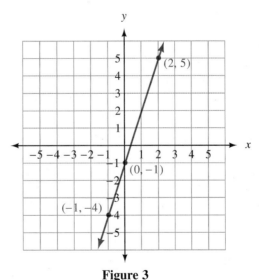

Figure 3

Example 2 again illustrates the connection between algebra and geometry that we mentioned previously. Descartes' rectangular coordinate system allows us to associate the equation $y = 3x - 1$ (an algebraic concept) with a specific straight line (a geometric concept). The study of the relationship between equations in algebra and their associated geometric figures is called **analytic geometry.** The rectangular coordinate system is often referred to as the **Cartesian coordinate system** in honor of Descartes.

▶ **EXAMPLE 3** Graph the equation $y = -\frac{1}{3}x + 2$.

Solution We need to find three ordered pairs that satisfy the equation. To do so, we can let x equal any numbers we choose and find corresponding values of y. But, since every value of x we substitute into the equation is going to be multiplied by $-\frac{1}{3}$, let's use numbers for x that are divisible by 3, like -3, 0, and 3. That way, when we multiply them by $-\frac{1}{3}$, the result will be an integer.

Let $x = -3$: $\quad y = -\dfrac{1}{3}(-3) + 2$

$$y = 1 + 2$$
$$y = 3$$

The ordered pair $(-3, 3)$ is one solution.

Let $x = 0$: $\quad y = -\dfrac{1}{3}(0) + 2$

$$y = 0 + 2$$
$$y = 2$$

The ordered pair $(0, 2)$ is a second solution.

Let $x = 3$: $\quad y = -\dfrac{1}{3}(3) + 2$

$$y = -1 + 2$$
$$y = 1$$

The ordered pair $(3, 1)$ is a third solution.

In table form

x	y
-3	3
0	2
3	1

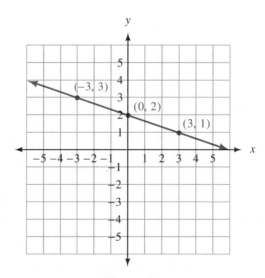

Figure 4

By graphing the ordered pairs $(-3, 3)$, $(0, 2)$, and $(3, 1)$ and by drawing a straight line through their graphs, we have the graph of the equation $y = -\frac{1}{3}x + 2$, as shown in Figure 4. ◀

Note In Example 3, the values of x we used, -3, 0, and 3, are referred to as convenient values of x because they are easier to work with than some other numbers. For instance, if we let $x = 2$ in our original equation, we would have to add $-\frac{2}{3}$ and 2 to find the corresponding value of y. Not only would the arithmetic be more difficult, but also the ordered pair we obtained would have a fraction for its y-coordinate, making it more difficult to graph accurately.

▶ **EXAMPLE 4** Graph the lines $x = 2$ and $y = -3$ on the same coordinate system.

Solution The line $x = 2$ is the set of all points whose x-coordinate is 2. When the variable y does not appear in the equation, it means that y can be any number. Therefore, the points $(2, 0)$, $(2, 3)$, $(2, -4)$, and $(2, 2)$ all satisfy the equation $x = 2$, simply because their x-coordinate is 2. The graph of $x = 2$, along with the points we named above, is given in Figure 5.

The line $y = -3$ is the set of all points whose y-coordinate is -3. The points $(0, -3)$, $(4, -3)$, and $(-4, -3)$ all satisfy the equation $y = -3$, because their y-coordinates are all -3. The graph of the line $y = -3$ is also shown in Figure 5.

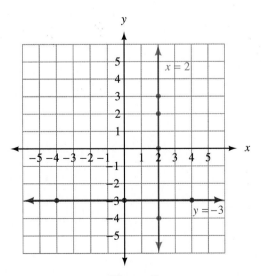

Figure 5 ◀

▶ **EXAMPLE 5**   Graph the solution set for $3x - 2y = 6$.

Solution   It will be easier to find convenient values of x to use in the equation if we first solve the equation for y. To do so, we add $-3x$ to each side and then we multiply each side by $-\frac{1}{2}$.

$$3x - 2y = 6 \qquad\qquad \text{Original equation}$$

$$-2y = -3x + 6 \qquad\qquad \text{Add } -3x \text{ to each side}$$

$$-\frac{1}{2}(-2y) = -\frac{1}{2}(-3x + 6) \qquad \text{Multiply each side by } -\frac{1}{2}$$

$$y = \frac{3}{2}x - 3 \qquad\qquad \text{Simplify each side}$$

Now, since each value of x will be multiplied by $\frac{3}{2}$, it will be to our advantage to choose values of x that are divisible by 2. That way, we will obtain values of y that do not contain fractions. This time, let's use 0, 2, and 4 for x.

When $x = 0$: $\qquad y = \dfrac{3}{2}(0) - 3$

$$y = 0 - 3$$

$$y = -3 \qquad\qquad (0, -3) \text{ is one solution}$$

When $x = 2$: $\qquad y = \dfrac{3}{2}(2) - 3$

$$y = 3 - 3$$

$$y = 0 \qquad\qquad (2, 0) \text{ is a second solution}$$

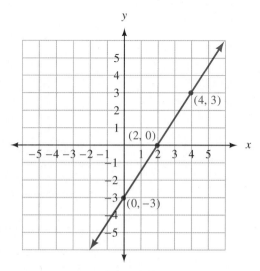

Figure 6

$$\text{When } x = 4: \qquad y = \frac{3}{2}(4) - 3$$

$$y = 6 - 3$$

$$y = 3 \qquad (4, 3) \text{ is a third solution}$$

Graphing the ordered pairs $(0, -3)$, $(2, 0)$, and $(4, 3)$, and drawing a line through them, we have the graph shown in Figure 6. ◄

Note After reading through Example 5, many students ask why we didn't use -2 for x when we were finding ordered pairs that were solutions to the original equation. The answer is, we could have. If we were to let $x = -2$, the corresponding value of y would have been -6. As you can see by looking at the graph in Figure 6, the ordered pair $(-2, -6)$ is on the graph.

P R O B L E M   S E T   3 . 3

For the following equations, complete the given ordered pairs and use the results to graph the solution set for the equation.

1. $x + y = 4$ $(0, \ )$, $(2, \ )$, $(\ , 0)$

2. $x - y = 3$ $(0, \ )$, $(2, \ )$, $(\ , 0)$

3. $x + y = 3$ $(0, \ )$, $(2, \ )$, $(\ , -1)$

4. $x - y = 4$ $(1, \ )$, $(-1, \ )$, $(\ , 0)$

5. $y = 2x$ $(0, \ )$, $(-2, \ )$, $(2, \ )$

6. $y = \dfrac{1}{2}x$ $(0, \ )$, $(-2, \ )$, $(2, \ )$

7. $y = \dfrac{1}{3}x$ $(-3, \ )$, $(0, \ )$, $(3, \ )$

8. $y = 3x$ $(-2, \ )$, $(0, \ )$, $(2, \ )$

9. $y = 2x + 1$ $(0, \ )$, $(-1, \ )$, $(1, \ )$

10. $y = -2x + 1$ $(0, \ )$, $(-1, \ )$, $(1, \ )$

11. $y = 4$ $(0, \ )$, $(-1, \ )$, $(2, \ )$

12. $x = 3$ $(\ , -2)$, $(\ , 0)$, $(\ , 5)$

13. $y = \dfrac{1}{2}x + 3$ $(-2, \ )$, $(0, \ )$, $(2, \ )$

14. $y = \dfrac{1}{2}x - 3$ $(-2, \ )$, $(0, \ )$, $(2, \ )$

15. $y = -\dfrac{2}{3}x + 1$ $(-3, \ )$, $(0, \ )$, $(3, \ )$

16. $y = -\dfrac{2}{3}x - 1$ $(-3, \ )$, $(0, \ )$, $(3, \ )$

Solve each equation for y. Then, complete the given ordered pairs and use them to draw the graph.

17. $2x + y = 3$ $(-1, \ )$, $(0, \ )$, $(1, \ )$

18. $3x + y = 2$ $(-1, \ )$, $(0, \ )$, $(1, \ )$

19. $3x + 2y = 6$ $(0, \ )$, $(2, \ )$, $(4, \ )$

20. $2x + 3y = 6$ $(0, \ )$, $(3, \ )$, $(6, \ )$

21. $-x + 2y = 6$ $(-2, \ )$, $(0, \ )$, $(2, \ )$

22. $-x + 3y = 6$ $(-3, \ )$, $(0, \ )$, $(3, \ )$

Find three solutions to each of the following equations and then graph the solution set.

23. $y = -\dfrac{1}{2}x$ **24.** $y = -2x$

25. $y = 3x - 1$ **26.** $y = -3x - 1$

27. $-2x + y = 1$ **28.** $-3x + y = 1$

29. $3x + 4y = 8$ **30.** $3x - 4y = 8$

31. $x = -2$ **32.** $y = 3$

33. $y = 2$ **34.** $x = -3$

35. If the perimeter of a rectangle is 10 inches, then the equation that describes the relationship between the length l and width w is

$$2l + 2w = 10$$

To graph this equation, we use a coordinate system in which the horizontal axis is labeled l and the vertical axis is labeled w.

The ordered pairs that satisfy the equation look like this: (l, w). Complete the ordered pairs $(1, \ )$, $(2, \ )$, and $(3, \ )$ for the equation $2l + 2w = 10$ and use the results to graph the equation. Compare your answer with Example 1 in this section.

36. The perimeter of a rectangle is 6 inches. Graph the equation that describes the relationship between the length l and width w.

37. The solutions to the equation $y = x$ are ordered pairs of the form (x, x). The x- and y-coordinates are equal. Graph the line $y = x$.

38. The solutions to a certain equation are ordered pairs of the form $(x, 2x)$. Find the equation and its graph.

39. The y-coordinate of each ordered pair that satisfies the equation $y = |x|$ will never be negative. Each value of y is the absolute value of the corresponding value of x. Fill in the following ordered pairs so that they are solutions to the equation $y = |x|$. Then graph them and connect their graphs in a way that makes the most sense to you.

$(-3, \ ) \ (-2, \ ) \ (-1, \ ) \ (0, \ ) \ (1, \ ) \ (2, \ )$
$(3, \ )$

40. Fill in the following ordered pairs so that they are solutions to the equation $y = |x - 2|$. Graph each point and then connect the graphs in a way that makes the most sense to you.

$(-1, \ ) \ (0, \ ) \ (1, \ ) \ (2, \ ) \ (3, \ ) \ (4, \ ) \ (5, \ )$

41. Graph the lines $y = x + 1$ and $y = x - 3$ on the same coordinate system. Can you tell from looking at these first two graphs where the graph of $y = x + 3$ would be?

42. Graph the lines $y = 2x + 2$ and $y = 2x - 1$ on the same coordinate system. Use the similarities between these two graphs to graph the line $y = 2x - 4$.

Review Problems

The problems below review material we covered in Sections 2.3 and 2.4.

Solve each equation.

43. $3(x - 2) = 9$
44. $-4(x - 3) = -16$
45. $2(3x - 1) + 4 = -10$
46. $-5(2x + 3) - 10 = 15$
47. $6 - 2(4x - 7) = -4$
48. $5 - 3(2 - 3x) = 8$
49. $\dfrac{1}{2}x + 4 = \dfrac{2}{3}x + 5$
50. $\dfrac{1}{4}x - 3 = \dfrac{3}{2}x + 7$

Research Project 5

As you know, the rectangular coordinate system is sometimes referred to as the Cartesian coordinate system in honor of its discoverer, René Descartes. In his book *The Closing of the American Mind*, the late Allan Bloom recalls being in a restaurant in France and overhearing a waiter call another waiter a "Cartesian." He goes on to say that many French people today define themselves in terms of the philosophy of Descartes, and as such may be referred to as Cartesians. Research the philosophy of Descartes and then write an essay in which you describe the characteristics and point of view a person considered a Cartesian would hold.

SECTION

3.4   ## More on Graphing: Intercepts

In this section we continue our work with graphing lines by finding the points where a line crosses the axes of our coordinate system. To do so, we use the fact that any point on the x-axis has a y-coordinate of 0, and any point on the y-axis has an x-coordinate of 0. We begin with the following definition.

> **DEFINITION** The **x-intercept** of a straight line is the *x*-coordinate of the point where the graph crosses the *x*-axis. The **y-intercept** is defined similarly. It is the *y*-coordinate of the point where the graph crosses the *y*-axis.

If the *x*-intercept is *a*, then the point $(a, 0)$ lies on the graph. (This is true because any point on the *x*-axis has a *y*-coordinate of 0.)

If the *y*-intercept is *b*, then the point $(0, b)$ lies on the graph. (This is true because any point on the *y*-axis has an *x*-coordinate of 0.)

Graphically, the relationship is shown in Figure 1.

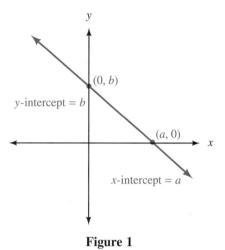

Figure 1

▶ **EXAMPLE 1** Find the *x*- and *y*-intercepts for $3x - 2y = 6$ and then use them to draw the graph.

Solution To find where the graph crosses the *x*-axis, we let $y = 0$. (The *y*-coordinate of any point on the *x*-axis is 0.)

x-intercept:

When $\qquad\qquad\qquad y = 0$

the equation $\qquad 3x - 2y = 6$

becomes $\qquad 3x - 2(0) = 6$

$$3x - 0 = 6$$

$$3x = 6$$

$$x = 2 \qquad \text{Multiply each side by } \tfrac{1}{3}$$

The graph crosses the *x*-axis at $(2, 0)$, which means the *x*-intercept is 2.

y-intercept:

When $x = 0$

the equation $3x - 2y = 6$

becomes $3(0) - 2y = 6$

$0 - 2y = 6$

$-2y = 6$

$y = -3$ Multiply each side by $-\frac{1}{2}$

The graph crosses the y-axis at $(0, -3)$, which means the y-intercept is -3.

Plotting the x- and y-intercepts and then drawing a line through them, we have the graph of $3x - 2y = 6$, as shown in Figure 2.

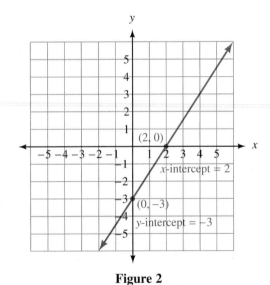

Figure 2   ◄

▶ **EXAMPLE 2**   Graph $-x + 2y = 4$ by finding the intercepts and using them to draw the graph.

Solution   Again, we find the x-intercept by letting $y = 0$ in the equation and solving for x. Similarly, we find the y-intercept by letting $x = 0$ and solving for y.

x-intercept:

When	$y = 0$
the equation	$-x + 2y = 4$
becomes	$-x + 2(0) = 4$
	$-x + 0 = 4$
	$-x = 4$
	$x = -4$ Multiply each side by -1

The x-intercept is -4, indicating that the point $(-4, 0)$ is on the graph of $-x + 2y = 4$.

y-intercept:

When	$x = 0$
the equation	$-x + 2y = 4$
becomes	$-0 + 2y = 4$
	$2y = 4$
	$y = 2$ Multiply each side by $\frac{1}{2}$

The y-intercept is 2, indicating that the point $(0, 2)$ is on the graph of $-x + 2y = 4$.

Plotting the intercepts and drawing a line through them, we have the graph of $-x + 2y = 4$, as shown in Figure 3.

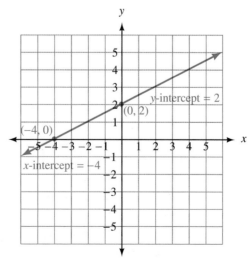

Figure 3

Graphing a line by finding the intercepts, as we have done in Examples 1 and 2, is an easy method of graphing if the equation has the form $ax + by = c$ and both the numbers a and b divide the number c evenly.

In our next example we use the intercepts to graph a line in which y is given in terms of x.

▶ **EXAMPLE 3**   Use the intercepts for $y = -\frac{1}{3}x + 2$ to draw its graph.

Solution   We graphed this line previously in Example 3 of Section 3.3 by substituting 3 different values of x into the equation and solving for y. This time we will graph the line by finding the intercepts.

x-intercept:

When $y = 0$

the equation $y = -\dfrac{1}{3}x + 2$

becomes $0 = -\dfrac{1}{3}x + 2$

$-2 = -\dfrac{1}{3}x$ Add -2 to each side

$6 = x$ Multiply each side by -3

The x-intercept is 6, which means the graph passes through the point $(6, 0)$.

y-intercept:

When $x = 0$

the equation $y = -\dfrac{1}{3}x + 2$

becomes $y = -\dfrac{1}{3}(0) + 2$

$y = 2$

The y-intercept is 2, which means the graph passes through the point $(0, 2)$.

The graph of $y = -\frac{1}{3}x + 2$ is shown in Figure 4. Compare this graph, and the method used to obtain it, with Example 3 in Section 3.3.

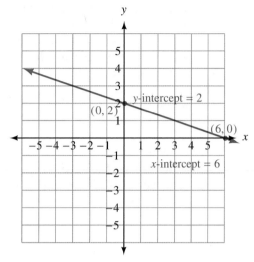

Figure 4

P R O B L E M   S E T   3 . 4

Find the x- and y-intercepts for the following equations. Then use the intercepts to graph each equation.

1. $2x + y = 4$ **2.** $2x + y = 2$

3. $-x + y = 3$ **4.** $-x + y = 4$

5. $-x + 2y = 2$ **6.** $-x + 2y = 4$

7. $5x + 2y = 10$ **8.** $2x + 5y = 10$

9. $4x - 2y = 8$ **10.** $2x - 4y = 8$

11. $-4x + 5y = 20$ **12.** $-5x + 4y = 20$

13. $y = 2x - 6$ **14.** $y = 2x + 6$

15. $y = 2x + 2$ **16.** $y = -2x + 2$

17. $y = 2x - 1$ **18.** $y = -2x - 1$

19. $y = \dfrac{1}{2}x + 3$ **20.** $y = \dfrac{1}{2}x - 3$

21. $y = -\dfrac{1}{3}x - 2$ **22.** $y = -\dfrac{1}{3}x + 2$

For each of the following lines the x-intercept and the y-intercept are both 0, which means the graph of each will go through the origin, (0, 0). Graph each line by finding a point on each, other than the origin, and then drawing a line through that point and the origin.

23. $y = -2x$ **24.** $y = \dfrac{1}{2}x$

25. $y = 2x$ **26.** $y = -\dfrac{1}{2}x$

27. $y = \dfrac{1}{3}x$ **28.** $y = 3x$

29. $y = -\dfrac{1}{3}x$ **30.** $y = -3x$

31. $y = \dfrac{2}{3}x$ **32.** $y = \dfrac{3}{2}x$

33. Graph the line that passes through the point $(-4, 4)$ and has an x-intercept of -2. What is the y-intercept of this line?

34. Graph the line that passes through the point $(-3, 4)$ and has a y-intercept of 3. What is the x-intercept of this line?

35. A line passes through the point (1, 4) and has a y-intercept of 3. Graph the line and name its x-intercept.

36. A line passes through the point $(3, 4)$ and has an x-intercept of 1. Graph the line and name its y-intercept.

37. Graph the line that passes through the points $(-2, 5)$ and $(5, -2)$. What are the x- and y-intercepts for this line?

38. Graph the line that passes through the points $(5, 3)$ and $(-3, -5)$. What are the x- and y-intercepts for this line?

39. The vertical line $x = 3$ has only one intercept. Graph $x = 3$ and name its intercept. (Remember, ordered pairs (x, y) that are solutions to the equation $x = 3$ are ordered pairs with an x-coordinate of 3 and any y-coordinate.)

40. Graph the vertical line $x = -2$. Then, name its intercept.

41. The horizontal line $y = 4$ has only one intercept. Graph $y = 4$ and name its intercept. (Ordered pairs (x, y) that are solutions to the equation $y = 4$ are ordered pairs with a y-coordinate of 4 and any x-coordinate.)

42. Graph the horizontal line $y = -3$. Then, name its intercept.

Review Problems

The problems below review material we covered in Section 1.6.

Simplify each expression.

43. $(-7)^2$

44. 2^3

45. $5 - 2 \cdot 6$

46. $7 - 4 \cdot 5$

47. $6 - 3(8 - 2)$

48. $7 - 2(3 - 5)$

49. $2(3)^2 - 4(3)^2$

50. $7(2)^3 - 3(3)^3$

51. $5 - 2[3 - 4(-2)]$

52. $6 - 3[5 - 2(-3)]$

3.5 The Slope of a Line

In defining the **slope** of a straight line, we are looking for a number to associate with a straight line that does two things. First of all, we want the slope of a line to measure the "steepness" of the line. That is, in comparing two lines, the slope of the steeper line should have the larger numerical value. Secondly, we want a line that *rises* going from left to right to have a **positive slope.** We want a line that *falls* going from left to right to have a **negative slope.** (A line that neither rises nor falls going from left to right must, therefore, have 0 slope.) These are illustrated in Figure 1.

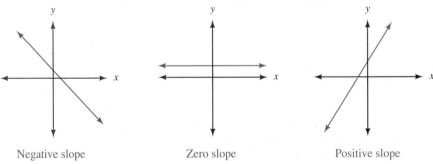

Negative slope   Zero slope   Positive slope

Figure 1

Suppose we know the coordinates of two points on a line. Since we are trying to develop a general formula for the slope of a line, we will use general points—call the two points P_1 (x_1, y_1) and P_2 (x_2, y_2). They represent the coordinates of any two different points on our line. We define the **slope** of our line to be the ratio of the vertical change to the horizontal change as we move from point (x_1, y_1) to point (x_2, y_2) on the line. (See Figure 2.)

Note   The 2 in x_2 is called a **subscript.** It is notation that allows us to distinguish between the variables x_1 and x_2, while still showing that they are both x-coordinates.

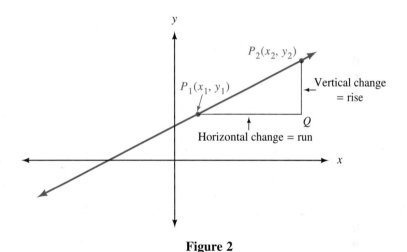

Figure 2

We call the vertical change the **rise** in the graph and the horizontal change the **run** in the graph. The slope, then, is

$$\text{Slope} = \frac{\text{vertical change}}{\text{horizontal change}} = \frac{\text{rise}}{\text{run}}$$

We would like to have a numerical value to associate with the rise in the graph and a numerical value to associate with the run in the graph. A quick study of Figure 2 shows that the coordinates of point Q must be (x_2, y_1), since Q is directly below point P_2 and right across from point P_1. We can draw our diagram again in the manner shown in Figure 3. It is apparent from this graph that the rise can be expressed as $(y_2 - y_1)$ and the run as $(x_2 - x_1)$. We usually denote the slope of a line by the letter m. The complete definition of slope follows along with a diagram (Figure 3) that illustrates the definition.

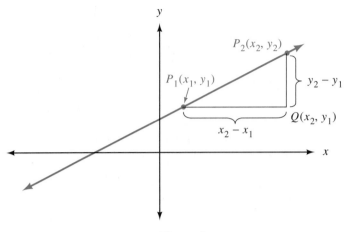

Figure 3

DEFINITION   If points (x_1, y_1) and (x_2, y_2) are any two different points, then the **slope** of the line on which they lie is

$$\text{Slope} = m = \frac{\text{rise}}{\text{run}} = \frac{y_2 - y_1}{x_2 - x_1}$$

This definition of the slope of a line does just what we want it to do. If the line rises going from left to right, the slope will be positive. If the line falls from left to right, the slope will be negative. Also, the steeper the line, the larger numerical value the slope will have.

▶ **EXAMPLE 1**   Find the slope of the line between the points (1, 2) and (3, 5).

Solution   We can let

$$(x_1, y_1) = (1, 2)$$

and

$$(x_2, y_2) = (3, 5)$$

then

$$m = \frac{y_2 - y_1}{x_2 - x_1} = \frac{5 - 2}{3 - 1} = \frac{3}{2}$$

The slope is $\frac{3}{2}$. For every vertical change of 3 units, there will be a corresponding horizontal change of 2 units. (See Figure 4.)

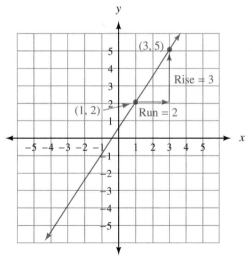

Figure 4

▶ **EXAMPLE 2**   Find the slope of the line through $(-2, 1)$ and $(5, -4)$.

Solution   It makes no difference which ordered pair we call (x_1, y_1) and which we call (x_2, y_2).

$$\text{Slope} = m = \frac{y_2 - y_1}{x_2 - x_1} = \frac{-4 - 1}{5 - (-2)} = \frac{-5}{7}$$

The slope is $-\frac{5}{7}$. Every vertical change of -5 units (down 5 units) is accompanied by a horizontal change of 7 units (to the right 7 units).

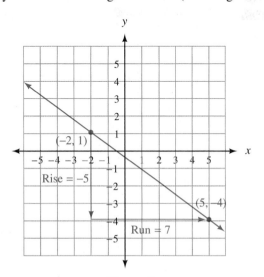

Figure 5

▶ **EXAMPLE 3** Graph the line with slope $\frac{3}{2}$ and y-intercept 1.

Solution Since the y-intercept is 1, we know that one point on the line is (0, 1). So, we begin by plotting the point (0, 1), as shown in Figure 6.

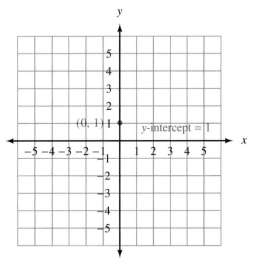

Figure 6

There are many lines that pass through the point shown in Figure 6, but only one of those lines has a slope of $\frac{3}{2}$. The slope, $\frac{3}{2}$, can be thought of as the rise in the graph divided by the run in the graph. Therefore, if we start at the point (0, 1) and move 3 units up (that's a rise of 3) and then 2 units to the right (a run of 2), we will be at another point on the graph.

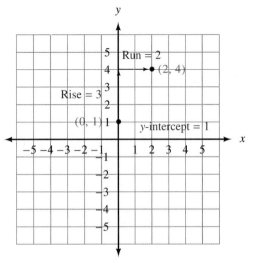

Figure 7

Figure 7 shows us that the point (2, 4) is another point on our line because it is 3 units up and 2 units over from the point (0, 1). To find the point (2, 4) we used the fact that our slope is given by

$$\text{Slope} = m = \frac{\text{rise}}{\text{run}} = \frac{3}{2}$$

To graph the line with slope $\frac{3}{2}$ and y-intercept 1, we simply draw a line through the two points in Figure 7 to obtain the graph shown in Figure 8.

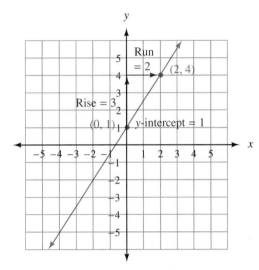

Figure 8

PROBLEM SET 3.5

Find the slope of the line through the following pairs of points. Then, plot each pair of points, draw a line through them, and indicate the rise and run in the graph in the same manner shown in Examples 1 and 2.

1. (2, 1), (4, 4)
2. (3, 1), (5, 4)
3. (1, 4), (5, 2)
4. (1, 3), (5, 2)
5. (1, −3), (4, 2)
6. (2, −3), (5, 2)
7. (−3, −2), (1, 3)
8. (−3, −1), (1, 4)
9. (−3, 2), (3, −2)
10. (−3, 3), (3, −1)

11. (2, −5), (3, −2)   **12.** (2, −4), (3, −1)

In each of the following problems, graph the line with the given slope and y-intercept, b.

13. $m = \dfrac{2}{3}, b = 1$ **14.** $m = \dfrac{3}{4}, b = -2$

15. $m = \dfrac{3}{2}, b = -3$ **16.** $m = \dfrac{4}{3}, b = 2$

17. $m = -\dfrac{4}{3}, b = 5$ **18.** $m = -\dfrac{3}{5}, b = 4$

19. $m = 2$, $b = 1$

20. $m = -2$, $b = 4$

21. $m = 3$, $b = -1$

22. $m = 3$, $b = -2$

23. Graph the line that has an x-intercept of 3 and a y-intercept of -2. What is the slope of this line?

24. Graph the line that has an x-intercept of 2 and a y-intercept of -3. What is the slope of this line?

25. Graph the line with x-intercept 4 and y-intercept 2. What is the slope of this line?

26. Graph the line with x-intercept -4 and y-intercept -2. What is the slope of this line?

27. Graph the line $y = 2x - 3$, then name the slope and y-intercept by looking at the graph.

28. Graph the line $y = -2x + 3$, then name the slope and y-intercept by looking at the graph.

29. Graph the line $y = \frac{1}{2}x + 1$, then name the slope and y-intercept by looking at the graph.

30. Graph the line $y = -\frac{1}{2}x - 2$, then name the slope and y-intercept by looking at the graph.

31. Find y if the line through $(4, 2)$ and $(6, y)$ has a slope of 2.

32. Find y if the line through $(1, y)$ and $(7, 3)$ has a slope of 6.

Review Problems

The following problems review material we covered in Section 2.5. Reviewing these problems will help you understand the material in the next section.

Solve each formula for the indicated variable.

33. $V = IR$ for I

34. $A = \frac{1}{2}bh$ for h

35. $P = 2l + 2w$ for l

36. $h = vt - 16t^2$ for v

37. $-2x + y = -4$ for y

38. $7x - y = 2$ for y

39. $5x + 4y = 8$ for y

40. $6x + 7y = 14$ for y

SECTION

3.6   **Finding the Equation of a Line**

To this point in the chapter, most of the problems we have worked have used the equation of a line to find different types of information about the line. For instance, given the equation of a line, we can find points on the line, the graph of the line, the intercepts, and the slope of the line. In this section we reverse things somewhat and move in the other direction; we will use information about a line, such as its slope and y-intercept, to find the equation of a line.

There are three main types of problems to solve in this section.

1. Find the equation of a line from the slope and y-intercept.
2. Find the equation of a line given one point on the line and the slope of the line.
3. Find the equation of a line given two points on the line.

Examples 1 and 2 illustrate the first type of problem. Example 5 solves the second type of problem. The third type of problem is solved in Example 6.

The Slope-Intercept Form of an Equation of a Straight Line

▶ **EXAMPLE 1**   Find the equation of the line with slope $\frac{3}{2}$ and y-intercept 1.

Solution   We graphed the line with slope $\frac{3}{2}$ and y-intercept 1 in the last example of the previous section. Figure 1 shows that graph.

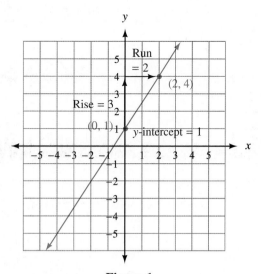

Figure 1

What we want to do now is find the equation of the line shown in Figure 1. To do so, we take any other point (x, y) on the line and apply our slope formula to that point and the point $(0, 1)$. We set that result equal to $\frac{3}{2}$, since $\frac{3}{2}$ is the slope of our line. The work is as follows, with a diagram of the situation on the next page.

$$\frac{y - 1}{x - 0} = \frac{3}{2} \qquad \text{Slope} = \frac{\text{vertical change}}{\text{horizontal change}}$$

$$\frac{y - 1}{x} = \frac{3}{2} \qquad x - 0 = x$$

$$y - 1 = \frac{3}{2}x \qquad \text{Multiply each side by } x$$

$$y = \frac{3}{2}x + 1 \qquad \text{Add 1 to each side}$$

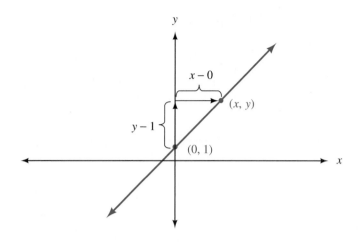

What is interesting and useful about the equation we have just found is that the number in front of x is the slope of the line and the constant term is the y-intercept. It is no coincidence that it turned out this way. Whenever an equation has the form $y = mx + b$, the graph is always a straight line with slope m and y-intercept b. To see that this is true in general, suppose we want the equation of a line with slope m and y-intercept b. Since the y-intercept is b, then the point $(0, b)$ is on the line. If (x, y) is any other point on the line, then we apply our slope formula to get

$$\frac{y - b}{x - 0} = m \qquad \text{Slope} = \frac{\text{vertical change}}{\text{horizontal change}}$$

$$\frac{y - b}{x} = m \qquad x - 0 = x$$

$$y - b = mx \qquad \text{Multiply each side by } x$$

$$y = mx + b \qquad \text{Add } b \text{ to each side}$$

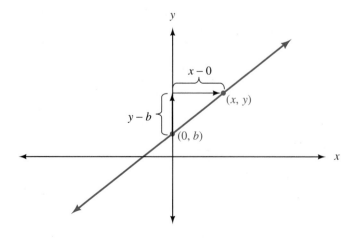

Here is a summary of what we have just found.

Slope-Intercept Form of the Equation of a Line

The equation of the line with slope m and y-intercept b is always given by

$$y = mx + b$$

▶ **EXAMPLE 2** Find the equation of the line with slope $-\frac{4}{3}$ and y-intercept 5. Then, graph the line.

Solution Substituting $m = -\frac{4}{3}$ and $b = 5$ into the equation $y = mx + b$, we have

$$y = -\frac{4}{3}x + 5$$

Finding the equation from the slope and y-intercept is just that easy. If the slope is m and the y-intercept is b, then the equation is always $y = mx + b$.

Since the y-intercept is 5, the graph goes through the point (0, 5). To find a second point on the graph, we start at (0, 5) and move 4 units down (that's a rise of -4) and 3 units to the right (a run of 3). The point we reach is (3, 1). Drawing a line that passes through (0, 5) and (3, 1), we have the graph of our equation. (Note that we could also let the rise $= 4$ and the run $= -3$ and obtain the same graph.) The graph is shown in Figure 2.

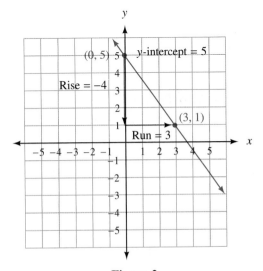

Figure 2

▶ **EXAMPLE 3** Find the slope and y-intercept for $-2x + y = -4$. Then, use them to draw the graph.

Solution To identify the slope and y-intercept from the equation, the equation must be in the form $y = mx + b$ (slope-intercept form). To write our equation in this form, we must solve the equation for y. To do so, we simply add $2x$ to each side of the equation.

$$-2x + y = -4 \qquad \text{Original equation}$$

$$y = 2x - 4 \qquad \text{Add } 2x \text{ to each side}$$

The equation is now in slope-intercept form, so the slope must be 2 and the y-intercept -4. The graph, therefore, crosses the y-axis at $(0, -4)$. Since the slope is 2, we can let the rise = 2 and the run = 1 and find a second point on the graph. The graph is shown in Figure 3.

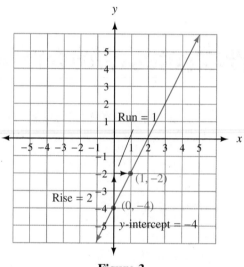

Figure 3   ◀

▶ **EXAMPLE 4** Find the slope and y-intercept for $3x - 2y = 6$.

Solution To find the slope and y-intercept from the equation, we must write the equation in the form $y = mx + b$. This means we must solve the equation $3x - 2y = 6$ for y.

$$3x - 2y = 6 \qquad \text{Original equation}$$

$$-2y = -3x + 6 \qquad \text{Add } -3x \text{ to each side}$$

$$-\frac{1}{2}(-2y) = -\frac{1}{2}(-3x + 6) \qquad \text{Multiply each side by } -\frac{1}{2}$$

$$y = \frac{3}{2}x - 3 \qquad \text{Simplify each side}$$

Now that the equation is written in slope-intercept form, we can identify the slope as $\frac{3}{2}$ and the y-intercept as -3. The graph is shown in Figure 4.

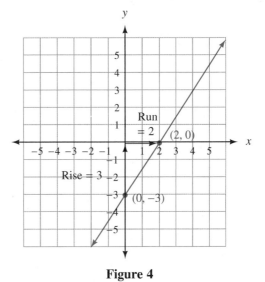

Figure 4   ◀

The Point-Slope Form of an Equation of a Straight Line

A second useful form of the equation of a straight line is the point-slope form.

Let line l contain the point (x_1, y_1) and have slope m. If (x, y) is any other point on l, then by the definition of slope, we have

$$\frac{y - y_1}{x - x_1} = m$$

Multiplying both sides by $(x - x_1)$ gives us

$$(x - x_1) \cdot \frac{y - y_1}{x - x_1} = m(x - x_1)$$

$$y - y_1 = m(x - x_1)$$

This last equation is known as the **point-slope form** of the equation of a straight line.

Point-Slope Form of the Equation of a Line

The equation of the line through (x_1, y_1) with slope m is given by

$$y - y_1 = m(x - x_1)$$

This form is used to find the equation of a line, either given one point on the line and the slope, or given two points on the line.

▶ **EXAMPLE 5** Find the equation of the line with slope -2 that contains the point $(-4, 3)$. Write the answer in slope-intercept form.

Solution

Using $(x_1, y_1) = (-4, 3)$ and $m = -2$

in $\quad y - y_1 = m(x - x_1) \qquad$ Point-slope form

gives us $\quad y - 3 = -2(x + 4) \qquad$ Note: $x - (-4) = x + 4$

$\qquad\qquad y - 3 = -2x - 8 \qquad$ Multiply out right side

$\qquad\qquad y = -2x - 5 \qquad$ Add 3 to each side

Figure 5 is the graph of the line that contains $(-4, 3)$ and has a slope of -2. Notice that the y-intercept on the graph matches that of the equation we found.

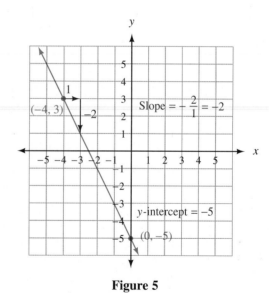

Figure 5

▶ **EXAMPLE 6** Find the equation of the line that passes through the points $(-3, 3)$ and $(3, -1)$.

Solution We begin by finding the slope of the line:

$$m = \frac{3 - (-1)}{-3 - 3} = \frac{4}{-6} = -\frac{2}{3}$$

Using $(x_1, y_1) = (3, -1)$ and $m = -\frac{2}{3}$ in $y - y_1 = m(x - x_1)$ yields

$$y + 1 = -\frac{2}{3}(x - 3) \qquad \text{Note: } y - (-1) = y + 1$$

$$y + 1 = -\frac{2}{3}x + 2 \qquad \text{Multiply out right side}$$

$$y = -\frac{2}{3}x + 1 \qquad \text{Add } -1 \text{ to each side}$$

Figure 6 shows the graph of the line that passes through the points $(-3, 3)$ and $(3, -1)$. As you can see, the slope and y-intercept are $-\frac{2}{3}$ and 1, respectively.

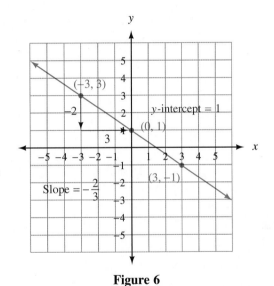

Figure 6

Note In Example 6, we could have used the point $(-3, 3)$ instead of $(3, -1)$ and obtained the same equation. That is, using $(x_1, y_1) = (-3, 3)$ and $m = -\frac{2}{3}$ in $y - y_1 = m(x - x_1)$ gives us

$$y - 3 = -\frac{2}{3}(x + 3)$$

$$y - 3 = -\frac{2}{3}x - 2$$

$$y = -\frac{2}{3}x + 1$$

which is the same result we obtained using $(3, -1)$.

Methods of Graphing Lines

1. Substitute convenient values of x into the equation and find the corresponding values of y. We used this method first for equations like $y = 2x - 3$. To use this method for equations that looked like $2x - 3y = 6$, we first solved them for y.
2. Find the x- and y-intercepts. This method works best for equations of the form $3x + 2y = 6$ where the numbers in front of x and y divide the constant term evenly.
3. Find the slope and y-intercept. This method works best when the equation has the form $y = mx + b$ and b is an integer.

PROBLEM SET 3.6

In each of the following problems, give the equation of the line with the given slope and y-intercept.

1. $m = \dfrac{2}{3}, b = 1$ **2.** $m = \dfrac{3}{4}, b = -2$

3. $m = \dfrac{3}{2}, b = -1$ **4.** $m = \dfrac{4}{3}, b = 2$

5. $m = -\dfrac{2}{5}, b = 3$ **6.** $m = -\dfrac{3}{5}, b = 4$

7. $m = 2, b = -4$ **8.** $m = -2, b = 4$
9. $m = -3, b = 2$ **10.** $m = 3, b = -2$

Find the slope and y-intercept for each of the following equations by writing them in the form $y = mx + b$. Then, graph each equation.

11. $-2x + y = 4$ **12.** $-2x + y = 2$
13. $3x + y = 3$ **14.** $3x + y = 6$
15. $3x + 2y = 6$ **16.** $2x + 3y = 6$
17. $4x - 5y = 20$ **18.** $2x - 5y = 10$
19. $-2x - 5y = 10$ **20.** $-4x + 5y = 20$

For each problem below, the slope and one point on a line are given. In each case, use the point-slope form to find the equation of that line. (Write your answers in slope-intercept form.)

21. $(-2, -5), m = 2$ **22.** $(-1, -5), m = 2$

23. $(-4, 1), m = -\dfrac{1}{2}$ **24.** $(-2, 1), m = -\dfrac{1}{2}$

25. $(2, -3), m = \dfrac{3}{2}$ **26.** $(3, -4), m = \dfrac{4}{3}$

27. $(-1, 4), m = -3$ **28.** $(-2, 5), m = -3$
29. $(2, 4), m = 1$ **30.** $(4, 2), m = -1$

Find the equation of the line that passes through each pair of points. Write your answers in slope-intercept form.

31. $(-2, -4), (1, -1)$ **32.** $(2, 4), (-3, -1)$
33. $(-1, -5), (2, 1)$ **34.** $(-1, 6), (1, 2)$
35. $(-3, -2), (3, 6)$ **36.** $(-3, 6), (3, -2)$
37. $(-3, -1), (3, -5)$ **38.** $(-3, -5), (3, 1)$
39. $(-2, 1), (1, -2)$ **40.** $(1, -2), (2, 1)$
41. Find the equation of the line with x-intercept 3 and y-intercept 2.
42. Find the equation of the line with x-intercept 2 and y-intercept 3.
43. Find the equation of the line with x-intercept -2 and y-intercept -5.
44. Find the equation of the line with x-intercept -3 and y-intercept -5.
45. The equation of the vertical line that passes through the points $(3, -2)$ and $(3, 4)$ is either $x = 3$ or $y = 3$. Which one is it?
46. Find the equation of the vertical line that passes through the points $(2, 5)$ and $(2, -3)$.
47. The equation of the horizontal line that passes through the points $(2, 3)$ and $(-1, 3)$ is either $x = 3$ or $y = 3$. Which one is it?
48. Find the equation of the horizontal line that passes through the points $(-4, -2)$ and $(3, -2)$.

Review Problems

The problems that follow review material we covered in Sections 2.5, 3.3, and 3.4.

Translate each of the following percent problems into an equation and then solve the equation. [2.5]

49. What number is 25% of 300?
50. What number is 40% of 250?
51. 25 is what percent of 125?

52. 30 is what percent of 120?
53. 60 is 15% of what number?
54. 75 is 30% of what number?

Graph each of the following lines. [3.3, 3.4]

55. $x + y = 4$
56. $x - y = -2$
57. $y = 2x - 3$
58. $y = 2x + 3$
59. $y = 2x$
60. $y = -2x$

SECTION

3.7   # Solving Linear Systems by Graphing

Two linear equations considered at the same time make up what is called a **system of linear equations.** Both equations contain two variables and, of course, have graphs that are straight lines. The following are systems of linear equations:

$$x + y = 3 \qquad y = 2x + 1 \qquad 2x - y = 1$$
$$3x + 4y = 2 \qquad y = 3x + 2 \qquad 3x - 2y = 6$$

The **solution set** for a *system* of linear equations is all ordered pairs that are solutions to both equations. Since each linear equation has a graph that is a straight line, we can expect the intersection of the graphs to be a point whose coordinates are solutions to the system. That is, if we graph both equations on the same coordinate system, we can read the coordinates of the point of intersection and have the solution to our system. Here is an example.

▶ **EXAMPLE 1** Solve the following system by graphing:

$$x + y = 4$$
$$x - y = -2$$

Solution On the same set of coordinate axes we graph each equation separately. Figure 1 shows both graphs, without showing the work necessary to get them. We can see from the graphs that they intersect at the point (1, 3). The point (1, 3) must therefore be the solution to our system, since it is the only ordered pair whose graph lies on both lines. Its coordinates satisfy both equations.

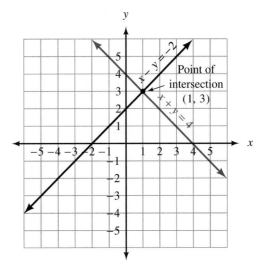

Figure 1

We can check our results by substituting the coordinates $x = 1$, $y = 3$ into both equations to see if they work.

When	$x = 1$	When	$x = 1$
and	$y = 3$	and	$y = 3$
the equation	$x + y = 4$	the equation	$x - y = -2$
becomes	$1 + 3 \overset{?}{=} 4$	becomes	$1 - 3 \overset{?}{=} -2$
or	$4 = 4$	or	$-2 = -2$

The point $(1, 3)$ satisfies both equations.   ◀

Here are some steps to follow in solving linear systems by graphing.

Strategy for Solving a Linear System by Graphing

Step 1: Graph the first equation by the methods described in Section 3.3 or 3.4.

Step 2: Graph the second equation on the same set of axes used for the first equation.

Step 3: Read the coordinates of the point of intersection of the two graphs. The ordered pair is the solution to the system.

Step 4: Check the solution in both equations, if necessary.

▶ **EXAMPLE 2** Solve the following system by graphing:

$$x + 2y = 8$$

$$2x - 3y = 2$$

Solution We graph each equation on the same set of coordinate axes and get the following:

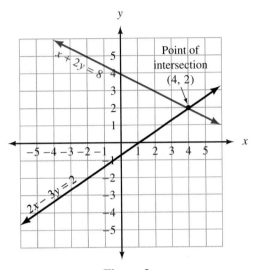

Figure 2

From the graph, we can see the solution for our system is (4, 2). We check this solution as follows.

When $\quad\quad\quad\quad x = 4$ When $\quad\quad\quad\quad\quad\quad\quad x = 4$

and $\quad\quad\quad\quad\quad y = 2$ and $\quad\quad\quad\quad\quad\quad\quad\quad y = 2$

the equation $\quad x + 2y = 8$ the equation $\quad\quad 2x - 3y = 2$

becomes $\quad 4 + 2(2) \overset{?}{=} 8$ becomes $\quad\quad 2(4) - 3(2) \overset{?}{=} 2$

$$4 + 4 \overset{?}{=} 8 \quad\quad\quad\quad\quad\quad\quad 8 - 6 \overset{?}{=} 2$$

$$8 = 8 \quad\quad\quad\quad\quad\quad\quad\quad\quad 2 = 2$$

The point (4, 2) satisfies both equations and, therefore, must be the solution to our system. ◀

▶ **EXAMPLE 3** Solve this system by graphing:

$$y = 2x - 3$$

$$x = 3$$

Solution Graphing both equations on the same set of axes, we have Figure 3.

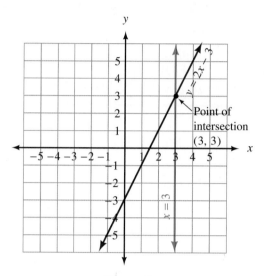

Figure 3

The solution to the system is the point (3, 3).   ◀

▶ **EXAMPLE 4** Solve by graphing:

$$y = x - 2$$
$$y = x + 1$$

Solution Graphing both equations, we have the following:

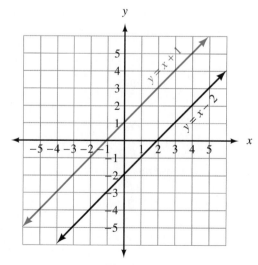

Figure 4

We can see in Figure 4 that the lines are parallel and therefore do not intersect. Our system has no ordered pair as a solution, since there is no ordered pair that satisfies both equations. We say the solution set is the empty set and write Ø.
◀

Example 4 is one example of two special cases associated with linear systems. The other special case happens when the two graphs coincide. Here is an example:

▶ **EXAMPLE 5**   Graph the system

$$2x + y = 4$$

$$4x + 2y = 8$$

Solution   Both graphs are shown in Figure 5.

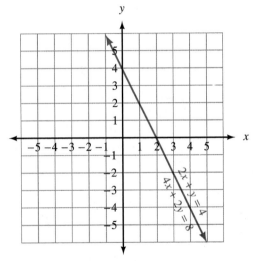

Figure 5

The two graphs coincide. The reason becomes apparent when we multiply both sides of the first equation by 2:

$$2x + y = 4$$

$$2(2x + y) = 2(4) \qquad \text{Multiply both sides by 2}$$

$$4x + 2y = 8$$

The equations have the same solution set. Any ordered pair that is a solution to one is a solution to the system. The system has an infinite number of solutions. (Any point on the line is a solution to the system.) ◀

The two special cases illustrated in the last two examples do not happen often. Usually a system has a single ordered pair as a solution. Solving a system of linear equations by graphing is useful only when the ordered pair in the solution set has integers for coordinates. There are two other solution methods that work well in all cases. We will develop the other two methods in the next two sections.

PROBLEM SET 3.7

Solve the following systems of linear equations by graphing.

1. $x + y = 3$
  $x - y = 1$

2. $x + y = 2$
  $x - y = 4$

3. $x + y = 1$
  $-x + y = 3$

4. $x + y = 1$
  $x - y = -5$

5. $x + y = 8$
  $-x + y = 2$

6. $x + y = 6$
  $-x + y = -2$

7. $3x - 2y = 6$
  $x - y = 1$

8. $5x - 2y = 10$
  $x - y = -1$

9. $6x - 2y = 12$
  $3x + y = -6$

10. $4x - 2y = 8$
  $2x + y = -4$

11. $4x + y = 4$
  $3x - y = 3$

12. $5x - y = 10$
  $2x + y = 4$

13. $x + 2y = 0$
  $2x - y = 0$

14. $3x + y = 0$
  $5x - y = 0$

15. $3x - 5y = 15$
  $-2x + y = 4$

16. $2x - 4y = 8$
  $2x - y = -1$

17. $y = 2x + 1$
  $y = -2x - 3$

18. $y = 3x - 4$
  $y = -2x + 1$

19. $x + 3y = 3$
  $y = x + 5$

20. $2x + y = -2$
  $y = x + 4$

21. $x + y = 2$
  $x = -3$

22. $x + y = 6$
  $y = 2$

23. $x = -4$
  $y = 6$

24. $x = 5$
  $y = -1$

25. $x + y = 4$
  $2x + 2y = -6$

26. $x - y = 3$
  $2x - 2y = 6$

27. $4x - 2y = 8$
  $2x - y = 4$

28. $3x - 6y = 6$
  $x - 2y = 4$

29. As you have probably guessed by now, it can be difficult to solve a system of equations by graphing if the solution to the system contains a fraction. The solution to the following system is $(\frac{1}{2}, 1)$. Solve the system by graphing.

$$y = -2x + 2$$
$$y = 4x - 1$$

30. The solution to the system below is $(\frac{1}{3}, -2)$. Solve the system by graphing.

$$y = 3x - 3$$
$$y = -3x - 1$$

31. A second difficulty can arise in solving a system of equations by graphing if one or both of the equations is difficult to graph. The solution to the system below is $(2, 1)$. Solve the system by graphing.

$$3x - 8y = -2$$
$$x - y = 1$$

32. The solution to the system below is $(-3, 2)$. Solve the system by graphing.

$$2x + 5y = 4$$
$$x - y = -5$$

Review Problems

The following problems review material we covered in Section 2.1.

Find the value of each expression when x is -3.

33. $2x - 9$

34. $-4x + 3$

35. $9 - 6x$

36. $7 - 5x$

37. $4(3x + 2) + 1$

38. $3(6x + 5) + 10$

39. $2x^2 + 3x + 4$

40. $4x^2 + 3x + 2$

The Elimination Method

The addition property of equality states that if equal quantities are added to both sides of an equation the solution set is unchanged. In the past we have used this property to help solve equations in one variable. We will now use it to solve systems of linear equations. Here is another way to state the addition property of equality.

Let A, B, C, and D represent algebraic expressions.

$$\text{If} \qquad A = B$$

$$\text{and} \qquad \underline{C = D}$$

$$\text{then} \quad A + C = B + D$$

Since C and D are equal (that is, they represent the same number), what we have done is added the same amount to both sides of the equation $A = B$. Let's see how we can use this form of the addition property of equality to solve a system of linear equations.

▶ **EXAMPLE 1** Solve the following system:

$$x + y = 4$$

$$x - y = 2$$

Solution The system is written in the form of the addition property of equality as written in this section. It looks like

$$A = B$$

$$C = D$$

where A is $x + y$, B is 4, C is $x - y$, and D is 2.

We use the addition property of equality to add the left sides together and the right sides together:

$$x + y = 4$$

$$\underline{x - y = 2}$$

$$2x + 0 = 6$$

We now solve the resulting equation for x:

$$2x + 0 = 6$$

$$2x = 6$$

$$x = 3$$

The value we get for x is the value of the x-coordinate of the point of intersection of the two lines $x + y = 4$ and $x - y = 2$. To find the y-coordinate we simply substitute $x = 3$ into either of the two original equations and get

$$3 + y = 4$$
$$y = 1$$

The solution to our system is the ordered pair $(3, 1)$. It satisfies both equations.

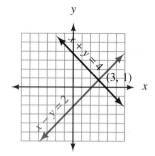

Figure 1

When	$x = 3$	When	$x = 3$
and	$y = 1$	and	$y = 1$
the equation	$x + y = 4$	the equation	$x - y = 2$
becomes	$3 + 1 \overset{?}{=} 4$	becomes	$3 - 1 \overset{?}{=} 2$
or	$4 = 4$	or	$2 = 2$

Figure 1 is visual evidence that the solution to our system is $(3, 1)$. ◀

The most important part of this method of solving linear systems is eliminating one of the variables when we add the left and right sides together. In our first example the equations were written so that the y variable was eliminated when we added the left and right sides together. If the equations are not set up this way to begin with, we have to work on one or both of them separately before we can add them together to eliminate one variable.

▶ **EXAMPLE 2** Solve the following system:

$$x + 2y = 4$$
$$x - y = -5$$

Solution Notice that if we were to add the equations together as they are, the resulting equation would have terms in both x and y. Let's eliminate the variable x by multiplying both sides of the second equation by -1 before we add the equations together. (As you will see, we can choose to eliminate either the x or the y variable.) Multiplying both sides of the second equation by -1 will not change its solution, so we do not need to be concerned that we have altered the system.

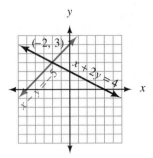

Figure 2

$$x + 2y = 4 \xrightarrow{\text{No change}} x + 2y = 4 \qquad \text{Add left and right}$$
$$x - y = -5 \xrightarrow[\text{Multiply by } -1]{} -x + y = 5 \qquad \text{sides to get}$$
$$\overline{}$$
$$0 + 3y = 9 \longleftarrow$$
$$3y = 9 \quad \left\{ \begin{array}{l} y\text{-coordinate of} \\ \text{the point of} \\ \text{intersection} \end{array} \right.$$
$$y = 3$$

Substituting $y = 3$ into either of the two original equations, we get $x = -2$. The solution to the system is $(-2, 3)$. It satisfies both equations. Figure 2 shows the solution to the system as the point where the two lines cross. ◄

▶ **EXAMPLE 3** Solve the following system:

$$2x - y = 6$$
$$x + 3y = 3$$

Solution Let's eliminate the y variable from the two equations. We can do this by multiplying the first equation by 3 and leaving the second equation unchanged:

$$2x - y = 6 \xrightarrow{\text{3 times both sides}} 6x - 3y = 18$$
$$x + 3y = 3 \xrightarrow[\text{No change}]{} x + 3y = 3$$

The important thing about our system now is that the coefficients (the numbers in front) of the y variables are opposites. When we add the terms on each side of the equal sign, then the terms in y will add to zero and be eliminated:

$$6x - 3y = 18$$
$$\underline{x + 3y = 3}$$
$$7x = 21 \qquad \text{Add corresponding terms}$$

which gives us $x = 3$. Using this value of x in the second equation of our original system, we have

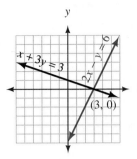

Figure 3

$$3 + 3y = 3$$
$$3y = 0$$
$$y = 0$$

We could substitute $x = 3$ into any of the equations with both x and y variables and also get $y = 0$. The solution to our system is the ordered pair $(3, 0)$. Figure 3 is a picture of the system of equations showing the solution $(3, 0)$. ◀

Note The graphs shown in the margin next to our first three examples are not part of the solution shown in each example. The graphs are there simply to show you that the results we obtain by the elimination method are consistent with the results we would obtain by graphing.

▶ **EXAMPLE 4** Solve the system:

$$2x + 3y = -1$$
$$3x + 5y = -2$$

Solution Let's eliminate x from the two equations. If we multiply the first equation by 3 and the second by -2, the coefficients of x will be 6 and -6, respectively. The x terms in the two equations will then add to zero:

$$2x + 3y = -1 \xrightarrow{\text{Multiply by 3}} 6x + 9y = -3$$
$$3x + 5y = -2 \xrightarrow[\text{Multiply by } -2]{} -6x - 10y = 4$$

We now add the left and right sides of our new system together:

$$\begin{array}{r} 6x + 9y = -3 \\ -6x - 10y = 4 \\ \hline -y = 1 \\ y = -1 \end{array}$$

Substituting $y = -1$ into the first equation in our original system, we have

$$2x + 3(-1) = -1$$
$$2x - 3 = -1$$
$$2x = 2$$
$$x = 1$$

The solution to our system is $(1, -1)$. It is the only ordered pair that satisfies both equations.   ◀

▶ **EXAMPLE 5**   Solve the system:

$$3x + 5y = -7$$

$$5x + 4y = 10$$

Solution   Let's eliminate y by multiplying the first equation by -4 and the second equation by 5:

$$3x + 5y = -7 \xrightarrow{\text{Multiply by } -4} -12x - 20y = 28$$

$$5x + 4y = 10 \xrightarrow[\text{Multiply by 5}]{} \underline{\quad 25x + 20y = 50}$$

$$13x \qquad = 78$$

$$x = 6$$

Substitute $x = 6$ into either equation in our original system, and the result will be $y = -5$. The solution is therefore $(6, -5)$.   ◀

▶ **EXAMPLE 6**   Solve the system:

$$\frac{1}{2}x - \frac{1}{3}y = 2$$

$$\frac{1}{4}x + \frac{2}{3}y = 6$$

Solution   Although we could solve this system without clearing the equations of fractions, there is probably less chance for error if we have only integer coefficients to work with. So let's begin by multiplying both sides of the top equation by 6, and both sides of the bottom equation by 12, to clear each equation of fractions:

$$\frac{1}{2}x - \frac{1}{3}y = 2 \xrightarrow{\text{Multiply by 6}} 3x - 2y = 12$$

$$\frac{1}{4}x + \frac{2}{3}y = 6 \xrightarrow[\text{Multiply by 12}]{} 3x + 8y = 72$$

Now we can eliminate x by multiplying the top equation by -1 and leaving the bottom equation unchanged:

$$3x - 2y = 12 \xrightarrow{\text{Multiply by } -1} -3x + 2y = -12$$

$$3x + 8y = 72 \xrightarrow[\text{No change}]{} \underline{\quad 3x + 8y = 72}$$

$$10y = 60$$

$$y = 6$$

We can substitute $y = 6$ into any equation that contains both x and y. Let's use $3x - 2y = 12$.

$$3x - 2(6) = 12$$
$$3x - 12 = 12$$
$$3x = 24$$
$$x = 8$$

The solution to the system is (8, 6).   ◀

Our next two examples will show what happens when we apply the elimination method to a system of equations consisting of parallel lines and to a system in which the lines coincide.

▶ **EXAMPLE 7**   Solve the system:

$$2x - y = 2$$
$$4x - 2y = 12$$

Solution   Let us choose to eliminate y from the system. We can do this by multiplying the first equation by -2 and leaving the second equation unchanged:

$$2x - y = 2 \xrightarrow{\text{Multiply by } -2} -4x + 2y = -4$$
$$4x - 2y = 12 \xrightarrow[\text{No change}]{} 4x - 2y = 12$$

If we add both sides of the resulting system, we have

$$-4x + 2y = -4$$
$$\underline{4x - 2y = 12}$$
$$0 + 0 = 8$$
$$\text{or} \quad 0 = 8 \quad \text{A false statement}$$

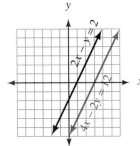

Figure 4

Both variables have been eliminated and we end up with the false statement $0 = 8$. We have tried to solve a system that consists of two parallel lines.

There is no solution, and that is the reason we end up with a false statement. Figure 4 is a visual representation of the situation and is conclusive evidence that there is no solution to our system.   ◀

▶ **EXAMPLE 8**   Solve the system:

$$4x - 3y = 2$$
$$8x - 6y = 4$$

Solution   Multiplying the top equation by -2 and adding, we can eliminate the variable x:

$$4x - 3y = 2 \xrightarrow{\text{Multiply by }-2} -8x + 6y = -4$$
$$8x - 6y = 4 \xrightarrow[\text{No change}]{} \underline{8x - 6y = 4}$$
$$0 = 0$$

Both variables have been eliminated and the resulting statement $0 = 0$ is true. In this case the lines coincide because the equations are equivalent. The solution set consists of all ordered pairs that satisfy either equation.   ◀

The preceding two examples illustrate the two special cases in which the graphs of the equations in the system either coincide or are parallel.

Here is a summary of our results from these two examples:

Both variables eliminated and the resulting statement false	↔	The lines are parallel and there is no solution to the system
Both variables eliminated and the resulting statement true	↔	The lines coincide and there is an infinite number of solutions to the system

The main idea in solving a system of linear equations by the elimination method is to use the multiplication property of equality on one or both of the original equations, if necessary, to make the coefficients of either variable opposites. On the following page are some steps to follow when solving a system of linear equations by the elimination method.

Strategy for Solving a System of Linear Equations by the Elimination Method

Step 1: Decide which variable to eliminate. (In some cases one variable will be easier to eliminate than the other. With some practice you will notice which one it is.)

Step 2: Use the multiplication property of equality on each equation separately to make the coefficients of the variable that is to be eliminated opposites.

Step 3: Add the respective left and right sides of the system together.

Step 4: Solve for the variable remaining.

Step 5: Substitute the value of the variable from step 4 into an equation containing both variables and solve for the other variable.

Step 6: Check your solution in both equations, if necessary.

PROBLEM SET 3.8

Solve the following systems of linear equations by elimination.

1. $x + y = 3$
$\quad x - y = 1$

2. $x + y = -2$
$\quad x - y = 6$

3. $\quad x + y = 10$
$\quad -x + y = 4$

4. $\quad x - y = 1$
$\quad -x - y = -7$

5. $\quad x - y = 7$
$\quad -x - y = 3$

6. $\quad x - y = 4$
$\quad 2x + y = 8$

7. $x + y = -1$
$\quad 3x - y = -3$

8. $\quad 2x - y = -2$
$\quad -2x - y = 2$

9. $\quad 3x + 2y = 1$
$\quad -3x - 2y = -1$

10. $-2x - 4y = 1$
$\quad 2x + 4y = -1$

Solve each of the following systems by eliminating the y variable.

11. $3x - y = 4$
$\quad 2x + 2y = 24$

12. $2x + y = 3$
$\quad 3x + 2y = 1$

13. $\quad 5x - 3y = -2$
$\quad 10x - y = 1$

14. $4x - y = -1$
$\quad 2x + 4y = 13$

15. $11x - 4y = 11$
$\quad 5x + y = 5$

16. $\quad 3x - y = 7$
$\quad 10x - 5y = 25$

Solve each of the following systems by eliminating the x variable.

17. $\quad 3x - 5y = 7$
$\quad -x + y = -1$

18. $4x + 2y = 32$
$\quad x + y = -2$

19. $\quad -x - 8y = -1$
$\quad -2x + 4y = 13$

20. $\quad -x + 10y = 1$
$\quad -5x + 15y = -9$

21. $-3x - y = 7$
$\quad 6x + 7y = 11$

22. $-5x + 2y = -6$
$\quad 10x + 7y = 34$

Solve each of the following systems of linear equations by the elimination method.

23. $6x - y = -8$
$\quad 2x + y = -16$

24. $5x - 3y = -3$
$\quad 3x + 3y = -21$

25. $\quad x + 3y = 9$
$\quad 2x - y = 4$

26. $\quad x + 2y = 0$
$\quad 2x - y = 0$

27. $\quad x - 6y = 3$
$\quad 4x + 3y = 21$

28. $8x + y = -1$
$\quad 4x - 5y = 16$

29. $2x + 9y = 2$
$\quad 5x + 3y = -8$

30. $5x + 2y = 11$
$\quad 7x + 8y = 7$

31. $\dfrac{1}{3}x + \dfrac{1}{4}y = \dfrac{7}{6}$

$\dfrac{3}{2}x - \dfrac{1}{3}y = \dfrac{7}{3}$

32. $\dfrac{7}{12}x - \dfrac{1}{2}y = \dfrac{1}{6}$

$\dfrac{2}{5}x - \dfrac{1}{3}y = \dfrac{11}{15}$

33. $3x + 2y = -1$
$6x + 4y = 0$

34. $8x - 2y = 2$
$4x - \ \ y = 2$

35. $11x + 6y = 17$
$5x - 4y = 1$

36. $\ \ 3x - 8y = 7$
$10x - 5y = 45$

37. $\dfrac{1}{2}x + \dfrac{1}{6}y = \dfrac{1}{3}$

$-x - \dfrac{1}{3}y = -\dfrac{1}{6}$

38. $-\dfrac{1}{3}x - \dfrac{1}{2}y = -\dfrac{2}{3}$

$-\dfrac{2}{3}x - \ \ y = -\dfrac{4}{3}$

39. For some systems of equations it is necessary to apply the addition property of equality to each equation to line up the x variables and y variables before trying to eliminate a variable. Solve the following system by first writing each equation so that the variable terms with x in them come first, the variable terms with y in them come second, and the constant terms are on the right side of each equation.

$$4x - 5y = 17 - 2x$$

$$5y = 3x + 4$$

40. Solve the following system by first writing each equation so that the variable terms with x in

them come first, the variable terms with y in them come second, and the constant terms are on the right side of each equation.

$$3x - 6y = -20 + 7x$$

$$4x = 3y - 34$$

41. Multiply both sides of the second equation in the following system by 100 and then solve as usual.

$$x + y = 22$$

$$0.05x + 0.10y = 1.70$$

42. Multiply both sides of the second equation in the following system by 100 and then solve as usual.

$$x + y = 15,000$$

$$0.06x + 0.07y = 980$$

Review Problems

The problems below review material we covered in Section 2.8.

Solve each inequality.

43. $x - 3 < 2$

44. $x + 4 \le 6$

45. $-3x \ge 12$

46. $-2x > 10$

47. $-\dfrac{x}{3} \le -1$

48. $-\dfrac{x}{5} < -2$

49. $-4x + 1 < 17$

50. $-3x + 2 \le -7$

SECTION

3.9

The Substitution Method

There is a third method of solving systems of equations. It is the substitution method, and, like the elimination method, it can be used on any system of linear equations. Some systems, however, lend themselves more to the substitution method than others do.

▶ **EXAMPLE 1** Solve the following system:

$$x + y = 2$$

$$y = 2x - 1$$

Solution If we were to solve this system by the methods used in the last section, we would have to rearrange the terms of the second equation so that similar terms would be in the same column. There is no need to do this, however, since the second equation tells us that y is $2x - 1$. We can replace the y variable in the first equation with the expression $2x - 1$ from the second equation. That is, we **substitute** $2x - 1$ from the second equation for y in the first equation. Here is what it looks like:

$$x + (2x - 1) = 2$$

The equation we end up with contains only the variable x. The y variable has been eliminated by substitution.

Solving the resulting equation, we have

$$x + (2x - 1) = 2$$

$$3x - 1 = 2$$

$$3x = 3$$

$$x = 1$$

This is the x-coordinate of the solution to our system. To find the y-coordinate, we substitute $x = 1$ into the second equation of our system. (We could substitute $x = 1$ into the first equation also and have the same result.)

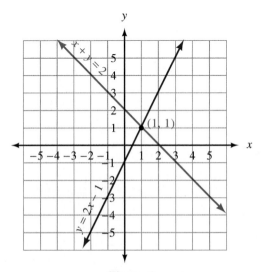

Figure 1

$$y = 2(1) - 1$$
$$y = 2 - 1$$
$$y = 1$$

The solution to our system is the ordered pair (1, 1). It satisfies both of the original equations. Figure 1 provides visual evidence that the substitution method yields the correct solution.

▶ **EXAMPLE 2**   Solve the following system by the substitution method:

$$2x - 3y = 12$$
$$y = 2x - 8$$

Solution   The second equation indicates that y is $2x - 8$. Since we are looking for the ordered pair that satisfies both equations, the y in the first equation must also be $2x - 8$. Substituting $2x - 8$ from the second equation for y in the first equation, we have

$$2x - 3(2x - 8) = 12$$

This equation can still be read as $2x - 3y = 12$, because $2x - 8$ is the same as y. Solving the equation, we have

$$2x - 3(2x - 8) = 12$$
$$2x - 6x + 24 = 12$$
$$-4x + 24 = 12$$
$$-4x = -12$$
$$x = 3$$

To find the y-coordinate of our solution, we substitute $x = 3$ into the second equation in the original system.

When $x = 3$

the equation $y = 2x - 8$

becomes $y = 2(3) - 8$

               $y = 6 - 8 = -2$

The solution to our system is (3, −2).

▶ **EXAMPLE 3**   Solve the following system by solving the first equation for x and then using the substitution method:

$$x - 3y = -1$$
$$2x - 3y = 4$$

Solution   We solve the first equation for x by adding $3y$ to both sides to get

$$x = 3y - 1$$

Using this value of x in the second equation, we have

$$2(3y - 1) - 3y = 4$$

$$6y - 2 - 3y = 4$$

$$3y - 2 = 4$$

$$3y = 6$$

$$y = 2$$

Next, we find x.

When $\qquad$ $y = 2$

the equation $\qquad$ $x = 3y - 1$

becomes $\qquad$ $x = 3(2) - 1$

$\qquad\qquad\qquad\quad$ $x = 6 - 1$

$\qquad\qquad\qquad\quad$ $x = 5$

The solution to our system is (5, 2). ◀

Here are the steps to use in solving a system of equations by the substitution method.

Strategy for Solving a System of Equations by the Substitution Method

Step 1: Solve either one of the equations for x or y. (This step is not necessary if one of the equations is already in the correct form, as in Examples 1 and 2.)

Step 2: Substitute the expression for the variable obtained in step 1 into the other equation and solve it.

Step 3: Substitute the solution from step 2 into any equation in the system that contains both variables and solve it.

Step 4: Check your results, if necessary.

▶ **EXAMPLE 4**   Solve by substitution:

$$-2x + 4y = 14$$

$$-3x + y = 6$$

Solution   We can solve either equation for either variable. If we look at the system closely, it becomes apparent that solving the second equation for y is the easiest way to go. If we add $3x$ to both sides of the second equation, we have

$$y = 3x + 6$$

Substituting the expression $3x + 6$ back into the first equation in place of y yields the following result:

$$-2x + 4(3x + 6) = 14$$
$$-2x + 12x + 24 = 14$$
$$10x + 24 = 14$$
$$10x = -10$$
$$x = -1$$

Substituting $x = -1$ into the equation $y = 3x + 6$ leaves us with

$$y = 3(-1) + 6$$
$$y = -3 + 6$$
$$y = 3$$

The solution to our system is $(-1, 3)$.   ◀

▶ **EXAMPLE 5**   Solve by substitution:

$$4x + 2y = 8$$
$$y = -2x + 4$$

Solution   Substituting the expression $-2x + 4$ for y from the second equation into the first equation, we have

$$4x + 2(-2x + 4) = 8$$
$$4x - 4x + 8 = 8$$
$$8 = 8 \qquad \text{A true statement}$$

Both variables have been eliminated and we are left with a true statement. Recall from the last section that a true statement in this situation tells us the lines coincide. That is, the equations $4x + 2y = 8$ and $y = -2x + 4$ have exactly the same graph. Any point on that graph has coordinates that satisfy both equations and is a solution to the system.   ◀

▶ **EXAMPLE 6**   Below are two monthly contract rates charged by GTE Mobilnet for cellular phone use in 1993. At how many minutes will the two rates cost the same amount?

	Flat Rate	plus	Per Minute Charge
Plan 1	$15		$1.50
Plan 2	$24.95		$0.75

Solution If we let y = the monthly charge for x minutes of phone use, then the equations for each plan are

$$\text{Plan 1: } y = 1.5x + 15$$

$$\text{Plan 2: } y = 0.75x + 24.95$$

We can solve this system by substitution by replacing the variable y in Plan 2 with the expression $1.5x + 15$ from Plan 1. If we do so we have

$$1.5x + 15 = 0.75x + 24.95$$

$$0.75x + 15 = 24.95$$

$$0.75x = 9.95$$

$$x = 13.27 \quad \text{to the nearest hundredth}$$

The monthly bill is based on the number of minutes you use the phone, with any fraction of a minute moving you up to the next minute. If you talk for a total of 13 minutes, you are billed for 13 minutes. If you talk for 13 minutes, 10 seconds, you are billed for 14 minutes. The number of minutes on your bill will always be a whole number. So, to calculate the cost for talking 13.27 minutes we would replace x with 14 and find y. Let's compare the two plans at $x = 13$ minutes and at $x = 14$ minutes.

$$\text{Plan 1: } y = 1.5x + 15$$

$$\text{When } x = 13, y = \$34.50$$

$$\text{When } x = 14, y = \$36.00$$

$$\text{Plan 2: } y = 0.75x + 24.95$$

$$\text{When } x = 13, y = \$34.70$$

$$\text{When } x = 14, y = \$35.45$$

The two plans will never give the same cost for talking x minutes. If you talk 13 or less minutes, Plan 1 will cost less. If you talk for more than 13 minutes, Plan 2 will cost less than Plan 1. ◀

PROBLEM SET 3.9

Solve the following systems by substitution. Substitute the expression in the second equation into the first equation and solve.

1. $x + y = 11$
$\quad y = 2x - 1$

2. $x - y = -3$
$\quad y = 3x + 5$

3. $x + y = 20$
$\quad y = 5x + 2$

4. $3x - y = -1$
$\quad x = 2y - 7$

5. $-2x + y = -1$
$\quad y = -4x + 8$

6. $4x - y = 5$
$\quad y = -4x + 1$

7. $3x - 2y = -2$
$x = -y + 6$
8. $2x - 3y = 17$
$x = -y + 6$
9. $5x - 4y = -16$ **10.** $6x + 2y = 18$
$y = 4$ $x = 3$
11. $5x + 4y = 7$ **12.** $10x + 2y = -6$
$y = -3x$ $y = -5x$

Solve the following systems by solving one of the equations for x or y and then using the substitution method.

13. $x + 3y = 4$ **14.** $x - y = 5$
$x - 2y = -1$ $x + 2y = -1$
15. $2x + y = 1$ **16.** $2x - 2y = 2$
$x - 5y = 17$ $x - 3y = -7$
17. $3x + 5y = -3$ **18.** $2x - 4y = -4$
$x - 5y = -5$ $x + 2y = 8$
19. $5x + 3y = 0$ **20.** $x - 3y = -5$
$x - 3y = -18$ $x - 2y = 0$
21. $-3x - 9y = 7$ **22.** $2x + 6y = -18$
$x + 3y = 12$ $x + 3y = -9$

Solve the following systems using the substitution method.

23. $5x - 8y = 7$
$y = 2x - 5$
24. $3x + 4y = 10$
$y = 8x - 15$
25. $7x - 6y = -1$
$x = 2y - 1$
26. $4x + 2y = 3$
$x = 4y - 3$

27. $-3x + 2y = 6$ **28.** $-2x - y = -3$
$y = 3x$ $y = -3x$
29. $5x - 6y = -4$ **30.** $2x - 4y = 0$
$x = y$ $y = x$
31. $3x + 3y = 9$
$y = 2x - 12$
32. $7x + 6y = -9$
$y = -2x + 1$
33. $7x - 11y = 16$
$y = 10$
34. $9x - 7y = -14$
$x = 7$

35. $-4x + 4y = -8$
$y = x - 2$
36. $-4x + 2y = -10$
$y = 2x - 5$

Solve each system by substitution. You can eliminate the decimals if you like, but you don't have to. The solution will be the same in either case.

37. $0.05x + 0.10y = 1.70$
$y = 22 - x$
38. $0.20x + 0.50y = 3.60$
$y = 12 - x$

39. Below are the rates charged by two bottled water companies for their monthly service, which includes rent on the water dispenser and the per gallon charge for water. If the water comes only in 5-gallon bottles, how many gallons must be used in a month for the two companies to charge the same amount?

	Flat Rate plus Per Gallon Charge	
Company 1	$7.00	$1.10
Company 2	$5.00	$1.15

40. The daily rates for renting a compact car from two rental car companies are shown below. How many miles would a car have to be driven in a *week* for the two companies to charge the same amount?

	Daily Rate plus Per Mile Charge	
Company 1	$21.99	$0.20
Company 2	$19.99	$0.22

Review Problems

The problems that follow review material we covered in Sections 2.6 and 2.7. Reviewing these problems will help you understand the next section.

41. A rectangle is 3 times as long as it is wide. If the perimeter is 24 meters, find the length and width.

42. The perimeter of a square is 48 inches. Find the length of its side.

43. A collection of coins consists of nickels and dimes and is worth $2.10. If there are 3 more dimes than nickels, how many of each coin are in the collection?

44. Mary has $5.75 in nickels and quarters. If she has 5 more quarters than nickels, how many of each coin does she have?

45. What number is 8% of 6,000?

46. 540 is 9% of what number?

47. A man invests twice as much money at 10% annual interest as he does at 8% annual interest. If his total interest for the year is $224, how much does he have invested at each rate?

48. A woman invests three times as much money at 8% annual interest as she does at 6% annual interest. If the total interest for the year is $66, how much does she have invested at each rate?

SECTION

···

3.10   # Applications

I have often heard students remark about the word problems in beginning algebra: ''What does this have to do with real life?'' Most of the word problems we will encounter don't have much to do with ''real life.'' We are actually just practicing. Ultimately, all problems requiring the use of algebra are word problems. That is, they are stated in words first, then translated to symbols. The problem then is solved by some system of mathematics, like algebra. Most real applications involve calculus or higher levels of mathematics. So, if the problems we solve are upsetting or frustrating to you, then you are probably taking them too seriously.

The word problems in this section have two unknown quantities. We will write two equations in two variables (each of which represents one of the unknown quantities), which of course is a system of equations. We then solve the system by one of the methods developed in the previous sections of this chapter. Here are the steps to follow in solving these word problems.

Blueprint for Problem Solving Using a System of Equations

Step 1: **Read** the problem and then mentally **list** the items that are known and the items that are unknown.

Step 2: **Assign variables** to each of the unknown items. That is, let $x =$ one of the unknown items and $y =$ the other unknown item. Then **translate** the other **information** in the problem to expressions involving the two variables.

Step 3: **Reread** the problem, and then **write a system of equations,** using the items and variables listed in steps 1 and 2, that describes the situation.

Step 4: **Solve the system** found in step 3.

Step 5: **Write** your **answers** using complete sentences.

Step 6: **Reread** the problem and **check** your solution with the original words in the problem.

Remember, the more problems you work, the more problems you will be able to work. If you have trouble getting started on the problem set, come back to the examples and work through them yourself. The examples are similar to the problems found in the problem set.

Number Problem

▶ **EXAMPLE 1**   One number is two more than five times another number. Their sum is 20. Find the two numbers.

Solution   Applying the steps in our blueprint we have

Step 1: We know that the two numbers have a sum of 20 and that one of them is two more than five times the other. We don't know what the numbers themselves are.

Step 2: Let x represent one of the numbers and y represent the other. "One number is two more than five times another" translates to

$$y = 5x + 2$$

"Their sum is 20" translates to

$$x + y = 20$$

Step 3: The system that describes the situation must be

$$x + y = 20$$
$$y = 5x + 2$$

Step 4: We can solve this system by substituting the expression $5x + 2$ in the second equation for y in the first equation:

$$x + 5x + 2 = 20$$
$$6x + 2 = 20$$
$$6x = 18$$
$$x = 3$$

Using $x = 3$ in either of the first two equations and then solving for y, we get $y = 17$.

Step 5: So 17 and 3 are the numbers for which we are looking.

Step 6: The number 17 is 2 more than 5 times 3, and the sum of 17 and 3 is 20.   ◀

Interest Problem

▶ **EXAMPLE 2**   Mr. Hicks had $15,000 to invest. He invested part at 6% and the rest at 7%. If he earns $980 in interest, how much did he invest at each rate?

Solution   Remember, step 1 is done mentally.

Step 1: We do not know the specific amounts invested in the two accounts. We do know that their sum is $15,000 and that the interest rates on the two accounts are 6% and 7%.

Step 2: Let x = the amount invested at 6% and y = the amount invested at 7%. Since Mr. Hicks invested a total of $15,000, we have

$$x + y = 15,000$$

The interest he earns comes from 6% of the amount invested at 6% and 7% of the amount invested at 7%. To find 6% of x we multiply x by 0.06, which gives us $0.06x$. To find 7% of y, we multiply 0.07 times y and get $0.07y$.

$$\begin{array}{ccc} \text{Interest} + & \text{interest} = & \text{total} \\ \text{at } 6\% & \text{at } 7\% & \text{interest} \\ 0.06x\ + & 0.07y\ = & 980 \end{array}$$

Step 3: The system is

$$x + \qquad y = 15,000$$
$$0.06x + 0.07y = 980$$

Step 4: We multiply the first equation by -6 and the second by 100 to eliminate x:

$$x + \qquad y = 15,000 \xrightarrow{\text{Multiply by } -6} -6x - 6y = -90,000$$
$$0.06x + 0.07y = 980 \xrightarrow[\text{Multiply by } 100]{} \quad \underline{6x + 7y = \quad 98,000}$$
$$y = \quad 8,000$$

Substituting $y = 8,000$ into the first equation and solving for x, we get $x = 7,000$.

Step 5: He invested $7,000 at 6% and $8,000 at 7%.

Step 6: Checking our solutions in the original problem, we have: The sum of $7,000 and $8,000 is $15,000, the total amount he invested. To complete our check, we find the total interest earned from the two accounts:

The interest on $7,000 at 6% is $0.06(7,000) = 420$

The interest on $8,000 at 7% is $\underline{0.07(8,000) = 560}$

The total interest is $980   ◀

Coin Problem

▶ **EXAMPLE 3** John has $1.70, all in dimes and nickels. He has a total of 22 coins. How many of each kind does he have?

Solution

Step 1: We know that John has 22 coins that are dimes and nickels. We know that a dime is worth 10 cents and a nickel is worth 5 cents. We do not know the specific number of dimes and nickels he has, but we do know that the total value of the coins is $1.70.

Step 2: Let x = the number of nickels and y = the number of dimes. The total number of coins is 22, so

$$x + y = 22$$

The total amount of money he has is $1.70, which comes from nickels and dimes:

$$\begin{array}{ccccc} \text{Amount of money} & + & \text{amount of money} & = & \text{total amount} \\ \text{in nickels} & & \text{in dimes} & & \text{of money} \\ 0.05x & + & 0.10y & = & 1.70 \end{array}$$

Step 3: The system that represents the situation is

$$\begin{aligned} x + \quad y &= 22 \qquad \text{The number of coins} \\ 0.05x + 0.10y &= 1.70 \qquad \text{The value of the coins} \end{aligned}$$

Step 4: We multiply the first equation by -5 and the second by 100 to eliminate the variable x:

$$\begin{aligned} x + \quad y = 22 \quad &\xrightarrow{\text{Multiply by } -5} \quad -5x - 5y = -110 \\ 0.05x + 0.10y = 1.70 \quad &\xrightarrow[\text{Multiply by 100}]{} \quad \underline{5x + 10y = 170} \\ & \qquad\qquad\qquad\qquad 5y = 60 \\ & \qquad\qquad\qquad\qquady = 12 \end{aligned}$$

Substituting $y = 12$ into our first equation, we get $x = 10$.

Step 5: John has 12 dimes and 10 nickels.

Step 6: Twelve dimes and 10 nickels total 22 coins.

$$12 \text{ dimes are worth } 12(0.10) = 1.20$$
$$\underline{10 \text{ nickels are worth } 10(0.05) = 0.50}$$
$$\text{The total value is} \qquad\qquad \$1.70 \qquad ◀$$

Mixture Problem

▶ **EXAMPLE 4** How much 20% alcohol solution and 50% alcohol solution must be mixed to get 12 gallons of 30% alcohol solution?

Solution To solve this problem we must first understand that a 20% alcohol solution is 20% alcohol and 80% water.

Step 1: We know there are two solutions that together must total 12 gallons. 20% of one of the solutions is alcohol and the rest is water, while the other solution is 50% alcohol and 50% water. We do not know how many gallons of each individual solution we need.

Step 2: Let x = the number of gallons of 20% alcohol solution needed, and y = the number of gallons of 50% alcohol solution needed. Since the total number of gallons we will end up with is 12, and this 12 gallons must come from the two solutions we are mixing, our first equation is

$$x + y = 12$$

To obtain our second equation we look at the amount of alcohol in our two original solutions and our final solution. The amount of alcohol in the x gallons of 20% solution is $0.20x$, while the amount of alcohol in y gallons of 50% solution is $0.50y$. The amount of alcohol in the 12 gallons of 30% solution is $0.30(12)$. Since the amount of alcohol we start with must equal the amount of alcohol we end up with, our second equation is

$$0.20x + 0.50y = 0.30(12)$$

The information we have so far can also be summarized with a table. Sometimes by looking at a table like the one that follows it is easier to see from where the equations come.

	20% Solution	50% Solution	Final Solution
Number of Gallons	x	y	12
Gallons of Alcohol	$0.20x$	$0.50y$	$0.30(12)$

Step 3: Our system of equations is

$$x + y = 12$$
$$0.20x + 0.50y = 0.30(12)$$

Step 4: We can solve this system by substitution. Solving the first equation for *y* and substituting the result into the second equation we have

$$0.20x + 0.50(12 - x) = 0.30(12)$$

Multiplying each side by 10 gives us an equivalent equation that is a little easier to work with:

$$2x + 5(12 - x) = 3(12)$$
$$2x + 60 - 5x = 36$$
$$-3x + 60 = 36$$
$$-3x = -24$$
$$x = 8$$

If *x* is 8, then *y* must be 4 because *x* + *y* = 12.

Step 5: It takes 8 gallons of 20% alcohol solution and 4 gallons of 50% alcohol solution to produce 12 gallons of 30% alcohol solution.

Step 6: Try it and see.   ◀

PROBLEM SET 3.10

Solve the following word problems. Be sure to show the equations used.

Number Problems

1. Two numbers have a sum of 25. One number is five more than the other. Find the numbers.
2. The difference of two numbers is 6. Their sum is 30. Find the two numbers.
3. The sum of two numbers is 15. One number is four times the other. Find the numbers.
4. The difference of two positive numbers is 28. One number is three times the other. Find the two numbers.
5. Two positive numbers have a difference of 5. The larger number is one more than twice the smaller. Find the two numbers.
6. One number is two more than three times another. Their sum is 26. Find the two numbers.
7. One number is five more than four times another. Their sum is 35. Find the two numbers.

8. The difference of two positive numbers is 8. If twice the smaller is decreased by 7, we arrive at the larger number. Find the two numbers.

Interest Problems

9. Mr. Wilson invested money in two accounts. His total investment was $20,000. If one account pays 6% in interest and the other pays 8% in interest, how much does he have in each account if he earned a total of $1,380 in interest in one year?
10. A total of $11,000 was invested. Part of the $11,000 was invested at 4% and the rest was invested at 7%. If the investments earn $680 in one year, how much was invested at each rate?
11. A woman invested four times as much at 5% as she did at 6%. The total amount of interest she earns in one year from both accounts is $520. How much did she invest at each rate?

12. Ms. Hagan invested twice as much money in an account that pays 7% interest as she did in an account that pays 6% in interest. Her total investment pays her $1,000 a year in interest. How much did she invest at each amount?

Coin Problems

13. Ron has 14 coins with a total value of $2.30. The coins are nickels and quarters. How many of each coin does he have?
14. Diane has $0.95 in dimes and nickels. She has a total of 11 coins. How many of each kind does she have?
15. Suppose Tom has 21 coins totaling $3.45. If he has only dimes and quarters, how many of each type does he have?
16. A coin collector has 31 dimes and nickels with a total face value of $2.40. (They are actually worth a lot more.) How many of each coin does she have?

Mixture Problems

17. How many liters of 50% alcohol solution and 20% alcohol solution must be mixed to obtain 18 liters of 30% alcohol solution?
18. How many liters of 10% alcohol solution and 5% alcohol solution must be mixed to obtain 40 liters of 8% alcohol solution?
19. A mixture of 8% disinfectant solution is to be made from 10% and 7% disinfectant solutions. How much of each solution should be used if 30 gallons of 8% solution are needed?
20. How much 50% antifreeze solution and 40% antifreeze solution should be combined to give 50 gallons of 46% antifreeze solution?

Miscellaneous Problems

21. For a Saturday matinee, adult tickets cost $5.50 while kids under 12 pay only $4.00. If 70 tickets are sold for a total of $310, how many of the tickets were adult tickets and how many were sold to kids under 12?
22. The Bishop's Peak 4H club is having its annual fundraising dinner. Adults pay $15 apiece and children pay $10 apiece. If the number of adult tickets sold is twice the number of children's tickets sold and the total income for the dinner is $1,600, how many of each kind of ticket did the 4H club sell?

23. A farmer has 96 feet of fence with which to make a corral. If he arranges it into a rectangle that is twice as long as it is wide, what are the dimensions?
24. If a 22-inch rope is to be cut into two pieces so that one piece is 3 inches longer than twice the other, how long is each piece?
25. A gambler finishes a session of blackjack with $5 chips and $25 chips. If he has 45 chips in all, with a total value of $465, how many of each kind of chip does the gambler have?
26. Tyler has been saving his winning lottery tickets. He has 23 tickets that are worth a total of $175. If each ticket is worth either $5 or $10, how many of each does he have?
27. Mary Jo spends $2,550 to buy stock in two companies. She pays $11 a share to one of the companies and $20 a share to the other. If she ends up with a total of 150 shares, how many shares did she buy at $11 a share and how many did she buy at $20 a share?
28. Kelly sells 62 shares of stock she owns for a total of $433. If the stock was in two different companies, one selling at $6.50 a share and the other at $7.25 a share, how many of each did she sell?

Review Problems

The following problems review material we covered in Section 2.8.

Solve each inequality.

29. $2x - 6 > 5x + 6$ 30. $4x - 3 \geq 8x + 1$
31. $3(2x + 4) \leq -6$ 32. $5(3x - 2) < 20$
33. $4(2 - x) \geq 12$ 34. $6(1 - x) > -12$
35. $6 - 2(x + 3) < -2$
36. $7 - 3(x - 5) \leq -2$

CHAPTER 3 SUMMARY

Examples

1. The equation $3x + 2y = 6$ is an example of a linear equation in two variables.

Linear Equations in Two Variables [3.1, 3.2, 3.3]

A *linear equation in two variables* is any equation that can be put in the form $ax + by = c$. The graph of every linear equation is a straight line.

2. To find the *x*-intercept for $3x + 2y = 6$ we let $y = 0$ and get

$$3x = 6$$

$$x = 2$$

In this case the *x*-intercept is 2, and the graph crosses the *x*-axis at (2, 0).

Intercepts [3.4]

The *x-intercept* of an equation is the *x*-coordinate of the point where the graph crosses the *x*-axis. The *y-intercept* is the *y*-coordinate of the point where the graph crosses the *y*-axis. We find the *y*-intercept by substituting $x = 0$ into the equation and solving for *y*. The *x*-intercept is found by letting $y = 0$ and solving for *x*.

3. The slope of the line through $(3, -5)$ and $(-2, 1)$ is

$$m = \frac{-5 - 1}{3 - (-2)} = \frac{-6}{5} = -\frac{6}{5}$$

Slope of a Line [3.5]

The *slope* of the line containing the points (x_1, y_1) and (x_2, y_2) is given by

$$\text{Slope} = m = \frac{y_2 - y_1}{x_2 - x_1} = \frac{\text{rise}}{\text{run}}$$

4. The equation of the line through $(1, 2)$ with a slope of 3 is

$$y - 2 = 3(x - 1)$$

$$y - 2 = 3x - 3$$

$$y = 3x - 1$$

Point-Slope Form of a Straight Line [3.6]

If a line has a slope of *m* and contains the point (x_1, y_1), the equation can be written as

$$y - y_1 = m(x - x_1)$$

5. The equation of the line with a slope of 2 and a *y*-intercept of 5 is

$$y = 2x + 5$$

Slope-Intercept Form of a Straight Line [3.6]

The equation of the line with a slope of *m* and a *y*-intercept *b* is

$$y = mx + b$$

6. The solution to the system

$$x + 2y = 4$$

$$x - y = 1$$

is the ordered pair (2, 1). It is the only ordered pair that satisfies both equations.

Definitions [3.7]

1. A *system of linear equations*, as the term is used in this book, is two linear equations that each contain the same two variables.
2. The *solution set* for a system of equations is the set of all ordered pairs that satisfy *both* equations. The solution set to a system of linear equations will contain:

Case I   One ordered pair when the graphs of the two equations intersect at only one point (this is the most common situation)

Case II   No ordered pairs when the graphs of the two equations are parallel lines

Case III   An infinite number of ordered pairs when the graphs of the two equations coincide (are the same line)

7. Solving the system in Example 6 by graphing looks like

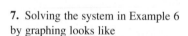
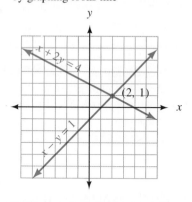

Strategy for Solving a System by Graphing [3.7]

Step 1: Graph the first equation.

Step 2: Graph the second equation on the same set of axes.

Step 3: The solution to the system consists of the coordinates of the point where the graphs cross each other (the coordinates of the point of intersection).

Step 4: Check the solution to see that it satisfies *both* equations, if necessary.

8. We can eliminate the y variable from the system in Example 6 by multiplying both sides of the second equation by 2 and adding the result to the first equation:

$$x + 2y = 4 \qquad\qquad x + 2y = 4$$
$$x - y = 1 \xrightarrow[\text{Multiply by 2}]{} 2x - 2y = 2$$
$$3x \quad\;\; = 6$$
$$x = 2$$

Substituting $x = 2$ into either of the original two equations gives $y = 1$. The solution is (2, 1).

Strategy for Solving a System by the Elimination Method [3.8]

Step 1: Look the system over to decide which variable will be easier to eliminate.

Step 2: Use the multiplication property of equality on each equation separately to ensure that the coefficients of the variable to be eliminated are opposites.

Step 3: Add the left and right sides of the system produced in step 2 and solve the resulting equation.

Step 4: Substitute the solution from step 3 back into any equation with both x and y variables and solve.

Step 5: Check your solution in both equations, if necessary.

9. We can apply the substitution method to the system in Example 6 by first solving the second equation for x to get $x = y + 1$.

Substituting this expression for x into the first equation, we have

$$y + 1 + 2y = 4$$
$$3y + 1 = 4$$
$$3y = 3$$
$$y = 1$$

Using $y = 1$ in either of the original equations gives $x = 2$.

Strategy for Solving a System by the Substitution Method [3.9]

Step 1: Solve either of the equations for one of the variables (this step is not necessary if one of the equations has the correct form already).

Step 2: Substitute the results of step 1 into the other equation and solve.

Step 3: Substitute the results of step 2 into an equation with both x and y variables and solve. (The equation produced in step 1 is usually a good one to use.)

Step 4: Check your solution, if necessary.

Special Cases [3.7, 3.8, 3.9]
In some cases, using the elimination or substitution method eliminates both variables. The situation is interpreted as follows:

1. If the resulting statement is *false*, then the lines are parallel and there is no solution to the system.
2. If the resulting statement is *true*, then the equations represent the same line (the lines coincide). In this case, any ordered pair that satisfies either equation is a solution to the system.

COMMON MISTAKES

The most common mistake encountered in solving linear systems is the failure to complete the problem. Here is an example:

$$x + y = 8$$
$$\underline{x - y = 4}$$
$$2x \quad\quad = 12$$
$$x = 6$$

This is only half the solution. To find the other half, we must substitute the 6 back into one of the original equations and then solve for y.

Remember, solutions to systems of linear equations always consist of ordered pairs. We need an x-coordinate and a y-coordinate; $x = 6$ can never be a solution to a system of linear equations.

CHAPTER 3 REVIEW

The numbers in brackets refer to the sections of the text in which similar problems can be found.

Graph the following ordered pairs. [3.1]

1. $(4, 2)$
2. $(-3, 1)$
3. $(0, 5)$
4. $(-2, -3)$
5. $(-3, 0)$
6. $\left(5, -\dfrac{3}{2}\right)$

For each equation complete the given ordered pairs. [3.2]

7. $3x + y = 6$ (4,), (0,), (, 3), (, 0)
8. $2x - 5y = 20$ (5,), (0,), (, 2), (, 0)

9. $y = 2x - 6$ (4,), (, -2), (, 3)
10. $y = 5x + 3$ (2,), (, 0), (, -3)
11. $y = -3$ (2,), (-1,), (-3,)
12. $x = 6$ (, 5), (, 0), (, -1)

For the following equations, tell which of the given ordered pairs are solutions. [3.2]

13. $3x - 4y = 12$ $\left(-2, \dfrac{9}{2}\right)$, $(0, 3)$, $\left(2, -\dfrac{3}{2}\right)$
14. $2x + 4y = 8$ $(6, -1)$, $(0, -2)$, $(8, 2)$
15. $y = 3x + 7$ $\left(-\dfrac{8}{3}, -1\right)$, $\left(\dfrac{7}{3}, 0\right)$, $(-3, -2)$
16. $y = 2x - 5$ $(3, -1)$, $(1, -3)$, $(-2, 9)$

For the following equations, complete the given ordered pairs and use the results to graph the solution set for the equations. [3.3]

17. $x + y = -2$ (, 0), (0,), (1,)
18. $x - y = 5$ (2,), (, −2), (0,)
19. $y = 3x$ (−1,), (1,), (, 0)
20. $y = \dfrac{1}{3}x$ (, 1), (, 0), (−3,)
21. $y = 2x - 1$ (1,), (0,), (, −3)
22. $y = -3x + 4$ (0,), (1,), (, −2)

Graph the following equations. [3.3, 3.4, 3.5]

23. $3x - y = 3$ **24.** $3x + 2y = 6$
25. $y = -\dfrac{1}{3}x$ **26.** $y = -3x$
27. $y = 2x + 1$ **28.** $y = -3x + 2$

Graph the following equations. [3.3]

29. $x = 5$ **30.** $y = 2$
31. $y = -3$ **32.** $x = 1$

Find the x- and y-intercepts for each equation. [3.4]

33. $3x - y = 6$ **34.** $2x + 5y = 10$
35. $y = x - 3$ **36.** $y = -x + 7$
37. $y = 3x - 6$ **38.** $y = -2x + 10$

Find the slope of the line through the given pair of points. [3.5]

39. (2, 3), (3, 5)   **40.** (3, −2), (1, 8)
41. (−2, 3), (6, −5)   **42.** (7, 3), (−2, −4)
43. (−1, −4), (−3, −8)
44. (−2, −5), (−4, −1)

Find the equation of the line with the given slope and y-intercept. [3.6]

45. $m = 3, b = 2$
46. $m = -5, b = -1$
47. $m = -1, b = 6$
48. $m = -4, b = -3$
49. $m = -\dfrac{1}{3}, b = \dfrac{3}{4}$
50. $m = -\dfrac{2}{5}, b = -\dfrac{2}{3}$

For each of the following equations, determine the slope and y-intercept. [3.6]

51. $y = 4x - 1$ **52.** $y = -3x + 2$
53. $2x + y = -5$ **54.** $3x - y = 4$
55. $6x + 3y = 9$ **56.** $8x - 4y = -12$
57. $5x + 2y = 8$ **58.** $3x - 4y = -16$

Find the equation of the line that contains the given point and has the given slope. Write your answers in slope-intercept form. [3.6]

59. (3, 1); $m = 4$ **60.** (−1, 4); $m = -2$
61. (4, 3); $m = \dfrac{1}{2}$ **62.** (−6, 4); $m = \dfrac{2}{3}$
63. (3, −2); $m = -\dfrac{3}{4}$
64. (−2, −1); $m = -\dfrac{2}{5}$

For each pair of points, find the equation of the line that passes through them. Write your answers in slope-intercept form. [3.6]

65. (2, 4), (1, 6)   **66.** (2, 5), (4, 8)
67. (2, 2), (4, 4)   **68.** (1, 3), (2, 6)
69. (3, −7), (−3, 1)   **70.** (−3, 5), (3, 3)

Solve the following systems by graphing. [3.7]

71. $x + y = 2$ **72.** $x + y = -1$
     $x - y = 6$ $-x + y = 5$
73. $2x - 3y = 12$ **74.** $4x - 2y = 8$
     $-2x + y = -8$ $3x + y = 6$
75. $y = 2x - 3$ **76.** $y = -x - 3$
     $y = \dfrac{1}{2}x$ $x = -1$

Solve the following systems by the elimination method. [3.8]

77. $x - y = 4$ **78.** $-x - y = -3$
     $x + y = -2$ $2x + y = 1$
79. $5x - 3y = 2$ **80.** $2x + 3y = -2$
     $-10x + 6y = -4$ $3x - 2y = 10$

81. $-3x + 4y = 1$
   $-4x + 5y = 1$

82. $-4x - 2y = 3$
   $6x + 3y = 1$

83. $-2x + 5y = -11$
   $7x - 3y = -5$

84. $-2x + 5y = -15$
   $3x - 4y = 19$

Solve the following systems by substitution. [3.9]

85. $x + y = 5$
   $y = -3x + 1$

86. $x - y = -2$
   $y = -2x - 10$

87. $4x - 3y = -16$
   $y = 3x + 7$

88. $5x + 2y = -2$
   $y = -8x + 10$

89. $-3x + 12y = -8$
   $x = 4y + 2$

90. $4x - 2y = 8$
   $y = -3x - 19$

91. $10x - 5y = 20$
   $x = -6y - 11$

92. $-6x + 2y = -4$
   $y = 3x - 2$

Solve the following applications. Be sure to show the equations used. [3.10]

93. The sum of two numbers is 18. If twice the smaller number is 6 more than the larger, find the two numbers.

94. The difference of two positive numbers is 16. One number is 3 times the other. Find the two numbers.

95. A total of $12,000 was invested. Part of the $12,000 was invested at 4% and the rest was invested at 5%. If the interest for one year is $560, how much was invested at each rate?

96. A total of $14,000 was invested. Part of the $14,000 was invested at 6% and the rest was invested at 8%. If the interest for one year is $1,060, how much was invested at each rate?

97. Barbara has $1.35 in dimes and nickels. She has a total of 17 coins. How many of each does she have?

98. Tom has $2.40 in dimes and quarters. He has a total of 15 coins. How many of each does he have?

99. How many liters of 20% alcohol solution and 10% alcohol solution must be mixed to obtain 50 liters of a 12% alcohol solution?

100. How many liters of 25% alcohol solution and 15% alcohol solution must be mixed to obtain 40 liters of a 20% alcohol solution?

101. Table 1 below gives the position and value of the first five terms of the sequence of even numbers. Construct a histogram and a scatter diagram from the table.

TABLE 1   Even Numbers

Position	Value
1	2
2	4
3	6
4	8
5	10

102. Table 2 below gives the interest that accumulates in an account when $100 has been deposited and earns interest that is compounded annually. Use the information in the table to construct a histogram and a line graph.

TABLE 2   Accumulating Interest

Time in Years	Interest in Dollars
0	0.00
1	5.00
2	10.25
3	15.76
4	21.55
5	27.63

CHAPTER 3 TEST

1. Fill in the following ordered pairs for the equation $2x - 5y = 10$. [3.2]

$$(0, \ ) \quad (\ , 0) \quad (10, \ ) \quad (\ , -3)$$

2. Which of the following ordered pairs are solutions to $y = 4x - 3$? [3.2]

$$(2, 5) \quad (0, -3) \quad (3, 0) \quad (-2, 11)$$

Graph each line. [3.3, 3.4, 3.5]

3. $y = 3x - 2$ **4.** $x = -2$

Give the slope, x-intercept, and y-intercept of the following straight lines and sketch their graphs. [3.4, 3.5, 3.6]

5. $3x - 2y = 6$ **6.** $2x - y = 5$

7. Find the equation of the line with a slope of 4 and y-intercept 8. [3.6]

8. Line l passes through the points $(-3, 1)$ and $(-2, 4)$. Find the equation of line l. [3.6]

Solve the system by graphing. [3.7]

9. $x + 2y = 5$
$\qquad y = 2x$

Solve each system by the elimination method. [3.8]

10. $x - y = 1$ **11.** $2x + \ y = 7$
$\quad\ 2x - 3y = 6$ $3x + 7y = -6$
12. $7x + 8y = -2$ **13.** $6x - 10y = 6$
$\quad\ 3x - 2y = 10$ $9x - 15y = 9$

Solve each system by the substitution method. [3.9]

14. $3x + 2y = 20$ **15.** $3x - 6y = -6$
$\qquad\quad y = 2x + 3$ $x = 3$

16. $2x - 7y = 2$
$\qquad\ x = 4y$

Solve the following applications. In each case, be sure to show the system of equations that describes the situation. [3.10]

17. The sum of two numbers is 12. Their difference is 2. Find the numbers.

18. The sum of two numbers is 15. One number is six more than twice the other. Find the two numbers.

19. Dr. Stork has $10,000 to invest. He would like to earn $980 per year in interest. How much should he invest at 9% if the rest is to be invested at 11%?

20. Diane has 12 coins that total $1.60. If the coins are all nickels and quarters, how many of each type does she have?

21. Table 1 below gives the position and value of the first five terms in a number sequence. Construct a histogram and a scatter diagram from the table.

TABLE 1 Number Sequence

Position	Value
1	12
2	14
3	18
4	24
5	32

4

EXPONENTS AND POLYNOMIALS

INTRODUCTION

If you were given a penny on the first day of September and then each day after that you were given twice the amount of money you received the day before, how much money would you receive on September 30? Here are two sequences that together model this situation:

Day of the month:	1	2	3	4	5	6	7	8	9	10
Money in cents:	1	2	4	8	16	32	64	128	256	512

On the tenth day of the month you would receive $5.12. To decide how much money you would receive on day 30, you could continue to multiply by 2, twenty more times. However, there is an easier way. If we were to write each number in the second row of the table as an exponential expression with a base of 2, then the number that would correspond to day 30 would be 2^{29} cents, which means you would receive $5,368,709.12; a little less than 5.4 million dollars. Since exponents represent repeated multiplication, it doesn't take much of a base and exponent to produce a very large number. Large numbers are often written in what we call **scientific notation.** The number 5.4 million dollars written in both scientific notation and expanded form is shown below.

Scientific Notation	Expanded Form
$\$5.4 \times 10^6$	$\$5,400,000.00$

Writing numbers in scientific notation can make them easier to work with. The same is true of numbers written with exponents. When you are finished with this chapter, you should have a good working knowledge of exponents and scientific notation, as well as some topics closely related to these two topics.

OVERVIEW

In this chapter we take a more detailed look at exponents. We begin by reviewing the definition for positive integer exponents. The rest of the chapter deals with developing properties associated with exponents.

To perform successfully in this chapter you should understand the following concepts. The concepts that are not familiar should be reviewed.

1. Subtraction is addition of the opposite
2. Division is multiplication by the reciprocal
3. Multiplication by positive integers is repeated addition
4. Addition and multiplication are commutative operations
5. Multiplication distributes over addition

STUDY SKILLS

The study skills for this chapter are about attitude. They are points of view that point toward success.

1. **Be Focused, Not Distracted**   I have students who begin their assignments by asking themselves, ''Why am I taking this class?'' Or, ''When am I ever going to use this stuff?'' If you are asking yourself similar questions, you may be distracting yourself away from doing the things that will produce the results you want in this course. Don't dwell on questions and evaluations of the class that can be used as excuses for not doing well. If you want to succeed in this course, focus your energy and efforts toward success, rather than distracting yourself away from your goals.

2. **Be Resilient**   Don't let setbacks keep you from your goals. You want to put yourself on the road to becoming a person who can succeed in this class or any class in college. Failing a test or quiz, or having a difficult time on some topics, is normal. No one goes through college without some setbacks. Don't let a temporary disappointment keep you from succeeding in this course. A low grade on a test or quiz is simply a signal that some reevaluation of your study habits needs to take place.

3. **Intend to Succeed**   I always have a few students who simply go through the motions of studying without intending to master the material. It is more important to them to look like they are studying than to actually study. You need to study with the intention of being successful in the course. Intend to master the material, no matter what it takes.

SECTION

4.1

Multiplication with Exponents

Recall that an **exponent** is a number written just above and to the right of another number, which is called the **base.** In the expression 5^2, for example, the exponent is 2 and the base is 5. The expression 5^2 is read ''5 to the second power'' or ''5 squared.'' The meaning of the expression is

$$5^2 = 5 \cdot 5 = 25$$

In the expression 5^3, the exponent is 3 and the base is 5. The expression 5^3 is read ''5 to the third power'' or ''5 cubed.'' The meaning of the expression is

$$5^3 = 5 \cdot 5 \cdot 5 = 125$$

Here are some further examples.

▶ **EXAMPLES**

1. $4^3 = 4 \cdot 4 \cdot 4 = 16 \cdot 4 = 64$ exponent 3, base 4

2. $-3^4 = -3 \cdot 3 \cdot 3 \cdot 3 = -81$ exponent 4, base 3

3. $(-2)^5 = (-2)(-2)(-2)(-2)(-2) = -32$ exponent 5, base -2

4. $\left(-\dfrac{3}{4}\right)^2 = \left(-\dfrac{3}{4}\right)\left(-\dfrac{3}{4}\right) = \dfrac{9}{16}$ exponent 2, base $-\dfrac{3}{4}$ ◀

QUESTION   In what way are $(-5)^2$ and -5^2 different?
ANSWER   In the first case, the base is -5. In the second case, the base is 5. The answer to the first is 25. The answer to the second is -25. Can you tell why?

We can simplify our work with exponents by developing some properties of exponents. We want to list the things we know are true about exponents and then use these properties to simplify expressions that contain exponents.

The first property of exponents applies to products with the same base. We can use the definition of exponents, as indicating repeated multiplication, to simplify expressions like $7^4 \cdot 7^2$.

$$7^4 \cdot 7^2 = (7 \cdot 7 \cdot 7 \cdot 7)(7 \cdot 7)$$
$$= (7 \cdot 7 \cdot 7 \cdot 7 \cdot 7 \cdot 7)$$
$$= 7^6 \qquad \text{Notice: } 4 + 2 = 6$$

As you can see, multiplication with the same base resulted in addition of exponents. We can summarize this result with the following property.

Property 1 for Exponents
If a is any real number and r and s are integers, then

$$a^r \cdot a^s = a^{r+s}$$

In words: To multiply two expressions with the same base, add exponents and use the common base.

Here are some examples using Property 1.

▶ **EXAMPLES**   Use Property 1 to simplify the following expressions. Leave your answers in terms of exponents.

5. $5^3 \cdot 5^6 = 5^{3+6} = 5^9$
6. $x^7 \cdot x^8 = x^{7+8} = x^{15}$
7. $3^4 \cdot 3^8 \cdot 3^5 = 3^{4+8+5} = 3^{17}$ ◀

Note   In Examples 5, 6, and 7, notice that in each case the base in the original problem is the same base that appears in the answer and that it is written only once in the answer. A very common mistake that people make when they first begin to use Property 1 is to write a 2 in front of the base in the answer. For example, people making this mistake would get $2x^{15}$ or $(2x)^{15}$ as the result in Example 6. To avoid this mistake, you must be sure you understand the meaning of Property 1 exactly as it is written.

Another common type of expression involving exponents is one in which an expression containing an exponent is raised to another power. The expression $(5^3)^2$ is an example:

$$(5^3)^2 = (5^3)(5^3)$$
$$= 5^{3+3}$$
$$= 5^6 \qquad \text{Notice } 3 \cdot 2 = 6$$

This result offers justification for the second property of exponents.

Property 2 for Exponents
If a is any real number and r and s are integers, then

$$(a^r)^s = a^{r \cdot s}$$

In words: A power raised to another power is the base raised to the product of the powers.

▶ **EXAMPLES**   Simplify the following expressions.

8. $(4^5)^6 = 4^{5 \cdot 6} = 4^{30}$
9. $(x^3)^5 = x^{3 \cdot 5} = x^{15}$ ◀

The third property of exponents applies to expressions in which the product of two or more numbers or variables is raised to a power. Let's look at how the expression $(2x)^3$ can be simplified:

$$(2x)^3 = (2x)(2x)(2x)$$
$$= (2 \cdot 2 \cdot 2)(x \cdot x \cdot x)$$
$$= 2^3 \cdot x^3 \qquad\qquad \text{Notice: The exponent 3 distributes}$$
$$= 8x^3 \qquad\qquad\qquad \text{over the product } 2x$$

We can generalize this result into a third property of exponents.

Property 3 for Exponents
If a and b are any two real numbers and r is an integer, then

$$(ab)^r = a^r b^r$$

In words: The power of a product is the product of the powers.

Here are some examples using Property 3 to simplify expressions.

▶ **EXAMPLES** Simplify the following expressions.

10. $(5x)^2 = 5^2 \cdot x^2$ Property 3
          $= 25x^2$

11. $(2xy)^3 = 2^3 \cdot x^3 \cdot y^3$ Property 3
            $= 8x^3y^3$

12. $(3x^2)^3 = 3^3(x^2)^3$ Property 3
           $= 27x^6$ Property 2

13. $\left(-\dfrac{1}{4}x^2y^3\right)^2 = \left(-\dfrac{1}{4}\right)^2 (x^2)^2(y^3)^2$ Property 3

$$= \frac{1}{16}x^4y^6 \qquad\qquad\qquad \text{Property 2}$$

14. $(x^4)^3(x^2)^5 = x^{12} \cdot x^{10}$ Property 2
             $= x^{22}$ Property 1

15. $(2y)^3(3y^2) = 2^3y^3(3y^2)$ Property 3
             $= 8 \cdot 3(y^3 \cdot y^2)$ Commutative and
                                    associative properties
             $= 24y^5$ Property 1

16. $(2x^2y^5)^3(3x^4y)^2 = 2^3(x^2)^3(y^5)^3 \cdot 3^2(x^4)^2y^2$ Property 3
                     $= 8x^6y^{15} \cdot 9x^8y^2$ Property 2
                     $= (8 \cdot 9)(x^6x^8)(y^{15}y^2)$ Commutative and
                                      associative properties
                     $= 72x^{14}y^{17}$ Property 1 ◀

Facts from Geometry: Volume of a Rectangular Solid

It is easy to see why the phrase "five squared" is associated with the expression 5^2. Simply find the area of the square shown in Figure 1 with a side of 5.

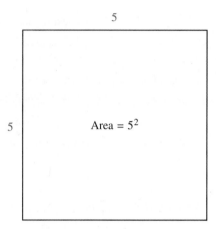

Figure 1

To see why the phrase "five cubed" is associated with the expression 5^3, we have to find the **volume** of a cube for which all three dimensions are 5 units long. The volume of a cube is a measure of the space occupied by the cube. To calculate the volume of the cube shown in Figure 2 we multiply the three dimensions together to get $5 \cdot 5 \cdot 5 = 5^3$.

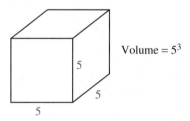

Figure 2

The cube shown in Figure 2 is a special case of a general category of three-dimensional geometric figures called **rectangular solids.** Rectangular solids have rectangles for sides, and all connecting sides meet at right angles. The three dimensions are length, width, and height. To find the volume of a rectangular solid we find the product of the three dimensions, as shown in Figure 3.

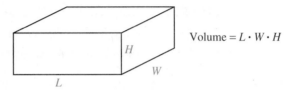

Figure 3

Note If we include units with the dimensions of the diagrams above, then the units for area will be square units and the units for volume will be cubic units. More specifically, if a square has a side 5 inches long, then its area will be

$$A = (5 \text{ inches})^2 = 25 \text{ inches}^2$$

where the unit inches^2 stands for square inches.

If a cube has a side 5 inches long, then its volume will be

$$V = (5 \text{ inches})^3 = 125 \text{ inches}^3$$

where the unit inches^3 stands for cubic inches.

If a rectangular solid has a length of 5 inches, a width of 4 inches, and a height of 3 inches, then its volume is

$$V = (5 \text{ inches})(4 \text{ inches})(3 \text{ inches})$$
$$= 60 \text{ inches}^3$$

Scientific Notation

Many branches of science require working with very large numbers. In astronomy, for example, distances commonly are given in light-years. A light-year is the distance light travels in a year. It is approximately

$$5{,}880{,}000{,}000{,}000 \text{ miles}$$

This number is difficult to use in calculations because of the number of zeros it contains. Scientific notation provides a way of writing very large numbers in a more manageable form.

DEFINITION A number is in **scientific notation** when it is written as the product of a number between 1 and 10 and an integer power of 10. A number written in scientific notation has the form

$$n \times 10^r$$

where $1 \le n < 10$ and $r =$ an integer.

▶ **EXAMPLE 17**   Write 376,000 in scientific notation.

Solution   We must rewrite 376,000 as the product of a number between 1 and 10 and a power of 10. To do so, we move the decimal point 5 places to the left so that it appears between the 3 and the 7. Then we multiply this number by 10^5. The number that results has the same value as our original number and is written in scientific notation.

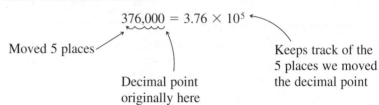

$$376,000 = 3.76 \times 10^5$$

Moved 5 places

Keeps track of the 5 places we moved the decimal point

Decimal point originally here

▶ **EXAMPLE 18**   Write 4.52×10^3 in expanded form.

Solution   Since 10^3 is 1,000, we can think of this as simply a multiplication problem. That is,

$$4.52 \times 10^3 = 4.52 \times 1,000 = 4,520$$

On the other hand, we can think of the exponent 3 as indicating the number of places we need to move the decimal point in order to write our number in expanded form. Since our exponent is positive 3, we move the decimal point three places to the right:

$$4.52 \times 10^3 = 4,520$$

P R O B L E M   S E T   4 . 1

Name the base and exponent in each of the following expressions. Then use the definition of exponents as repeated multiplication to simplify.

1. 4^2

2. 6^2

3. $(0.3)^2$

4. $(0.03)^2$

5. 4^3

6. 10^3

7. $(-5)^2$

8. -5^2

9. -2^3

10. $(-2)^3$

11. 3^4

12. $(-3)^4$

13. $\left(\dfrac{2}{3}\right)^2$

14. $\left(\dfrac{2}{3}\right)^3$

15. $\left(\dfrac{1}{2}\right)^4$

16. $\left(\dfrac{4}{5}\right)^2$

Use Property 1 for exponents to simplify each expression. Leave all answers in terms of exponents.

17. $x^4 \cdot x^5$

18. $x^7 \cdot x^3$

19. $7^6 \cdot 7^1$

20. $8^9 \cdot 8^1$

21. $2^5 \cdot 2^4 \cdot 2^3$

22. $4^2 \cdot 4^3 \cdot 4^4$

23. $x^4 \cdot x^6 \cdot x^8 \cdot x^{10}$

24. $x^{20} \cdot x^{18} \cdot x^{16} \cdot x^{14}$

Use Property 2 for exponents to write each of the following problems with a single exponent. (Assume all variables are positive numbers.)

25. $(x^2)^5$

26. $(x^5)^2$

27. $(5^4)^3$

28. $(5^3)^4$

29. $(2^5)^{10}$

30. $(10^5)^2$

31. $(a^3)^x$ **32.** $(a^5)^x$
33. $(b^x)^y$ **34.** $(b^r)^s$

Use Property 3 for exponents to simplify each of the following expressions.

35. $(4x)^2$ **36.** $(2x)^4$
37. $(-3x)^4$ **38.** $(-3x)^3$
39. $(0.5ab)^2$ **40.** $(0.4ab)^2$
41. $(4xyz)^3$ **42.** $(5xyz)^3$

Simplify the following expressions by using the properties of exponents.

43. $(2x^4)^3$ **44.** $(3x^5)^2$
45. $(x^2)^3(x^4)^2$ **46.** $(x^5)^2(x^3)^5$
47. $(2x)^3(2x)^4$ **48.** $(3x)^2(3x)^3$
49. $(3x^2)^3(2x)^4$ **50.** $(3x)^3(2x^3)^2$
51. $(4x^2y^3)^2$ **52.** $(9x^3y^5)^2$

53. $\left(\dfrac{2}{3}a^4b^5\right)^3$ **54.** $\left(\dfrac{3}{4}ab^7\right)^3$

55. $(3x^2)(2x^3)(5x^4)$ **56.** $(x^3)(2x)(5x^5)$

57. $(4x^2y)^3\left(\dfrac{1}{8}xy\right)^2$

58. $\left(\dfrac{1}{4}x^3y^2\right)^2(2x^4y^5)^4$

Write each number in scientific notation.

59. 43,200   **60.** 432,000
61. 570   **62.** 5,700
63. 238,000   **64.** 2,380,000

Write each number in expanded form.

65. 2.49×10^3 **66.** 2.49×10^4
67. 3.52×10^2 **68.** 3.52×10^5
69. 2.8×10^4 **70.** 2.8×10^3

Applying the Concepts

The volume of a cube with side s is $V = s^3$. Likewise, the volume of a rectangular solid with length L, width W, and height H is $V = LWH$.

71. Find the volume of a cube if each side is 3 inches long.
72. Find the volume of a cube if each side is 3 feet long.

73. A bottle of perfume is packaged in a box that is in the shape of a cube. Find the volume of the box if each side is 2.5 inches long. Round to the nearest tenth.
74. A television set is packaged in a box that is in the shape of a cube. Find the volume of the box if each side is 18 inches long.
75. A rented videotape is in a plastic container that has the shape of a rectangular solid. Find the volume of the container if the length is 8 inches, its width is 4.5 inches, and the height is 1 inch.
76. Your textbook is in the shape of a rectangular solid. Find the volume in cubic inches.
77. If a box has a volume of 42 cubic feet, is it possible for you to fit inside the box? Explain your answer.
78. A box has a volume of 45 cubic inches. Will a can of soup fit inside the box? Explain your answer.
79. If you are 21 years old, you have been alive for over 650,000,000 seconds. Write this last number in scientific notation.
80. The distance around the earth at the equator is more than 130,000,000 feet. Write this number in scientific notation.
81. If you earn at least $12 an hour and work full time for 30 years, you will make at least 7.4×10^5 dollars. Write this last number in expanded form.
82. If your pulse is 72, then in one year your heart will beat at least 3.78×10^7 times. Write this last number in expanded form.
83. If you put $1,000 into a savings account every year from the time you are 25 years old until you are 55 years old, you will have more than 1.8×10^5 dollars in the account when you reach 55 years of age (assuming 10% annual interest). Write 1.8×10^5 in expanded form.
84. If you put $20 into a savings account every month from the time you are 20 years old until you are 30 years old, you will have over 3.27×10^3 dollars in the account when you reach 30 years of age (assuming 6% annual interest compounded monthly). Write 3.27×10^3 in expanded form.

85. If you were to use the definition of exponents as repeated multiplication to evaluate 10^2, 10^3, and 10^4, you would get $10^2 = 100$, $10^3 = 1,000$, and $10^4 = 10,000$. Without actually multiplying, what is 10^5? What is the relationship between the exponent on each of these powers of 10 and the number of zeros in the result?

86. Use the results found in Problem 85 to evaluate 10^1 and 10^0.

87. In this section we evaluated expressions like $(2^3)^2$, but not expressions like 2^{3^2}. What is the difference in the meanings of the two expressions? Evaluate each of them.

88. Evaluate the expressions $(2^2)^3$ and 2^{2^3}. (The answers in both cases are a little larger than usual, so be careful when you multiply.)

89. Here is an example of an analogy. These kinds of questions are common on entrance exams and aptitude tests.

x^4 is to x^2 as 16 is to

(a) 8   (b) 4   (c) 2   (d) 1

90. Here is an analogy problem that is a little more difficult than the first one:

x^2 is to x as Superman is to

(a) Mr. White   (b) Jimmy

(c) Clark Kent   (d) Lois Lane

Review Problems

The following problems review material on subtraction with negative numbers that we covered in Section 1.4. Reviewing these problems will help you understand the material in the next section.

Subtract.

91. $4 - 7$ **92.** $-4 - 7$

93. $4 - (-7)$ **94.** $-4 - (-7)$

95. $15 - 20$ **96.** $15 - (-20)$

97. $-15 - (-20)$ **98.** $-15 - 20$

SECTION

··

4.2   **Division with Exponents**

In Section 4.1 we found that multiplication with the same base results in addition of exponents, that is, $a^r \cdot a^s = a^{r+s}$. Since division is the inverse operation of multiplication, we can expect division with the same base to result in subtraction of exponents.

To develop the properties for exponents under division we again apply the definition of exponents:

$$\frac{x^5}{x^3} = \frac{x \cdot x \cdot x \cdot x \cdot x}{x \cdot x \cdot x}$$

$$= \frac{x \cdot x \cdot x}{x \cdot x \cdot x}(x \cdot x)$$

$$= 1(x \cdot x)$$

$$= x^2 \quad \text{Notice: } 5 - 3 = 2$$

$$\frac{2^4}{2^7} = \frac{2 \cdot 2 \cdot 2 \cdot 2}{2 \cdot 2 \cdot 2 \cdot 2 \cdot 2 \cdot 2 \cdot 2}$$

$$= \frac{2 \cdot 2 \cdot 2 \cdot 2}{2 \cdot 2 \cdot 2 \cdot 2} \cdot \frac{1}{2 \cdot 2 \cdot 2}$$

$$= \frac{1}{2 \cdot 2 \cdot 2}$$

$$= \frac{1}{2^3} \quad \text{Notice: } 7 - 4 = 3$$

In both cases, division with the same base resulted in subtraction of the smaller exponent from the larger. The problem is deciding whether the answer is a fraction or not. The problem is resolved quite easily by the following definition.

DEFINITION If r is a positive integer, then $a^{-r} = \dfrac{1}{a^r} = \left(\dfrac{1}{a}\right)^r$ $(a \neq 0)$

The following examples illustrate how we use this definition to simplify expressions that contain negative exponents.

▶ **EXAMPLES** Write each expression with a positive exponent and then simplify.

1. $2^{-3} = \dfrac{1}{2^3} = \dfrac{1}{8}$ Notice: Negative exponents do not indicate negative numbers. They indicate reciprocals.

2. $5^{-2} = \dfrac{1}{5^2} = \dfrac{1}{25}$

3. $3x^{-6} = 3 \cdot \dfrac{1}{x^6} = \dfrac{3}{x^6}$ ◀

Now let us look back to one of our original problems and try to work it again with the help of a negative exponent. We know that $2^4/2^7 = 1/2^3$. Let us decide now that with division of the same base, we will always subtract the exponent in the denominator from the exponent in the numerator and see if this conflicts with what we know is true.

$$\frac{2^4}{2^7} = 2^{4-7} \qquad \text{Subtracting the bottom exponent from the top exponent}$$

$$= 2^{-3} \qquad \text{Subtraction}$$

$$= \frac{1}{2^3} \qquad \text{Definition of negative exponents}$$

Subtracting the exponent in the denominator from the exponent in the numerator and then using the definition of negative exponents gives us the same result we obtained previously. We can now continue the list of properties of exponents we started in Section 4.1.

Property 4 for Exponents

If a is any real number and r and s are integers, then

$$\frac{a^r}{a^s} = a^{r-s} \quad (a \neq 0)$$

In words: To divide with the same base, subtract the exponent in the denominator from the exponent in the numerator and raise the base to the exponent that results.

The following examples show how we use Property 4 and the definition for negative exponents to simplify expressions involving division.

▶ **EXAMPLES**   Simplify the following expressions.

4. $\dfrac{x^9}{x^6} = x^{9-6} = x^3$

5. $\dfrac{x^4}{x^{10}} = x^{4-10} = x^{-6} = \dfrac{1}{x^6}$

6. $\dfrac{2^{15}}{2^{20}} = 2^{15-20} = 2^{-5} = \dfrac{1}{2^5} = \dfrac{1}{32}$ ◀

Our final property of exponents is similar to Property 3 from Section 4.1, but it involves division instead of multiplication. After we have stated the property, we will give a proof of it. The proof shows why this property is true.

Property 5 for Exponents

If a and b are any two real numbers ($b \neq 0$) and r is an integer, then

$$\left(\frac{a}{b}\right)^r = \frac{a^r}{b^r}$$

In words: A quotient raised to a power is the quotient of the powers.

PROOF

$$\left(\frac{a}{b}\right)^r = \left(a \cdot \frac{1}{b}\right)^r \qquad \text{By the definition of division}$$

$$= a^r \cdot \left(\frac{1}{b}\right)^r \qquad \text{By Property 3}$$

$$= a^r \cdot b^{-r} \qquad \text{By the definition of negative exponents}$$

$$= a^r \cdot \frac{1}{b^r} \qquad \text{By the definition of negative exponents}$$

$$= \frac{a^r}{b^r} \qquad \text{By the definition of division}$$

▶ **Examples** Simplify these expressions.

7. $\left(\dfrac{x}{2}\right)^3 = \dfrac{x^3}{2^3} = \dfrac{x^3}{8}$

8. $\left(\dfrac{5}{y}\right)^2 = \dfrac{5^2}{y^2} = \dfrac{25}{y^2}$

9. $\left(\dfrac{2}{3}\right)^4 = \dfrac{2^4}{3^4} = \dfrac{16}{81}$ ◀

Zero and One as Exponents

We have two special exponents left to deal with before our rules for exponents are complete: 0 and 1. To obtain an expression for x^1, we will solve a problem two different ways:

$$\left. \begin{array}{l} \dfrac{x^3}{x^2} = \dfrac{x \cdot x \cdot x}{x \cdot x} = x \\[2em] \dfrac{x^3}{x^2} = x^{3-2} = x^1 \end{array} \right\} \quad \text{Hence } x^1 = x$$

Stated generally, this rule says that $a^1 = a$. This seems reasonable and we will use it, since it is consistent with our property of division using the same base.

We use the same procedure to obtain an expression for x^0:

$$\left. \begin{array}{l} \dfrac{5^2}{5^2} = \dfrac{25}{25} = 1 \\[2em] \dfrac{5^2}{5^2} = 5^{2-2} = 5^0 \end{array} \right\} \quad \text{Hence } 5^0 = 1$$

It seems, therefore, that the best definition of x^0 is 1 for all x except $x = 0$. In the case of $x = 0$, we have 0^0, which we will not define. The definition $x^0 = 1$ will probably seem awkward at first. Most people would like to define x^0 as 0 when they first encounter it. Remember, the zero in this expression is an exponent, so x^0 does not mean to multiply by zero.

Here are some examples involving the exponents 0 and 1.

▶ **EXAMPLES**   Simplify the following expressions.

10. $8^0 = 1$
11. $8^1 = 8$
12. $4^0 + 4^1 = 1 + 4 = 5$
13. $(2x^2y)^0 = 1$ ◀

Here is a summary of the definitions and properties of exponents we have developed so far. For each definition or property in the list, a and b are real numbers, and r and s are integers.

Definitions	Properties
$$a^{-r} = \frac{1}{a^r} = \left(\frac{1}{a}\right)^r \quad a \neq 0$$ $$a^1 = a$$ $$a^0 = 1 \quad a \neq 0$$	1. $a^r \cdot a^s = a^{r+s}$ 2. $(a^r)^s = a^{rs}$ 3. $(ab)^r = a^r b^r$ 4. $\dfrac{a^r}{a^s} = a^{r-s} \qquad a \neq 0$ 5. $\left(\dfrac{a}{b}\right)^r = \dfrac{a^r}{b^r} \qquad b \neq 0$

Here are some additional examples. These examples use a combination of the preceding properties and definitions.

▶ **EXAMPLES**   Simplify each expression. Write all answers with positive exponents only.

14. $\dfrac{(5x^3)^2}{x^4} = \dfrac{25x^6}{x^4}$ Properties 2 and 3

$\qquad\qquad = 25x^2$ Property 4

15. $\dfrac{x^{-8}}{(x^2)^3} = \dfrac{x^{-8}}{x^6}$ Property 2

$\qquad\qquad = x^{-8-6}$ Property 4
$\qquad\qquad = x^{-14}$ Subtraction

$\qquad\qquad = \dfrac{1}{x^{14}}$ Definition of negative exponents

16. $\left(\dfrac{y^5}{y^3}\right)^2 = \dfrac{(y^5)^2}{(y^3)^2}$ Property 5

$\qquad\qquad = \dfrac{y^{10}}{y^6}$ Property 2

$\qquad\qquad = y^4$ Property 4

Notice in Example 16 that we could have simplified inside the parentheses first and then raised the result to the second power:

$$\left(\frac{y^5}{y^3}\right)^2 = (y^2)^2 = y^4$$

17. $(3x^5)^{-2} = \dfrac{1}{(3x^5)^2}$ Definition of negative exponents

       $= \dfrac{1}{9x^{10}}$ Properties 2 and 3

18. $x^{-8} \cdot x^5 = x^{-8+5}$ Property 1
       $= x^{-3}$ Addition

       $= \dfrac{1}{x^3}$ Definition of negative exponents

19. $\dfrac{(a^3)^2 a^{-4}}{(a^{-4})^3} = \dfrac{a^6 a^{-4}}{a^{-12}}$ Property 2

       $= \dfrac{a^2}{a^{-12}}$ Property 1

       $= a^{14}$ Property 4   ◄

In the next two examples we use division to compare the area and volume of geometric figures.

▶ **EXAMPLE 20** Suppose you have two squares, one of which is larger than the other. If the length of a side of the larger square is 3 times as long as the length of a side of the smaller square, how many of the smaller squares will it take to cover up the larger square?

Solution If we let x represent the length of a side of the smaller square, then the length of a side of the larger square is $3x$. The area of each square, along with a diagram of the situation, is given in Figure 1.

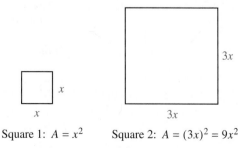

Square 1: $A = x^2$ Square 2: $A = (3x)^2 = 9x^2$

Figure 1

To find out how many smaller squares it will take to cover up the larger square, we divide the area of the larger square by the area of the smaller square.

$$\frac{\text{Area of square 2}}{\text{Area of square 1}} = \frac{9x^2}{x^2} = 9$$

It will take 9 of the smaller squares to cover the larger square. ◀

▶ **EXAMPLE 21**   Suppose you have two boxes, each of which is a cube. If the length of a side in the second box is 3 times as long as the length of a side of the first box, how many of the smaller boxes will fit inside the larger box?

Solution   If we let x represent the length of a side of the smaller box, then the length of a side of the larger box is $3x$. The volume of each box, along with a diagram of the situation, is given in Figure 2.

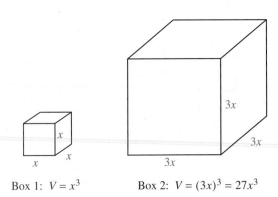

Box 1: $V = x^3$ Box 2: $V = (3x)^3 = 27x^3$

Figure 2

To find out how many smaller boxes will fit inside the larger box, we divide the volume of the larger box by the volume of the smaller box.

$$\frac{\text{Volume of box 2}}{\text{Volume of box 1}} = \frac{27x^3}{x^3} = 27$$

We can fit 27 of the smaller boxes inside the larger box. ◀

More on Scientific Notation

Now that we have completed our list of definitions and properties of exponents, we can expand the work we did previously with scientific notation.

Recall that a number is in scientific notation when it is written in the form

$$n \times 10^r$$

where $1 \leq n < 10$ and r is an integer.

Since negative exponents give us reciprocals, we can use negative exponents to write very small numbers in scientific notation. For example, the number 0.00057, when written in scientific notation, is equivalent to 5.7×10^{-4}. Here's why:

$$5.7 \times 10^{-4} = 5.7 \times \frac{1}{10^4} = 5.7 \times \frac{1}{10,000} = \frac{5.7}{10,000} = 0.00057$$

The table below lists some other numbers in both scientific notation and expanded form.

▶ **EXAMPLE 22**

Number Written the Long Way		Number Written Again in Scientific Notation
376,000	=	3.76×10^5
49,500	=	4.95×10^4
3,200	=	3.2×10^3
591	=	5.91×10^2
46	=	4.6×10^1
8	=	8×10^0
0.47	=	4.7×10^{-1}
0.093	=	9.3×10^{-2}
0.00688	=	6.88×10^{-3}
0.0002	=	2×10^{-4}
0.000098	=	9.8×10^{-5}

◀

Notice that in each case, when the number is written in scientific notation, the decimal point in the first number is placed so that the number is between 1 and 10. The exponent on 10 in the second number keeps track of the number of places we moved the decimal point in the original number to get a number between 1 and 10:

$$376,000 = 3.76 \times 10^5$$

Moved 5 places

Decimal point originally here

Keeps track of the 5 places we moved the decimal point

$$0.00688 = 6.88 \times 10^{-3}$$

Moved 3 places

Keeps track of the 3 places we moved the decimal point

PROBLEM SET 4.2

Write each of the following with positive exponents and then simplify, when possible.

1. 3^{-2} **2.** 3^{-3}
3. 6^{-2} **4.** 2^{-6}
5. 5^{-3} **6.** 9^{-2}
7. $2x^{-3}$ **8.** $5x^{-1}$
9. $(2x)^{-3}$ **10.** $(5x)^{-1}$
11. 10^{-2} **12.** 10^{-3}

Use Property 4 to simplify each of the following expressions. Write all answers that contain exponents with positive exponents only.

13. $\dfrac{5^3}{5^1}$ **14.** $\dfrac{7^8}{7^6}$

15. $\dfrac{5^1}{5^3}$ **16.** $\dfrac{7^6}{7^8}$

17. $\dfrac{x^{10}}{x^4}$ **18.** $\dfrac{x^4}{x^{10}}$

19. $\dfrac{6^{11}}{6}$ **20.** $\dfrac{8^7}{8}$

21. $\dfrac{6}{6^{11}}$ **22.** $\dfrac{8}{8^7}$

23. $\dfrac{2^{-5}}{2^3}$ **24.** $\dfrac{2^{-5}}{2^{-3}}$

25. $\dfrac{2^5}{2^{-3}}$ **26.** $\dfrac{2^{-3}}{2^{-5}}$

Simplify the following expressions. Any answers that contain exponents should contain positive exponents only.

27. $(3xy)^4$ **28.** $(4xy)^3$
29. 10^0 **30.** 10^1
31. $(2a^2b)^1$ **32.** $(2a^2b)^0$
33. $(7y^3)^{-2}$ **34.** $(5y^4)^{-2}$
35. $x^{-3}x^{-5}$ **36.** $x^{-6} \cdot x^8$
37. $y^7 \cdot y^{-10}$ **38.** $y^{-4} \cdot y^{-6}$

39. $\dfrac{(x^2)^3}{x^4}$ **40.** $\dfrac{(x^5)^3}{x^{10}}$

41. $\dfrac{(a^4)^3}{(a^3)^2}$ **42.** $\dfrac{(a^5)^3}{(a^5)^2}$

43. $\left(\dfrac{y^7}{y^2}\right)^8$ **44.** $\left(\dfrac{y^2}{y^3}\right)^4$

45. $\dfrac{(x^{-2})^3}{x^{-5}}$ **46.** $\dfrac{(x^2)^{-3}}{x^{-5}}$

47. $\left(\dfrac{x^{-2}}{x^{-5}}\right)^3$ **48.** $\left(\dfrac{x^2}{x^{-5}}\right)^{-3}$

49. $\dfrac{(a^3)^2(a^4)^5}{(a^5)^2}$ **50.** $\dfrac{(a^4)^8(a^2)^5}{(a^3)^4}$

51. $\dfrac{(a^{-2})^3(a^4)^2}{(a^{-3})^{-2}}$ **52.** $\dfrac{(a^{-5})^{-3}(a^7)^{-1}}{(a^{-3})^5}$

53. $\dfrac{(x^{-7})^3(x^4)^5}{(x^3)^2(x^{-1})^8}$ **54.** $\dfrac{(x^{-9})^0(x^4)^{-3}}{(x^{-1})^5(x^5)^{-2}}$

Write each of the following numbers in scientific notation.

55. 0.000357   **56.** 0.00357
57. 35,700   **58.** 3,570
59. 0.0048   **60.** 0.000048
61. 0.000009   **62.** 0.0009

Write each of the following numbers in expanded form.

63. 4.23×10^{-3} **64.** 4.23×10^3
65. 5.6×10^4 **66.** 5.6×10^{-4}
67. 8×10^{-5} **68.** 8×10^5
69. 7.89×10^1 **70.** 7.89×10^0

Applying the Concepts

71. Some home computers can do a calculation in 2×10^{-3} seconds. Write this number in expanded form.

72. Some of the cells in the human body have a radius of 3×10^{-5} inches. Write this number in expanded form.

73. One of the smallest ants in the world has a length of only 0.006 inches. Write this number in scientific notation.

74. Some cameras used in scientific research can take 1 picture every 0.000000167 second. Write this number in scientific notation.

75. The number 25×10^3 is not in scientific notation because 25 is larger than 10. Write 25×10^3 in scientific notation.

76. The number 0.25×10^3 is not in scientific notation because 0.25 is less than 1. Write 0.25×10^3 in scientific notation.

77. The number 23.5×10^4 is not in scientific notation because 23.5 is not between 1 and 10. Rewrite 23.5×10^4 in scientific notation.

78. The number 375×10^3 is not in scientific notation because 375 is not between 1 and 10. Rewrite 375×10^3 in scientific notation.

79. The number 0.82×10^{-3} is not in scientific notation because 0.82 is not between 1 and 10. Rewrite 0.82×10^{-3} in scientific notation.

80. The number 0.93×10^{-2} is not in scientific notation because 0.93 is not between 1 and 10. Rewrite 0.93×10^{-2} in scientific notation.

Suppose you have two squares, one of which is larger than the other. Suppose further that the side of the larger square is twice as long as the side of the smaller square.

81. If the length of the side of the smaller square is 10 inches, give the area of each square. Then find the number of smaller squares it will take to cover the larger square.

82. How many smaller squares will it take to cover the larger square if the length of the side of the smaller square is 1 foot?

83. If the length of the side of the smaller square is x, find the area of each square. Then find the number of smaller squares it will take to cover the larger square.

84. Suppose the length of the side of the larger square is 1 foot. How many smaller squares will it take to cover the larger square?

Suppose you have two boxes, each of which is a cube. Suppose further that the length of a side of the second box is twice as long as the length of a side of the first box.

85. If the length of a side of the first box is 6 inches, give the volume of each box. Then find the number of smaller boxes that will fit inside the larger box.

86. How many smaller boxes can be placed inside the larger box if the length of a side of the second box is 1 foot?

87. If the length of a side of the first box is x, find the volume of each box. Then find the number of smaller boxes that will fit inside the larger box.

88. Suppose the length of a side of the larger box is 12 inches. How many smaller boxes will fit inside the larger box?

Review Problems

The following problems review material we covered in Sections 2.1 and 3.4. Reviewing the problems from Section 2.1 will help you understand some of the material in the next section.

Simplify the following expressions. [2.1]

89. $4x + 3x$ **90.** $9x + 7x$
91. $5a - 3a$ **92.** $10a - 2a$
93. $4y + 5y + y$ **94.** $6y - y + 2y$

Find the x- and y-intercepts for each equation and then use the intercepts to draw the graph of the equation. [3.4]

95. $3x - 2y = 6$ **96.** $2x - 3y = 6$
97. $2x + y = 4$ **98.** $x + 2y = 4$

Operations with Monomials

We have developed all the tools necessary to perform the four basic operations on the simplest of polynomials: monomials.

> **DEFINITION** A **monomial** is a one-term expression that is either a constant (number) or the product of a constant and one or more variables raised to whole number exponents.

The following are examples of monomials:

$$-3 \qquad 15x \qquad -23x^2y \qquad 49x^4y^2z^4 \qquad \frac{3}{4}a^2b^3$$

The numerical part of each monomial is called the **numerical coefficient,** or just **coefficient.** Monomials are also called **terms.**

Multiplication and Division of Monomials

There are two basic steps involved in the multiplication of monomials. First, we rewrite the products using the commutative and associative properties. Then, we simplify by multiplying coefficients and adding exponents of like bases.

▶ **EXAMPLES** Multiply.

1. $(-3x^2)(4x^3) = (-3 \cdot 4)(x^2 \cdot x^3)$ — Commutative and associative properties

   $\qquad\qquad\qquad = -12x^5$ — Multiply coefficients, add exponents

2. $\left(\frac{4}{5}x^5y^2\right)(10x^3y) = \left(\frac{4}{5} \cdot 10\right)(x^5 \cdot x^3)(y^2 \cdot y)$ — Commutative and associative properties

   $\qquad\qquad\qquad = 8x^8y^3$ — Multiply coefficients, add exponents ◀

You can see that in each case the work was the same—multiply coefficients and add exponents of the same base. We can expect division of monomials to proceed in a similar way. Since our properties are consistent, division of monomials will result in division of coefficients and subtraction of exponents of like bases.

▶ **EXAMPLES** Divide.

3. $\dfrac{15x^3}{3x^2} = \dfrac{15}{3} \cdot \dfrac{x^3}{x^2}$ — Write as separate fractions

   $\qquad = 5x$ — Divide coefficients, subtract exponents

4. $\dfrac{39x^2y^3}{3xy^5} = \dfrac{39}{3} \cdot \dfrac{x^2}{x} \cdot \dfrac{y^3}{y^5}$ Write as separate fractions

$\qquad\qquad = 13x \cdot \dfrac{1}{y^2}$ Divide coefficients, subtract exponents

$\qquad\qquad = \dfrac{13x}{y^2}$ Write answer as a single fraction   ◀

In Example 4, the expression y^3/y^5 simplifies to $1/y^2$ because of Property 4 for exponents and the definition of negative exponents. If we were to show all the work in this simplification process, it would look like this:

$$\dfrac{y^3}{y^5} = y^{3-5}$$ Property 4 for exponents

$$\qquad = y^{-2}$$ Subtraction

$$\qquad = \dfrac{1}{y^2}$$ Definition of negative exponents

The point of this explanation is this: Even though we may not show all the steps when simplifying an expression involving exponents, the result we obtain can still be justified using the properties of exponents. We have not introduced any new properties in Example 4; we have just not shown the details of each simplification.

▶ **EXAMPLE 5**   Divide.

$$\dfrac{25a^5b^3}{50a^2b^7} = \dfrac{25}{50} \cdot \dfrac{a^5}{a^2} \cdot \dfrac{b^3}{b^7}$$ Write as separate fractions

$$\qquad = \dfrac{1}{2} \cdot a^3 \cdot \dfrac{1}{b^4}$$ Divide coefficients, subtract exponents

$$\qquad = \dfrac{a^3}{2b^4}$$ Write answer as a single fraction   ◀

Notice in Example 5 that dividing 25 by 50 results in $\frac{1}{2}$. This is the same result we would obtain if we reduced the fraction $\frac{25}{50}$ to lowest terms, and there is no harm in thinking of it that way. Also, notice that the expression b^3/b^7 simplifies to $1/b^4$ by Property 4 for exponents and the definition of negative exponents, even though we have not shown the steps involved in doing so.

Multiplication and Division of Numbers Written in Scientific Notation

We multiply and divide numbers written in scientific notation using the same steps we used to multiply and divide monomials.

▶ **EXAMPLE 6**   Multiply $(4 \times 10^7)(6 \times 10^{-4})$. Write your answer in scientific notation.

Solution   Since multiplication is commutative and associative, we can rearrange the order of these numbers and group them as follows:

$$(4 \times 10^7)(6 \times 10^{-4}) = (4 \times 6)(10^7 \times 10^{-4})$$
$$= 24 \times 10^3$$

The number 24×10^3 is not in scientific notation because 24 is larger than 10. To rewrite 24×10^3 in scientific notation we change 24 to 2.4, then we change 10^3 to 10^4. The answer in scientific notation is

$$2.4 \times 10^4$$

Notice that we add exponents, $7 + (-4) = 3$, when we multiply with the same base. ◀

▶ **EXAMPLE 7**   Divide $\dfrac{9.6 \times 10^{12}}{3 \times 10^4}$.

Solution   We group the numbers between 1 and 10 separately from the powers of 10 and proceed as we did in Example 6:

$$\frac{9.6 \times 10^{12}}{3 \times 10^4} = \frac{9.6}{3} \times \frac{10^{12}}{10^4}$$
$$= 3.2 \times 10^8$$

Notice that the procedure we used in both of these examples is very similar to multiplication and division of monomials, for which we multiplied or divided coefficients and added or subtracted exponents. ◀

Addition and Subtraction of Monomials

Addition and subtraction of monomials will be almost identical, since subtraction is defined as addition of the opposite. With multiplication and division of monomials, the key was rearranging the numbers and variables using the commutative and associative properties. With addition, the key is application of the distributive property. We sometimes use the phrase *combine monomials* to describe addition and subtraction of monomials.

> **DEFINITION**   Two terms (monomials) with the same variable part (same variables raised to the same powers) are called **similar** (or *like*) **terms.**

You can add only similar terms. This is because the distributive property (which is the key to addition of monomials) cannot be applied to terms that are not similar.

▶ **EXAMPLES** Combine the following monomials.

8. $-3x^2 + 15x^2 = (-3 + 15)x^2$ Distributive property
$$= 12x^2$$ Add coefficients

9. $9x^2y - 20x^2y = (9 - 20)x^2y$ Distributive property
$$= -11x^2y$$ Add coefficients

10. $5x^2 + 8y^2$ In this case we cannot apply the distributive property, so we cannot add the monomials ◀

The next examples show how we simplify expressions containing monomials when more than one operation is involved.

▶ **EXAMPLE 11** Simplify $\dfrac{(6x^4y)(3x^7y^5)}{9x^5y^2}$.

Solution We begin by multiplying the two monomials in the numerator:

$$\frac{(6x^4y)(3x^7y^5)}{9x^5y^2} = \frac{18x^{11}y^6}{9x^5y^2} \qquad \text{Simplify numerator}$$

$$= 2x^6y^4 \qquad \text{Divide} \qquad ◀$$

▶ **EXAMPLE 12** Simplify $\dfrac{(6.8 \times 10^5)(3.9 \times 10^{-7})}{7.8 \times 10^{-4}}$.

Solution We group the numbers between 1 and 10 separately from the powers of 10:

$$\frac{(6.8)(3.9)}{7.8} \times \frac{(10^5)(10^{-7})}{10^{-4}} = 3.4 \times 10^{5 + (-7) - (-4)}$$

$$= 3.4 \times 10^2 \qquad ◀$$

▶ **EXAMPLE 13** Simplify $\dfrac{14x^5}{2x^2} + \dfrac{15x^8}{3x^5}$.

Solution Simplifying each expression separately and then combining similar terms gives

$$\frac{14x^5}{2x^2} + \frac{15x^8}{3x^5} = 7x^3 + 5x^3 \qquad \text{Divide}$$

$$= 12x^3 \qquad \text{Add} \qquad ◀$$

▶ **EXAMPLE 14** A rectangular solid is twice as long as it is wide and one-half as high as it is wide. Write an expression for the volume.

Solution We begin by making a diagram of the object (Figure 1) with the dimensions labeled as given in the problem.

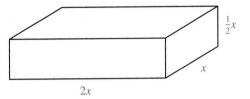

Figure 1

The volume is the product of the three dimensions.

$$V = 2x \cdot x \cdot \frac{1}{2}x = x^3$$

The box has the same volume as a cube with side x, as shown in Figure 2.

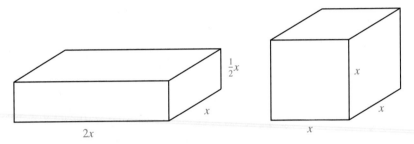

Equal Volumes

Figure 2

More about Graphing

In Example 2 of Section 3.1, we compared the numbers in the sequence of squares with their position in the sequence using a table and a scatter diagram, reproduced below as Table 1 and Figure 3.

TABLE 1 Squares

Position	Value
1	1
2	4
3	9
4	16
5	25

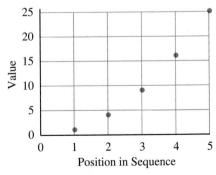

Figure 3

If we let x represent the position of a number in the sequence of squares and y represent the value of the number in the sequence, then each pair of numbers in the table satisfies the equation $y = x^2$. There are many other pairs of numbers that satisfy the equation $y = x^2$ besides those shown in the table. If we expand the numbers available for x to include negative numbers, we begin to see the relationship between the variables in the equation $y = x^2$, as Example 15 illustrates.

▶ **EXAMPLE 15**   Graph the equation $y = x^2$.

Solution   Just as we did when we first began graphing straight lines, we start by finding some ordered pairs that are solutions to the equation. In this case, we can use a table to help us find these ordered pairs. The table we will use looks like this:

x	$y = x^2$	y

We can choose any convenient numbers for x and then use the equation $y = x^2$ to find the corresponding values for y. Let's use the values -3, -2, $-1, 0, 1, 2$, and 3 for x and find corresponding values for y. Here is how the table looks when we let x have these values:

x	$y = x^2$	y
-3	$y = (-3)^2 = 9$	9
-2	$y = (-2)^2 = 4$	4
-1	$y = (-1)^2 = 1$	1
0	$y = 0^2 = 0$	0
1	$y = 1^2 = 1$	1
2	$y = 2^2 = 4$	4
3	$y = 3^2 = 9$	9

The table gives us the solutions $(-3, 9)$, $(-2, 4)$, $(-1, 1)$, $(0, 0)$, $(1, 1)$, $(2, 4)$, and $(3, 9)$ for the equation $y = x^2$. We plot each of the points on a rectangular coordinate system and draw a smooth curve between them, as shown in Figure 4 on the next page.

This graph is called a **parabola**. All equations of the form $y = ax^2$ $(a \neq 0)$ produce parabolas when graphed. The point $(0, 0)$ is called the **vertex** of the parabola in Figure 4.

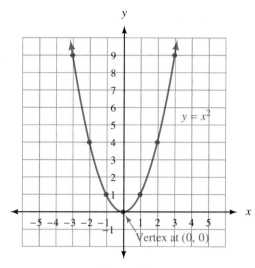

Figure 4

PROBLEM SET 4.3

Multiply.

1. $(3x^4)(4x^3)$

2. $(6x^5)(-2x^2)$

3. $(-2y^4)(8y^7)$

4. $(5y^{10})(2y^5)$

5. $(8x)(4x)$

6. $(7x)(5x)$

7. $(10a^3)(10a)(2a^2)$

8. $(5a^4)(10a)(10a^4)$

9. $(6ab^2)(-4a^2b)$

10. $(-5a^3b)(4ab^4)$

11. $(4x^2y)(3x^3y^3)(2xy^4)$

12. $(5x^6)(-10xy^4)(-2x^2y^6)$

Divide. Write all answers with positive exponents only.

13. $\dfrac{15x^3}{5x^2}$

14. $\dfrac{25x^5}{5x^4}$

15. $\dfrac{18y^9}{3y^{12}}$

16. $\dfrac{24y^4}{-8y^7}$

17. $\dfrac{32a^3}{64a^4}$

18. $\dfrac{25a^5}{75a^6}$

19. $\dfrac{21a^2b^3}{-7ab^5}$

20. $\dfrac{32a^5b^6}{8ab^5}$

21. $\dfrac{3x^3y^2z}{27xy^2z^3}$

22. $\dfrac{5x^5y^4z}{30x^3yz^2}$

23. $\dfrac{144x^9y^2}{-12x^{10}y^8}$

24. $\dfrac{256x^9y^2}{32x^4}$

Find each product. Write all answers in scientific notation.

25. $(3 \times 10^3)(2 \times 10^5)$

26. $(4 \times 10^8)(1 \times 10^6)$

27. $(3.5 \times 10^4)(5 \times 10^{-6})$

28. $(7.1 \times 10^5)(2 \times 10^{-8})$

29. $(5.5 \times 10^{-3})(2.2 \times 10^{-4})$

30. $(3.4 \times 10^{-2})(4.5 \times 10^{-6})$

Find each quotient. Write all answers in scientific notation.

31. $\dfrac{8.4 \times 10^5}{2 \times 10^2}$

32. $\dfrac{9.6 \times 10^{20}}{3 \times 10^6}$

33. $\dfrac{6 \times 10^8}{2 \times 10^{-2}}$

34. $\dfrac{8 \times 10^{12}}{4 \times 10^{-3}}$

35. $\dfrac{2.5 \times 10^{-6}}{5 \times 10^{-4}}$

36. $\dfrac{4.5 \times 10^{-8}}{9 \times 10^{-4}}$

Combine by adding or subtracting as indicated.

37. $3x^2 + 5x^2$

38. $4x^3 + 8x^3$

39. $8x^5 - 19x^5$

40. $75x^6 - 50x^6$

41. $2a + a - 3a$

42. $5a + a - 6a$

43. $10x^3 - 8x^3 + 2x^3$

44. $7x^5 + 8x^5 - 12x^5$

45. $20ab^2 - 19ab^2 + 30ab^2$

46. $18a^3b^2 - 20a^3b^2 + 10a^3b^2$

47. $-4abc - 9abc - abc$

48. $-7abc - abc - abc$

Simplify. Write all answers with positive exponents only.

49. $\dfrac{(3x^2)(8x^5)}{6x^4}$

50. $\dfrac{(7x^3)(6x^8)}{14x^5}$

51. $\dfrac{(9a^2b)(2a^3b^4)}{18a^5b^7}$

52. $\dfrac{(21a^5b)(2a^8b^4)}{14ab}$

53. $\dfrac{(4x^3y^2)(9x^4y^{10})}{(3x^5y)(2x^6y)}$

54. $\dfrac{(5x^4y^4)(10x^3y^3)}{(25xy^5)(2xy^7)}$

Simplify each expression and write all answers in scientific notation.

55. $\dfrac{(6 \times 10^8)(3 \times 10^5)}{9 \times 10^7}$

56. $\dfrac{(8 \times 10^4)(5 \times 10^{10})}{2 \times 10^7}$

57. $\dfrac{(5 \times 10^3)(4 \times 10^{-5})}{2 \times 10^{-2}}$

58. $\dfrac{(7 \times 10^6)(4 \times 10^{-4})}{1.4 \times 10^{-3}}$

59. $\dfrac{(2.8 \times 10^{-7})(3.6 \times 10^4)}{2.4 \times 10^3}$

60. $\dfrac{(5.4 \times 10^2)(3.5 \times 10^{-9})}{4.5 \times 10^6}$

Simplify.

61. $\dfrac{18x^4}{3x} + \dfrac{21x^7}{7x^4}$

62. $\dfrac{24x^{10}}{6x^4} + \dfrac{32x^7}{8x}$

63. $\dfrac{45a^6}{9a^4} - \dfrac{50a^8}{2a^6}$

64. $\dfrac{16a^9}{4a} - \dfrac{28a^{12}}{4a^4}$

65. $\dfrac{6x^7y^4}{3x^2y^2} + \dfrac{8x^5y^8}{2y^6}$

66. $\dfrac{40x^{10}y^{10}}{8x^2y^5} + \dfrac{10x^8y^8}{5y^3}$

Use your knowledge of the properties and definitions of exponents to find x in each of the following.

67. $4^x \cdot 4^5 = 4^7$

68. $\dfrac{5^x}{5^3} = 5^4$

69. $(7^3)^x = 7^{12}$

70. $\dfrac{3^x}{3^4} = 9$

71. The statement $(a + b)^2 = a^2 + b^2$ looks similar to Property 3 for exponents. However, it is not a property of exponents because almost every time we replace a and b with numbers, this expression becomes a false statement. Let $a = 4$ and $b = 5$ in the expressions $(a + b)^2$ and $a^2 + b^2$ and see what each simplifies to.

72. Show that the statement $(a - b)^2 = a^2 - b^2$ is not, in general, true by substituting 3 for a and 5 for b in each of the expressions and then simplifying each result.

73. Show that the expressions $(a + b)^2$ and $a^2 + 2ab + b^2$ are equal when $a = 3$ and $b = 4$.

74. Show that the expressions $(a - b)^2$ and $a^2 - 2ab + b^2$ are equal when $a = 7$ and $b = 5$.

Applying the Concepts

75. A rectangle is twice as long as it is wide. Write an expression for the perimeter and an expression for the area of the rectangle.

76. A rectangle is three times as long as it is wide. Write an expression for the perimeter and an expression for the area of the rectangle.

77. A rectangular solid is twice as long as it is wide. If the height is 4 inches, find an expression for the volume.

78. A rectangular solid is four times as long as it is wide. If the height is 2 inches, find an expression for the volume.

79. A box is four times as long as it is wide and half as high as it is wide. Find an expression for the volume.

80. The length of a box is twice the height, while the width is half the height. Find an expression for the volume.

Graphing

Each of the following equations will have a graph that has a shape similar to the graph shown in Figure 4 of this section. For each equation, fill in the given ordered pairs and use them to draw the graph of the equation.

81. $y = x^2$ $(-4, \ ), (-2, \ ), (-1, \ ), (0, \ ),$ $(1, \ ), (2, \ ), (4, \ )$

82. $y = x^2$ $(-2, \ ), (-1.5, \ ), (-\frac{1}{2}, \ ),$ $(0, \ ), (\frac{1}{2}, \ ), (1.5, \ ), (2, \ )$

83. $y = 2x^2$ $(-3, \ ), (-2, \ ), (-1, \ ),$ $(0, \ ), (1, \ ), (2, \ ), (3, \ )$

84. $y = 3x^2$ $(-3, \ ), (-2, \ ), (-1, \ ),$ $(0, \ ), (1, \ ), (2, \ ), (3, \ )$

85. $y = \frac{1}{2}x^2$ $(-4, \ ), (-2, \ ), (-1, \ ),$ $(0, \ ), (1, \ ), (2, \ ), (4, \ )$

86. $y = \frac{1}{3}x^2$ $(-3, \ ), (-1, \ ), (0, \ ), (1, \ ),$ $(3, \ )$

87. $y = \frac{1}{4}x^2$ $(-4, \ ), (-2, \ ), (-1, \ ),$ $(0, \ ), (1, \ ), (2, \ ), (4, \ )$

88. $y = 4x^2$ $(-2, \ ), (-1, \ ), (0, \ ), (1, \ ),$ $(2, \ )$

Review Problems

The problems that follow review material we covered in Sections 2.1 and 3.3. Reviewing the problems from Section 2.1 will help you understand some of the material in the next section.

Find the value of each expression when $x = -2$. [2.1]

89. $4x$

90. $-3x$

91. $-2x + 5$

92. $-4x - 1$

93. $x^2 + 5x + 6$

94. $x^2 - 5x + 6$

For each equation below, complete the given ordered pairs so each is a solution to the equation and then use the ordered pairs to graph the equation. [3.3]

95. $y = 2x + 2$ $(-2, \ ), (0, \ ), (2, \ )$

96. $y = 2x - 3$ $(-1, \ ), (0, \ ), (2, \ )$

97. $y = \frac{1}{3}x + 1$ $(-3, \ ), (0, \ ), (3, \ )$

98. $y = \frac{1}{2}x - 2$ $(-2, \ ), (0, \ ), (2, \ )$

Research Project 6

Two numbers that are used frequently in astronomy and sometimes written in scientific notation are

1. a light-year
2. the speed of light

Give a definition for each of these numbers using both scientific notation and expanded form, then show how the two numbers are related to one another. Finally, give two examples from astronomy that involve light-years. Organize your work into an essay that flows smoothly from one idea to the next.

SECTION

4.4 **Addition and Subtraction of Polynomials**

In this section we will extend what we learned in Section 4.3 to expressions called polynomials. We begin this section with the definition of a polynomial.

DEFINITION A **polynomial** is a finite sum of monomials (terms).

EXAMPLES The following are polynomials:

$$3x^2 + 2x + 1 \qquad 15x^2y + 21xy^2 - y^2 \qquad 3a - 2b + 4c - 5d$$

Polynomials can be further classified by the number of terms they contain. A polynomial with two terms is called a **binomial.** If it has three terms, it is a **trinomial.** As stated before, a **monomial** has only one term.

> **DEFINITION** The **degree** of a polynomial in one variable is the highest power to which the variable is raised.

EXAMPLES

$3x^5 + 2x^3 + 1$ A trinomial of degree 5
$2x + 1$ A binomial of degree 1
$3x^2 + 2x + 1$ A trinomial of degree 2
$3x^5$ A monomial of degree 5
-9 A monomial of degree 0

There are no new rules for adding one or more polynomials. We rely only on our previous knowledge. Here are some examples.

▶ **EXAMPLE 1**   Add $(2x^2 - 5x + 3) + (4x^2 + 7x - 8)$.

Solution   We use the commutative and associative properties to group similar terms together and then apply the distributive property to add

$(2x^2 - 5x + 3) + (4x^2 + 7x - 8)$
$= (2x^2 + 4x^2) + (-5x + 7x) + (3 - 8)$ Commutative and associative properties
$= (2 + 4)x^2 + (-5 + 7)x + (3 - 8)$ Distributive property
$= 6x^2 + 2x - 5$ Addition

The results here indicate that to add two polynomials, we add coefficients of similar terms.   ◀

▶ **EXAMPLE 2**   Add $x^2 + 3x + 2x + 6$.

Solution   The only similar terms here are the two middle terms. We combine them as usual to get

$$x^2 + 3x + 2x + 6 = x^2 + 5x + 6$$ ◀

You will recall from Chapter 1 the definition of subtraction: $a - b = a + (-b)$. To subtract one expression from another, we simply add its opposite. The letters a and b in the definition can each represent polynomials. The opposite of a polynomial is the opposite of each of its terms. When you subtract one polynomial from another you subtract each of its terms.

▶ **EXAMPLE 3**   Subtract $(3x^2 + x + 4) - (x^2 + 2x + 3)$.

Solution   To subtract $x^2 + 2x + 3$, we change the sign of each of its terms and add. If you are having trouble remembering why we do this, remember that we can think of $-(x^2 + 2x + 3)$ as $-1(x^2 + 2x + 3)$. If we distribute the -1 across $x^2 + 2x + 3$, we get $-x^2 - 2x - 3$:

$(3x^2 + x + 4) - (x^2 + 2x + 3)$

$$= 3x^2 + x + 4 - x^2 - 2x - 3 \quad \text{Take the opposite of each}$$
$$\text{term in the second polynomial}$$

$$= (3x^2 - x^2) + (x - 2x) + (4 - 3)$$

$$= 2x^2 - x + 1 \qquad \blacktriangleleft$$

▶ **EXAMPLE 4**   Subtract $-4x^2 + 5x - 7$ from $x^2 - x - 1$.

Solution   The polynomial $x^2 - x - 1$ comes first, then the subtraction sign, and finally the polynomial $-4x^2 + 5x - 7$ in parentheses.

$(x^2 - x - 1) - (-4x^2 + 5x - 7)$

$$= x^2 - x - 1 + 4x^2 - 5x + 7 \quad \text{We take the opposite of each}$$
$$\text{term in the second polynomial}$$

$$= (x^2 + 4x^2) + (-x - 5x) + (-1 + 7)$$

$$= 5x^2 - 6x + 6 \qquad \blacktriangleleft$$

There are two important points to remember when adding or subtracting polynomials. First, to add or subtract two polynomials you always add or subtract *coefficients* of similar terms. Second, the exponents never increase in value when you are adding or subtracting similar terms.

The last topic we want to consider in this section is finding the value of a polynomial for a given value of the variable.

To find the value of the polynomial $3x^2 + 1$ when x is 5, we replace x with 5 and simplify the result:

When	$x = 5$
the polynomial	$3x^2 + 1$
becomes	$3(5)^2 + 1 = 3(25) + 1$
	$= 75 + 1$
	$= 76$

▶ **EXAMPLE 5**   Find the value of $3x^2 - 5x + 4$ when $x = -2$.

Solution

When	$x = -2$
the polynomial	$3x^2 - 5x + 4$

becomes

$$3(-2)^2 - 5(-2) + 4 = 3(4) + 10 + 4$$
$$= 12 + 10 + 4$$
$$= 26 \quad \blacktriangleleft$$

P R O B L E M   S E T   4 . 4

Identify each of the following polynomials as a trinomial, binomial, or monomial, and give the degree in each case.

1. $2x^3 - 3x^2 + 1$
2. $4x^2 - 4x + 1$
3. $5 + 8a - 9a^3$
4. $6 + 12x^3 + x^4$
5. $2x - 1$
6. $4 + 7x$
7. $45x^2 - 1$
8. $3a^3 + 8$
9. $7a^2$
10. $90x$
11. -4
12. 56

Perform the following additions and subtractions.

13. $(2x^2 + 3x + 4) + (3x^2 + 2x + 5)$
14. $(x^2 + 5x + 6) + (x^2 + 3x + 4)$
15. $(3a^2 - 4a + 1) + (2a^2 - 5a + 6)$
16. $(5a^2 - 2a + 7) + (4a^2 - 3a + 2)$
17. $x^2 + 4x + 2x + 8$ **18.** $x^2 + 5x - 3x - 15$
19. $6x^2 - 3x - 10x + 5$
20. $10x^2 + 30x - 2x - 6$
21. $x^2 - 3x + 3x - 9$
22. $x^2 - 5x + 5x - 25$
23. $3y^2 - 5y - 6y + 10$
24. $y^2 - 18y + 2y - 12$
25. $(6x^3 - 4x^2 + 2x) + (9x^2 - 6x + 3)$
26. $(5x^3 + 2x^2 + 3x) + (2x^2 + 5x + 1)$
27. $\left(\dfrac{2}{3}x^2 - \dfrac{1}{5}x - \dfrac{3}{4}\right) + \left(\dfrac{4}{3}x^2 - \dfrac{4}{5}x + \dfrac{7}{4}\right)$
28. $\left(\dfrac{3}{8}x^3 - \dfrac{5}{7}x^2 - \dfrac{2}{5}\right) + \left(\dfrac{5}{8}x^3 - \dfrac{2}{7}x^2 + \dfrac{7}{5}\right)$
29. $(a^2 - a - 1) - (-a^2 + a + 1)$
30. $(5a^2 - a - 6) - (-3a^2 - 2a + 4)$

31. $\left(\dfrac{5}{9}x^3 + \dfrac{1}{3}x^2 - 2x + 1\right) - \left(\dfrac{2}{3}x^3 + x^2 + \dfrac{1}{2}x - \dfrac{3}{4}\right)$
32. $\left(4x^3 - \dfrac{2}{5}x^2 + \dfrac{3}{8}x - 1\right) - \left(\dfrac{9}{2}x^3 + \dfrac{1}{4}x^2 - x + \dfrac{5}{6}\right)$
33. $(4y^2 - 3y + 2) + (5y^2 + 12y - 4) - (13y^2 - 6y + 20)$
34. $(2y^2 - 7y - 8) - (6y^2 + 6y - 8) + (4y^2 - 2y + 3)$
35. Subtract $10x^2 + 23x - 50$ from $11x^2 - 10x + 13$.
36. Subtract $2x^2 - 3x + 5$ from $4x^2 - 5x + 10$.
37. Subtract $3y^2 + 7y - 15$ from $11y^2 + 11y + 11$.
38. Subtract $15y^2 - 8y - 2$ from $3y^2 - 3y + 2$.
39. Add $50x^2 - 100x - 150$ to $25x^2 - 50x + 75$.
40. Add $7x^2 - 8x + 10$ to $-8x^2 + 2x - 12$.
41. Subtract $2x + 1$ from the sum of $3x - 2$ and $11x + 5$.
42. Subtract $3x - 5$ from the sum of $5x + 2$ and $9x - 1$.
43. Find the value of the polynomial $x^2 - 2x + 1$ when x is 3.
44. Find the value of the polynomial $(x - 1)^2$ when x is 3.
45. Find the value of the polynomial $(y - 5)^2$ when y is 10.
46. Find the value of the polynomial $y^2 - 10y + 25$ when y is 10.
47. Find the value of $a^2 + 4a + 4$ when a is 2.
48. Find the value of $(a + 2)^2$ when a is 2.

Review Problems

The problems below review material we covered in Section 4.3. Reviewing these problems will help you understand the next section.

Multiply. [4.3]

49. $2x(5x)$
50. $2x(-2x)$
51. $3x(-5x)$
52. $-3x(-7x)$
53. $2x(3x^2)$
54. $x^2(3x)$
55. $3x^2(2x^2)$
56. $4x^2(2x^2)$

SECTION

4.5 **Multiplication with Polynomials**

We begin our discussion of multiplication of polynomials by finding the product of a monomial and a trinomial.

▶ **EXAMPLE 1** Multiply $3x^2(2x^2 + 4x + 5)$.

Solution Applying the distributive property gives us

$$3x^2(2x^2 + 4x + 5) = 3x^2(2x^2) + 3x^2(4x) + 3x^2(5) \quad \text{Distributive} \\ \text{property}$$

$$= 6x^4 + 12x^3 + 15x^2 \qquad \text{Multiplication} \ \blacktriangleleft$$

The distributive property is the key to multiplication of polynomials. We can use it to find the product of any two polynomials. There are some shortcuts we can use in certain situations, however. Let's look at an example that involves the product of two binomials.

▶ **EXAMPLE 2** Multiply $(3x - 5)(2x - 1)$.

Solution

$$(3x - 5)(2x - 1) = 3x(2x - 1) - 5(2x - 1)$$

$$= 3x(2x) + 3x(-1) + (-5)(2x) + (-5)(-1)$$

$$= 6x^2 - 3x - 10x + 5$$

$$= 6x^2 - 13x + 5 \qquad\qquad\qquad \blacktriangleleft$$

If we look closely at the second and third lines of work in this example, we can see that the terms in the answer come from all possible products of terms in the first binomial with terms in the second binomial. This result is generalized as follows.

> **Rule**
> To multiply any two polynomials, multiply each term in the first with each term in the second.

There are two ways we can put this rule to work.

FOIL Method

If we look at the original problem in Example 2 and then to the answer, we see that the first term in the answer came from multiplying the first terms in each binomial:

$$3x \cdot 2x = 6x^2 \qquad \text{FIRST}$$

The middle term in the answer came from adding the products of the two outside terms with the two inside terms in each binomial:

$$3x(-1) = -\ 3x \qquad \text{OUTSIDE}$$
$$-5(2x) = \underline{-\ 10x} \qquad \text{INSIDE}$$
$$-\ 13x$$

The last term in the answer came from multiplying the two last terms:

$$-5(-1) = 5 \qquad \text{LAST}$$

To summarize the FOIL method we will multiply another two binomials.

▶ **EXAMPLE 3**   Multiply $(2x + 3)(5x - 4)$.

Solution

$$(2x + 3)(5x - 4) = 2x(5x) + 2x(-4) + 3(5x) + 3(-4)$$

First

Outside

Inside

Last

$$= 10x^2 - 8x + 15x - 12$$
$$= 10x^2 + 7x - 12$$

With practice $-8x + 15x = 7x$ can be done mentally.   ◀

COLUMN Method

The FOIL method can be applied only when multiplying two binomials. To find products of polynomials with more than two terms we use what is called the COLUMN method.

The COLUMN method of multiplying two polynomials is very similar to long multiplication with whole numbers. It is just another way of finding all possible products of terms in one polynomial with terms in another polynomial.

▶ **EXAMPLE 4** Multiply $(2x + 3)(3x^2 - 2x + 1)$.

Solution

$$
\begin{array}{r}
3x^2 - 2x + 1 \\
2x + 3 \\
\hline
6x^3 - 4x^2 + 2x \qquad \leftarrow 2x(3x^2 - 2x + 1) \\
9x^2 - 6x + 3 \leftarrow 3(3x^2 - 2x + 1) \\
\hline
6x^3 + 5x^2 - 4x + 3 \leftarrow \text{Add similar terms}
\end{array}
$$
◄

It will be to your advantage to become very fast and accurate at multiplying polynomials. You should be comfortable using either method. The following examples illustrate the three types of multiplication.

▶ **EXAMPLES** Multiply.

5. $4a^2(2a^2 - 3a + 5) = 4a^2(2a^2) + 4a^2(-3a) + 4a^2(5)$
$$= 8a^4 - 12a^3 + 20a^2$$

6. $(x - 2)(y + 3) = \underset{\text{F}}{x(y)} + \underset{\text{O}}{x(3)} + \underset{\text{I}}{(-2)(y)} + \underset{\text{L}}{(-2)(3)}$

$$= xy + 3x - 2y - 6$$

7. $(x + y)(a - b) = \underset{\text{F}}{x(a)} + \underset{\text{O}}{x(-b)} + \underset{\text{I}}{y(a)} + \underset{\text{L}}{y(-b)}$

$$= xa - xb + ya - yb$$

8. $(5x - 1)(2x + 6) = \underset{\text{F}}{5x(2x)} + \underset{\text{O}}{5x(6)} + \underset{\text{I}}{(-1)(2x)} + \underset{\text{L}}{(-1)(6)}$

$$= 10x^2 + 30x + (-2x) + (-6)$$
$$= 10x^2 + 28x - 6$$

9. $(3x + 2)(x^2 - 5x + 6)$

$$
\begin{array}{r}
x^2 - 5x + 6 \\
3x + 2 \\
\hline
3x^3 - 15x^2 + 18x \\
2x^2 - 10x + 12 \\
\hline
3x^3 - 13x^2 + 8x + 12
\end{array}
$$
◄

▶ **EXAMPLE 10** The length of a rectangle is 3 more than twice the width. Write an expression for the area of the rectangle.

Solution We begin by drawing a rectangle and labeling the width with x. Since the length is 3 more than twice the width, we label the length with $2x + 3$.

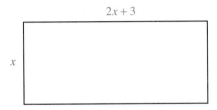

Since the area, A, of a rectangle is the product of the length and width, we write our formula for the area of this rectangle as

$$A = x(2x + 3)$$
$$A = 2x^2 + 3x \qquad \text{Multiply} \qquad \blacktriangleleft$$

▶ **EXAMPLE 11**   A piece of cardboard that is 17 inches long and 11 inches wide is used to construct a box. The box is constructed by cutting a square from each corner of the cardboard (each square is the same size) and then folding the flaps up, as shown in Figure 1.

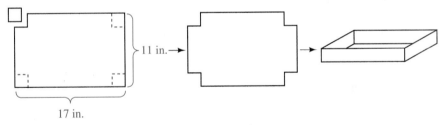

Figure 1

Make a table that gives the dimensions and volume of the box if the corner squares that are removed have sides of 1, 2, 3, 4, and 5 inches. Then find a formula that gives the volume of the box that is formed if a square of side x inches is cut from each corner.

Solution   Table 1 shows the dimensions and volume for the different boxes that are constructed when the squares are removed from each corner of the 11-by-17-inch piece of cardboard. To understand where the numbers in the table come from, look at the third row in the table (it begins with the number 3). The 3 is the length of the side of each square that is cut from the corners of the original rectangle. If the length of the original rectangle is 17 inches and a 3-inch square is cut from each corner, then the length of the box that is formed is $17 - 2 \cdot 3$. Likewise, if the width of the original rectangle is 11 inches, then the width of the resulting box is $11 - 2 \cdot 3$. The height of the box is equal to the side of the square that is removed from each corner.

TABLE 1 Constructing a Box from a 17-inch-by-11-inch Rectangle

| Corner Square | Dimensions and Volume of the Box | | | |
Side	Length	Width	Height	Volume
1	15	9	1	135 in.^{3}
2	13	7	2	182 in.^{3}
3	11	5	3	165 in.^{3}
4	9	3	4	108 in.^{3}
5	7	1	5	35 in.^{3}

The volume of each box is found by multiplying the length, width, and height. If the length of the side of the square that is cut from each corner is x, then the length of the box will be $17 - 2x$. The width of the box will be $11 - 2x$, and the height of the box will be x. The volume is the product of these dimensions, or

$$V = (17 - 2x)(11 - 2x)x$$

If we were to multiply this out we would obtain a third-degree polynomial. However, it may be easier to work with this formula in the form it is in, so multiplication is unnecessary. For example, substituting 3 for x in the formula gives us

$$V = (17 - 2 \cdot 3)(11 - 2 \cdot 3)(3) = (11)(5)(3) = 165$$

A histogram that shows the relationship between the length of the side of the square that is removed and the volume of the resulting box is shown in Figure 2.

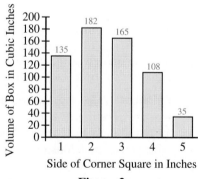

Figure 2

Suppose that a store sells x items at p dollars per item. The total amount of money obtained by selling the items is called the *revenue*. It can be found by multiplying the number of items sold, x, by the price per item, p. For example, if 100 items are sold for $6 each, the revenue is $100(6) = \$600$. Similarly, if 500 items are sold for $8 each, the total revenue is 500(8) = $4,000. If we denote the revenue with the letter R, then the formula that relates R, x, and p is

$$\text{Revenue} = (\text{number of items sold})(\text{price of each item})$$

In symbols: $R = xp$.

▶ **EXAMPLE 12**   A store selling diskettes for home computers knows from past experience that it can sell x diskettes each day at a price of p dollars per diskette, according to the equation $x = 800 - 100p$. Write a formula for the daily revenue that involves only the variables R and p.

Solution   From our discussion above, we know that the revenue R is given by the formula

$$R = xp$$

But, since $x = 800 - 100p$, we can substitute $800 - 100p$ for x in the revenue equation to obtain

$$R = (800 - 100p)p$$
$$R = 800p - 100p^2$$

This last formula gives the revenue, R, in terms of the price, p.   ◀

P R O B L E M   S E T   4 . 5

Multiply the following by applying the distributive property.

1. $2x(3x + 1)$ **2.** $4x(2x - 3)$
3. $2x^2(3x^2 - 2x + 1)$ **4.** $5x(4x^3 - 5x^2 + x)$
5. $2ab(a^2 - ab + 1)$
6. $3a^2b(a^3 + a^2b^2 + b^3)$
7. $y^2(3y^2 + 9y + 12)$ **8.** $5y(2y^2 - 3y + 5)$
9. $4x^2y(2x^3y + 3x^2y^2 + 8y^3)$
10. $6xy^3(2x^2 + 5xy + 12y^2)$

Multiply the binomials below. You should do about half the problems using the FOIL method and the other half using the COLUMN method. Remember, you want to be comfortable using both methods.

11. $(x + 3)(x + 4)$ **12.** $(x + 2)(x + 5)$
13. $(x + 6)(x + 1)$ **14.** $(x + 1)(x + 4)$

15. $\left(x + \dfrac{1}{2}\right)\left(x + \dfrac{3}{2}\right)$

16. $\left(x + \dfrac{3}{5}\right)\left(x + \dfrac{2}{5}\right)$

17. $(a + 5)(a - 3)$ **18.** $(a - 8)(a + 2)$
19. $(x - a)(y + b)$ **20.** $(x + a)(y - b)$
21. $(x + 6)(x - 6)$ **22.** $(x + 3)(x - 3)$

23. $\left(y + \dfrac{5}{6}\right)\left(y - \dfrac{5}{6}\right)$

24. $\left(y - \dfrac{4}{7}\right)\left(y + \dfrac{4}{7}\right)$

25. $(2x - 3)(x - 4)$ **26.** $(3x - 5)(x - 2)$
27. $(a + 2)(2a - 1)$ **28.** $(a - 6)(3a + 2)$
29. $(2x - 5)(3x - 2)$ **30.** $(3x + 6)(2x - 1)$
31. $(2x + 3)(a + 4)$ **32.** $(2x - 3)(a - 4)$
33. $(5x - 4)(5x + 4)$ **34.** $(6x + 5)(6x - 5)$

35. $\left(2x - \dfrac{1}{2}\right)\left(x + \dfrac{3}{2}\right)$

36. $\left(4x - \dfrac{3}{2}\right)\left(x + \dfrac{1}{2}\right)$

37. $(1 - 2a)(3 - 4a)$ **38.** $(1 - 3a)(3 + 2a)$
39. $(7 - 6x)(8 - 5x)$ **40.** $(7 - 4x)(8 - 3x)$

Multiply the following.

41. $(x + 1)(x^2 + 3x - 4)$
42. $(x - 2)(x^2 + 3x - 4)$
43. $(a - 3)(a^2 - 3a + 2)$
44. $(a + 5)(a^2 + 2a + 3)$
45. $(x + 2)(x^2 - 2x + 4)$
46. $(x + 3)(x^2 - 3x + 9)$
47. $(2x + 1)(x^2 + 8x + 9)$
48. $(3x - 2)(x^2 - 7x + 8)$
49. $(5x^2 + 2x + 1)(x^2 - 3x + 5)$
50. $(2x^2 + x + 1)(x^2 - 4x + 3)$

Multiply.

51. $(x^2 + 3)(2x^2 - 5)$
52. $(4x^3 - 8)(5x^3 + 4)$
53. $(3a^4 + 2)(2a^2 + 5)$
54. $(7a^4 - 8)(4a^3 - 6)$
55. $(x + 3)(x + 4)(x + 5)$
56. $(x - 3)(x - 4)(x - 5)$
57. The length of a rectangle is 5 more than twice the width. Write an expression for the area of the rectangle.
58. The length of a rectangle is 2 more than three times the width. Write an expression for the area of the rectangle.
59. The width and length of a rectangle are given by two consecutive integers. Write an expression for the area of the rectangle.
60. The width and length of a rectangle are given by two consecutive even integers. Write an expression for the area of the rectangle.
61. A piece of cardboard 17 inches long and 11 inches wide is used to construct a box by cutting a square from each corner of the cardboard, as shown in Example 11 in this section. Table 2 is an extension of Table 1 from that example.
(a) Fill in the missing numbers in the table.
(b) Use the data from the table and the coordinate system below it to construct a line graph that shows the relationship between the length of the side of the corner square and the volume of the resulting box.

TABLE 2 Constructing a Box from a 17-inch-by-11-inch Rectangle

Corner Square	Dimensions and Volume of the Box			
Side	Length	Width	Height	Volume
0				
1	15	9	1	135 in.^{3}
1.5				
2	13	7	2	182 in.^{3}
2.5				
3	11	5	3	165 in.^{3}
3.5				
4	9	3	4	108 in.^{3}
4.5				
5	7	1	5	35 in.^{3}
5.5				

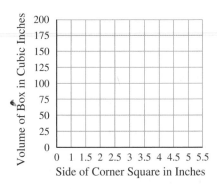

62. A piece of cardboard 11 inches long and 11 inches wide is used to construct a box by cutting a square from each corner of the cardboard. Table 3 has data that correspond to that problem.
(a) Fill in the missing numbers in the table.
(b) Use the data from the table and the coordinate system below it to construct a line graph that shows the relationship between the length of the side of the corner square and the volume of the resulting box.

TABLE 3   Constructing a Box from an 11-inch-by-11-inch Square

Corner Square	Dimensions and Volume of the Box			
Side	Length	Width	Height	Volume
0				
1	9	9	1	81 in.^{3}
1.5				
2	7	7	2	98 in.^{3}
2.5				
3	5	5	3	75 in.^{3}
3.5				
4	3	3	4	36 in.^{3}
4.5				
5	1	1	5	5 in.^{3}
5.5				

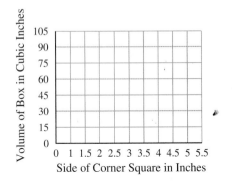

63. A store selling typewriter ribbons knows that the number of ribbons it can sell each week, x, is related to the price per ribbon, p, by the equation $x = 1,200 - 100p$. Write an expression for the weekly revenue that involves only the variables R and p. (Remember: The equation for revenue is $R = xp$.)

64. A store selling small portable radios knows from past experience that the number of radios it can sell each week, x, is related to the price per radio, p, by the equation $x = 1,300 - 100p$. Write an expression for the weekly revenue that involves only the variables R and p.

65. The relationship between the number of calculators a company sells per day, x, and the price of each calculator, p, is given by the equation $x = 1,700 - 100p$. Write an expression for the daily revenue that involves only the variables R and p.

66. The relationship between the number of pencil sharpeners a company can sell each day, x, and the price of each sharpener, p, is given by the equation $x = 1,800 - 100p$. Write an expression for the daily revenue that involves only the variables R and p.

Review Problems

The problems that follow review material we covered in Section 3.6.

Find the slope and y-intercept for each of the following equations by first writing them in the form $y = mx + b$.

67. $-2x + 4y = 8$ **68.** $5x - 3y = 15$

For each problem below, the slope and one point on a line are given. Find the equation.

69. $(-2, -6)$, $m = 3$ **70.** $(4, 2)$, $m = \frac{1}{2}$

Find the equation of the line that passes through each pair of points.

71. $(-3, -5)$, $(3, 1)$ **72.** $(-1, -5)$, $(2, 1)$

SECTION

4.6   Binomial Squares and Other Special Products

In this section we will combine the results of the last section with our definition of exponents to find some special products.

▶ **EXAMPLE 1**   Find the square of $(3x - 2)$.

Solution   To square $(3x - 2)$, we multiply it by itself:

$$
\begin{aligned}
(3x - 2)^2 &= (3x - 2)(3x - 2) && \text{Definition of exponents} \\
&= 9x^2 - 6x - 6x + 4 && \text{FOIL method} \\
&= 9x^2 - 12x + 4 && \text{Combine similar terms} \quad ◀
\end{aligned}
$$

Notice that the first and last terms in the answer are the square of the first and last terms in the original binomial and that the middle term is twice the product of the two terms in the original binomial.

▶ **EXAMPLES**

2. $(a + b)^2 = (a + b)(a + b)$
$\qquad\qquad = a^2 + 2ab + b^2$

3. $(a - b)^2 = (a - b)(a - b)$
$\qquad\qquad = a^2 - 2ab + b^2$ ◀

Binomial squares having the form of Examples 2 and 3 occur very frequently in algebra. It will be to your advantage to memorize the following rule for squaring a binomial.

Rule

The square of a binomial is the sum of the square of the first term, the square of the last term, and twice the product of the two original terms. In symbols, this rule is written as follows:

$$(x + y)^2 = \underset{\substack{\text{Square} \\ \text{of} \\ \text{first} \\ \text{term}}}{x^2} + \underset{\substack{\text{Twice} \\ \text{product} \\ \text{of the} \\ \text{two terms}}}{2xy} + \underset{\substack{\text{Square} \\ \text{of} \\ \text{last} \\ \text{term}}}{y^2}$$

Note   A very common mistake when squaring binomials is to write

$$(a + b)^2 = a^2 + b^2$$

which just isn't true. The mistake becomes obvious when we substitute 2 for a and 3 for b:

$$(2 + 3)^2 \neq 2^2 + 3^2$$

$$25 \neq 13$$

Exponents do not distribute over addition or subtraction.

▶ **EXAMPLES** Multiply using the preceding rule.

	First term squared	Twice their product	Last term squared	Answer
4. $(x - 5)^2 =$	x^2	$+ 2(x)(-5) +$	25	$= x^2 - 10x + 25$
5. $(x + 2)^2 =$	x^2	$+ 2(x)(2) +$	4	$= x^2 + 4x + 4$
6. $(2x - 3)^2 =$	$4x^2$	$+ 2(2x)(-3) +$	9	$= 4x^2 - 12x + 9$
7. $(5x - 4)^2 =$	$25x^2$	$+ 2(5x)(-4) +$	16	$= 25x^2 - 40x + 16$ ◀

Another special product that occurs frequently is $(a + b)(a - b)$. The only difference in the two binomials is the sign between the two terms. The interesting thing about this type of product is that the middle term is always zero. Here are some examples.

▶ **EXAMPLES** Multiply using the FOIL method.

8. $(2x - 3)(2x + 3) = 4x^2 + 6x - 6x - 9$ FOIL method
$= 4x^2 - 9$

9. $(x - 5)(x + 5) = x^2 + 5x - 5x - 25$ FOIL method
$= x^2 - 25$

10. $(3x - 1)(3x + 1) = 9x^2 + 3x - 3x - 1$ FOIL method
$= 9x^2 - 1$ ◀

Notice that in each case the middle term is zero and therefore doesn't appear in the answer. The answers all turn out to be the difference of two squares. Here is a rule to help you memorize the result.

Rule

When multiplying two binomials that differ only in the sign between their terms, subtract the square of the last term from the square of the first term.

$$(a - b)(a + b) = a^2 - b^2$$

Here are some problems that result in the difference of two squares.

▶ **EXAMPLES** Multiply using the preceding rule.

11. $(x + 3)(x - 3) = x^2 - 9$
12. $(a + 2)(a - 2) = a^2 - 4$
13. $(9a + 1)(9a - 1) = 81a^2 - 1$
14. $(2x - 5y)(2x + 5y) = 4x^2 - 25y^2$
15. $(3a - 7b)(3a + 7b) = 9a^2 - 49b^2$ ◀

Although all the problems in this section can be worked correctly using the methods in the last section, they can be done much faster if the two rules are *memorized.* Here is a summary of the two rules:

$$(a + b)^2 = (a + b)(a + b) = a^2 + 2ab + b^2$$

$$(a - b)^2 = (a - b)(a - b) = a^2 - 2ab + b^2$$

$$(a - b)(a + b) = a^2 - b^2$$

▶ **EXAMPLE 16**   Write an expression in symbols for the sum of the squares of three consecutive even integers. Then, simplify that expression.

Solution   If we let $x =$ the first of the even integers, then $x + 2$ is the next consecutive even integer, and $x + 4$ is the one after that. An expression for the sum of their squares is

$$x^2 + (x + 2)^2 + (x + 4)^2 \qquad \text{Sum of squares}$$
$$= x^2 + (x^2 + 4x + 4) + (x^2 + 8x + 16) \qquad \text{Expand squares}$$
$$= 3x^2 + 12x + 20 \qquad \text{Add similar terms}$$

◀

PROBLEM SET 4.6

Perform the indicated operations.

1. $(x - 2)^2$
2. $(x + 2)^2$
3. $(a + 3)^2$
4. $(a - 3)^2$
5. $(x - 5)^2$
6. $(x - 4)^2$
7. $\left(a - \dfrac{1}{2}\right)^2$
8. $\left(a + \dfrac{1}{2}\right)^2$
9. $(x + 10)^2$
10. $(x - 10)^2$
11. $(a + 0.8)^2$
12. $(a - 0.4)^2$
13. $(2x - 1)^2$
14. $(3x + 2)^2$
15. $(4a + 5)^2$
16. $(4a - 5)^2$
17. $(3x - 2)^2$
18. $(2x - 3)^2$
19. $(3a + 5b)^2$
20. $(5a - 3b)^2$
21. $(4x - 5y)^2$
22. $(5x + 4y)^2$
23. $(7m + 2n)^2$
24. $(2m - 7n)^2$

25. $(6x - 10y)^2$
26. $(10x + 6y)^2$
27. $(x^2 + 5)^2$
28. $(x^2 + 3)^2$
29. $(a^2 + 1)^2$
30. $(a^2 - 2)^2$
31. $(x^3 - 7)^2$
32. $(x^3 + 4)^2$
33. $(x - 3)(x + 3)$
34. $(x + 4)(x - 4)$
35. $(a + 5)(a - 5)$
36. $(a - 6)(a + 6)$
37. $(y - 1)(y + 1)$
38. $(y - 2)(y + 2)$
39. $(9 + x)(9 - x)$
40. $(10 - x)(10 + x)$
41. $(2x + 5)(2x - 5)$
42. $(3x + 5)(3x - 5)$
43. $\left(4x + \dfrac{1}{3}\right)\left(4x - \dfrac{1}{3}\right)$
44. $\left(6x + \dfrac{1}{4}\right)\left(6x - \dfrac{1}{4}\right)$
45. $(2a + 7)(2a - 7)$

46. $(3a + 10)(3a - 10)$
47. $(6 - 7x)(6 + 7x)$ **48.** $(7 - 6x)(7 + 6x)$
49. $(x^2 + 3)(x^2 - 3)$ **50.** $(x^2 + 2)(x^2 - 2)$
51. $(a^2 + 4)(a^2 - 4)$
52. $(a^2 + 9)(a^2 - 9)$
53. $(5y^4 - 8)(5y^4 + 8)$
54. $(7y^5 + 6)(7y^5 - 6)$

Multiply and simplify.

55. $(x + 3)(x - 3) + (x - 5)(x + 5)$
56. $(x - 7)(x + 7) + (x - 4)(x + 4)$
57. $(2x + 3)^2 - (4x - 1)^2$
58. $(3x - 5)^2 - (2x + 3)^2$
59. $(a + 1)^2 - (a + 2)^2 + (a + 3)^2$
60. $(a - 1)^2 + (a - 2)^2 - (a - 3)^2$
61. $(2x + 3)^3$
62. $(3x - 2)^3$
63. The formula for the difference of two squares can be used as a shortcut to multiplying certain whole numbers if they have the correct form. Use the difference of two squares formula to multiply 49(51) by first writing 49 as $(50 - 1)$ and 51 as $(50 + 1)$.
64. Use the difference of two squares formula to multiply 101(99) by first writing 101 as $(100 + 1)$ and 99 as $(100 - 1)$.
65. Evaluate the expression $(x + 3)^2$ and the expression $x^2 + 6x + 9$ for $x = 2$.
66. Evaluate the expression $x^2 - 25$ and the expression $(x - 5)(x + 5)$ for $x = 6$.
67. Write an expression for the sum of the squares of two consecutive integers. Then, simplify that expression.
68. Write an expression for the sum of the squares of two consecutive odd integers. Then, simplify that expression.
69. Write an expression for the sum of the squares of three consecutive integers. Then, simplify that expression.
70. Write an expression for the sum of the squares of three consecutive odd integers. Then, simplify that expression.
71. We can use the concept of area to further justify our rule for squaring a binomial. The length of each side of the square shown in the figure is

$a + b$. (The longer line segment has length a and the shorter line segment has length b.) The area of the whole square is $(a + b)^2$. On the other hand, the whole area is the sum of the areas of the two smaller squares and the two smaller rectangles that make it up. Write the area of the two smaller squares and the two smaller rectangles and then add them together to verify the formula $(a + b)^2 = a^2 + 2ab + b^2$.

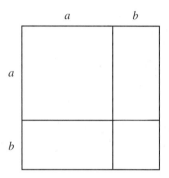

72. The length of each side of the large square shown in the figure is $x + 5$. Therefore, its area is $(x + 5)^2$. Find the area of the two smaller squares and the two smaller rectangles that make up the large square, then add them together to verify the formula $(x + 5)^2 = x^2 + 10x + 25$.

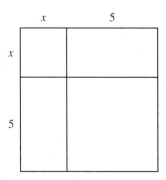

Review Problems

The problems below review material we covered in Sections 4.3 and 3.7. Reviewing the problems from Section 4.3 will help you understand the next section.

Simplify each expression (divide). [4.3]

73. $\dfrac{10x^3}{5x}$ **74.** $\dfrac{-15x^2}{5x}$

75. $\dfrac{15x^2y}{3xy}$ **76.** $\dfrac{21xy^2}{3xy}$

77. $\dfrac{35a^6b^8}{70a^2b^{10}}$ **78.** $\dfrac{75a^2b^6}{25a^4b^3}$

Solve each system by graphing. [3.7]

79. $x + y = 2$ **80.** $x + y = 1$
$\quad\ x - y = 4$ $\quad\ x - y = -3$
81. $y = 2x + 3$ **82.** $y = 3x - 2$
$\quad\ y = -2x - 1$ $\quad\ y = -2x + 3$

Research Project 7

The title on the diagram below is *Binomial Expansions* because each line gives the expansion of the binomial $x + y$ raised to a whole number power.

Binomial Expansions

$$(x + y)^0 = \qquad\qquad 1$$
$$(x + y)^1 = \qquad\qquad x + y$$
$$(x + y)^2 = \qquad\quad x^2 + 2xy + y^2$$
$$(x + y)^3 = \qquad x^3 + 3x^2y + 3xy^2 + y^3$$
$$(x + y)^4 =$$
$$(x + y)^5 =$$

Complete the diagram above by expanding the binomials $(x + y)^4$ and $(x + y)^5$ using the multiplication procedures you have learned in this section. Then look for patterns in the completed diagram that will allow you to continue the diagram one more row, without using multiplication. When you are finished, write an essay in which you describe what you have done and the results you have obtained.

SECTION

4.7   Dividing a Polynomial by a Monomial

To divide a polynomial by a monomial, we will use the definition of division and apply the distributive property. Follow the steps in this example closely.

▶ **EXAMPLE 1** Divide $10x^3 - 15x^2$ by $5x$.

Solution

$$\frac{10x^3 - 15x^2}{5x} = (10x^3 - 15x^2)\frac{1}{5x}$$
Division by $5x$ is the same as multiplication by $\frac{1}{5x}$

$$= 10x^3\left(\frac{1}{5x}\right) - 15x^2\left(\frac{1}{5x}\right)$$
Distribute $\frac{1}{5x}$ to both terms

$$= \frac{10x^3}{5x} - \frac{15x^2}{5x}$$
Multiplication by $\frac{1}{5x}$ is the same as division by $5x$

$$= 2x^2 - 3x$$
Division of monomials as done in Section 4.3 ◀

If we were to leave out the first steps, the problem would look like this:

$$\frac{10x^3 - 15x^2}{5x} = \frac{10x^3}{5x} - \frac{15x^2}{5x}$$

$$= 2x^2 - 3x$$

The problem is much shorter and clearer this way. You may leave out the first two steps from Example 1 when working problems in this section. They are part of Example 1 only to help show you why the following rule is true.

> **Rule**
>
> To divide a polynomial by a monomial, simply divide each term in the polynomial by the monomial.

   Here are some further examples using our rule for division of a polynomial by a monomial.

▶ **EXAMPLE 2**   Divide $\dfrac{3x^2 - 6}{3}$.

Solution   We begin by writing the 3 in the denominator under each term in the numerator. Then we simplify the result:

$$\frac{3x^2 - 6}{3} = \frac{3x^2}{3} - \frac{6}{3} \qquad \text{Divide each term in the numerator by 3}$$

$$= x^2 - 2 \qquad \text{Simplify} \qquad \blacktriangleleft$$

▶ **EXAMPLE 3**   Divide $\dfrac{4x^2 - 2}{2}$.

Solution   Dividing each term in the numerator by 2 we have

$$\frac{4x^2 - 2}{2} = \frac{4x^2}{2} - \frac{2}{2} \qquad \text{Divide each term in the numerator by 2}$$

$$= 2x^2 - 1 \qquad \text{Simplify} \qquad \blacktriangleleft$$

▶ **EXAMPLE 4**   Find the quotient of $27x^3 - 9x^2$ and $3x$.

Solution   We are again asked to divide the first polynomial by the second one:

$$\frac{27x^3 - 9x^2}{3x} = \frac{27x^3}{3x} - \frac{9x^2}{3x} \qquad \text{Divide each term by } 3x$$

$$= 9x^2 - 3x \qquad \text{Simplify} \qquad \blacktriangleleft$$

▶ **EXAMPLE 5** Divide $(15x^2y - 21xy^2) \div (-3xy)$.

Solution This is the same type of problem we have shown in the first four examples; it is just worded a little differently. Note that when we divide each term in the first polynomial by $-3xy$, the negative sign must be taken into account:

$$\frac{15x^2y - 21xy^2}{-3xy} = \frac{15x^2y}{-3xy} - \frac{21xy^2}{-3xy} \qquad \text{Divide each term by } -3xy$$

$$= -5x - (-7y) \qquad \text{Simplify}$$

$$= -5x + 7y \qquad \text{Simplify} \qquad ◀$$

▶ **EXAMPLE 6** Divide $\dfrac{10x^3 - 5x^2 + 20x}{10x^3}$.

Solution Proceeding as we have in the first five examples, we distribute the $10x^3$ under each term in the numerator:

$$\frac{10x^3 - 5x^2 + 20x}{10x^3} = \frac{10x^3}{10x^3} - \frac{5x^2}{10x^3} + \frac{20x}{10x^3} \qquad \begin{array}{c}\text{Divide each term}\\\text{by } 10x^3\end{array}$$

$$= 1 - \frac{1}{2x} + \frac{2}{x^2} \qquad \text{Simplify}$$

Note that our answer is not a polynomial. ◀

▶ **EXAMPLE 7** Divide $\dfrac{24x^3y^2 + 16x^2y^2 - 4x^2y^3}{8x^2y}$.

Solution Writing $8x^2y$ under each term in the numerator and then simplifying, we have

$$\frac{24x^3y^2 + 16x^2y^2 - 4x^2y^3}{8x^2y} = \frac{24x^3y^2}{8x^2y} + \frac{16x^2y^2}{8x^2y} - \frac{4x^2y^3}{8x^2y}$$

$$= 3xy + 2y - \frac{y^2}{2} \qquad ◀$$

From the examples in this section, it is clear that to divide a polynomial by a monomial, we must divide each term in the polynomial by the monomial. Often, students taking algebra for the first time will make the following mistake:

$$\frac{x + \cancel{2}}{\cancel{2}} = x + 1 \qquad \text{Mistake}$$

The mistake here is in not dividing both terms in the numerator by 2. The correct way to divide $x + 2$ by 2 looks like this:

$$\frac{x + 2}{2} = \frac{x}{2} + \frac{2}{2} = \frac{x}{2} + 1 \qquad \text{Correct}$$

P R O B L E M   S E T   4 . 7

Divide the following polynomials by $5x$.

1. $5x^2 - 10x$

2. $10x^3 - 15x$

3. $15x - 10x^3$

4. $50x^3 - 20x^2$

5. $25x^2y - 10xy$

6. $15xy^2 + 20x^2y$

7. $35x^5 - 30x^4 + 25x^3$

8. $40x^4 - 30x^3 + 20x^2$

9. $50x^5 - 25x^3 + 5x$

10. $75x^6 + 50x^3 - 25x$

Divide the following by $-2a$.

11. $8a^2 - 4a$

12. $a^3 - 6a^2$

13. $16a^5 + 24a^4$

14. $30a^6 + 20a^3$

15. $8ab + 10a^2$

16. $6a^2b - 10ab^2$

17. $12a^3b - 6a^2b^2 + 14ab^3$

18. $4ab^3 - 16a^2b^2 - 22a^3b$

19. $a^2 + 2ab + b^2$

20. $a^2b - 2ab^2 + b^3$

Perform the following divisions (find the following quotients).

21. $\dfrac{6x + 8y}{2}$

22. $\dfrac{9x - 3y}{3}$

23. $\dfrac{7y - 21}{-7}$

24. $\dfrac{14y - 12}{2}$

25. $\dfrac{10xy - 8x}{2x}$

26. $\dfrac{12xy^2 - 18x}{-6x}$

27. $\dfrac{x^2y - x^3y^2}{x}$

28. $\dfrac{x^2y - x^3y^2}{x^2}$

29. $\dfrac{x^2y - x^3y^2}{-x^2y}$

30. $\dfrac{ab + a^2b^2}{ab}$

31. $\dfrac{a^2b^2 - ab^2}{-ab^2}$

32. $\dfrac{a^2b^2c + ab^2c^2}{abc}$

33. $\dfrac{x^3 - 3x^2y + xy^2}{x}$

34. $\dfrac{x^2 - 3xy^2 + xy^3}{x}$

35. $\dfrac{10a^2 - 15a^2b + 25a^2b^2}{5a^2}$

36. $\dfrac{11a^2b^2 - 33ab}{-11ab}$

37. $\dfrac{26x^2y^2 - 13xy}{-13xy}$

38. $\dfrac{6x^2y^2 - 3xy}{6xy}$

39. $\dfrac{4x^2y^2 - 2xy}{4xy}$

40. $\dfrac{6x^2a + 12x^2b - 6x^2c}{36x^2}$

41. $\dfrac{5a^2x - 10ax^2 + 15a^2x^2}{20a^2x^2}$

42. $\dfrac{12ax - 9bx + 18cx}{6x^2}$

43. $\dfrac{16x^5 + 8x^2 + 12x}{12x^3}$

44. $\dfrac{27x^2 - 9x^3 - 18x^4}{-18x^3}$

Divide. Assume all variables represent positive numbers.

45. $\dfrac{9a^{5m} - 27a^{3m}}{3a^{2m}}$

46. $\dfrac{26a^{3m} - 39a^{5m}}{13a^{3m}}$

47. $\dfrac{10x^{5m} - 25x^{3m} + 35x^m}{5x^m}$

48. $\dfrac{18x^{2m} + 24x^{4m} - 30x^{6m}}{6x^{2m}}$

Simplify each numerator, and then divide.

49. $\dfrac{2x^3(3x + 2) - 3x^2(2x - 4)}{2x^2}$

50. $\dfrac{5x^2(6x - 3) + 6x^3(3x - 1)}{3x}$

51. $\dfrac{(x + 2)^2 - (x - 2)^2}{2x}$

52. $\dfrac{(x - 3)^2 - (x + 3)^2}{3x}$

53. $\dfrac{(x + 5)^2 + (x + 5)(x - 5)}{2x}$

54. $\dfrac{(x - 4)^2 + (x + 4)(x - 4)}{2x}$

55. Evaluate the expression $\dfrac{10x + 15}{5}$ and the expression $2x + 3$ when $x = 2$.

56. Evaluate the expression $\dfrac{6x^2 + 4x}{2x}$ and the expression $3x + 2$ when $x = 5$.

57. Show that the expression $\dfrac{3x + 8}{2}$ is not the same as the expression $3x + 4$ by replacing x with 10 in both expressions and simplifying the results.

58. Show that the expression $\dfrac{x + 10}{x}$ is not equal to 10 by replacing x with 5 and simplifying.

Review Problems

The following problems review material we covered in Sections 3.8 and 3.9.

Solve each system of equations by the elimination method. [3.8]

59. $x + y = 6$
$x - y = 8$

60. $2x + y = 5$
$-x + y = -4$

61. $2x - 3y = -5$
$x + y = 5$

62. $2x - 4y = 10$
$3x - 2y = -1$

Solve each system by the substitution method. [3.9]

63. $x + y = 2$
$y = 2x - 1$

64. $2x - 3y = 4$
$x = 3y - 1$

65. $4x + 2y = 8$
$y = -2x + 4$

66. $4x + 2y = 8$
$y = -2x + 5$

Dividing a Polynomial by a Polynomial

Since long division for polynomials is very similar to long division with whole numbers, we will begin by reviewing a division problem with whole numbers. You may realize when looking at Example 1 that you don't have a very good idea why you proceed as you do with long division. What you do know is that the process always works. We are going to approach the explanations in this section in much the same manner. That is, we won't always be sure why the steps we will use are important, only that they always produce the correct result.

▶ **EXAMPLE 1** Divide $27\overline{)3{,}962}$.

Solution

$$
\begin{array}{r}
1 \\
27\overline{)3{,}962} \\
2\,7 \\
\hline
1\,2
\end{array}
$$
← Estimate 27 into 39
← Multiply $1 \times 27 = 27$
← Subtract $39 - 27 = 12$

$$
\begin{array}{r}
1 \\
27\overline{)3{,}962} \\
2\,7\!\downarrow \\
\hline
1\,26
\end{array}
$$
← Bring down the 6

These are the four basic steps in long division. Estimate, multiply, subtract, and bring down the next term. To finish the problem, we simply perform the same four steps again:

$$\begin{array}{r} 14 \\ 27\overline{)3{,}962} \end{array}$$ ← 4 is the estimate

$$\begin{array}{r} 27 \\ \hline 126 \\ 108 \\ \hline 182 \end{array}$$ ← Multiply to get 108
← Subtract to get 18,
then bring down the 2

One more time.

$$\begin{array}{r} 146 \\ 27\overline{)3{,}962} \end{array}$$ ← 6 is the estimate

$$\begin{array}{r} 2\,7 \\ \hline 1\,26 \\ 1\,08 \\ \hline 182 \\ 162 \\ \hline 20 \end{array}$$ ← Multiply to get 162
← Subtract to get 20

Since there is nothing left to bring down we have our answer.

$$\frac{3{,}962}{27} = 146 + \frac{20}{27} \quad \text{or} \quad 146\,\frac{20}{27}$$ ◄

Here is how it works with polynomials.

► **EXAMPLE 2**   Divide $\dfrac{x^2 - 5x + 8}{x - 3}$.

Solution

$$\begin{array}{r} x \\ x-3\overline{)\;x^2 - 5x + 8} \end{array}$$ ← Estimate $x^2 \div x = x$

$$\begin{array}{r} \overset{-}{\cancel{x^2}}\;\overset{+}{\cancel{3x}} \\ \hline -\,2x \end{array}$$ ← Multiply $x(x-3) = x^2 - 3x$
← Subtract $(x^2 - 5x) - (x^2 - 3x) = -2x$

$$\begin{array}{r} x \\ x-3\overline{)\;x^2 - 5x + 8} \end{array}$$

$$\begin{array}{r} \overset{-}{\cancel{x^2}}\;\overset{+}{\cancel{3x}} \\ \hline -\,2x + 8 \end{array}$$ ← Bring down the 8

Notice that to subtract one polynomial from another, we add its opposite. That is why we change the signs on $x^2 - 3x$ and add what we get to $x^2 - 5x$. (To subtract the second polynomial, simply change the signs and add.)

We perform the same four steps again:

$$
\begin{array}{r}
x - 2 \leftarrow -2 \text{ is the estimate } (-2x \div x = -2) \\
x - 3\overline{)\ x^2 - 5x + 8} \\
\end{array}
$$

$$
\begin{array}{r}
- \quad + \\
\cancel{+}\,x^2 \,\cancel{/}\, 3x \quad \downarrow \\
\hline
- 2x + 8 \\
+ \quad - \\
\cancel{/}\, 2x \,\cancel{+}\, 6 \leftarrow \text{Multiply } -2(x - 3) = -2x + 6 \\
\hline
2 \leftarrow \text{Subtract } (-2x + 8) - (-2x + 6) = 2
\end{array}
$$

Since there is nothing left to bring down, we have our answer:

$$
\frac{x^2 - 5x + 8}{x - 3} = x - 2 + \frac{2}{x - 3}
$$

To check our answer, we multiply $(x - 3)(x - 2)$ to get $x^2 - 5x + 6$. Then, adding on the remainder, 2, we have $x^2 - 5x + 8$. ◀

▶ **EXAMPLE 3**   Divide $\dfrac{6x^2 - 11x - 14}{2x - 5}$.

Solution

$$
\begin{array}{r}
3x + \ \ 2 \\
2x - 5\overline{)\ 6x^2 - 11x - 14} \\
\end{array}
$$

$$
\begin{array}{r}
- \quad + \\
\cancel{+}\, 6x^2 \,\cancel{/}\, 15x \quad \downarrow \\
\hline
+ \ 4x - 14 \\
- \quad + \\
\cancel{/}\ \ 4x \,\cancel{/}\, 10 \\
\hline
- \ 4
\end{array}
$$

$$
\frac{6x^2 - 11x - 14}{2x - 5} = 3x + 2 + \frac{-4}{2x - 5}
$$
◀

One last step is sometimes necessary. The two polynomials in a division problem must both be in descending powers of the variable and cannot skip any powers from the highest power down to the constant term.

▶ **EXAMPLE 4**   Divide $\dfrac{2x^3 - 3x + 2}{x - 5}$.

Solution   The problem will be much less confusing if we write $2x^3 - 3x + 2$ as $2x^3 + 0x^2 - 3x + 2$. Adding $0x^2$ does not change our original problem.

$$\begin{array}{r} 2x^2 \quad\quad\quad\quad \\ x - 5 \overline{)\quad 2x^3 + 0x^2 - 3x + 2} \end{array}$$ ← Estimate $2x^3 \div x = 2x^2$

$$\begin{array}{r} -\quad+\quad\quad \\ \cancel{+}\ 2x^3 \cancel{-} 10x^2 \\ \hline + 10x^2 - 3x \end{array}$$

← Multiply $2x^2(x - 5) = 2x^3 - 10x^2$
← Subtract
$(2x^3 + 0x^2) - (2x^3 - 10x^2) = 10x^2$
Bring down the next term

Adding the term $0x^2$ gives us a column in which to write $10x^2$. (Remember, you can add and subtract only similar terms.)

Here is the completed problem:

$$\begin{array}{r} 2x^2 + 10x + 47 \\ x - 5 \overline{)\quad 2x^3 + 0x^2 - 3x + 2} \\ -\quad+\quad\quad\quad\quad\quad \\ \cancel{+}\ 2x^3 \cancel{-} 10x^2 \\ \hline + 10x^2 - 3x \\ -\quad+\quad\quad \\ \cancel{+}\ 10x^2 \cancel{-} 50x \\ \hline + 47x + 2 \\ -\quad+\quad \\ \cancel{+}\ 47x \cancel{-} 235 \\ \hline 237 \end{array}$$

Our answer is $\dfrac{2x^3 - 3x + 2}{x - 5} = 2x^2 + 10x + 47 + \dfrac{237}{x - 5}$. ◀

As you can see, long division with polynomials is a mechanical process. Once you have done it correctly a couple of times, it becomes very easy to produce the correct answer.

P R O B L E M   S E T   4 . 8

Divide.

1. $\dfrac{x^2 - 5x + 6}{x - 3}$

2. $\dfrac{x^2 - 5x + 6}{x - 2}$

11. $\dfrac{x^2 + 5x + 8}{x + 3}$

12. $\dfrac{x^2 + 5x + 4}{x + 3}$

3. $\dfrac{a^2 + 9a + 20}{a + 5}$

4. $\dfrac{a^2 + 9a + 20}{a + 4}$

13. $\dfrac{a^2 + 3a + 2}{a + 5}$

14. $\dfrac{a^2 + 4a + 3}{a + 5}$

5. $\dfrac{x^2 - 6x + 9}{x - 3}$

6. $\dfrac{x^2 + 10x + 25}{x + 5}$

15. $\dfrac{x^2 + 2x + 1}{x - 2}$

16. $\dfrac{x^2 + 6x + 9}{x - 3}$

7. $\dfrac{2x^2 + 5x - 3}{2x - 1}$

8. $\dfrac{4x^2 + 4x - 3}{2x - 1}$

17. $\dfrac{x^2 + 5x - 6}{x + 1}$

18. $\dfrac{x^2 - x - 6}{x + 1}$

9. $\dfrac{2a^2 - 9a + 5}{2a + 1}$

10. $\dfrac{4a^2 - 8a - 5}{2a + 1}$

19. $\dfrac{a^2 + 3a + 1}{a + 2}$

20. $\dfrac{a^2 - a + 3}{a + 1}$

21. $\dfrac{2x^2 - 2x + 5}{2x + 4}$

22. $\dfrac{15x^2 + 19x - 4}{3x + 8}$

23. $\dfrac{6a^2 + 5a + 1}{2a + 3}$

24. $\dfrac{4a^2 + 4a + 3}{2a + 1}$

25. $\dfrac{6a^3 - 13a^2 - 4a + 15}{3a - 5}$

26. $\dfrac{2a^3 - a^2 + 3a + 2}{2a + 1}$

Fill in the missing terms in the numerator and then use long division to find the quotients (see Example 4).

27. $\dfrac{x^3 + 4x + 5}{x + 1}$

28. $\dfrac{x^3 + 4x^2 - 8}{x + 2}$

29. $\dfrac{x^3 - 1}{x - 1}$

30. $\dfrac{x^3 + 1}{x + 1}$

31. $\dfrac{x^3 - 8}{x - 2}$

32. $\dfrac{x^3 + 27}{x + 3}$

Review Problems

The problems that follow review material we covered in Section 3.10.

Use systems of equations to solve the following applications.

33. The sum of two numbers is 25. One of the numbers is four times the other. Find the numbers.

34. The sum of two numbers is 24. One of the numbers is 3 more than twice the other. Find the numbers.

35. Suppose you have a total of $1,200 invested in two accounts. One of the accounts pays 8% annual interest and the other pays 9% annual interest. If your total interest for the year is $100, how much money did you invest in each of the accounts?

36. If you invest twice as much money in an account that pays 12% annual interest as you do in an account that pays 11% annual interest, how much do you have in each account if your total interest for a year is $210?

37. If you have a total of $160 in $5 bills and $10 bills, how many of each type of bill do you have if you have four more $10 bills than $5 bills?

38. Suppose you have 20 coins worth a total of $2.80. If the coins are all nickels and quarters, how many of each type do you have?

39. How many gallons of 20% antifreeze solution and 60% antifreeze solution must be mixed to get 16 gallons of 35% antifreeze solution?

40. A chemist wants to obtain 80 liters of a solution that is 12% hydrochloric acid. How many liters of 10% hydrochloric acid solution and 20% hydrochloric acid solution should he mix to do so?

CHAPTER 4 SUMMARY

Examples

Exponents: Definitions and Properties [4.1, 4.2]

Integer exponents indicate repeated multiplications.

1. (a) $2^3 = 2 \cdot 2 \cdot 2 = 8$

(b) $x^5 \cdot x^3 = x^{5+3} = x^8$

(c) $\dfrac{x^5}{x^3} = x^{5-3} = x^2$

(d) $(3x)^2 = 3^2 \cdot x^2 = 9x^2$

$a^r \cdot a^s = a^{r+s}$ — To multiply with the same base you add exponents.

$\dfrac{a^r}{a^s} = a^{r-s}$ — To divide with the same base you subtract exponents.

$(ab)^r = a^r \cdot b^r$ — Exponents distribute over multiplication.

(e) $\left(\dfrac{2}{3}\right)^3 = \dfrac{2^3}{3^3} = \dfrac{8}{27}$

$\left(\dfrac{a}{b}\right)^r = \dfrac{a^r}{b^r}$ Exponents distribute over division.

(f) $(x^5)^3 = x^{5 \cdot 3} = x^{15}$

$(a^r)^s = a^{r \cdot s}$ A power of a power is the product of the powers.

(g) $3^{-2} = \dfrac{1}{3^2} = \dfrac{1}{9}$

$a^{-r} = \dfrac{1}{a^r}$ Negative exponents imply reciprocals.

2. $(5x^2)(3x^4) = 15x^6$

Multiplication of Monomials [4.3]
To multiply two monomials, multiply coefficients and add exponents.

3. $\dfrac{12x^9}{4x^5} = 3x^4$

Division of Monomials [4.3]
To divide two monomials, divide coefficients and subtract exponents.

4. $768{,}000 = 7.68 \times 10^5$

$0.00039 = 3.9 \times 10^{-4}$

Scientific Notation [4.1, 4.2, 4.3]
A number is in scientific notation when it is written as the product of a number between 1 and 10 and an integer power of 10.

5. $(3x^2 - 2x + 1) + (2x^2 + 7x - 3)$
$= 5x^2 + 5x - 2$

Addition of Polynomials [4.4]
To add two polynomials, add coefficients of similar terms.

6. $(3x + 5) - (4x - 3)$
$= 3x + 5 - 4x + 3$
$= -x + 8$

Subtraction of Polynomials [4.4]
To subtract one polynomial from another, add the opposite of the second to the first.

7. (a) $2a^2(5a^2 + 3a - 2)$
$= 10a^4 + 6a^3 - 4a^2$
(b) $(x + 2)(3x - 1)$
$= 3x^2 - x + 6x - 2$
$= 3x^2 + 5x - 2$
(c) $2x^2 - 3x + 4$
             $3x - 2$

$6x^3 - 9x^2 + 12x$
          $- 4x^2 + 6x - 8$

$6x^3 - 13x^2 + 18x - 8$

Multiplication of Polynomials [4.5]
To multiply a polynomial by a monomial, we apply the distributive property. To multiply two binomials we use the FOIL method. In other situations we use the COLUMN method. Each method achieves the same result: To multiply any two polynomials, we multiply each term in the first polynomial by each term in the second polynomial.

8. $(x + 3)^2 = x^2 + 6x + 9$

$(x - 3)^2 = x^2 - 6x + 9$

$(x + 3)(x - 3) = x^2 - 9$

Special Products [4.6]

$(a + b)^2 = a^2 + 2ab + b^2$ ⎫
$(a - b)^2 = a^2 - 2ab + b^2$ ⎬ Binomial squares

$(a + b)(a - b) = a^2 - b^2$ Difference of two squares

9. $\dfrac{12x^3 - 18x^2}{6x}$

$= 2x^2 - 3x$

Dividing a Polynomial by a Monomial [4.7]
To divide a polynomial by a monomial, divide each term in the polynomial by the monomial.

10.
$$
\begin{array}{r}
x - 2 \\
x - 3 \overline{)\ x^2 - 5x + 8\ } \\
\end{array}
$$

$$
\begin{array}{r}
-\quad + \\
\not{x^2}\ \not{+}\ 3x \\
\hline
-2x + 8 \\
+\quad - \\
\not{-}\ 2x\ \not{-}\ 6 \\
\hline
2
\end{array}
$$

Long Division with Polynomials [4.8]
Division with polynomials is similar to long division with whole numbers. The steps in the process are: estimate, multiply, subtract, and bring down the next term. The divisors in all the long-division problems in this chapter were binomials.

COMMON MISTAKES
1. If a term contains a variable that is raised to a power, then the exponent on the variable is associated only with that variable, unless there are parentheses. That is, the expression $3x^2$ means $3 \cdot x \cdot x$, not $3x \cdot 3x$. It is a mistake to write $3x^2$ as $9x^2$. The only way to end up with $9x^2$ is to start with $(3x)^2$.
2. It is a mistake to add nonsimilar terms. For example, $2x$ and $3x^2$ are nonsimilar terms and therefore cannot be combined. That is, $2x + 3x^2 \neq 5x^3$. If you were to substitute 10 for x in the preceding expression, you would see that the two sides are not equal.
3. It is a mistake to distribute exponents over sums and differences. That is, $(a + b)^2 \neq a^2 + b^2$. Convince yourself of this by letting $a = 2$ and $b = 3$ and then simplifying both sides.
4. Another common mistake can occur when dividing a polynomial by a monomial. Here is an example:

$$\frac{x + \not{2}}{\not{2}} = x + 1 \qquad \text{Mistake}$$

The mistake here is in not dividing both terms in the numerator by 2. The correct way to divide $x + 2$ by 2 looks like this:

$$\frac{x + 2}{2} = \frac{x}{2} + \frac{2}{2} \qquad \text{Correct}$$

$$= \frac{x}{2} + 1$$

CHAPTER 4 REVIEW

The numbers in brackets refer to the sections of the text in which similar problems can be found.

Simplify. [4.1]

1. $(-1)^3$

2. $(-4)^3$

3. -8^2

4. -2^4

5. $\left(\dfrac{3}{7}\right)^2$

6. $\left(\dfrac{1}{3}\right)^3$

Simplify each expression. [4.1]

7. $y^3 \cdot y^9$

8. $y^{10} \cdot y^6$

9. $x^{15} \cdot x^7 \cdot x^5 \cdot x^3$

10. $x^{21} \cdot x^{13} \cdot x^9 \cdot x^7$

Write each problem with a single exponent. [4.1]

11. $(x^7)^5$

12. $(x^4)^{10}$

13. $(2^6)^4$

14. $(3^8)^5$

Simplify each expression. [4.1]

15. $(3y)^3$

16. $(2y)^4$

17. $(-2xyz)^3$

18. $\left(-\dfrac{2}{3}\,a^2b^3c^4\right)^2$

Write each of the following with positive exponents and then simplify, when possible. [4.2]

19. 7^{-2}

20. 9^{-2}

21. $4x^{-5}$

22. $7x^{-6}$

23. $(3y)^{-3}$

24. $(6y)^{-2}$

Simplify each expression. Any answers that contain exponents should contain positive exponents only. [4.2]

25. $\dfrac{a^9}{a^3}$

26. $\dfrac{a^{10}}{a^2}$

27. $\left(\dfrac{x^3}{x^5}\right)^2$

28. $\left(\dfrac{x^5}{x^3}\right)^{-2}$

29. $\dfrac{x^9}{x^{-6}}$

30. $\dfrac{x^4}{x^{-8}}$

31. $\dfrac{x^{-7}}{x^{-2}}$

32. $\dfrac{x^{-9}}{x^{-13}}$

Simplify each expression. [4.2]

33. $(-3xy)^0$

34. $(9xy)^0$

35. $3^0 - 5^1 + 5^0$

36. $4^1 + 9^0 + (-7)^0$

Simplify. Any answers that contain exponents should contain positive exponents only. [4.1, 4.2]

37. $(3x^3y^2)^2$

38. $(-2x^4y^3)^3$

39. $(2a^3b^2)^4(2a^5b^6)^2$

40. $(3a^2b^5)^3(2a^6b^7)^2$

41. $(-3xy^2)^{-3}$

42. $(-2x^3y^5)^{-4}$

43. $\dfrac{(b^3)^4(b^2)^5}{(b^7)^3}$

44. $\dfrac{(b^6)^2(b^3)^4}{(b^{10})^3}$

45. $\dfrac{(x^{-3})^3(x^6)^{-1}}{(x^{-5})^{-4}}$

46. $\dfrac{(x^{-4})^{-3}(x^{-3})^4}{x^0}$

Simplify. Write all answers with positive exponents only. [4.3]

47. $\dfrac{(2x^4)(15x^9)}{(6x^6)}$

48. $\dfrac{(3x^5)(20x^3)}{15x^{10}}$

49. $\dfrac{(10x^3y^5)(21x^2y^6)}{(7xy^3)(5x^9y)}$

50. $\dfrac{(12xy^5)(16x^2y^2)}{(8x^3y^3)(3x^5y)}$

51. $\dfrac{21a^{10}}{3a^4} - \dfrac{18a^{17}}{6a^{11}}$

52. $\dfrac{24a^{12}}{6a^3} + \dfrac{30a^{24}}{10a^{15}}$

53. $\dfrac{8x^8y^3}{2x^3y} - \dfrac{10x^6y^9}{5xy^7}$

54. $\dfrac{50x^8y^8}{25x^4y^2} + \dfrac{28x^7y^7}{14x^3y}$

Simplify and write all answers in scientific notation. [4.3]

55. $(3.2 \times 10^3)(2 \times 10^4)$

56. $(5 \times 10^{-5})(2.1 \times 10^3)$

57. $\dfrac{4.6 \times 10^5}{2 \times 10^{-3}}$

58. $\dfrac{3.5 \times 10^{-7}}{7 \times 10^{-3}}$

59. $\dfrac{(4 \times 10^6)(6 \times 10^5)}{3 \times 10^8}$

60. $\dfrac{(6 \times 10^5)(6 \times 10^{-3})}{9 \times 10^{-4}}$

Perform the following additions and subtractions. [4.4]

61. $(3a^2 - 5a + 5) + (5a^2 - 7a - 8)$
62. $(4a^3 - 10a^2 + 6) - (6a^3 + 5a - 7)$
63. $(-7x^2 + 3x - 6) - (8x^2 - 4x + 7) +$
    $(3x^2 - 2x - 1)$
64. $(-4x^2 - 5x + 2) + (3x^2 - 6x + 1) -$
    $(-x^2 + 2x - 7)$
65. Subtract $8x^2 + 3x - 2$ from $4x^2 - 3x - 2$.
66. Subtract $4x + 3$ from the sum of $2x - 7$ and
    $9x + 5$.

Find the value of the following. [4.4]

67. $2x^2 - 3x + 5$ when $x = 3$
68. $(y - 3)^2$ when $y = -8$

Multiply. [4.5]

69. $3x(4x - 7)$ **70.** $5x(8x + 3)$
71. $8x^3 y(3x^2 y - 5xy^2 + 4y^3)$
72. $5xy^2(3x^3 + 4x^2 y + 10xy^2)$
73. $(a + 1)(a^2 + 5a - 4)$
74. $(a - 2)(a^2 - 6a + 7)$
75. $(x + 5)(x^2 - 5x + 25)$
76. $(x - 1)(x^2 + x + 1)$

Multiply the following binomials. [4.5]

77. $(3x - 7)(2x - 5)$ **78.** $(5x - 2)(3x + 4)$

79. $\left(5y + \dfrac{1}{5}\right)\left(5y - \dfrac{1}{5}\right)$

80. $\left(\dfrac{1}{2}y + 2\right)\left(\dfrac{1}{2}y - 2\right)$

81. $(a^2 - 3)(a^2 + 3)$ **82.** $(a^2 + 7)(a^2 - 7)$

Perform the indicated operations. [4.6]

83. $(a - 5)^2$ **84.** $(a + 6)^2$
85. $(3x + 4)^2$ **86.** $(5x - 1)^2$
87. $(y^2 + 3)^2$ **88.** $(y^2 - 2)^2$

Work the following division problems. [4.7]

89. Divide $10ab + 20a^2$ by $-5a$
90. Divide $15a^3 b - 10a^2 b^2 - 20ab^3$ by $5ab$
91. Divide $40x^5 y^4 - 32x^3 y^3 - 16x^2 y$ by $-8xy$
92. Divide $28x^4 y^4 - 14x^2 y^3 + 21xy^2$ by $-7xy^2$

Perform the following divisions. [4.7]

93. $\dfrac{16xy^2 - 10xy}{-2xy}$

94. $\dfrac{40x^3 y^2 + 20x^2 y^2}{-10xy^2}$

95. $\dfrac{20a^2 - 16a^2 b + 24a^2 b^2}{4a^2}$

96. $\dfrac{15x^5 - 10x^2 + 20x}{5x^5}$

Divide using long division. [4.8]

97. $\dfrac{x^2 + 15x + 54}{x + 6}$ **98.** $\dfrac{x^2 - 3x - 28}{x + 4}$

99. $\dfrac{6x^2 + 13x - 5}{3x - 1}$ **100.** $\dfrac{8x^2 + 10x + 3}{4x + 3}$

101. $\dfrac{x^3 + 64}{x + 4}$ **102.** $\dfrac{x^3 - 27}{x - 3}$

103. $\dfrac{3x^2 - 7x + 10}{3x + 2}$ **104.** $\dfrac{4x^2 + 8x - 10}{2x - 3}$

105. $\dfrac{2x^3 - 7x^2 + 6x + 10}{2x + 1}$

106. $\dfrac{3x^3 - 7x^2 + 14x - 10}{3x - 1}$

A box is in the shape of a rectangular solid and is three times as long as it is wide. The height and the width are equal. [4.1, 4.2, 4.3]

107. Write an expression for the volume of this box.
108. If the width of this box is 8.6 inches, will this book fit inside the box?
109. If the width of this box is 2 feet, will it hold as much food as your refrigerator?

110. Approximately what width would the box have to have in order for you to fit inside it?

111. If the length of the side of a cube is the same as the width of the box, how many cubes will fit inside the box?

112. If the length of the side of a cube is half as long as the width of the box, how many cubes will fit inside the box?

CHAPTER 4 TEST

Simplify each of the following expressions. [4.1]

1. $(-3)^4$

2. $\left(\dfrac{3}{4}\right)^2$

3. $(3x^3)^2(2x^4)^3$

Simplify each expression. Write all answers with positive exponents only. [4.2]

4. 3^{-2}

5. $(3a^4b^2)^0$

6. $\dfrac{a^{-3}}{a^{-5}}$

7. $\dfrac{(x^{-2})^3(x^{-3})^{-5}}{(x^{-4})^{-2}}$

8. Write 0.0278 in scientific notation. [4.2]

9. Write 2.43×10^5 in expanded form. [4.2]

Simplify. Write all answers with positive exponents only. [4.3]

10. $\dfrac{35x^2y^4z}{70x^6y^2z}$

11. $\dfrac{(6a^2b)(9a^3b^2)}{18a^4b^3}$

12. $\dfrac{24x^7}{3x^2} + \dfrac{14x^9}{7x^4}$

13. $\dfrac{(2.4 \times 10^5)(4.5 \times 10^{-2})}{1.2 \times 10^{-6}}$

Add and subtract as indicated. [4.4]

14. $8x^2 - 4x + 6x + 2$

15. $(5x^2 - 3x + 4) - (2x^2 - 7x - 2)$

16. Subtract $3x - 4$ from $6x - 8$. [4.4]

17. Find the value of $2y^2 - 3y - 4$ when y is -2. [4.4]

Multiply. [4.5]

18. $2a^2(3a^2 - 5a + 4)$

19. $\left(x + \dfrac{1}{2}\right)\left(x + \dfrac{1}{3}\right)$

20. $(4x - 5)(2x + 3)$

21. $(x - 3)(x^2 + 3x + 9)$

Multiply. [4.6]

22. $(x + 5)^2$

23. $(3a - 2b)^2$

24. $(3x - 4y)(3x + 4y)$

25. $(a^2 - 3)(a^2 + 3)$

26. Divide $10x^3 + 15x^2 - 5x$ by $5x$. [4.7]

Divide. [4.8]

27. $\dfrac{8x^2 - 6x - 5}{2x - 3}$

28. $\dfrac{3x^3 - 2x + 1}{x - 3}$

29. Find the volume of a cube if the length of a side is 2.5 centimeters. [4.1]

30. Find the volume of a rectangular solid if the length is five times the width, and the height is one fifth the width. [4.3]

FACTORING

INTRODUCTION

One of the most important theorems in the history of mathematics, and also one of the oldest, is the Pythagorean Theorem. This theorem gives the relationship among the sides in *any* right triangle. It states that in any right triangle, the square of the longest side is equal to the sum of the squares of the other two sides.

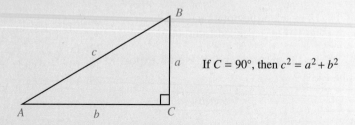

If $C = 90°$, then $c^2 = a^2 + b^2$

You will find some problems based on the Pythagorean Theorem toward the end of this chapter.

OVERVIEW

In Chapter 1 we used the diagram below to illustrate the relationship between multiplication and factoring.

Multiplication

Factors $\rightarrow 3 \cdot 5 = 15 \leftarrow$ Product

Factoring

A similar relationship holds for multiplication of polynomials. Reading the diagram below from left to right, we say the product of the binomials $x + 2$ and $x + 3$ is the trinomial $x^2 + 5x + 6$. On the other hand, if we read in the other direction, we can say that $x^2 + 5x + 6$ factors into the product of $x + 2$ and $x + 3$.

Multiplication

Factors $\rightarrow (x + 2)(x + 3) = x^2 + 5x + 6 \leftarrow$ Product

Factoring

In this chapter we develop a systematic method of factoring polynomials. The more familiar you are with the idea of the distributive property and the process of multiplying binomials, the more easily you will understand factoring.

STUDY SKILLS

The study skills for this chapter have to do with the way you approach new situations in mathematics. The first study skill has to do with your natural instincts for what does and doesn't work in mathematics. The second study skill gives you a way of testing your instincts.

1. **Don't Let Your Intuition Fool You**   As you become more experienced and more successful in mathematics you will be able to trust your mathematical intuition. For now though, it can get in the way of success. For example, if you ask a beginning algebra student to "subtract 3 from -5" many will answer -2 or 2. Both answers are incorrect, even though they may seem intuitively true.
2. **Test Properties About Which You Are Unsure**   You can always test a property or statement by substituting numbers for variables. For instance, I always have students that rewrite $(x + 3)^2$ as $x^2 + 9$, thinking that the two expressions are equivalent. The fact that the two expressions are not equivalent becomes obvious when we substitute 10 for x in each one.

    When $x = 10$, the expression $(x + 3)^2$ is $(10 + 3)^2 = 13^2 = 169$

    When $x = 10$, the expression $x^2 + 9 = 10^2 + 9 = 100 + 9 = 109$

Along the same lines, there may come a time when you are wondering if $\sqrt{x^2 + 25}$ is the same as $x + 5$. If you try $x = 10$ in each expression, you will find out quickly that the two expressions are not the same.

When $x = 10$, the expression $\sqrt{x^2 + 25} = \sqrt{10^2 + 25} = \sqrt{125} \approx 11.2$

When $x = 10$, the expression $x + 5 = 10 + 5 = 15$

When you test the equivalence of expressions by substituting numbers for the variable, make it easy on yourself by choosing numbers that are easy to work with, such as 10. Also, don't try to verify the equivalence of expressions by substituting 0, 1, or 2 for the variable, as these numbers sometimes make expressions seem equivalent when in fact they are not equivalent.

It is not unusual, nor is it wrong, to try occasionally to apply a property that doesn't exist. If you have any doubt about generalizations you are making, test them by replacing variables with numbers and simplifying.

SECTION

5.1   The Greatest Common Factor and Factoring by Grouping

In this section we will apply the distributive property to polynomials to factor from them what is called the **greatest common factor.**

DEFINITION   The **greatest common factor** for a polynomial is the largest monomial that divides (is a factor of) each term of the polynomial.

We use the term largest monomial to mean the monomial with the greatest coefficient and highest power of the variable.

▶ **EXAMPLE 1**   Find the greatest common factor for the polynomial

$$3x^5 + 12x^2$$

Solution   The terms of the polynomial are $3x^5$ and $12x^2$. The largest number that divides the coefficients is 3, and the highest power of x that is a factor of x^5 and x^2 is x^2. Therefore, the greatest common factor for $3x^5 + 12x^2$ is $3x^2$. That is, $3x^2$ is the largest monomial that divides each term of $3x^5 + 12x^2$.   ◀

▶ **EXAMPLE 2**   Find the greatest common factor for

$$8a^3b^2 + 16a^2b^3 + 20a^3b^3$$

Solution   The largest number that divides each of the coefficients is 4. The highest power of the variable that is a factor of a^3b^2, a^2b^3, and a^3b^3 is a^2b^2. The greatest common factor for $8a^3b^2 + 16a^2b^3 + 20a^3b^3$ is $4a^2b^2$. It is the largest monomial that is a factor of each term.   ◀

Once we have recognized the greatest common factor of a polynomial, we can apply the distributive property and factor it out of each term. We rewrite the polynomial as the product of its greatest common factor with the polynomial that remains after the greatest common factor has been factored from each term in the original polynomial.

▶ **EXAMPLE 3**   Factor the greatest common factor from $3x - 15$.

Solution   The greatest common factor for the terms $3x$ and 15 is 3. We can rewrite both $3x$ and 15 so that the greatest common factor 3 is showing in each term. It is important to realize that $3x$ means $3 \cdot x$. The 3 and the x are not ''stuck'' together.

$$3x - 15 = 3 \cdot x - 3 \cdot 5$$

Now, applying the distributive property, we have

$$3 \cdot x - 3 \cdot 5 = 3(x - 5)$$

To check a factoring problem like this, we can multiply 3 and $x - 5$ to get $3x - 15$, which is what we started with. Factoring is simply a procedure by which we change sums and differences into products. In this case, we changed the difference $3x - 15$ into the product $3(x - 5)$. Note, however, that we have not changed the meaning or value of the expression. The expression we end up with is equal to the expression we started with.   ◀

▶ **EXAMPLE 4**   Factor the greatest common factor from

$$5x^3 - 15x^2$$

Solution   The greatest common factor is $5x^2$. We rewrite the polynomial as

$$5x^3 - 15x^2 = 5x^2 \cdot x - 5x^2 \cdot 3$$

Then we apply the distributive property to get

$$5x^2 \cdot x - 5x^2 \cdot 3 = 5x^2(x - 3)$$

To check our work, we simply multiply $5x^2$ and $x - 3$ to get $5x^3 - 15x^2$, which is our original polynomial.   ◀

▶ **EXAMPLE 5**   Factor the greatest common factor from

$$16x^5 - 20x^4 + 8x^3$$

Solution   The greatest common factor is $4x^3$. We rewrite the polynomial so we can see the greatest common factor $4x^3$ in each term; then we apply the distributive property to factor it out:

$$16x^5 - 20x^4 + 8x^3 = 4x^3 \cdot 4x^2 - 4x^3 \cdot 5x + 4x^3 \cdot 2$$
$$= 4x^3(4x^2 - 5x + 2)$$ ◀

▶ **EXAMPLE 6**   Factor the greatest common factor from

$$6x^3y - 18x^2y^2 + 12xy^3$$

Solution   The greatest common factor is $6xy$. We rewrite the polynomial in terms of $6xy$ and then apply the distributive property as follows:

$$6x^3y - 18x^2y^2 + 12xy^3 = 6xy \cdot x^2 - 6xy \cdot 3xy + 6xy \cdot 2y^2$$
$$= 6xy(x^2 - 3xy + 2y^2)$$ ◀

▶ **EXAMPLE 7**   Factor the greatest common factor from

$$3a^2b - 6a^3b^2 + 9a^3b^3$$

Solution   The greatest common factor is $3a^2b$:

$$3a^2b - 6a^3b^2 + 9a^3b^3 = 3a^2b(1) - 3a^2b(2ab) + 3a^2b(3ab^2)$$
$$= 3a^2b(1 - 2ab + 3ab^2)$$ ◀

Factoring by Grouping

To develop our next method of factoring, called **factoring by grouping,** we start by examining the polynomial $xc + yc$. The greatest common factor for the two terms is c. Factoring c from each term we have

$$xc + yc = c(x + y)$$

But suppose that c itself was a more complicated expression, such as $a + b$, so that the expression we were trying to factor was $x(a + b) + y(a + b)$, instead of $xc + yc$. The greatest common factor for $x(a + b) + y(a + b)$ is $(a + b)$. Factoring this common factor from each term looks like this:

$$x(a + b) + y(a + b) = (a + b)(x + y)$$

To see how all of this applies to factoring polynomials, consider the polynomial

$$xy + 3x + 2y + 6$$

There is no greatest common factor other than the number 1. However, if we group the terms together two at a time, we can factor an x from the first two terms and a 2 from the last two terms:

$$xy + 3x + 2y + 6 = x(y + 3) + 2(y + 3)$$

The expression on the right can be thought of as having two terms: $x(y + 3)$ and $2(y + 3)$. Each of these expressions contains the common factor $y + 3$, which can be factored out using the distributive property:

$$x(y + 3) + 2(y + 3) = (y + 3)(x + 2)$$

This last expression is in factored form. The process we used to obtain it is called factoring by grouping. Here are some additional examples.

▶ **EXAMPLE 8**   Factor $ax + bx + ay + by$.

Solution   We begin by factoring x from the first two terms and y from the last two terms:

$$ax + bx + ay + by = x(a + b) + y(a + b)$$
$$= (a + b)(x + y)$$

To convince yourself that this is factored correctly, multiply the two factors $(a + b)$ and $(x + y)$.   ◀

▶ **EXAMPLE 9**   Factor by grouping: $3ax - 2a + 15x - 10$.

Solution   First, we factor a from the first two terms and 5 from the last two terms. Then, we factor $3x - 2$ from the remaining two expressions:

$$3ax - 2a + 15x - 10 = a(3x - 2) + 5(3x - 2)$$
$$= (3x - 2)(a + 5)$$

Again, multiplying $(3x - 2)$ and $(a + 5)$ will convince you that these are the correct factors.   ◀

▶ **EXAMPLE 10**   Factor $2x^2 + 5ax - 2xy - 5ay$.

Solution   From the first two terms we factor x. From the second two terms we must factor $-y$ so that the binomial that remains after we do so matches the binomial produced by the first two terms.

$$2x^2 + 5ax - 2xy - 5ay = x(2x + 5a) - y(2x + 5a)$$
$$= (2x + 5a)(x - y)$$

Another way to accomplish the same result is to use the commutative property to interchange the middle two terms, and then factor by grouping:

$$2x^2 + 5ax - 2xy - 5ay$$
$$= 2x^2 - 2xy + 5ax - 5ay \qquad \text{Commutative property}$$
$$= 2x(x - y) + 5a(x - y)$$
$$= (x - y)(2x + 5a)$$

This is the same result we obtained previously.   ◀

▶ **EXAMPLE 11**   Factor $6x^2 - 3x - 4x + 2$ by grouping.

Solution   The first two terms have $3x$ in common, while the last two terms have either a 2 or a -2 in common. Suppose we factor $3x$ from the first two terms and 2 from the last two terms. We get

$$6x^2 - 3x - 4x + 2 = 3x(2x - 1) + 2(-2x + 1)$$

We can't go any further because there is no common factor that will allow us to factor further. However, if we factor -2, instead of 2, from the last two terms, our problem is solved.

$$6x^2 - 3x - 4x + 2 = 3x(2x - 1) - 2(2x - 1)$$
$$= (2x - 1)(3x - 2)$$

In this case, factoring -2 from the last two terms gives us an expression that can be factored further.   ◀

PROBLEM SET 5.1

Factor the following by taking out the greatest common factor.

1. $15x + 25$ **2.** $14x + 21$
3. $6a + 9$ **4.** $8a + 10$
5. $4x - 8y$ **6.** $9x - 12y$
7. $3x^2 - 6x - 9$ **8.** $2x^2 + 6x + 4$
9. $3a^2 - 3a - 60$ **10.** $2a^2 - 18a + 28$
11. $24y^2 - 52y + 24$ **12.** $18y^2 + 48y + 32$
13. $9x^2 - 8x^3$ **14.** $7x^3 - 4x^2$
15. $13a^2 - 26a^3$ **16.** $5a^2 - 10a^3$
17. $21x^2y - 28xy^2$ **18.** $30xy^2 - 25x^2y$
19. $22a^2b^2 - 11ab^2$
20. $15x^3 - 25x^2 + 30x$
21. $7x^3 + 21x^2 - 28x$
22. $16x^4 - 20x^2 - 16x$
23. $121y^4 - 11x^4$ **24.** $25a^4 - 5b^4$

Factor the following by taking out the greatest common factor.

25. $100x^4 - 50x^3 + 25x^2$
26. $36x^5 + 72x^3 - 81x^2$
27. $8a^2 + 16b^2 + 32c^2$ **28.** $9a^2 - 18b^2 - 27c^2$
29. $4a^2b - 16ab^2 + 32a^2b^2$
30. $5ab^2 + 10a^2b^2 + 15a^2b$
31. $121a^3b^2 - 22a^2b^3 + 33a^3b^3$
32. $20a^4b^3 - 18a^3b^4 + 22a^4b^4$
33. $12x^2y^3 - 72x^5y^3 - 36x^4y^4$
34. $49xy - 21x^2y^2 + 35x^3y^3$

Factor by grouping.

35. $xy + 5x + 3y + 15$
36. $xy + 2x + 4y + 8$
37. $xy + 6x + 2y + 12$
38. $xy + 2y + 6x + 12$
39. $ab + 7a - 3b - 21$
40. $ab + 3b - 7a - 21$
41. $ax - bx + ay - by$
42. $ax - ay + bx - by$
43. $2ax + 6x - 5a - 15$
44. $3ax + 21x - a - 7$
45. $3xb - 4b - 6x + 8$
46. $3xb - 4b - 15x + 20$
47. $x^2 + ax + 2x + 2a$ **48.** $x^2 + ax + 3x + 3a$

49. $x^2 - ax - bx + ab$ **50.** $x^2 + ax - bx - ab$

Factor by grouping. You can group the terms together two at a time or three at a time. Either way will produce the same result.

51. $ax + ay + bx + by + cx + cy$
52. $ax + bx + cx + ay + by + cy$

Factor the polynomials below by grouping the terms together two at a time.

53. $6x^2 + 9x + 4x + 6$
54. $6x^2 - 9x - 4x + 6$
55. $20x^2 - 2x + 50x - 5$
56. $20x^2 + 25x + 4x + 5$
57. $20x^2 + 4x + 25x + 5$
58. $20x^2 + 4x - 25x - 5$
59. $x^3 + 2x^2 + 3x + 6$
60. $x^3 - 5x^2 - 4x + 20$
61. $6x^3 - 4x^2 + 15x - 10$
62. $8x^3 - 12x^2 + 14x - 21$
63. The greatest common factor of the binomial $3x + 6$ is 3. The greatest common factor of the binomial $2x + 4$ is 2. What is the greatest common factor of their product, $(3x + 6)(2x + 4)$, when it has been multiplied out?
64. The greatest common factors of the binomials $4x + 2$ and $5x + 10$ are 2 and 5, respectively. What is the greatest common factor of their product, $(4x + 2)(5x + 10)$, when it has been multiplied out?
65. The following factorization is incorrect. Find the mistake and correct the right-hand side:

$$12x^2 + 6x + 3 = 3(4x^2 + 2x)$$

66. Find the mistake in the following factorization and then rewrite the right-hand side correctly:

$$10x^2 + 2x + 6 = 2(5x^2 + 3)$$

Applying the Concepts

67. If you invest $1,000 in an account with an annual interest rate of r compounded annually, the amount of money you have in the account after one year is

$$A = 1{,}000 + 1{,}000r$$

Write this formula again with the right side in factored form. Then, find the amount of money in this account at the end of one year if the interest rate is 12%.

68. If you invest P dollars in an account with an annual interest rate of 8% compounded annually, then the amount of money in that account after one year is given by the formula

$$A = P + 0.08P$$

Rewrite this formula with the right side in factored form and then find the amount of money in the account at the end of one year if $500 was the initial investment.

Review Problems

The problems below review material we covered in Section 4.5. Reviewing these problems will help you with the next section.

Multiply using the FOIL method.

69. $(x + 3)(x + 4)$	**70.** $(x - 5)(x + 2)$
71. $(x + 7)(x - 2)$	**72.** $(x + 7)(x + 2)$
73. $(x - 7)(x + 2)$	**74.** $(x - 7)(x - 2)$
75. $(x - 3)(x + 2)$	**76.** $(x + 3)(x - 2)$

77. $(x + 3)(x^2 - 3x + 9)$
78. $(x - 2)(x^2 + 2x + 4)$
79. $(2x + 1)(x^2 + 4x - 3)$
80. $(3x + 2)(x^2 - 2x - 4)$

SECTION

5.2   Factoring Trinomials

In this section we will factor trinomials in which the coefficient of the squared term is 1. The more familiar we are with multiplication of binomials the easier factoring trinomials will be.

Recall multiplication of binomials from Chapter 4:

$$(x + 3)(x + 4) = x^2 + 7x + 12$$

$$(x - 5)(x + 2) = x^2 - 3x - 10$$

The first term in the answer is the product of the first terms in each binomial. The last term in the answer is the product of the last terms in each binomial. The middle term in the answer comes from adding the product of the outside terms with the product of the inside terms.

Let's have a and b represent real numbers and look at the product of $(x + a)$ and $(x + b)$:

$$(x + a)(x + b) = x^2 + ax + bx + ab$$
$$= x^2 + (a + b)x + ab$$

The coefficient of the middle term is the sum of a and b. The last term is the product of a and b. Writing this as a factoring problem, we have

$$x^2 + \underset{\text{Sum}}{(a + b)}x + \underset{\text{Product}}{ab} = (x + a)(x + b)$$

To factor a trinomial in which the coefficient of x^2 is 1, we need only find the numbers a and b whose sum is the coefficient of the middle term and whose product is the constant term (last term).

▶ **EXAMPLE 1** Factor $x^2 + 8x + 12$.

Solution The coefficient of x^2 is 1. We need two numbers whose sum is 8 and whose product is 12. The numbers are 6 and 2:

$$x^2 + 8x + 12 = (x + 6)(x + 2)$$

We can easily check our work by multiplying $(x + 6)$ and $(x + 2)$:

Check:
$$(x + 6)(x + 2) = x^2 + 6x + 2x + 12$$
$$= x^2 + 8x + 12 \qquad ◀$$

▶ **EXAMPLE 2** Factor $x^2 - 2x - 15$.

Solution The coefficient of x^2 is again 1. We need to find a pair of numbers whose sum is -2 and whose product is -15. Here are all the possibilities for products that are -15.

Products	Sums
$-1(15) = -15$	$-1 + 15 = 14$
$1(-15) = -15$	$1 + (-15) = -14$
$-5(3) = -15$	$-5 + 3 = -2$
$5(-3) = -15$	$5 + (-3) = 2$

The third line gives us what we want. The factors of $x^2 - 2x - 15$ are $(x - 5)$ and $(x + 3)$.

$$x^2 - 2x - 15 = (x - 5)(x + 3) \qquad ◀$$

▶ **EXAMPLE 3** Factor $2x^2 + 10x - 28$.

Solution The coefficient of x^2 is 2. We begin by factoring out the greatest common factor, which is 2:

$$2x^2 + 10x - 28 = 2(x^2 + 5x - 14)$$

Now we factor the remaining trinomial by finding a pair of numbers whose sum is 5 and whose product is -14. Here are the possibilities.

Products	Sums
$-1(14) = -14$	$-1 + 14 = 13$
$1(-14) = -14$	$1 + (-14) = -13$
$-7(2) = -14$	$-7 + 2 = -5$
$7(-2) = -14$	$7 + (-2) = 5$

From the last line we see that the factors of $x^2 + 5x - 14$ are $(x + 7)$ and $(x - 2)$. Here is the complete problem:

$$2x^2 + 10x - 28 = 2(x^2 + 5x - 14)$$
$$= 2(x + 7)(x - 2) \quad \blacktriangleleft$$

Note In Example 3 we began by factoring out the greatest common factor. The first step in factoring any trinomial is to look for the greatest common factor. If the trinomial in question has a greatest common factor other than 1, we factor it out first and then try to factor the trinomial that remains.

▶ **EXAMPLE 4** Factor $3x^3 - 3x^2 - 18x$.

Solution We begin by factoring out the greatest common factor, which is $3x$. Then we factor the remaining trinomial. Without showing the table of products and sums as we did in Examples 2 and 3, here is the complete problem:

$$3x^3 - 3x^2 - 18x = 3x(x^2 - x - 6)$$
$$= 3x(x - 3)(x + 2) \quad \blacktriangleleft$$

▶ **EXAMPLE 5** Factor $x^2 + 8xy + 12y^2$.

Solution This time we need two expressions whose product is $12y^2$ and whose sum is $8y$. The two expressions are $6y$ and $2y$ (see Example 1 in this section):

$$x^2 + 8xy + 12y^2 = (x + 6y)(x + 2y)$$

You should convince yourself that these factors are correct by finding their product. $\quad \blacktriangleleft$

▶ **EXAMPLE 6** A ball is tossed into the air with an upward velocity of 16 feet per second from the top of a building 32 feet high. The equation that gives the height of the ball above the ground at any time t is

$$h = 32 + 16t - 16t^2$$

Factor the right side of this equation and then find h when t is 2.

Solution We begin by factoring out the greatest common factor, 16. Then, we factor the trinomial that remains:

$$h = 32 + 16t - 16t^2$$

$$h = 16(2 + t - t^2)$$

$$h = 16(2 - t)(1 + t)$$

Letting $t = 2$ in the equation, we have

$$h = 16(2 - 2)(1 + 2)$$

$$= 16(0)(3)$$

$$= 0$$

When t is 2, h is 0. ◀

P R O B L E M   S E T   5 . 2

Factor the following trinomials.

1. $x^2 + 7x + 12$　　　　**2.** $x^2 + 7x + 10$
3. $x^2 + 3x + 2$　　　　**4.** $x^2 + 7x + 6$
5. $a^2 + 10a + 21$　　　**6.** $a^2 - 7a + 12$
7. $x^2 - 7x + 10$　　　　**8.** $x^2 - 3x + 2$
9. $y^2 - 10y + 21$　　　**10.** $y^2 - 7y + 6$
11. $x^2 - x - 12$　　　　**12.** $x^2 - 4x - 5$
13. $y^2 + y - 12$　　　　**14.** $y^2 + 3y - 18$
15. $x^2 + 5x - 14$　　　**16.** $x^2 - 5x - 24$
17. $r^2 - 8r - 9$　　　　**18.** $r^2 - r - 2$
19. $x^2 - x - 30$　　　　**20.** $x^2 + 8x + 12$
21. $a^2 + 15a + 56$　　　**22.** $a^2 - 9a + 20$
23. $y^2 - y - 42$　　　　**24.** $y^2 + y - 42$
25. $x^2 + 13x + 42$　　　**26.** $x^2 - 13x + 42$

Factor the following problems completely. First, factor out the greatest common factor, and then factor the remaining trinomial.

27. $2x^2 + 6x + 4$　　　　**28.** $3x^2 - 6x - 9$
29. $3a^2 - 3a - 60$　　　**30.** $2a^2 - 18a + 28$
31. $100x^2 - 500x + 600$
32. $100x^2 - 900x + 2,000$
33. $100p^2 - 1,300p + 4,000$
34. $100p^2 - 1,200p + 3,200$
35. $x^4 - x^3 - 12x^2$　　　**36.** $x^4 - 11x^3 + 24x^2$
37. $2r^3 + 4r^2 - 30r$　　　**38.** $5r^3 + 45r^2 + 100r$

39. $2y^4 - 6y^3 - 8y^2$　　　**40.** $3r^3 - 3r^2 - 6r$
41. $x^5 + 4x^4 + 4x^3$　　　**42.** $x^5 + 13x^4 + 42x^3$
43. $3y^4 - 12y^3 - 15y^2$　　**44.** $5y^4 - 10y^3 + 5y^2$
45. $4x^4 - 52x^3 + 144x^2$
46. $3x^3 - 3x^2 - 18x$

Factor the following trinomials.

47. $x^2 + 5xy + 6y^2$　　　**48.** $x^2 - 5xy + 6y^2$
49. $x^2 - 9xy + 20y^2$　　　**50.** $x^2 + 9xy + 20y^2$
51. $a^2 + 2ab - 8b^2$　　　**52.** $a^2 - 2ab - 8b^2$
53. $a^2 - 10ab + 25b^2$　　**54.** $a^2 + 6ab + 9b^2$
55. $a^2 + 10ab + 25b^2$　　**56.** $a^2 - 6ab + 9b^2$
57. $x^2 + 2xa - 48a^2$　　　**58.** $x^2 - 3xa - 10a^2$
59. $x^2 - 5xb - 36b^2$　　　**60.** $x^2 - 13xb + 36b^2$

Factor completely.

61. $x^4 - 5x^2 + 6$　　　　**62.** $x^6 - 2x^3 - 15$
63. $x^2 - 80x - 2,000$
64. $x^2 - 190x - 2,000$

65. $x^2 - x + \dfrac{1}{4}$　　　**66.** $x^2 - \dfrac{2}{3}x + \dfrac{1}{9}$

67. $x^2 + 0.6x + 0.08$　　**68.** $x^2 + 0.8x + 0.15$
69. If one of the factors of $x^2 + 24x + 128$ is $x + 8$, what is the other factor?
70. If one factor of $x^2 + 260x + 2,500$ is $x + 10$, what is the other factor?

71. What polynomial, when factored, gives $(4x + 3)(x - 1)$?

72. What polynomial factors to $(4x - 3)(x + 1)$?

Applying the Concepts

73. If an arrow is shot into the air with a velocity of 48 feet per second from the top of a building 64 feet high, the equation that gives the height of the arrow at any time t is

$$h = 64 + 48t - 16t^2$$

Factor the right side of this equation and then find h when t is 4.

74. A bullet is fired into the air with an initial upward velocity of 64 feet per second from the top of a building 80 feet high. The equation that gives the height of the bullet at any time t is

$$h = 80 + 64t - 16t^2$$

Factor the right side of this equation and then find h when t is 5.

Review Problems

The problems that follow review material we covered in Sections 4.4 and 4.5. Reviewing the problems from Section 4.5 will help you with the next section.

Multiply using the FOIL method. [4.5]

75. $(6a + 1)(a + 2)$ **76.** $(6a - 1)(a - 2)$

77. $(3a + 2)(2a + 1)$ **78.** $(3a - 2)(2a - 1)$

79. $(6a + 2)(a + 1)$ **80.** $(3a + 1)(2a + 2)$

Subtract. [4.4]

81. $(5x^2 + 5x - 4) - (3x^2 - 2x + 7)$

82. $(7x^4 - 4x^2 - 5) - (2x^4 - 4x^2 + 5)$

83. Subtract $4x - 5$ from $7x + 3$.

84. Subtract $3x + 2$ from $-6x + 1$.

85. Subtract $2x^2 - 4x$ from $5x^2 - 5$.

86. Subtract $6x^2 + 3$ from $2x^2 - 4x$.

SECTION 5.3   More Trinomials to Factor

We will now consider trinomials whose greatest common factor is 1 and whose leading coefficient (the coefficient of the squared term) is a number other than 1. We present two methods for factoring trinomials of this type. The first method involves listing possible factors until the correct pair of factors is found. This requires a certain amount of trial and error. The second method is based on the factoring by grouping process that we covered in Section 5.1. Either method can be used to factor trinomials whose leading coefficient is a number other than 1. Here is our first method:

Method 1 **Factoring $ax^2 + bx + c$ by trial and error**

Suppose we want to factor the trinomial $2x^2 - 5x - 3$. We know the factors (if they exist) will be a pair of binomials. The product of their first terms is $2x^2$ and the product of their last terms is -3. Let us list all the possible factors along with the trinomial that would result if we were to multiply them together. Remember, the middle term comes from the product of the inside terms plus the product of the outside terms.

Binomial Factors	First Term	Middle Term	Last Term
$(2x - 3)(x + 1)$	$2x^2$	$-x$	-3
$(2x + 3)(x - 1)$	$2x^2$	$+x$	-3
$(2x - 1)(x + 3)$	$2x^2$	$+5x$	-3
$(2x + 1)(x - 3)$	$2x^2$	$-5x$	-3

We can see from the last line that the factors of $2x^2 - 5x - 3$ are $(2x + 1)$ and $(x - 3)$. Using this method, there is no straightforward way, as there was in the last section, to find the factors, other than by trial and error or by simply listing all the possibilities. We look for possible factors that, when multiplied, will give the correct first and last terms, and then we see if we can adjust them to give the correct middle term.

▶ **EXAMPLE 1**   Factor $6a^2 + 7a + 2$.

Solution   We list all the possible pairs of factors that, when multiplied together, give a trinomial whose first term is $6a^2$ and whose last term is $+2$.

Binomial Factors	First Term	Middle Term	Last Term
$(6a + 1)(a + 2)$	$6a^2$	$+13a$	$+2$
$(6a - 1)(a - 2)$	$6a^2$	$-13a$	$+2$
$(3a + 2)(2a + 1)$	$6a^2$	$+7a$	$+2$
$(3a - 2)(2a - 1)$	$6a^2$	$-7a$	$+2$

The factors of $6a^2 + 7a + 2$ are $(3a + 2)$ and $(2a + 1)$.

Check: $(3a + 2)(2a + 1) = 6a^2 + 7a + 2$ ◀

Notice that in the preceding list we did not include the factors $(6a + 2)$ and $(a + 1)$. We do not need to try these, since the first factor has a 2 common to each term and so could be factored again, giving $2(3a + 1)(a + 1)$. Since our original trinomial, $6a^2 + 7a + 2$, did *not* have a greatest common factor of 2, neither of its factors will.

▶ **EXAMPLE 2**   Factor $4x^2 - x - 3$.

Solution   We list all the possible factors that, when multiplied, give a trinomial whose first term is $4x^2$ and whose last term is -3.

Binomial Factors	First Term	Middle Term	Last Term
$(4x + 1)(x - 3)$	$4x^2$	$-11x$	-3
$(4x - 1)(x + 3)$	$4x^2$	$+11x$	-3
$(4x + 3)(x - 1)$	$4x^2$	$-x$	-3
$(4x - 3)(x + 1)$	$4x^2$	$+x$	-3
$(2x + 1)(2x - 3)$	$4x^2$	$-4x$	-3
$(2x - 1)(2x + 3)$	$4x^2$	$+4x$	-3

The third line shows that the factors are $(4x + 3)$ and $(x - 1)$.

Check: $(4x + 3)(x - 1) = 4x^2 - x - 3$ ◀

You will find that the more practice you have at factoring this type of trinomial, the faster you will get the correct factors. You will pick up some shortcuts along the way or maybe come across a system of eliminating some factors as possibilities. Whatever works best for you is the method you should use. Factoring is a very important tool and you must be good at it.

▶ **EXAMPLE 3**   Factor $12y^3 + 10y^2 - 12y$.

Solution   We begin by factoring out the greatest common factor, $2y$:

$$12y^3 + 10y^2 - 12y = 2y(6y^2 + 5y - 6)$$

We now list all possible factors of a trinomial with the first term $6y^2$ and last term -6, along with the associated middle terms.

Possible Factors	Middle Term When Multiplied
$(3y + 2)(2y - 3)$	$-5y$
$(3y - 2)(2y + 3)$	$+5y$
$(6y + 1)(y - 6)$	$-35y$
$(6y - 1)(y + 6)$	$+35y$

The second line gives the correct factors. The complete problem is

$$12y^3 + 10y^2 - 12y = 2y(6y^2 + 5y - 6)$$
$$= 2y(3y - 2)(2y + 3)$$ ◀

▶ **EXAMPLE 4**   Factor $30x^2y - 5xy^2 - 10y^3$.

Solution   The greatest common factor is $5y$:

$$30x^2y - 5xy^2 - 10y^3 = 5y(6x^2 - xy - 2y^2)$$
$$= 5y(2x + y)(3x - 2y) \qquad \blacktriangleleft$$

Method 2 **Factoring $ax^2 + bx + c$ by grouping**

Recall from Section 5.1 that we can use factoring by grouping to factor the polynomial $6x^2 - 3x - 4x + 2$. We begin by factoring $3x$ from the first two terms and -2 from the last two terms. For review, here is the complete problem.

$$6x^2 - 3x - 4x + 2 = 3x(2x - 1) - 2(2x - 1)$$
$$= (2x - 1)(3x - 2)$$

Now, let's back up a little and notice that our original polynomial $6x^2 - 3x - 4x + 2$ can be simplified to $6x^2 - 7x + 2$ by adding $-3x$ and $-4x$. This means that $6x^2 - 7x + 2$ can be factored to $(2x - 1)(3x - 2)$ by the grouping method shown in Section 5.1. The key to using this process is to rewrite the middle term, $-7x$, as $-3x - 4x$.

To generalize this discussion, here are the steps we use to factor trinomials by grouping.

Strategy for Factoring $ax^2 + bx + c$ by Grouping

> *Step 1:* Form the product ac.
> *Step 2:* Find a pair of numbers whose product is ac and whose sum is b.
> *Step 3:* Rewrite the polynomial to be factored so that the middle term bx is written as the sum of two terms whose coefficients are the two numbers found in step 2.
> *Step 4:* Factor by grouping.

▶ **EXAMPLE 5** Factor $3x^2 - 10x - 8$ using these steps.

Solution The trinomial $3x^2 - 10x - 8$ has the form $ax^2 + bx + c$, where $a = 3$, $b = -10$, and $c = -8$.

> *Step 1:* The product ac is $3(-8) = -24$.
> *Step 2:* We need to find two numbers whose product is -24 and whose sum is -10. Let's systematically begin to list all the pairs of numbers whose product is -24 to find the pair whose sum is -10.

Product	*Sum*
$-24(1) = -24$	$-24 + 1 = -23$
$-12(2) = -24$	$-12 + 2 = -10$

We stop here because we have found the pair of numbers whose product is -24, and whose sum is -10. The numbers are -12 and 2.

Step 3: We now rewrite our original trinomial so the middle term $-10x$ is written as the sum of $-12x$ and $2x$:

$$3x^2 - 10x - 8 = 3x^2 - 12x + 2x - 8$$

Step 4: Factoring by grouping, we have

$$3x^2 - 12x + 2x - 8 = 3x(x - 4) + 2(x - 4)$$
$$= (x - 4)(3x + 2)$$

We can check our work by multiplying $x - 4$ and $3x + 2$ to get $3x^2 - 0x - 8$. ◀

▶ **EXAMPLE 6**   Factor $4x^2 - x - 3$.

Solution   In this case, $a = 4$, $b = -1$, and $c = -3$. The product ac is $4(-3) = -12$. We need a pair of numbers whose product is -12 and whose sum is -1. We begin listing pairs of numbers whose product is -12 and whose sum is -1.

Product	Sum
$-12(1) = -12$	$-12 + 1 = -11$
$-6(2) = -12$	$-6 + 2 = -4$
$-4(3) = -12$	$-4 + 3 = -1$

We stop here because we have found the pair of numbers for which we are looking. They are -4 and 3. Next, we rewrite the middle term, $-x$, as the sum $-4x + 3x$ and proceed to factor by grouping.

$$4x^2 - x - 3 = 4x^2 - 4x + 3x - 3$$
$$= 4x(x - 1) + 3(x - 1)$$
$$= (x - 1)(4x + 3)$$

Compare this procedure and the result with those shown in Example 2 of this section. ◀

▶ **EXAMPLE 7**   Factor $8x^2 - 2x - 15$.

Solution   The product ac is $8(-15) = -120$. There are many pairs of numbers whose product is -120. We are looking for the pair whose sum is also -2. The numbers are -12 and 10. Writing $-2x$ as $-12x + 10x$ and then factoring by grouping, we have

$$8x^2 - 2x - 15 = 8x^2 - 12x + 10x - 15$$
$$= 4x(2x - 3) + 5(2x - 3)$$
$$= (2x - 3)(4x + 5)$$ ◀

PROBLEM SET 5.3

Factor the following trinomials.

1. $2x^2 + 7x + 3$ **2.** $2x^2 + 5x + 3$
3. $2a^2 - a - 3$ **4.** $2a^2 + a - 3$
5. $3x^2 + 2x - 5$ **6.** $3x^2 - 2x - 5$
7. $3y^2 - 14y - 5$ **8.** $3y^2 + 14y - 5$
9. $6x^2 + 13x + 6$ **10.** $6x^2 - 13x + 6$
11. $4x^2 - 12xy + 9y^2$ **12.** $4x^2 + 12xy + 9y^2$
13. $4y^2 - 11y - 3$ **14.** $4y^2 + y - 3$
15. $20x^2 - 41x + 20$ **16.** $20x^2 + 9x - 20$
17. $20a^2 + 48ab - 5b^2$
18. $20a^2 + 29ab + 5b^2$
19. $20x^2 - 21x - 5$ **20.** $20x^2 - 48x - 5$
21. $12m^2 + 16m - 3$ **22.** $12m^2 + 20m + 3$
23. $20x^2 + 37x + 15$ **24.** $20x^2 + 13x - 15$
25. $12a^2 - 25ab + 12b^2$
26. $12a^2 + 7ab - 12b^2$
27. $3x^2 - xy - 14y^2$
28. $3x^2 + 19xy - 14y^2$
29. $14x^2 + 29x - 15$ **30.** $14x^2 + 11x - 15$
31. $6x^2 - 43x + 55$ **32.** $6x^2 - 7x - 55$
33. $15t^2 - 67t + 38$ **34.** $15t^2 - 79t - 34$

Factor each of the following completely. Look first for the greatest common factor.

35. $4x^2 + 2x - 6$ **36.** $6x^2 - 51x + 63$
37. $24a^2 - 50a + 24$ **38.** $18a^2 + 48a + 32$
39. $10x^3 - 23x^2 + 12x$
40. $10x^4 + 7x^3 - 12x^2$
41. $6x^4 - 11x^3 - 10x^2$
42. $6x^3 + 19x^2 + 10x$
43. $10a^3 - 6a^2 - 4a$ **44.** $6a^3 + 15a^2 + 9a$

45. $15x^3 - 102x^2 - 21x$
46. $2x^4 - 24x^3 + 64x^2$
47. $35y^3 - 60y^2 - 20y$
48. $14y^4 - 32y^3 + 8y^2$
49. $15a^4 - 2a^3 - a^2$
50. $10a^5 - 17a^4 + 3a^3$
51. $24x^2y - 6xy - 45y$
52. $8x^2y^2 + 26xy^2 + 15y^2$
53. $12x^2y - 34xy^2 + 14y^3$
54. $12x^2y - 46xy^2 + 14y^3$
55. Evaluate the expression $2x^2 + 7x + 3$ and the expression $(2x + 1)(x + 3)$ for $x = 2$.
56. Evaluate the expression $2a^2 - a - 3$ and the expression $(2a - 3)(a + 1)$ for $a = 5$.
57. What polynomial factors to $(2x + 3)(2x - 3)$?
58. What polynomial factors to $(5x + 4)(5x - 4)$?
59. What polynomial factors to $(x + 3)(x - 3)(x^2 + 9)$?
60. What polynomial factors to $(x + 2)(x - 2)(x^2 + 4)$?

Review Problems

The following problems review material we covered in Section 4.6. Reviewing these problems will help you understand the next section.

Multiply.

61. $(x + 3)(x - 3)$ **62.** $(x + 5)(x - 5)$
63. $(6a + 1)(6a - 1)$ **64.** $(4a + 5)(4a - 5)$
65. $(x + 4)^2$ **66.** $(x - 5)^2$
67. $(2x + 3)^2$ **68.** $(2x - 3)^2$

SECTION

5.4   The Difference of Two Squares

In Chapter 4 we listed the following three special products:

$$(a + b)^2 = (a + b)(a + b) = a^2 + 2ab + b^2$$

$$(a - b)^2 = (a - b)(a - b) = a^2 - 2ab + b^2$$

$$(a + b)(a - b) = a^2 - b^2$$

Since factoring is the reverse of multiplication, we can also consider the three special products as three special factorizations:

$$a^2 + 2ab + b^2 = (a + b)^2$$

$$a^2 - 2ab + b^2 = (a - b)^2$$

$$a^2 - b^2 = (a + b)(a - b)$$

Any trinomial of the form $a^2 + 2ab + b^2$ or $a^2 - 2ab + b^2$ can be factored by the methods of Section 5.3. The last line is the factorization for the difference of two squares. The difference of two squares always factors in this way. Again, these are patterns you must be able to recognize on sight.

▶ **EXAMPLE 1**   Factor $16x^2 - 25$.

Solution   We can see that the first term is a perfect square and the last term is also. This fact becomes even more obvious if we rewrite the problem as

$$16x^2 - 25 = (4x)^2 - (5)^2$$

The first term is the square of the quantity $4x$ and the last term is the square of 5. The completed problem looks like this:

$$16x^2 - 25 = (4x)^2 - (5)^2$$
$$= (4x + 5)(4x - 5)$$

To check our results we multiply:

$$(4x + 5)(4x - 5) = 16x^2 + 20x - 20x - 25$$
$$= 16x^2 - 25 \qquad ◀$$

▶ **EXAMPLE 2**   Factor $36a^2 - 1$.

Solution   We rewrite the two terms to show they are perfect squares and then factor. Remember, 1 is its own square, $1^2 = 1$.

$$36a^2 - 1 = (6a)^2 - (1)^2$$
$$= (6a + 1)(6a - 1)$$

To check our results we multiply:

$$(6a + 1)(6a - 1) = 36a^2 + 6a - 6a - 1$$
$$= 36a^2 - 1 \qquad ◀$$

▶ **EXAMPLE 3**   Factor $x^4 - y^4$.

Solution x^4 is the perfect square $(x^2)^2$, and y^4 is $(y^2)^2$.

$$x^4 - y^4 = (x^2)^2 - (y^2)^2$$
$$= (x^2 - y^2)(x^2 + y^2)$$

The factor $(x^2 - y^2)$ is itself the difference of two squares and therefore can be factored again. The factor $(x^2 + y^2)$ is the *sum* of two squares and cannot be factored again. The complete problem is this:

$$x^4 - y^4 = (x^2)^2 - (y^2)^2$$
$$= (x^2 - y^2)(x^2 + y^2)$$
$$= (x + y)(x - y)(x^2 + y^2)$$ ◀

Note If you think the sum of two squares, $x^2 + y^2$, factors, you should try it. Write down the factors you think it has and then multiply them using the FOIL method. You won't get $x^2 + y^2$.

▶ **EXAMPLE 4** Factor $25x^2 - 60x + 36$.

Solution Although this trinomial can be factored by the method we used in Section 5.3, we notice that the first and last terms are the perfect squares $(5x)^2$ and $(6)^2$. Before going through the method for factoring trinomials by listing all possible factors, we can check to see if $25x^2 - 60x + 36$ factors to $(5x - 6)^2$. We need only multiply to check:

$$(5x - 6)^2 = (5x - 6)(5x - 6)$$
$$= 25x^2 - 30x - 30x + 36$$
$$= 25x^2 - 60x + 36$$

The trinomial $25x^2 - 60x + 36$ factors to $(5x - 6)(5x - 6) = (5x - 6)^2$.

◀

▶ **EXAMPLE 5** Factor $m^2 + 14m + 49$.

Solution Since the first and last terms are perfect squares, we can try the factors $(m + 7)(m + 7)$:

$$(m + 7)^2 = (m + 7)(m + 7)$$
$$= m^2 + 7m + 7m + 49$$
$$= m^2 + 14m + 49$$

The factors of $m^2 + 14m + 49$ are $(m + 7)(m + 7) = (m + 7)^2$. ◀

Note As we have indicated before, perfect square trinomials like the ones in Examples 4 and 5 can be factored by the methods developed in previous sections. Recognizing that they factor to binomial squares simply saves time in factoring.

▶ **EXAMPLE 6** Factor $5x^2 + 30x + 45$.

Solution We begin by factoring out the greatest common factor, which is 5. Then we notice that the trinomial that remains is a perfect square trinomial:

$$5x^2 + 30x + 45 = 5(x^2 + 6x + 9)$$
$$= 5(x + 3)^2 \qquad \blacktriangleleft$$

▶ **EXAMPLE 7**   Factor $(x - 3)^2 - 25$.

Solution   This example has the form $a^2 - b^2$ where a is $x - 3$ and b is 5. We factor it according to the formula for the difference of two squares:

$$(x - 3)^2 - 25 = (x - 3)^2 - 5^2 \qquad \text{Write 25 as } 5^2$$
$$= [(x - 3) - 5][(x - 3) + 5] \qquad \text{Factor}$$
$$= (x - 8)(x + 2) \qquad \text{Simplify}$$

Notice in this example we could have expanded $(x - 3)^2$, subtracted 25, and then factored to obtain the same result:

$$(x - 3)^2 - 25 = x^2 - 6x + 9 - 25 \qquad \text{Expand } (x - 3)^2$$
$$= x^2 - 6x - 16 \qquad \text{Simplify}$$
$$= (x - 8)(x + 2) \qquad \text{Factor} \qquad \blacktriangleleft$$

▶ **EXAMPLE 8**   Factor $x^2 - 10x + 25 - y^2$.

Solution   Notice the first three terms form a perfect square trinomial. That is, $x^2 - 10x + 25 = (x - 5)^2$. If we replace the first three terms by $(x - 5)^2$, the expression that results has the form $a^2 - b^2$. We can then factor as we did in Example 7:

$$x^2 - 10x + 25 - y^2$$
$$= (x^2 - 10x + 25) - y^2 \qquad \text{Group first 3 terms together}$$
$$= (x - 5)^2 - y^2 \qquad \text{This has the form } a^2 - b^2$$
$$= [(x - 5) - y][(x - 5) + y] \qquad \text{Factor}$$
$$= (x - 5 - y)(x - 5 + y) \qquad \text{Simplify}$$

We could check this result by multiplying the two factors together.   ◀

P R O B L E M   S E T   5 . 4

Factor the following.

1. $x^2 - 9$
2. $x^2 - 25$
3. $a^2 - 36$
4. $a^2 - 64$
5. $x^2 - 49$
6. $x^2 - 121$
7. $4a^2 - 16$
8. $4a^2 + 16$
9. $9x^2 + 25$
10. $16x^2 - 36$
11. $25x^2 - 169$
12. $x^2 - y^2$
13. $9a^2 - 16b^2$
14. $49a^2 - 25b^2$
15. $9 - m^2$
16. $16 - m^2$
17. $25 - 4x^2$
18. $36 - 49y^2$
19. $2x^2 - 18$
20. $3x^2 - 27$
21. $32a^2 - 128$
22. $3a^3 - 48a$
23. $8x^2y - 18y$
24. $50a^2b - 72b$
25. $a^4 - b^4$
26. $a^4 - 16$
27. $16m^4 - 81$
28. $81 - m^4$
29. $3x^3y - 75xy^3$
30. $2xy^3 - 8x^3y$

Factor the following.

31. $x^2 - 2x + 1$ **32.** $x^2 - 6x + 9$

33. $x^2 + 2x + 1$ **34.** $x^2 + 6x + 9$

35. $a^2 - 10a + 25$ **36.** $a^2 + 10a + 25$

37. $y^2 + 4y + 4$ **38.** $y^2 - 8y + 16$

39. $x^2 - 4x + 4$ **40.** $x^2 + 8x + 16$

41. $m^2 - 12m + 36$ **42.** $m^2 + 12m + 36$

43. $4a^2 + 12a + 9$ **44.** $9a^2 - 12a + 4$

45. $49x^2 - 14x + 1$ **46.** $64x^2 - 16x + 1$

47. $9y^2 - 30y + 25$ **48.** $25y^2 + 30y + 9$

49. $x^2 + 10xy + 25y^2$ **50.** $25x^2 + 10xy + y^2$

51. $9a^2 + 6ab + b^2$ **52.** $9a^2 - 6ab + b^2$

Factor the following by first factoring out the greatest common factor.

53. $3a^2 + 18a + 27$ **54.** $4a^2 - 16a + 16$

55. $2x^2 + 20xy + 50y^2$

56. $3x^2 + 30xy + 75y^2$

57. $5x^3 + 30x^2y + 45xy^2$

58. $12x^2y - 36xy^2 + 27y^3$

Factor by grouping the first three terms together (see Example 8).

59. $x^2 + 6x + 9 - y^2$

60. $x^2 + 10x + 25 - y^2$

61. $x^2 + 2xy + y^2 - 9$

62. $a^2 + 2ab + b^2 - 25$

63. Find the value of b for which the polynomial $x^2 + bx + 49$ factors to $(x + 7)^2$.

64. Find the value of b for which the polynomial $x^2 + bx + 81$ factors to $(x + 9)^2$.

65. Find the value of c for which the polynomial $x^2 + 10x + c$ factors to $(x + 5)^2$.

66. Find the value of a for which the polynomial $ax^2 + 12x + 9$ factors to $(2x + 3)^2$.

Review Problems

The following problems review material we covered in Section 4.8.

Use long division to divide.

67. $\dfrac{x^2 - 5x + 8}{x - 3}$ **68.** $\dfrac{x^2 + 7x + 12}{x + 4}$

69. $\dfrac{6x^2 + 5x + 3}{2x + 3}$ **70.** $\dfrac{x^3 + 27}{x + 3}$

SECTION

5.5   Factoring: A General Review

In this section we will review the different methods of factoring that we have presented in the previous sections of the chapter. This section is important because it will give you an opportunity to factor a variety of polynomials. Prior to this section, the polynomials you have worked with have been grouped together according to the method used to factor them. That is, in Section 5.4 all the polynomials you factored were either the difference of two squares or perfect square trinomials. What usually happens in a situation like this is that you become proficient at factoring the kind of polynomial you are working with at the time, but have trouble when given a variety of polynomials to factor.

We begin this section with a checklist that can be used in factoring polynomials of any type. When you have finished this section and the problem set that follows, you want to be proficient enough at factoring that the checklist is second nature to you.

Strategy for Factoring a Polynomial

Step 1: If the polynomial has a greatest common factor other than 1, then factor out the greatest common factor.

Step 2: If the polynomial has two terms (it is a binomial), then see if it is the difference of two squares. Remember, if it is the sum of two squares, it will not factor.

Step 3: If the polynomial has three terms (a trinomial), then either it is a perfect square trinomial, which will factor into the square of a binomial, or it is not a perfect square trinomial, in which case you use the methods developed in Section 5.3.

Step 4: If the polynomial has more than three terms, try to factor it by grouping.

Step 5: As a final check, see if any of the factors you have written can be factored further. If you have overlooked a common factor, you can catch it here.

Here are some examples illustrating how we use the checklist.

▶ **EXAMPLE 1**   Factor $2x^5 - 8x^3$.

Solution   First we check to see if the greatest common factor is other than 1. Since the greatest common factor is $2x^3$, we begin by factoring it out. Once we have done so, we notice that the binomial that remains is the difference of two squares:

$$2x^5 - 8x^3 = 2x^3(x^2 - 4) \qquad \text{Factor out the greatest common factor, } 2x^3$$

$$= 2x^3(x + 2)(x - 2) \qquad \text{Factor the difference of two squares}$$ ◀

Note that the greatest common factor $2x^3$ that we factored from each term in the first step of Example 1 remains as part of the answer to the problem. That is because it is one of the factors of the original binomial. Remember, the expression we end up with when factoring must be equal to the expression we start with. We can't just drop a factor and expect the resulting expression to equal the original expression.

▶ **EXAMPLE 2**   Factor $3x^4 - 18x^3 + 27x^2$.

Solution   Step 1 is to factor out the greatest common factor, $3x^2$. After we have done so, we notice that the trinomial that remains is a perfect square trinomial, which will factor as the square of a binomial:

$$3x^4 - 18x^3 + 27x^2 = 3x^2(x^2 - 6x + 9) \qquad \text{Factor out } 3x^2$$
$$= 3x^2(x - 3)^2 \qquad x^2 - 6x + 9 \text{ is the}$$
$$\text{square of } x - 3 \quad \blacktriangleleft$$

▶ **EXAMPLE 3**   Factor $y^3 + 25y$.

Solution   We begin by factoring out the y that is common to both terms. The binomial that remains after we have done so is the sum of two squares, which does not factor, so after the first step we are finished:

$$y^3 + 25y = y(y^2 + 25) \qquad \text{Factor out the greatest common factor,}$$
$$y; \text{ then notice that } y^2 + 25 \text{ cannot be}$$
$$\text{factored further} \quad \blacktriangleleft$$

▶ **EXAMPLE 4**   Factor $6a^2 - 11a + 4$.

Solution   Here we have a trinomial that does not have a greatest common factor other than 1. Since it is not a perfect square trinomial, we factor it by trial and error. That is, we look for binomial factors the product of whose first terms is $6a^2$ and the product of whose last terms is 4. Then we look for the combination of these types of binomials whose product gives us a middle term of $-11a$. Without showing all the different possibilities, here is the answer:

$$6a^2 - 11a + 4 = (3a - 4)(2a - 1) \quad \blacktriangleleft$$

▶ **EXAMPLE 5**   Factor $6x^3 - 12x^2 - 48x$.

Solution   This trinomial has a greatest common factor of $6x$. The trinomial that remains after the $6x$ has been factored from each term must be factored by trial and error:

$$6x^3 - 12x^2 - 48x = 6x(x^2 - 2x - 8)$$
$$= 6x(x - 4)(x + 2) \quad \blacktriangleleft$$

▶ **EXAMPLE 6**   Factor $2ab^5 + 8ab^4 + 2ab^3$.

Solution   The greatest common factor is $2ab^3$. We begin by factoring it from each term. After that we find that the trinomial that remains cannot be factored further:

$$2ab^5 + 8ab^4 + 2ab^3 = 2ab^3(b^2 + 4b + 1) \quad \blacktriangleleft$$

▶ **EXAMPLE 7**   Factor $xy + 8x + 3y + 24$.

Solution   Since our polynomial has four terms, we try factoring by grouping:

$$xy + 8x + 3y + 24 = x(y + 8) + 3(y + 8)$$
$$= (y + 8)(x + 3) \quad \blacktriangleleft$$

PROBLEM SET 5.5

Factor each of the following polynomials completely. That is, once you are finished factoring, none of the factors you obtain should be factorable. Also, note that the even-numbered problems are not necessarily similar to the odd-numbered problems that precede them in this problem set.

1. $x^2 - 81$
2. $x^2 - 18x + 81$
3. $x^2 + 2x - 15$
4. $15x^2 + 11x - 6$
5. $x^2 + 6x + 9$
6. $12x^2 - 11x + 2$
7. $y^2 - 10y + 25$
8. $21y^2 - 25y - 4$
9. $2a^3b + 6a^2b + 2ab$
10. $6a^2 - ab - 15b^2$
11. $x^2 + x + 1$
12. $2x^2 - 4x + 2$
13. $12a^2 - 75$
14. $18a^2 - 50$
15. $9x^2 - 12xy + 4y^2$
16. $x^3 - x^2$
17. $4x^3 + 16xy^2$
18. $16x^2 + 49y^2$
19. $2y^3 + 20y^2 + 50y$
20. $3y^2 - 9y - 30$
21. $a^6 + 4a^4b^2$
22. $5a^2 - 45b^2$
23. $xy + 3x + 4y + 12$
24. $xy + 7x + 6y + 42$
25. $x^4 - 16$
26. $x^4 - 81$
27. $xy - 5x + 2y - 10$
28. $xy - 7x + 3y - 21$
29. $5a^2 + 10ab + 5b^2$
30. $3a^3b^2 + 15a^2b^2 + 3ab^2$
31. $x^2 + 49$
32. $16 - x^4$
33. $3x^2 + 15xy + 18y^2$
34. $3x^2 + 27xy + 54y^2$
35. $2x^2 + 15x - 38$
36. $2x^2 + 7x - 85$
37. $100x^2 - 300x + 200$
38. $100x^2 - 400x + 300$
39. $x^2 - 64$
40. $9x^2 - 4$
41. $x^2 + 3x + ax + 3a$
42. $x^2 + 4x + bx + 4b$
43. $49a^7 - 9a^5$
44. $a^4 - 1$
45. $49x^2 + 9y^2$
46. $12x^4 - 62x^3 + 70x^2$
47. $25a^3 + 20a^2 + 3a$
48. $36a^4 - 100a^2$
49. $xa - xb + ay - by$
50. $xy - bx + ay - ab$
51. $48a^4b - 3a^2b$
52. $18a^4b^2 - 12a^3b^3 + 8a^2b^4$
53. $20x^4 - 45x^2$
54. $16x^3 + 16x^2 + 3x$
55. $3x^2 + 35xy - 82y^2$
56. $3x^2 + 37xy - 86y^2$
57. $16x^5 - 44x^4 + 30x^3$
58. $16x^2 + 16x - 1$
59. $2x^2 + 2ax + 3x + 3a$
60. $2x^2 + 2ax + 5x + 5a$
61. $y^4 - 1$
62. $25y^7 - 16y^5$
63. $12x^4y^2 + 36x^3y^3 + 27x^2y^4$
64. $16x^3y^2 - 4xy^2$

Review Problems

The problems that follow review material we covered in Sections 2.3 and 4.1. Reviewing the problems from Section 2.3 will help you understand the next section.

Solve each equation. [2.3]

65. $3x - 6 = 9$
66. $5x - 1 = 14$
67. $2x + 3 = 0$
68. $4x - 5 = 0$
69. $4x + 3 = 0$
70. $3x - 1 = 0$

Simplify, using the properties of exponents. [4.1]

71. $x^8 \cdot x^7$
72. $(x^5)^2$
73. $(3x^3)^2(2x^4)^3$
74. $(5x^2y)^2(4xy^3)^2$
75. Write the number 57,600 in scientific notation.
76. Write the number 4.3×10^5 in expanded form.

SECTION

5.6 Solving Equations by Factoring

In this section we will use the methods of factoring developed in previous sections, along with a special property of 0, to solve quadratic equations.

DEFINITION   Any equation that can be put in the form $ax^2 + bx + c = 0$, where a, b, and c are real numbers ($a \neq 0$), is called a **quadratic equation**. The equation $ax^2 + bx + c = 0$ is called *standard form* for a quadratic equation:

an x^2 term   an x term   and a constant term

$a(\text{variable})^2 + b(\text{variable}) + (\text{absence of the variable}) = 0$

The number zero has a special property that is the key to solving quadratic equations. If we multiply two numbers and the product is 0, then one or both of the original two numbers must be 0. In symbols, this property looks like this.

Zero-Factor Property

Let a and b represent real numbers. If $a \cdot b = 0$, then $a = 0$ or $b = 0$.

Suppose we want to solve the quadratic equation $x^2 + 5x + 6 = 0$. We can factor the left side into $(x + 2)(x + 3)$. Then we have

$$x^2 + 5x + 6 = 0$$

$$(x + 2)(x + 3) = 0$$

Now $(x + 2)$ and $(x + 3)$ both represent real numbers. Their product is 0; therefore, either $(x + 3)$ is 0 or $(x + 2)$ is 0. Either way we have a solution to our equation. We use the property of zero stated above to finish the problem:

$$x^2 + 5x + 6 = 0$$

$$(x + 2)(x + 3) = 0$$

$$x + 2 = 0 \quad \text{or} \quad x + 3 = 0$$

$$x = -2 \quad \text{or} \quad x = -3$$

Our solution set is $\{-2, -3\}$. Our equation has two solutions. To check our solutions we have to check each one separately to see that they both produce a true statement when used in place of the variable:

When $x = -3$

the equation $x^2 + 5x + 6 = 0$

becomes $(-3)^2 + 5(-3) + 6 \stackrel{?}{=} 0$

$$9 + (-15) + 6 \stackrel{?}{=} 0$$

$$0 = 0$$

When	$x = -2$
the equation	$x^2 + 5x + 6 = 0$
becomes	$(-2)^2 + 5(-2) + 6 \overset{?}{=} 0$
	$4 + (-10) + 6 \overset{?}{=} 0$
	$0 = 0$

We have solved a quadratic equation by replacing it with two linear equations in one variable.

Strategy for Solving a Quadratic Equation by Factoring

Step 1: Put the equation in standard form, that is, 0 on one side and decreasing powers of the variable on the other.

Step 2: Factor completely.

Step 3: Use the zero-factor property to set each variable factor from step 2 to 0.

Step 4: Solve each equation produced in step 3.

Step 5: Check each solution, if necessary.

▶ **EXAMPLE 1**   Solve the equation $2x^2 - 5x = 12$.

Solution

Step 1: We begin by adding -12 to both sides, so the equation is in standard form:

$$2x^2 - 5x = 12$$
$$2x^2 - 5x - 12 = 0$$

Step 2: We factor the left side completely:

$$(2x + 3)(x - 4) = 0$$

Step 3: We set each factor to 0:

$$2x + 3 = 0 \quad \text{or} \quad x - 4 = 0$$

Step 4: Solve each of the equations from step 3:

$$2x + 3 = 0 \qquad x - 4 = 0$$
$$2x = -3 \qquad\quad x = 4$$
$$x = -\frac{3}{2}$$

Step 5: Substitute each solution into $2x^2 - 5x = 12$ to check:

$$\text{Check: } -\frac{3}{2} \qquad\qquad\qquad \text{Check: } 4$$

$$2\left(-\frac{3}{2}\right)^2 - 5\left(-\frac{3}{2}\right) \overset{?}{=} 12 \qquad\qquad 2(4)^2 - 5(4) \overset{?}{=} 12$$

$$2\left(\frac{9}{4}\right) + 5\left(\frac{3}{2}\right) \overset{?}{=} 12 \qquad\qquad 2(16) - 20 \overset{?}{=} 12$$

$$\frac{9}{2} + \frac{15}{2} \overset{?}{=} 12 \qquad\qquad 32 - 20 \overset{?}{=} 12$$

$$\frac{24}{2} \overset{?}{=} 12 \qquad\qquad 12 = 12$$

$$12 = 12 \qquad\qquad\qquad \blacktriangleleft$$

▶ **EXAMPLE 2**   Solve for a: $16a^2 - 25 = 0$.

Solution   The equation is already in standard form.

$$16a^2 - 25 = 0$$

$$(4a - 5)(4a + 5) = 0 \qquad\qquad \text{Factor left side}$$

$$4a - 5 = 0 \quad \text{or} \quad 4a + 5 = 0 \qquad \text{Set each factor to 0}$$

$$4a = 5 \qquad\qquad 4a = -5 \qquad \text{Solve the resulting equations}$$

$$a = \frac{5}{4} \qquad\qquad a = -\frac{5}{4} \qquad\qquad\qquad \blacktriangleleft$$

▶ **EXAMPLE 3**   Solve $4x^2 = 8x$.

Solution   We begin by adding $-8x$ to each side of the equation to put it in standard form. Then, we factor the left side of the equation by factoring out the greatest common factor:

$$4x^2 = 8x$$

$$4x^2 - 8x = 0 \qquad \text{Add } -8x \text{ to each side}$$

$$4x(x - 2) = 0 \qquad \text{Factor the left side}$$

$$4x = 0 \quad \text{or} \quad x - 2 = 0 \qquad \text{Set each factor to 0}$$

$$x = 0 \quad \text{or} \qquad x = 2 \qquad \text{Solve the resulting equations}$$

The solutions are 0 and 2. $\qquad\qquad\qquad\qquad\qquad\qquad \blacktriangleleft$

▶ **EXAMPLE 4**   Solve $x(2x + 3) = 44$.

Solution   We must multiply out the left side first and then put the equation in standard form:

$$x(2x + 3) = 44$$
$$2x^2 + 3x = 44 \qquad \text{Multiply out the left side}$$
$$2x^2 + 3x - 44 = 0 \qquad \text{Add } -44 \text{ to each side}$$
$$(2x + 11)(x - 4) = 0 \qquad \text{Factor the left side}$$
$$2x + 11 = 0 \quad \text{or} \quad x - 4 = 0 \qquad \text{Set each factor to 0}$$
$$2x = -11 \quad \text{or} \quad x = 4 \qquad \text{Solve the resulting equations}$$
$$x = -\frac{11}{2} \qquad \qquad x = 4 \qquad \text{Solve the resulting equations}$$

The two solutions are $-\frac{11}{2}$ and 4.   ◀

▶ **EXAMPLE 5**   Solve for x: $5^2 = x^2 + (x + 1)^2$.

Solution   Before we can put this equation in standard form we must square the binomial. Remember, to square a binomial, we use the formula $(a + b)^2 = a^2 + 2ab + b^2$.

$$5^2 = x^2 + (x + 1)^2$$
$$25 = x^2 + x^2 + 2x + 1 \qquad \text{Expand } 5^2 \text{ and } (x + 1)^2$$
$$25 = 2x^2 + 2x + 1 \qquad \text{Simplify the right side}$$
$$0 = 2x^2 + 2x - 24 \qquad \text{Add } -25 \text{ to each side}$$
$$0 = 2(x^2 + x - 12) \qquad \text{Begin factoring}$$
$$0 = 2(x + 4)(x - 3) \qquad \text{Factor completely}$$
$$x + 4 = 0 \quad \text{or} \quad x - 3 = 0 \qquad \text{Set each variable factor to 0}$$
$$x = -4 \quad \text{or} \quad x = 3$$

Note, in the second to the last line, that we do not set 2 equal to 0. That is because 2 can never be 0. It is always 2. We only use the zero-factor property to set variable factors to 0 because they are the only factors that can possibly be 0.

Also notice that it makes no difference which side of the equation is 0 when we write the equation in standard form.   ◀

Although the equation in the next example is not a quadratic equation, it can be solved by the method shown in the first five examples.

▶ **EXAMPLE 6** Solve $24x^3 = -10x^2 + 6x$ for x.

Solution First we write the equation in standard form:

$$24x^3 + 10x^2 - 6x = 0 \qquad \text{Standard form}$$

$$2x(12x^2 + 5x - 3) = 0 \qquad \text{Factor out } 2x$$

$$2x(3x - 1)(4x + 3) = 0 \qquad \text{Factor remaining trinomial}$$

$$2x = 0 \quad \text{or} \quad 3x - 1 = 0 \quad \text{or} \quad 4x + 3 = 0 \qquad \text{Set factors to 0}$$

$$x = 0 \quad \text{or} \qquad x = \frac{1}{3} \quad \text{or} \qquad x = -\frac{3}{4} \qquad \text{Solutions} \qquad ◀$$

PROBLEM SET 5.6

The following equations are already in factored form. Use the special property with zero to set the factors to zero and solve.

1. $(x + 2)(x - 1) = 0$ **2.** $(x + 3)(x + 2) = 0$

3. $(a - 4)(a - 5) = 0$

4. $(a + 6)(a - 1) = 0$

5. $x(x + 1)(x - 3) = 0$

6. $x(2x + 1)(x - 5) = 0$

7. $(3x + 2)(2x + 3) = 0$

8. $(4x - 5)(x - 6) = 0$

9. $m(3m + 4)(3m - 4) = 0$

10. $m(2m - 5)(3m - 1) = 0$

11. $2y(3y + 1)(5y + 3) = 0$

12. $3y(2y - 3)(3y - 4) = 0$

Solve the following equations.

13. $x^2 + 3x + 2 = 0$ **14.** $x^2 - x - 6 = 0$

15. $x^2 - 9x + 20 = 0$ **16.** $x^2 + 2x - 3 = 0$

17. $a^2 - 2a - 24 = 0$

18. $a^2 - 11a + 30 = 0$

19. $100x^2 - 500x + 600 = 0$

20. $100x^2 - 300x + 200 = 0$

21. $x^2 = -6x - 9$ **22.** $x^2 = 10x - 25$

23. $a^2 - 16 = 0$ **24.** $a^2 - 36 = 0$

25. $2x^2 + 5x - 12 = 0$

26. $3x^2 + 14x - 5 = 0$

27. $9x^2 + 12x + 4 = 0$

28. $12x^2 - 24x + 9 = 0$

29. $a^2 + 25 = 10a$ **30.** $a^2 + 16 = 8a$

31. $2x^2 = 3x + 20$ **32.** $6x^2 = x + 2$

33. $3m^2 = 20 - 7m$

34. $2m^2 = -18 + 15m$

35. $4x^2 - 49 = 0$ **36.** $16x^2 - 25 = 0$

37. $x^2 + 6x = 0$ **38.** $x^2 - 8x = 0$

39. $x^2 - 3x = 0$ **40.** $x^2 + 5x = 0$

41. $2x^2 = 8x$ **42.** $2x^2 = 10x$

43. $3x^2 = 15x$ **44.** $5x^2 = 15x$

45. $1,400 = 400 + 700x - 100x^2$

46. $2,700 = 700 + 900x - 100x^2$

47. $6x^2 = -5x + 4$ **48.** $9x^2 = 12x - 4$

49. $x(2x - 3) = 20$ **50.** $x(3x - 5) = 12$

51. $t(t + 2) = 80$ **52.** $t(t + 2) = 99$

53. $4,000 = (1,300 - 100p)p$

54. $3,200 = (1,200 - 100p)p$

55. $x(14 - x) = 48$ **56.** $x(12 - x) = 32$

57. $(x + 5)^2 = 2x + 9$

58. $(x + 7)^2 = 2x + 13$

59. $(y - 6)^2 = y - 4$ **60.** $(y + 4)^2 = y + 6$

61. $10^2 = (x + 2)^2 + x^2$

62. $15^2 = (x + 3)^2 + x^2$

63. $2x^3 + 11x^2 + 12x = 0$

64. $3x^3 + 17x^2 + 10x = 0$

65. $4y^3 - 2y^2 - 30y = 0$

66. $9y^3 + 6y^2 - 24y = 0$

67. $8x^3 + 16x^2 = 10x$

68. $24x^3 - 22x^2 = -4x$

69. $20a^3 = -18a^2 + 18a$

70. $12a^3 = -2a^2 + 10a$

Use factoring by grouping to solve the following equations.

71. $x^3 + 3x^2 - 4x - 12 = 0$
72. $x^3 + 5x^2 - 9x - 45 = 0$
73. $x^3 + x^2 - 16x - 16 = 0$
74. $4x^3 + 12x^2 - 9x - 27 = 0$

Review Problems

The following problems review material we covered in Sections 3.10 and 4.2.

The following applications are taken from the book *Academic Algebra,* written by William J. Milne and published by the American Book Company in 1901. Solve each problem. [3.10]

75. A bicycle and a suit cost $90. How much did each cost, if the bicycle cost five times as much as the suit?

76. A man bought a cow and a calf for $36, paying eight times as much for the cow as for the calf. What was the cost of each?

77. A house and a lot cost $3,000. If the house cost four times as much as the lot, what was the cost of each?

78. A plumber and two helpers together earned $7.50 per day. How much did each earn per day, if the plumber earned four times as much as each helper?

Use the properties of exponents to simplify each expression. [4.2]

79. 2^{-3}

80. 5^{-2}

81. $\dfrac{x^5}{x^{-3}}$

82. $\dfrac{x^{-2}}{x^{-5}}$

83. $\dfrac{(x^2)^3}{(x^{-3})^4}$

84. $\dfrac{(x^2)^{-4}(x^{-2})^3}{(x^{-3})^{-5}}$

85. Write the number 0.0056 in scientific notation.
86. Write the number 2.34×10^{-4} in expanded form.

SECTION
5.7   Applications

In this section we will look at some application problems, the solutions to which require solving a quadratic equation. We will also introduce the Pythagorean Theorem, one of the oldest theorems in the history of mathematics. The person whose name we associate with the theorem, Pythagoras, was a Greek philosopher and mathematician who lived from 580 B.C. to 500 B.C. According to the British philosopher Bertrand Russell, Pythagoras was "intellectually one of the most important men that ever lived."

Also in this section, the solutions to the examples show only the essential steps from our Blueprint for Problem Solving. Recall that step 1 is done mentally; we read the problem and mentally list the items that are known and the items that are unknown. This is an essential part of problem solving. However, now that you have had experience with application problems, you are doing step 1 automatically.

Number Problems

▶ **EXAMPLE 1**   The product of two consecutive odd integers is 63. Find the integers.

Solution   Let x = the first odd integer, then $x + 2$ = the second odd integer. An equation that describes the situation is

$$x(x + 2) = 63 \qquad \text{(Their product is 63)}$$

We solve the equation:

$$x(x + 2) = 63$$
$$x^2 + 2x = 63$$
$$x^2 + 2x - 63 = 0$$
$$(x - 7)(x + 9) = 0$$
$$x - 7 = 0 \quad \text{or} \quad x + 9 = 0$$
$$x = 7 \quad \text{or} \qquad x = -9$$

If the first odd integer is 7, the next odd integer is $7 + 2 = 9$. If the first odd integer is -9, the next consecutive odd integer is $-9 + 2 = -7$. We have two pairs of consecutive odd integers that are solutions. They are 7, 9 and -9, -7.

We check to see that their products are 63:

$$7(9) = 63$$
$$-7(-9) = 63 \qquad \blacktriangleleft$$

Suppose we know that the sum of two numbers is 50. We want to find a way to represent each number using only one variable. If we let x represent one of the two numbers, how can we represent the other? Let's suppose for a moment that x turns out to be 30. Then the other number will be 20, because their sum is 50. That is, if two numbers add up to 50 and one of them is 30, then the other must be $50 - 30 = 20$. Generalizing this to any number x, we see that if two numbers have a sum of 50 and one of the numbers is x, then the other must be $50 - x$. The table that follows shows some additional examples.

If two numbers have a sum of	and one of them is	then the other must be
50	x	$50 - x$
100	x	$100 - x$
10	y	$10 - y$
12	n	$12 - n$

Now let's look at an example that uses this idea.

▶ **EXAMPLE 2**   The sum of two numbers is 13. Their product is 40. Find the numbers.

Solution   If we let x represent one of the numbers, then $13 - x$ must be the other number because their sum is 13. Since their product is 40 we can write

$$x(13 - x) = 40 \qquad \text{The product of the two numbers is 40}$$

$$13x - x^2 = 40 \qquad \text{Multiply left side}$$

$$x^2 - 13x = -40 \qquad \text{Multiply both sides by } -1 \text{ and reverse} \\ \text{order of terms of left side}$$

$$x^2 - 13x + 40 = 0 \qquad \text{Add 40 to each side}$$

$$(x - 8)(x - 5) = 0 \qquad \text{Factor left side}$$

$$x - 8 = 0 \quad \text{or} \quad x - 5 = 0$$

$$x = 8 \qquad\qquad x = 5$$

The two solutions are 8 and 5. If x is 8, then the other number is $13 - x = 13 - 8 = 5$. Likewise, if x is 5, the other number is $13 - x = 13 - 5 = 8$. Therefore, the two numbers we are looking for are 8 and 5. Their sum is 13 and their product is 40.   ◀

Geometry Problems

Many applications dealing with area can best be described algebraically by quadratic equations.

▶ **EXAMPLE 3**   The length of a rectangle is three more than twice the width. The area is 44 square inches (sq. in.). Find the dimensions (find the length and width).

Solution   As shown in Figure 1, let $x =$ the width of the rectangle. Then $2x + 3 =$ the length of the rectangle, because the length is three more than twice the width.

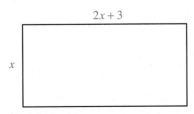

Figure 1

Since the area is 44 sq. in., an equation that describes the situation is

$$x(2x + 3) = 44 \qquad \text{(Width} \cdot \text{length} = \text{area)}$$

We now solve the equation:

$$x(2x + 3) = 44$$
$$2x^2 + 3x = 44$$
$$2x^2 + 3x - 44 = 0$$
$$(2x + 11)(x - 4) = 0$$
$$2x + 11 = 0 \qquad \text{or} \quad x - 4 = 0$$
$$x = -\frac{11}{2} \quad \text{or} \qquad x = 4$$

The solution $x = -\frac{11}{2}$ cannot be used, since length and width are always given in positive units. The width is 4. The length is three more than twice the width or $2(4) + 3 = 11$.

$$\text{Width} = 4 \text{ inches}$$
$$\text{Length} = 11 \text{ inches}$$

The solutions check in the original problem, since $4(11) = 44$. ◄

▶ **EXAMPLE 4** The numerical value of the area of a square is twice its perimeter. What is the length of its side?

Solution   As shown in Figure 2, let $x =$ the length of its side. Then $x^2 =$ the area of the square and $4x =$ the perimeter of the square.
    An equation that describes the situation is

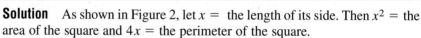

$$x^2 = 2(4x) \qquad \text{The area is 2 times the perimeter}$$
$$x^2 = 8x$$
$$x^2 - 8x = 0$$
$$x(x - 8) = 0$$
$$x = 0 \quad \text{or} \quad x = 8$$

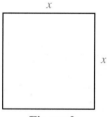

x

x

Figure 2

Since $x = 0$ does not make sense in our original problem, we use $x = 8$. If the side has length 8, then the perimeter is $4(8) = 32$ and the area is $8^2 = 64$. Since 64 is twice 32, our solution is correct. ◄

Fact from Geometry: The Pythagorean Theorem

Next we will work some problems involving the Pythagorean Theorem, which we mentioned in the introduction to this chapter. It may interest you to know that Pythagoras formed a secret society around the year 540 B.C. Known as the

Pythagoreans, members kept no written record of their work; everything was handed down by spoken word. They influenced not only mathematics, but religion, science, medicine, and music as well. Among other things, they discovered the correlation between musical notes and the reciprocals of counting numbers, $\frac{1}{2}, \frac{1}{3}, \frac{1}{4}$, and so on. In their daily lives they followed strict dietary and moral rules to achieve a higher rank in future lives.

Pythagorean Theorem
In any right triangle (Figure 3), the square of the longest side (called the hypotenuse) is equal to the sum of the squares of the other two sides (called legs).

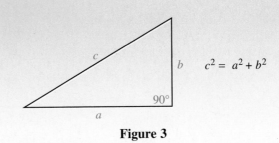

$$c^2 = a^2 + b^2$$

Figure 3

▶ **EXAMPLE 5** The three sides of a right triangle are three consecutive integers. Find the lengths of the three sides.

Solution Let $x =$ the first integer (shortest side)

then $x + 1 =$ the next consecutive integer

and $x + 2 =$ the last consecutive integer (longest side)

A diagram of the triangle is shown in Figure 4.

The Pythagorean Theorem tells us that the square of the longest side, $(x + 2)^2$, is equal to the sum of the squares of the two shorter sides, $(x + 1)^2 + x^2$. Here is the equation:

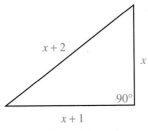

Figure 4

$$(x + 2)^2 = (x + 1)^2 + x^2$$

$$x^2 + 4x + 4 = x^2 + 2x + 1 + x^2 \qquad \text{Expand squares}$$

$$x^2 - 2x - 3 = 0 \qquad \text{Standard form}$$

$$(x - 3)(x + 1) = 0 \qquad \text{Factor}$$

$$x - 3 = 0 \quad \text{or} \quad x + 1 = 0 \qquad \text{Set factors to 0}$$

$$x = 3 \quad \text{or} \qquad x = -1$$

Since a triangle cannot have a side with a negative number for its length, we must not use -1 for a solution to our original problem. Therefore the shortest side is 3. The other two sides are the next two consecutive integers, 4 and 5. ◀

▶ **EXAMPLE 6** The hypotenuse of a right triangle is 5 inches, while the lengths of the two legs (the other two sides) are given by two consecutive integers. Find the lengths of the two legs.

Solution If we let $x =$ the length of the shortest side, then the other side must be $x + 1$. A diagram of the triangle is shown in Figure 5.

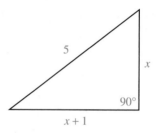

Figure 5

The Pythagorean Theorem tells us that the square of the longest side, 5^2, is equal to the sum of the squares of the two shorter sides, $x^2 + (x + 1)^2$. Here is the equation:

$$5^2 = x^2 + (x + 1)^2 \qquad \text{Pythagorean Theorem}$$
$$25 = x^2 + x^2 + 2x + 1 \qquad \text{Expand } 5^2 \text{ and } (x + 1)^2$$
$$25 = 2x^2 + 2x + 1 \qquad \text{Simplify the right side}$$
$$0 = 2x^2 + 2x - 24 \qquad \text{Add } -25 \text{ to each side}$$
$$0 = 2(x^2 + x - 12) \qquad \text{Begin factoring}$$
$$0 = 2(x + 4)(x - 3) \qquad \text{Factor completely}$$
$$x + 4 = 0 \quad \text{or} \quad x - 3 = 0 \qquad \text{Set variable factors to 0}$$
$$x = -4 \quad \text{or} \qquad x = 3$$

Since a triangle cannot have a side with a negative number for its length, we cannot use -4. Therefore, the shortest side must be 3 inches. The next side is $x + 1 = 3 + 1 = 4$ inches. Since the hypotenuse is 5, we can check our solutions with the Pythagorean Theorem as shown in Figure 6.

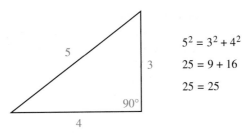

Figure 6 ◀

Business Problems

Our next two examples are from the world of business. If you are planning on taking finite mathematics, statistics, or business calculus in the future, these examples will give you a head start on some of the problems you will see in those classes.

▶ **EXAMPLE 7**   A company can manufacture x hundred items for a total cost of $C = 300 + 500x - 100x^2$. How many items were manufactured if the total cost is $900?

Solution   We are looking for x when C is 900. We begin by substituting 900 for C in the cost equation. Then we solve for x:

When $\qquad\qquad\qquad C = 900$

the equation $\qquad\quad C = 300 + 500x - 100x^2$

becomes $\qquad\quad 900 = 300 + 500x - 100x^2$

We can write this equation in standard form by adding -300, $-500x$, and $100x^2$ to each side. The result looks like this:

$$100x^2 - 500x + 600 = 0$$

$100(x^2 - 5x + 6) = 0 \qquad$ Begin factoring

$100(x - 2)(x - 3) = 0 \qquad$ Factor completely

$x - 2 = 0 \quad$ or $\quad x - 3 = 0 \qquad$ Set variable factors to 0

$x = 2 \quad$ or $\qquad x = 3$

Our solutions are 2 and 3, which means that the company can manufacture 2 hundred items or 3 hundred items for a total cost of $900. ◀

▶ **EXAMPLE 8**   A manufacturer of small portable radios knows that the number of radios she can sell each week is related to the price of the radios by the equation $x = 1{,}300 - 100p$ (x is the number of radios and p is the price per radio). What price should she charge for the radios in order to have a weekly revenue of $4,000?

Solution First, we must find the revenue equation. The equation for total revenue is $R = xp$ where x is the number of units sold and p is the price per unit. Since we want R in terms of p, we substitute $1{,}300 - 100p$ for x in the equation $R = xp$:

$$\text{If} \qquad R = xp$$

$$\text{and} \qquad x = 1{,}300 - 100p$$

$$\text{then} \qquad R = (1{,}300 - 100p)p$$

We want to find p when R is 4,000. Substituting 4,000 for R in the equation gives us

$$4{,}000 = (1{,}300 - 100p)p$$

If we multiply out the right side we have

$$4{,}000 = 1{,}300p - 100p^2$$

To write this equation in standard form we add $100p^2$ and $-1{,}300p$ to each side.

$$100p^2 - 1{,}300p + 4{,}000 = 0 \qquad \text{Add } 100p^2 \text{ and } -1{,}300p \text{ to each side}$$

$$100(p^2 - 13p + 40) = 0 \qquad \text{Begin factoring}$$

$$100(p - 5)(p - 8) = 0 \qquad \text{Factor completely}$$

$$p - 5 = 0 \quad \text{or} \quad p - 8 = 0 \qquad \text{Set variable factors to 0}$$

$$p = 5 \quad \text{or} \qquad p = 8$$

If she sells the radios for \$5 each or for \$8 each she will have a weekly revenue of \$4,000. ◀

PROBLEM SET 5.7

Solve the following applications. Be sure to show the equation used.

Number Problems

1. The product of two consecutive even integers is 80. Find the two integers.
2. The product of two consecutive integers is 72. Find the two integers.
3. The product of two consecutive odd integers is 99. Find the two integers.
4. The product of two consecutive integers is 132. Find the two integers.
5. The product of two consecutive even integers is ten less than five times their sum. Find the two integers.
6. The product of two consecutive odd integers is one less than four times their sum. Find the two integers.
7. The sum of two numbers is 14. Their product is 48. Find the numbers.
8. The sum of two numbers is 12. Their product is 32. Find the numbers.
9. One number is two more than five times another. Their product is 24. Find the numbers.
10. One number is one more than twice another. Their product is 55. Find the numbers.

11. One number is four times another. Their product is four times their sum. Find the numbers.

12. One number is two more than twice another. Their product is two more than twice their sum. Find the numbers.

Geometry Problems

13. The length of a rectangle is one more than the width. The area is 12 sq. in. Find the dimensions.

14. The length of a rectangle is three more than twice the width. The area is 44 sq. in. Find the dimensions.

15. The height of a triangle is twice the base. The area is 9 sq. in. Find the base.

16. The height of a triangle is two more than twice the base. The area is 20 sq. ft. Find the base.

17. The hypotenuse of a right triangle is 10 inches. The lengths of the two legs are given by two consecutive even integers. Find the lengths of the two legs.

18. The hypotenuse of a right triangle is 15 inches. One of the legs is 3 inches more than the other. Find the lengths of the two legs.

19. The shorter leg of a right triangle is 5 meters. The hypotenuse is 1 meter longer than the longer leg. Find the length of the longer leg.

20. The shorter leg of a right triangle is 12 yards. If the hypotenuse is 20 yards, how long is the other leg?

Business Problems

21. A company can manufacture x hundred items for a total cost of $C = 400 + 700x - 100x^2$. Find x if the total cost is $1,400.

22. If the total cost, C, of manufacturing x hundred items is given by the equation $C = 700 + 900x - 100x^2$, find x when C is $2,700.

23. The total cost, C, of manufacturing x hundred videotapes is given by the equation

$$C = 600 + 1,000x - 100x^2$$

Find x if the total cost is $2,200.

24. The total cost, C, of manufacturing x hundred pen and pencil sets is given by the equation

$C = 500 + 800x - 100x^2$. Find x when C is $1,700.

25. A company that manufactures typewriter ribbons knows that the number of ribbons it can sell each week, x, is related to the price per ribbon, p, by the equation $x = 1,200 - 100p$. At what price should the company sell the ribbons if it wants the weekly revenue to be $3,200? (*Remember:* The equation for revenue is $R = xp$.)

26. A company manufactures diskettes for home computers. It knows from past experience that the number of diskettes it can sell each day, x, is related to the price per diskette, p, by the equation $x = 800 - 100p$. At what price should the company sell the diskettes if it wants the daily revenue to be $1,200?

27. The relationship between the number of calculators a company sells per week, x, and the price of each calculator, p, is given by the equation $x = 1,700 - 100p$. At what price should the calculators be sold if the weekly revenue is to be $7,000?

28. The relationship between the number of pencil sharpeners a company can sell each week, x, and the price of each sharpener, p, is given by the equation $x = 1,800 - 100p$. At what price should the sharpeners be sold if the weekly revenue is to be $7,200?

Review Problems

The problems that follow review material we covered in Section 4.3.

Multiply.

29. $(6a^2b)(7a^3b^2)$

30. $(2a^3b^2)(6ab)$

Divide.

31. $\dfrac{12x^3y^5}{6xy^3}$

32. $\dfrac{25x^4y^8}{5x^3y}$

Simplify.

33. $\dfrac{(5x^4y^4)(10x^3y^3)}{2x^2y^7}$

34. $\dfrac{(4x^3y^2)(9x^4y^{10})}{3x^5y^2}$

35. $(2 \times 10^5)(3 \times 10^{-8})$

36. $\dfrac{8 \times 10^{-8}}{2 \times 10^{-3}}$ **37.** $\dfrac{45a^6}{9a^3} - \dfrac{15a^8}{5a^5}$

38. $\dfrac{27a^{10}}{9a^6} + \dfrac{12a^{12}}{6a^8}$

Research Project 8

The Pythagorean Theorem, which we introduced in this section, is one of the most widely proven theorems in the history of mathematics. Many people have given a proof of this theorem, including one of the presidents of the United States. One of the most common proofs is based on the diagram in Figure 7 and the formula for the area of a triangle. Figure 7 is constructed by taking the right triangle in the lower right corner and repeating it three times so that the final diagram is a square in which each side has a

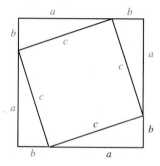

Figure 7

length $a + b$. Write an essay in which you use this diagram to derive the formula $c^2 = a^2 + b^2$. Your essay should be written so that any beginning algebra student can understand it.

CHAPTER 5 SUMMARY

Examples

1. $8x^4 - 10x^3 + 6x^2$

$\quad = 2x^2 \cdot 4x^2 - 2x^2 \cdot 5x + 2x^2 \cdot 3$

$\quad = 2x^2(4x^2 - 5x + 3)$

Greatest Common Factor [5.1]

The largest monomial that divides each term of a polynomial is called the **greatest common factor** for that polynomial. We begin all factorizations by factoring out the greatest common factor.

2. $x^2 + 5x + 6 = (x + 2)(x + 3)$

$\quad x^2 - 5x + 6 = (x - 2)(x - 3)$

$\quad 6x^2 - x - 2 = (2x + 1)(3x - 2)$

$\quad 6x^2 + 7x + 2 = (2x + 1)(3x + 2)$

Factoring Trinomials [5.2, 5.3]

One method of factoring a trinomial of the form $ax^2 + bx + c$ is to list all pairs of binomials the product of whose first terms gives the first term of the trinomial and the product of whose last terms gives the last term of the trinomial. We then choose the pair that gives the correct middle term for the original trinomial. Or, we can use the factoring by grouping method in which we form the product ac and then look for a pair of numbers whose product is ac and whose sum is b.

3. $x^2 + 10x + 25 = (x + 5)^2$

$\quad x^2 - 10x + 25 = (x - 5)^2$

$\quad x^2 - 25 = (x + 5)(x - 5)$

Special Factorizations [5.4]

$$a^2 + 2ab + b^2 = (a + b)^2$$

$$a^2 - 2ab + b^2 = (a - b)^2$$

$$a^2 - b^2 = (a + b)(a - b)$$

4. (a) $2x^5 - 8x^3$

$$= 2x^3(x^2 - 4)$$

$$= 2x^3(x + 2)(x - 2)$$

(b) $3x^4 - 18x^3 + 27x^2$

$$= 3x^2(x^2 - 6x + 9)$$

$$= 3x^2(x - 3)^2$$

(c) $6x^3 - 12x^2 - 48x$

$$= 6x(x^2 - 2x - 8)$$

$$= 6x(x - 4)(x + 2)$$

(d) $x^2 + ax + bx + ab$

$$= x(x + a) + b(x + a)$$

$$= (x + a)(x + b)$$

Strategy for Factoring a Polynomial [5.5]

Step 1: If the polynomial has a greatest common factor other than 1, then factor out the greatest common factor.

Step 2: If the polynomial has two terms (it is a binomial), then see if it is the difference of two squares. Remember, if it is the sum of two squares, it will not factor.

Step 3: If the polynomial has three terms (a trinomial), then either it is a perfect square trinomial that will factor into the square of a binomial, or it is not a perfect square trinomial, in which case you use the methods developed in Section 5.3.

Step 4: If the polynomial has more than three terms, then try to factor it by grouping.

Step 5: As a final check, see if any of the factors you have written can be factored further. If you have overlooked a common factor, you can catch it here.

5. Solve $x^2 - 6x = -8$.

$$x^2 - 6x + 8 = 0$$

$$(x - 4)(x - 2) = 0$$

$$x - 4 = 0 \quad \text{or} \quad x - 2 = 0$$

$$x = 4 \quad \text{or} \quad x = 2$$

Both solutions check.

Strategy for Solving a Quadratic Equation [5.6]

Step 1: Write the equation in standard form: $ax^2 + bx + c = 0$.

Step 2: Factor completely.

Step 3: Set each variable factor equal to zero.

Step 4: Solve the equations found in step 3.

Step 5: Check solutions, if necessary.

6. The hypotenuse of a right triangle is 5 inches, while the lengths of the two legs (the other two sides) are given by two consecutive integers. Find the lengths of the two legs.

If we let $x =$ the length of the shortest side, then the other side must be $x + 1$. The Pythagorean Theorem tells us that

$$5^2 = x^2 + (x + 1)^2$$

$$25 = x^2 + x^2 + 2x + 1$$

$$25 = 2x^2 + 2x + 1$$

The Pythagorean Theorem [5.7]

In any right triangle, the square of the longest side (called the hypotenuse) is equal to the sum of the squares of the other two sides (called legs).

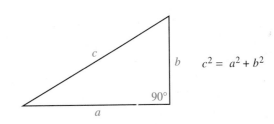

$0 = 2x^2 + 2x - 24$

$0 = 2(x^2 + x - 12)$

$0 = 2(x + 4)(x - 3)$

$x + 4 = 0$ or $x - 3 = 0$

$x = -4$ or $x = 3$

Since a triangle cannot have a side with a negative number for its length, we cannot use -4. The shortest side is $x = 3$ and the other side is $x + 1 = 3 + 1 = 4$.

COMMON MISTAKE

Trying to apply the zero-factor property to other numbers. For example, consider the equation $(x - 3)(x + 4) = 18$. A fairly common mistake is to attempt to solve it with the following steps:

$$(x - 3)(x + 4) = 18$$
$$x - 3 = 18 \quad \text{or} \quad x + 4 = 18 \quad \leftarrow \text{Mistake}$$
$$x = 21 \quad \text{or} \quad x = 14$$

These are obviously not solutions, as a quick check will verify:

Check $x = 21$

$(21 - 3)(21 + 4) \overset{?}{=} 18$

$18 \cdot 25 \overset{?}{=} 18$

$450 = 18$ $\overset{\text{False statements}}{\longleftrightarrow}$

Check $x = 14$

$(14 - 3)(14 + 4) \overset{?}{=} 18$

$11 \cdot 18 \overset{?}{=} 18$

$198 = 18$

The mistake is in setting each factor equal to 18. It is not necessarily true, when the product of two numbers is 18, that either one of them is itself 18. The correct solution looks like this:

$$(x - 3)(x + 4) = 18$$
$$x^2 + x - 12 = 18$$
$$x^2 + x - 30 = 0$$
$$(x + 6)(x - 5) = 0$$
$$x + 6 = 0 \quad \text{or} \quad x - 5 = 0$$
$$x = -6 \quad \text{or} \quad x = 5$$

To avoid this mistake, remember that before you factor a quadratic equation, you must write it in standard form. It is in standard form only when 0 is on one side and decreasing powers of the variable are on the other.

CHAPTER 5 REVIEW

The numbers in brackets refer to the sections of the text in which similar problems can be found.

Factor the following by factoring out the greatest common factor. [5.1]

1. $10x - 20$

2. $4x^3 - 9x^2$

3. $5x - 5y$

4. $7x^3 + 2x$

5. $8x + 4$

6. $2x^2 + 14x + 6$

7. $24y^2 - 40y + 48$

8. $18a^4 - 30a^3$

9. $30xy^3 - 45x^3y^2$

10. $8y^4 - 20y^2 - 32y$

11. $49a^3 - 14b^3$

12. $48x^5 - 36x^3 - 60x^2$

13. $6ab^2 + 18a^3b^3 - 24a^2b$
14. $33a^4b^5 - 48a^5b^3 + 18a^4b^4$

Factor by grouping. [5.1]

15. $xy + bx + ay + ab$
16. $xy + 5x + ay + 5a$
17. $xy + 4x - 5y - 20$
18. $xy - 4x + 5y - 20$
19. $2xy + 10x - 3y - 15$
20. $3xy + 15x - 2y - 10$
21. $5x^2 - 4ax - 10bx + 8ab$
22. $15x^2 - 10bx - 3bx + 2b^2$

Factor the following trinomials. [5.2]

23. $y^2 + 9y + 14$ **24.** $w^2 + 15w + 50$
25. $a^2 - 14a + 48$ **26.** $r^2 + r - 20$
27. $r^2 - 18r + 72$ **28.** $a^2 + 2a - 35$
29. $y^2 + 20y + 99$ **30.** $x^2 - 5x - 24$
31. $y^2 + 8y + 12$ **32.** $n^2 - 5n - 36$

Factor the following trinomials. [5.3]

33. $2x^2 + 13x + 15$ **34.** $4y^2 - 12y + 5$
35. $5y^2 + 11y + 6$
36. $8n^2 - 6mn - 5m^2$
37. $20a^2 - 27a + 9$
38. $14x^2 + 31xy - 10y^2$
39. $6r^2 + 5rt - 6t^2$ **40.** $20y^2 - 36y + 9$
41. $10x^2 - 29x - 21$
42. $12x^2 - 11xy - 15y^2$

Factor the following if possible. [5.4]

43. $n^2 - 81$ **44.** $4y^2 - 9$
45. $x^2 + 49$ **46.** $4r^2 - 9t^2$
47. $36y^2 - 121x^2$ **48.** $25 - y^2$
49. $64a^2 - 121b^2$ **50.** $y^4 - 81$
51. $64 - 9m^2$ **52.** $16 - a^4$

Factor the following. [5.4]

53. $y^2 + 20y + 100$ **54.** $m^2 - 16m + 64$
55. $64t^2 + 16t + 1$ **56.** $49x^2 - 14x + 1$
57. $16n^2 - 24n + 9$
58. $49w^2 + 112w + 64$
59. $4r^2 - 12rt + 9t^2$

60. $16x^2 + 72xy + 81y^2$
61. $9m^2 + 30mn + 25n^2$
62. $25x^2 - 30xy + 9y^2$

Factor the following. [5.2]

63. $2x^2 + 20x + 48$ **64.** $a^3 - 10a^2 + 21a$
65. $3m^3 - 18m^2 - 21m$
66. $x^5 - x^4 - 30x^3$
67. $5y^4 + 10y^3 - 40y^2$
68. $6y^4 + 30y^3 - 36y^2$

Factor the following trinomials. [5.3]

69. $8x^2 + 16x + 6$ **70.** $3a^3 - 14a^2 - 5a$
71. $20m^3 - 34m^2 + 6m$
72. $18x^3 - 3x^2y - 3xy^2$
73. $30x^2y - 55xy^2 + 15y^3$
74. $32x^2y - 56xy^2 + 12y^3$

Factor the following. [5.4]

75. $4x^2 + 40x + 100$ **76.** $4x^3 + 12x^2 + 9x$
77. $5x^2 - 45$
78. $45x^2y - 30xy^2 + 5y^3$
79. $12x^3 - 27xy^2$ **80.** $80x^3 - 5xy^2$

Factor the following polynomials completely. [5.5]

81. $6a^3b + 33a^2b^2 + 15ab^3$
82. $x^5 - x^3$ **83.** $4y^6 + 9y^4$
84. $12x^5 + 20x^4y - 8x^3y^2$
85. $30a^4b + 35a^3b^2 - 15a^2b^3$
86. $18a^3b^2 + 3a^2b^3 - 6ab^4$

Solve. [5.6]

87. $(x - 5)(x + 2) = 0$
88. $x(3x + 2)(x - 4) = 0$
89. $3(2y + 5)(2y - 5) = 0$
90. $4m(m - 7)(2m - 7) = 0$
91. $m^2 + 3m = 10$ **92.** $3x^2 = 7x + 20$
93. $a^2 - 49 = 0$ **94.** $16x^2 - 81 = 0$
95. $m^2 - 9m = 0$ **96.** $98r^2 - 18 = 0$
97. $6y^2 = -13y - 6$ **98.** $5x^2 = -15x$
99. $9x^4 + 9x^3 = 10x^2$
100. $6x^4 = 33x^3 - 42x^2$

Solve the following applications. [5.7]

101. The product of two consecutive even integers is 120. Find the two integers.

102. The product of two consecutive integers is 110. Find the two integers.

103. The product of two consecutive odd integers is one less than three times their sum. Find the integers.

104. The sum of two numbers is 20. Their product is 75. Find the numbers.

105. One number is one less than twice another. Their product is 66. Find the numbers.

106. The length of a rectangle is one more than five times the width. The area is 48 sq. in. Find the dimensions.

107. The height of a triangle is eight times the base. The area is 16 sq. in. Find the base.

108. The shorter leg of a right triangle is 6 meters. The hypotenuse is 2 meters longer than the longer leg. Find the length of the longer leg.

CHAPTER 5 TEST

Factor out the greatest common factor. [5.1]

1. $5x - 10$

2. $18x^2y - 9xy - 36xy^2$

Factor by grouping. [5.1]

3. $x^2 + 2ax - 3bx - 6ab$

4. $xy + 4x - 7y - 28$

Factor the following completely. [5.2–5.5]

5. $x^2 - 5x + 6$ **6.** $x^2 - x - 6$

7. $a^2 - 16$ **8.** $x^2 + 25$

9. $x^4 - 81$ **10.** $27x^2 - 75y^2$

11. $x^3 + 5x^2 - 9x - 45$

12. $x^2 - bx + 5x - 5b$ **13.** $4a^2 + 22a + 10$

14. $3m^2 - 3m - 18$ **15.** $6y^2 + 7y - 5$

16. $12x^3 - 14x^2 - 10x$

Solve the following equations. [5.6]

17. $x^2 + 7x + 12 = 0$ **18.** $x^2 - 4x + 4 = 0$

19. $x^2 - 36 = 0$ **20.** $x^2 = x + 20$

21. $x^2 - 11x = -30$ **22.** $y^3 = 16y$

23. $2a^2 = a + 15$ **24.** $30x^3 - 20x^2 = 10x$

Solve the following word problems. Be sure to show the equation used. [5.7]

25. Two numbers have a sum of 20. Their product is 64. Find the numbers.

26. The product of two consecutive odd integers is seven more than their sum. Find the integers.

27. The length of a rectangle is five more than three times the width. The area is 42 sq. ft. Find the dimensions.

28. One leg of a right triangle is 2 more than twice the other. The hypotenuse is 13 meters. Find the lengths of the two legs.

29. A company can manufacture x hundred items for a total cost C of $C = 200 + 500x - 100x^2$. How many items can be manufactured if the total cost is to be $800?

30. A manufacturer knows that the number of items he can sell each week, x, is related to the price of each item, p, by the equation $x = 900 - 100p$. What price should he charge for each item in order to have a weekly revenue of $1,800? (Remember: $R = xp$.)

RATIONAL EXPRESSIONS

Introduction

In previous chapters we have worked with number sequences. The sequence below is a more complicated sequence than those we have seen before.

$$1 + \frac{1}{1+1}, \, 1 + \frac{1}{1 + \dfrac{1}{1+1}}, \, 1 + \frac{1}{1 + \dfrac{1}{1 + \dfrac{1}{1+1}}}, \, \ldots$$

What makes this sequence more complicated than the sequences we have worked with previously is that each of the expressions above must be simplified before the actual numbers in the sequence appear. Later in this chapter we will find out how to simplify expressions such as the ones in this sequence. When we do, you will see that this sequence is connected very closely to the numbers in the Fibonacci sequence.

The rate equation is also an important part of this chapter. The most common form of this equation is

$$\text{Distance} = \text{rate} \cdot \text{time}$$

We will use the rate equation, along with something called unit analysis, to find the speed in miles per hour of a chair lift at a ski area.

Overview

Rational expressions are to polynomials what fractions are to integers. As you will see, the problems you will encounter in this chapter are very similar to the problems you have already worked with fractions. The methods used in working with fractions are the same methods we use to work with rational expressions. The first few sections in the chapter are concerned with reducing, adding, subtracting, multiplying, and dividing rational expressions. We conclude the chapter with a look at some application problems, including problems involving proportion.

Although there are many properties and principles from previous chapters that you will need in order to solve the problems in this chapter, the three most essential are

1. the ability to factor polynomials
2. division as multiplication by the reciprocal
3. the multiplication property of equality

Number 1 is by far the most important. You must know how to factor polynomials to understand the material in this chapter. There are many problems in this chapter that are impossible to solve without factoring.

STUDY SKILLS

This is the last chapter in which we will mention study skills. You know by now what works best for you, and what you have to do to achieve your goals for this course. From now on it is simply a matter of sticking with the things that work for you and avoiding the things that do not work. It seems simple, but as with anything that takes effort, it is up to you to see that you maintain the skills that get you where you want to be in the course.

If you intend to take more classes in mathematics, and you want to ensure your success in those classes, then you can work towards this goal: *Become the type of student who can learn mathematics on their own.* Most people who have degrees in mathematics were students who could learn mathematics on their own. This doesn't mean that you have to learn it all on your own; it simply means that if you have to, you can learn it on your own. Attaining this goal gives you independence, and puts you in control of your success in any math class you take.

SECTION

6.1

Reducing Rational Expressions to Lowest Terms

In Chapter 1 we defined the set of rational numbers to be the set of all numbers that could be put in the form $\dfrac{a}{b}$, where a and b are integers ($b \neq 0$):

$$\text{Rational numbers} = \left\{ \frac{a}{b} \,\middle|\, a \text{ and } b \text{ are integers, } b \neq 0 \right\}$$

A **rational expression** is any expression that can be put in the form $\dfrac{P}{Q}$, where P and Q are polynomials and $Q \neq 0$:

$$\text{Rational expressions} = \left\{ \frac{P}{Q} \,\middle|\, P \text{ and } Q \text{ are polynomials, } Q \neq 0 \right\}$$

Each of the following is an example of a rational expression:

$$\frac{2x + 3}{x} \qquad \frac{x^2 - 6x + 9}{x^2 - 4} \qquad \frac{5}{x^2 + 6} \qquad \frac{2x^2 + 3x + 4}{2}$$

For the rational expression

$$\frac{x^2 - 6x + 9}{x^2 - 4}$$

the polynomial on top, $x^2 - 6x + 9$, is called the numerator, and the polynomial on the bottom, $x^2 - 4$, is called the denominator. The same is true of the other rational expressions.

We must be careful that we do not use a value of the variable that will give us a denominator of zero. Remember, division by zero is not defined.

▶ **EXAMPLES** State the restrictions on the variable in the following rational expressions:

1. $\dfrac{x + 2}{x - 3}$

Solution The variable x can be any real number except $x = 3$, since, when $x = 3$, the denominator is $3 - 3 = 0$. We state this restriction by writing $x \neq 3$.

2. $\dfrac{5}{x^2 - x - 6}$

Solution If we factor the denominator, we have $x^2 - x - 6 = (x - 3)(x + 2)$. If either of the factors is zero, the whole denominator is zero. Our restrictions are $x \neq 3$ and $x \neq -2$, since either one makes $x^2 - x - 6$ equal to 0. ◀

We will not always list each restriction on a rational expression, but we should be aware of them and keep in mind that no rational expression can have a denominator of zero.

The two fundamental properties of rational expressions are listed next. We will use these two properties many times in this chapter.

Properties of Rational Expressions

Property 1
Multiplying the numerator and denominator of a rational expression by the same nonzero quantity will not change the value of the rational expression.

Property 2
Dividing the numerator and denominator of a rational expression by the same nonzero quantity will not change the value of the rational expression.

We can use Property 2 to reduce rational expressions to lowest terms. Since this process is almost identical to the process of reducing fractions to lowest terms, let's recall how the fraction $\frac{6}{15}$ is reduced to lowest terms:

$$\frac{6}{15} = \frac{2 \cdot 3}{5 \cdot 3} \qquad \text{Factor numerator and denominator}$$

$$= \frac{2 \cdot \cancel{3}}{5 \cdot \cancel{3}} \qquad \text{Divide out the common factor, 3}$$

$$= \frac{2}{5} \qquad \text{Reduce to lowest terms}$$

The same procedure applies to reducing rational expressions to lowest terms. The process is summarized in the following rule.

Rule

To reduce a rational expression to lowest terms, first factor the numerator and denominator completely and then divide both the numerator and denominator by any factors they have in common.

▶ **EXAMPLE 3**   Reduce $\dfrac{x^2 - 9}{x^2 + 5x + 6}$ to lowest terms.

Solution   We begin by factoring:

$$\frac{x^2 - 9}{x^2 + 5x + 6} = \frac{(x - 3)(x + 3)}{(x + 2)(x + 3)}$$

Notice that both polynomials contain the factor $(x + 3)$. If we divide the numerator by $(x + 3)$, we are left with $(x - 3)$. If we divide the denominator by $(x + 3)$, we are left with $(x + 2)$. The complete problem looks like this:

$$\frac{x^2 - 9}{x^2 + 5x + 6} = \frac{(x - 3)\cancel{(x + 3)}}{(x + 2)\cancel{(x + 3)}} \qquad \begin{array}{l}\text{Factor the numerator and}\\ \text{denominator completely}\end{array}$$

$$= \frac{x - 3}{x + 2} \qquad \begin{array}{l}\text{Divide out the common}\\ \text{factor, } x + 3\end{array} \qquad ◀$$

It is convenient to draw a line through the factors as we divide them out. It is especially helpful when the problems become longer.

▶ **EXAMPLE 4**   Reduce $\dfrac{10a + 20}{5a^2 - 20}$ to lowest terms.

Solution   We begin by factoring out the greatest common factor from the numerator and denominator:

$$\frac{10a + 20}{5a^2 - 20} = \frac{10(a + 2)}{5(a^2 - 4)}$$

Factor the greatest common factor from the numerator and denominator

$$= \frac{10(\cancel{a + 2})}{5(\cancel{a + 2})(a - 2)}$$

Factor the denominator as the difference of two squares

$$= \frac{2}{a - 2}$$

Divide out the common factors 5 and $a + 2$ ◀

▶ **EXAMPLE 5** Reduce $\dfrac{2x^3 + 2x^2 - 24x}{x^3 + 2x^2 - 8x}$ to lowest terms.

Solution We begin by factoring the numerator and denominator completely. Then we divide out all factors common to the numerator and denominator. Here is what it looks like:

$$\frac{2x^3 + 2x^2 - 24x}{x^3 + 2x^2 - 8x} = \frac{2x(x^2 + x - 12)}{x(x^2 + 2x - 8)}$$

Factor out the greatest common factor first

$$= \frac{2\cancel{x}(x - 3)\cancel{(x + 4)}}{\cancel{x}(x - 2)\cancel{(x + 4)}}$$

Factor the remaining trinomials

$$= \frac{2(x - 3)}{x - 2}$$

Divide out the factors common to the numerator and denominator ◀

▶ **EXAMPLE 6** Reduce $\dfrac{x - 5}{x^2 - 25}$ to lowest terms.

Solution

$$\frac{x - 5}{x^2 - 25} = \frac{\cancel{x - 5}}{\cancel{(x - 5)}(x + 5)}$$

Factor numerator and denominator completely

$$= \frac{1}{x + 5}$$

Divide out the common factor, $x - 5$ ◀

Ratios

For the rest of this section we will concern ourselves with ratio, a topic closely related to reducing fractions and rational expressions to lowest terms. Let's start with a definition.

DEFINITION If a and b are any two numbers, $b \neq 0$, then the **ratio** of a to b is

$$\frac{a}{b}$$

As you can see, ratios are another name for fractions or rational numbers. They are a way of comparing quantities. Since we can also think of a/b as the quotient of a and b, ratios are also quotients. The following table gives some ratios in words and as fractions.

Ratio	As a Fraction	In Lowest Terms
25 to 75	$\frac{25}{75}$	$\frac{1}{3}$
8 to 2	$\frac{8}{2}$	$\frac{4}{1}$
20 to 16	$\frac{20}{16}$	$\frac{5}{4}$

▶ **EXAMPLE 7** A solution of hydrochloric acid (HCl) and water contains 49 milliliters of water and 21 milliliters of HCl. Find the ratio of HCl to water and of HCl to the total volume of the solution.

Solution The ratio of HCl to water is 21 to 49 or

$$\frac{21}{49} = \frac{3}{7}$$

The amount of total solution volume is $49 + 21 = 70$ milliliters. Therefore the ratio of HCl to total solution is 21 to 70, or

$$\frac{21}{70} = \frac{3}{10} \qquad \blacktriangleleft$$

Many of the problems in this chapter will use what is called the **rate equation.** You use this equation on an intuitive level when you are estimating how long it will take you to drive long distances. For example, if you drive at 50 miles per hour for 2 hours, you will travel 100 miles. Here is the rate equation.

$$\text{distance} = \text{rate} \cdot \text{time}$$

$$d = r \cdot t$$

The rate equation has two equivalent forms, the most common of which is obtained by solving for r. Here it is.

$$r = \frac{d}{t}$$

The rate, r, in the rate equation is the ratio of distance to time, and is also referred to as *average speed*. The units for rate are miles per hour, feet per second, kilometers per hour, and so on.

▶ **EXAMPLE 8**   The Forest chair lift at the Northstar ski resort in Lake Tahoe is 5,603 feet long. If a ride on this chair lift takes 11 minutes, what is the average speed of the lift in feet per minute?

Solution   To find the speed of the lift we find the ratio of distance covered to time. (Our answer is rounded to the nearest whole number.)

$$\text{Rate} = \frac{\text{distance}}{\text{time}} = \frac{5{,}603 \text{ feet}}{11 \text{ minute}} = \frac{5{,}603}{11} \text{ feet/minute} = 509 \text{ ft/min}$$

Note how we separate the numerical part of the problem from the units. In the next section we will convert this rate to miles per hour. ◀

P R O B L E M   S E T   6 . 1

Reduce the following rational expressions to lowest terms, if possible. Also, specify any restrictions on the variable in Problems 1–10.

1. $\dfrac{5}{5x - 10}$

2. $\dfrac{-4}{2x - 8}$

3. $\dfrac{a - 3}{a^2 - 9}$

4. $\dfrac{a + 4}{a^2 - 16}$

5. $\dfrac{x + 5}{x^2 - 25}$

6. $\dfrac{x - 2}{x^2 - 4}$

7. $\dfrac{2x^2 - 8}{4}$

8. $\dfrac{5x - 10}{x - 2}$

9. $\dfrac{2x - 10}{3x - 6}$

10. $\dfrac{4x - 8}{x - 2}$

11. $\dfrac{10a + 20}{5a + 10}$

12. $\dfrac{11a + 33}{6a + 18}$

13. $\dfrac{5x^2 - 5}{4x + 4}$

14. $\dfrac{7x^2 - 28}{2x + 4}$

15. $\dfrac{x - 3}{x^2 - 6x + 9}$

16. $\dfrac{x^2 - 10x + 25}{x - 5}$

17. $\dfrac{3x + 15}{3x^2 + 24x + 45}$

18. $\dfrac{5x + 15}{5x^2 + 40x + 75}$

19. $\dfrac{a^2 - 3a}{a^3 - 8a^2 + 15a}$

20. $\dfrac{a^2 + 3a}{a^3 - 2a^2 - 15a}$

21. $\dfrac{3x - 2}{9x^2 - 4}$

22. $\dfrac{2x - 3}{4x^2 - 9}$

23. $\dfrac{x^2 + 8x + 15}{x^2 + 5x + 6}$

24. $\dfrac{x^2 - 8x + 15}{x^2 - x - 6}$

25. $\dfrac{2m^3 - 2m^2 - 12m}{m^2 - 5m + 6}$

26. $\dfrac{2m^3 + 4m^2 - 6m}{m^2 - m - 12}$

27. $\dfrac{x^3 + 3x^2 - 4x}{x^3 - 16x}$

28. $\dfrac{3a^2 - 8a + 4}{9a^3 - 4a}$

29. $\dfrac{4x^3 - 10x^2 + 6x}{2x^3 + x^2 - 3x}$

30. $\dfrac{3a^3 - 8a^2 + 5a}{4a^3 - 5a^2 + a}$

31. $\dfrac{4x^2 - 12x + 9}{4x^2 - 9}$ **32.** $\dfrac{5x^2 + 18x - 8}{5x^2 + 13x - 6}$

33. $\dfrac{x + 3}{x^4 - 81}$ **34.** $\dfrac{x^2 + 9}{x^4 - 81}$

35. $\dfrac{3x^2 + x - 10}{x^4 - 16}$ **36.** $\dfrac{5x^2 - 26x + 24}{x^4 - 64}$

37. $\dfrac{42x^3 - 20x^2 - 48x}{6x^2 - 5x - 4}$

38. $\dfrac{36x^3 + 132x^2 - 135x}{6x^2 + 25x - 9}$

To reduce each of the following rational expressions to lowest terms, you will have to use factoring by grouping. Be sure to factor each numerator and denominator completely before dividing out any common factors. (*Remember:* Factoring by grouping takes two steps.)

39. $\dfrac{xy + 3x + 2y + 6}{xy + 3x + 5y + 15}$

40. $\dfrac{xy + 7x + 4y + 28}{xy + 3x + 4y + 12}$

41. $\dfrac{x^2 - 3x + ax - 3a}{x^2 - 3x + bx - 3b}$

42. $\dfrac{x^2 - 6x + ax - 6a}{x^2 - 7x + ax - 7a}$

43. $\dfrac{xy + bx + ay + ab}{xy + bx + 3y + 3b}$

44. $\dfrac{x^2 + 5x + ax + 5a}{x^2 + 5x + bx + 5b}$

Write each ratio as a fraction in lowest terms.

45. 8 to 6 **46.** 6 to 8
47. 200 to 250 **48.** 250 to 200
49. 32 to 4 **50.** 4 to 32
51. $15x^2$ to $3x^3$ **52.** $8x^4$ to $4x^7$

Applying the Concepts

53. A car travels 122 miles in 3 hours. Find the average speed of the car in miles per hour. Round to the nearest tenth.

54. A bullet fired from a gun travels a distance of 4,500 feet in 3 seconds. Find the average speed of the bullet in feet per second.

55. The first Ferris wheel was designed and built by George Ferris in 1893. It was a large wheel with a circumference of 785 feet. If one trip around the circumference of the wheel took 20 minutes, find the average speed of a rider in feet per minute.

56. In 1897, a large Ferris wheel was built in Vienna, which is still in operation today. Known as *The Great Wheel,* it has a circumference of 618 feet. If one trip around the wheel takes 15 minutes, find the average speed of a rider on this wheel in feet per minute.

57. A Ferris wheel called *Colossus* was built in St. Louis in 1986. It has a circumference of 518 feet. If a trip around the circumference of *Colossus* takes 40 seconds, find the average speed of a rider in feet per second.

58. A person riding a ferris wheel at the county fair travels once around the wheel, a distance of 188 feet, in 30 seconds. What is the average speed of the rider in feet per second? Round to the nearest tenth.

59. Tina is training for a biathlon. As part of her training she runs an 8-mile course, 2 miles of which is on level ground and 6 miles of which is downhill. It takes her 20 minutes to run the level part of the course and 40 minutes to run the downhill part of the course. Find her average speed in minutes per mile and in miles per minute for each part of the course. Round to the nearest hundredth, if rounding is necessary.

60. A jogger covers a distance of 3 miles in 24 minutes. Find the average speed of the jogger in miles per minute.

61. An economy car travels 168 miles on 3.5 gallons of gas. Give the average fuel consumption of the car in miles per gallon.

62. A luxury car travels 100 miles on 8 gallons of gas. Give the average fuel consumption of the car in miles per gallon.

63. Replace x with 5 and y with 4 in the expression

$$\frac{x^2 - y^2}{x - y}$$

and simplify the result. Is the result equal to $5 - 4$ or $5 + 4$?

64. Replace x with 2 in the expression

$$\frac{x^3 - 1}{x - 1}$$

and simplify the result. Your answer should be equal to what you would get if you replaced x with 2 in $x^2 + x + 1$.

Replace a with 10 in the following expressions and then simplify. The result in each case will be -1.

65. $\dfrac{a - 7}{7 - a}$ **66.** $\dfrac{a^2 - 4}{4 - a^2}$

67. Let $x = 3$ and $y = 1$ and simplify $\dfrac{x^4 - y^4}{x^2 + y^2}$.

68. Let $x = 5$ and $y = 3$ and simplify

$$\frac{x^3 - y^3}{x^2 + xy + y^2}.$$

Review Problems

The problems below review material we covered in Sections 4.3 and 4.7.

Simplify. [4.3]

69. $\dfrac{27x^5}{9x^2} - \dfrac{45x^8}{15x^5}$ **70.** $\dfrac{36x^9}{4x} - \dfrac{45x^3}{5x^{-5}}$

71. $\dfrac{72a^3b^7}{9ab^5} + \dfrac{64a^5b^3}{8a^3b}$ **72.** $\dfrac{80a^5b^{11}}{10a^2b} + \dfrac{33a^6b^{12}}{11a^3b^2}$

Divide. [4.7]

73. $\dfrac{38x^7 + 42x^5 - 84x^3}{2x^3}$

74. $\dfrac{49x^6 - 63x^4 - 35x^2}{7x^2}$

75. $\dfrac{28a^5b^5 + 36ab^4 - 44a^4b}{4ab}$

76. $\dfrac{30a^3b - 12a^2b^2 + 6ab^3}{6ab}$

SECTION

6.2

Multiplication and Division of Rational Expressions

Recall that to multiply two fractions we simply multiply numerators and multiply denominators and then reduce to lowest terms, if possible:

$$\frac{3}{4} \cdot \frac{10}{21} = \frac{30}{84} \leftarrow \text{Multiply numerators} \atop \leftarrow \text{Multiply denominators}$$

$$= \frac{5}{14} \leftarrow \text{Reduce to lowest terms}$$

Recall also that the same result can be achieved by factoring numerators and denominators first and then dividing out the factors they have in common:

$$\frac{3}{4} \cdot \frac{10}{21} = \frac{3}{2 \cdot 2} \cdot \frac{2 \cdot 5}{3 \cdot 7} \qquad \text{Factor}$$

$$= \frac{\cancel{3} \cdot \cancel{2} \cdot 5}{\cancel{2} \cdot 2 \cdot \cancel{3} \cdot 7} \qquad \begin{array}{l} \text{Multiply numerators} \\ \text{Multiply denominators} \end{array}$$

$$= \frac{5}{14} \qquad \text{Divide out common factors}$$

We can apply the second process to the product of two rational expressions, as the following example illustrates.

▶ **EXAMPLE 1** Multiply $\dfrac{x - 2}{x + 3} \cdot \dfrac{x^2 - 9}{2x - 4}$.

Solution We begin by factoring numerators and denominators as much as possible. Then we multiply the numerators and denominators. The last step consists of dividing out all factors common to the numerator and denominator:

$$\frac{x - 2}{x + 3} \cdot \frac{x^2 - 9}{2x - 4} = \frac{x - 2}{x + 3} \cdot \frac{(x - 3)(x + 3)}{2(x - 2)} \qquad \text{Factor completely}$$

$$= \frac{(x - 2)(x - 3)(x + 3)}{(x + 3)(2)(x - 2)} \qquad \begin{array}{l} \text{Multiply numerators} \\ \text{and denominators} \end{array}$$

$$= \frac{x - 3}{2} \qquad \begin{array}{l} \text{Divide out common} \\ \text{factors} \end{array} \qquad ◀$$

In Chapter 1 we defined division as the equivalent of multiplication by the reciprocal. This is how it looks with fractions:

$$\frac{4}{5} \div \frac{8}{9} = \frac{4}{5} \cdot \frac{9}{8} \qquad \text{Division as multiplication by the reciprocal}$$

$$\left. \begin{array}{l} = \dfrac{\cancel{2} \cdot \cancel{2} \cdot 3 \cdot 3}{5 \cdot \cancel{2} \cdot \cancel{2} \cdot 2} \\[2em] = \dfrac{9}{10} \end{array} \right\} \text{Factor and divide out common factors}$$

The same idea holds for division with rational expressions. The rational expression that follows the division symbol is called the **divisor;** to divide, we multiply by the reciprocal of the divisor.

▶ **EXAMPLE 2** Divide $\dfrac{3x - 9}{x^2 - x - 20} \div \dfrac{x^2 + 2x - 15}{x^2 - 25}$.

Solution   We begin by taking the reciprocal of the divisor and writing the problem again in terms of multiplication. We then factor, multiply, and, finally, divide out all factors common to the numerator and denominator of the resulting expression. The complete problem looks like this:

$$\frac{3x - 9}{x^2 - x - 20} \div \frac{x^2 + 2x - 15}{x^2 - 25}$$

$$= \frac{3x - 9}{x^2 - x - 20} \cdot \frac{x^2 - 25}{x^2 + 2x - 15} \qquad \text{Multiply by the reciprocal of the divisor}$$

$$= \frac{3(x - 3)}{(x + 4)(x - 5)} \cdot \frac{(x - 5)(x + 5)}{(x + 5)(x - 3)} \qquad \text{Factor}$$

$$= \frac{3(\cancel{x - 3})(\cancel{x - 5})(\cancel{x + 5})}{(x + 4)(\cancel{x - 5})(\cancel{x + 5})(\cancel{x - 3})} \qquad \text{Divide out common factors}$$

$$= \frac{3}{x + 4} \qquad \blacktriangleleft$$

As you can see, factoring is the single most important tool we use in working with rational expressions. Most of the work we have done or will do with rational expressions is most easily accomplished if the rational expressions are in factored form. Here are some more examples of multiplication and division with rational expressions.

▶ **EXAMPLES**

3. Multiply $\dfrac{3a + 6}{a^2} \cdot \dfrac{a}{2a + 4}$.

Solution

$$\frac{3a + 6}{a^2} \cdot \frac{a}{2a + 4}$$

$$= \frac{3(a + 2)}{a^2} \cdot \frac{a}{2(a + 2)} \qquad \text{Factor completely}$$

$$= \frac{3(\cancel{a + 2})\cancel{a}}{\cancel{a^2}(2)(\cancel{a + 2})} \qquad \text{Multiply}$$

$$= \frac{3}{2a} \qquad \text{Divide numerator and denominator by common factors } a(a + 2)$$

4. Divide $\dfrac{x^2 + 7x + 12}{x^2 - 16} \div \dfrac{x^2 + 6x + 9}{2x - 8}$.

Solution

$$\frac{x^2 + 7x + 12}{x^2 - 16} \div \frac{x^2 + 6x + 9}{2x - 8}$$

$$= \frac{x^2 + 7x + 12}{x^2 - 16} \cdot \frac{2x - 8}{x^2 + 6x + 9}$$ Division is multiplication
by the reciprocal

$$= \frac{(x + 3)(x + 4)(2)(x - 4)}{(x - 4)(x + 4)(x + 3)(x + 3)}$$ Factor and multiply

$$= \frac{2}{x + 3}$$ Divide out common
factors   ◀

In Example 4 we factored and multiplied the two expressions in a single step. This saves writing the problem one extra time.

▶ **EXAMPLE 5**   Multiply $(x^2 - 49) \left(\dfrac{x + 4}{x + 7} \right)$.

Solution   We can think of the polynomial $x^2 - 49$ as having a denominator of 1. Thinking of $x^2 - 49$ in this way allows us to proceed as we did in previous examples:

$$(x^2 - 49) \left(\frac{x + 4}{x + 7} \right) = \frac{x^2 - 49}{1} \cdot \frac{x + 4}{x + 7}$$ Write $x^2 - 49$ with
denominator 1

$$= \frac{(x + 7)(x - 7)(x + 4)}{x + 7}$$ Factor and multiply

$$= (x - 7)(x + 4)$$ Divide out common
factors

We can leave the answer in this form or multiply to get $x^2 - 3x - 28$. In this section let's agree to leave our answers in factored form.   ◀

▶ **EXAMPLE 6**   Multiply $a(a + 5)(a - 5) \left(\dfrac{a + 4}{a^2 + 5a} \right)$.

Solution   We can think of the expression $a(a + 5)(a - 5)$ as having a denominator of 1:

$$a(a + 5)(a - 5) \left(\frac{a + 4}{a^2 + 5a} \right)$$

$$= \frac{a(a + 5)(a - 5)}{1} \cdot \frac{a + 4}{a^2 + 5a}$$

$$= \frac{a(a + 5)(a - 5)(a + 4)}{a(a + 5)}$$ Factor and multiply

$$= (a - 5)(a + 4)$$ Divide out common
factors   ◀

Unit Analysis

Unit analysis is a method of converting between units of measure by multiplying by the number 1. Here is our first illustration: Suppose you are flying in a commercial airliner and the pilot tells you the plane has reached its cruising altitude of 35,000 feet. How many miles is the plane above the ground?

If you know that 1 mile is 5,280 feet, then it is simply a matter of deciding what to do with the two numbers, 5,280 and 35,000. By using unit analysis, this decision is unnecessary.

$$35,000 \text{ feet} = \frac{35,000 \text{ feet}}{1} \cdot \frac{1 \text{ mile}}{5,280 \text{ feet}}$$

We treat the units common to the numerator and denominator in the same way we treat factors common to the numerator and denominator; common units can be divided out, just as common factors are. In the expression above, we have feet common to the numerator and denominator. Dividing them out leaves us with miles only. Here is the complete problem.

$$35,000 \text{ feet} = \frac{35,000 \text{ feet}}{1} \cdot \frac{1 \text{ mile}}{5,280 \text{ feet}}$$

$$= \frac{35,000}{5,280} \text{ miles}$$

$$= 6.6 \text{ miles to the nearest tenth of a mile}$$

The expression $\dfrac{1 \text{ mile}}{5,280 \text{ feet}}$ is called a **conversion factor.** It is simply the number 1 written in a convenient form. Because it is the number 1, we can multiply any other number by it and always be sure we have not changed that number. The key to unit analysis is choosing the right conversion factors.

▶ **EXAMPLE 7** The Mall of America in the Twin Cities covers 78 acres of land. If 1 square mile = 640 acres, how many square miles does the Mall of America cover? Round your answer to the nearest hundredth of a square mile.

Solution We are starting with acres and want to end up with square miles. We need to multiply by a conversion factor that will allow acres to divide out and leave us with square miles.

$$78 \text{ acres} = \frac{78 \text{ acres}}{1} \cdot \frac{1 \text{ square mile}}{640 \text{ acres}}$$

$$= \frac{78}{640} \text{ square miles}$$

$$= 0.12 \text{ square miles to the nearest hundredth} \quad ◀$$

The next example is a continuation of Example 8 from Section 6.1.

▶ **EXAMPLE 8** The Forest chair lift at the Northstar ski resort in Lake Tahoe is 5,603 feet long. If a ride on this chair lift takes 11 minutes, what is the average speed of the lift in miles per hour?

Solution First we find the speed of the lift in feet per minute, as we did in Example 8 of Section 6.1, by taking the ratio of distance to time.

$$\text{Rate} = \frac{\text{distance}}{\text{time}} = \frac{5{,}603 \text{ feet}}{11 \text{ minutes}} = \frac{5{,}603}{11} \text{ feet/minute} = 509 \text{ feet/minute}$$

Next, we convert feet per minute to miles per hour. To do this, we need to know that

1 mile = 5,280 feet
1 hour = 60 minutes

$$\text{Speed} = 509 \text{ feet/minute} = \frac{509 \text{ feet}}{1 \text{ minute}} \cdot \frac{1 \text{ mile}}{5{,}280 \text{ feet}} \cdot \frac{60 \text{ minutes}}{1 \text{ hour}}$$

$$= \frac{509 \cdot 60}{5{,}280} \text{ miles/hour}$$

$$= 5.8 \text{ miles/hour to the nearest tenth} \qquad ◀$$

PROBLEM SET 6.2

Multiply or divide as indicated. Be sure to reduce all answers to lowest terms. (That is, the numerator and denominator of the answer should not have any factors in common.)

1. $\dfrac{x + y}{3} \cdot \dfrac{6}{x + y}$

2. $\dfrac{x - 1}{x + 1} \cdot \dfrac{5}{x - 1}$

3. $\dfrac{2x + 10}{x^2} \cdot \dfrac{x^3}{4x + 20}$

4. $\dfrac{3x^4}{3x - 6} \cdot \dfrac{x - 2}{x^2}$

5. $\dfrac{9}{2a - 8} \div \dfrac{3}{a - 4}$

6. $\dfrac{8}{a^2 - 25} \div \dfrac{16}{a + 5}$

7. $\dfrac{x + 1}{x^2 - 9} \div \dfrac{2x + 2}{x + 3}$

8. $\dfrac{11}{x - 2} \div \dfrac{22}{2x^2 - 8}$

9. $\dfrac{a^2 + 5a}{7a} \cdot \dfrac{4a^2}{a^2 + 4a}$

10. $\dfrac{4a^2 + 4a}{a^2 - 25} \cdot \dfrac{a^2 - 5a}{8a}$

11. $\dfrac{y^2 - 5y + 6}{2y + 4} \div \dfrac{2y - 6}{y + 2}$

12. $\dfrac{y^2 - 7y}{3y^2 - 48} \div \dfrac{y^2 - 9}{y^2 - 7y + 12}$

13. $\dfrac{2x - 8}{x^2 - 4} \cdot \dfrac{x^2 + 6x + 8}{x - 4}$

14. $\dfrac{x^2 + 5x + 1}{7x - 7} \cdot \dfrac{x - 1}{x^2 + 5x + 1}$

15. $\dfrac{x - 1}{x^2 - x - 6} \cdot \dfrac{x^2 + 5x + 6}{x^2 - 1}$

16. $\dfrac{x^2 - 3x - 10}{x^2 - 4x + 3} \cdot \dfrac{x^2 - 5x + 6}{x^2 - 3x - 10}$

17. $\dfrac{a^2 + 10a + 25}{a + 5} \div \dfrac{a^2 - 25}{a - 5}$

18. $\dfrac{a^2 + a - 2}{a^2 + 5a + 6} \div \dfrac{a - 1}{a}$

19. $\dfrac{y^3 - 5y^2}{y^4 + 3y^3 + 2y^2} \div \dfrac{y^2 - 5y + 6}{y^2 - 2y - 3}$

20. $\dfrac{y^2 - 5y}{y^2 + 7y + 12} \div \dfrac{y^3 - 7y^2 + 10y}{y^2 + 9y + 18}$

21. $\dfrac{2x^2 + 17x + 21}{x^2 + 2x - 35} \cdot \dfrac{x^2 - 25}{2x^2 - 7x - 15}$

22. $\dfrac{x^2 - 13x + 42}{4x^2 + 31x + 21} \cdot \dfrac{4x^2 - 5x - 6}{x^2 - 4}$

23. $\dfrac{2x^2 + 10x + 12}{4x^2 + 24x + 32} \cdot \dfrac{2x^2 + 18x + 40}{x^2 + 8x + 15}$

24. $\dfrac{3x^2 - 3}{6x^2 + 18x + 12} \cdot \dfrac{2x^2 - 8}{x^2 - 3x + 2}$

25. $\dfrac{2a^2 + 7a + 3}{a^2 - 16} \div \dfrac{4a^2 + 8a + 3}{2a^2 - 5a - 12}$

26. $\dfrac{3a^2 + 7a - 20}{a^2 + 3a - 4} \div \dfrac{3a^2 - 2a - 5}{a^2 - 2a + 1}$

27. $\dfrac{4y^2 - 12y + 9}{y^2 - 36} \div \dfrac{2y^2 - 5y + 3}{y^2 + 5y - 6}$

28. $\dfrac{5y^2 - 6y + 1}{y^2 - 1} \div \dfrac{16y^2 - 9}{4y^2 + 7y + 3}$

29. $\dfrac{x^2 - 1}{6x^2 + 42x + 60} \cdot \dfrac{7x^2 + 17x + 6}{x + 1} \cdot \dfrac{6x + 30}{7x^2 - 11x - 6}$

30. $\dfrac{4x^2 - 1}{3x - 15} \cdot \dfrac{4x^2 - 17x - 15}{4x^2 - 9x - 9} \cdot \dfrac{3x - 3}{x^2 - 9}$

31. $\dfrac{18x^3 + 21x^2 - 60x}{21x^2 - 25x - 4} \cdot \dfrac{28x^2 - 17x - 3}{16x^3 + 28x^2 - 30x}$

32. $\dfrac{56x^3 + 54x^2 - 20x}{8x^2 - 2x - 15} \cdot \dfrac{6x^2 + 5x - 21}{63x^3 + 129x^2 - 42x}$

Multiply the following expressions using the method shown in Examples 5 and 6 in this section.

33. $(x^2 - 9)\left(\dfrac{2}{x + 3}\right)$

34. $(x^2 - 9)\left(\dfrac{-3}{x - 3}\right)$

35. $a(a + 5)(a - 5)\left(\dfrac{2}{a^2 - 25}\right)$

36. $a(a^2 - 4)\left(\dfrac{a}{a + 2}\right)$

37. $(x^2 - x - 6)\left(\dfrac{x + 1}{x - 3}\right)$

38. $(x^2 - 2x - 8)\left(\dfrac{x + 3}{x - 4}\right)$

39. $(x^2 - 4x - 5)\left(\dfrac{-2x}{x + 1}\right)$

40. $(x^2 - 6x + 8)\left(\dfrac{4x}{x - 2}\right)$

Each one of the following problems involves some factoring by grouping. Remember, before you can divide out factors common to the numerators and denominators of a product, you must factor completely.

41. $\dfrac{x^2 - 9}{x^2 - 3x} \cdot \dfrac{2x + 10}{xy + 5x + 3y + 15}$

42. $\dfrac{x^2 - 16}{x^2 - 4x} \cdot \dfrac{3x + 18}{xy + 6x + 4y + 24}$

43. $\dfrac{2x^2 + 4x}{x^2 - y^2} \cdot \dfrac{x^2 + 3x + xy + 3y}{x^2 + 5x + 6}$

44. $\dfrac{x^2 - 25}{3x^2 + 3xy} \cdot \dfrac{x^2 + 4x + xy + 4y}{x^2 + 9x + 20}$

45. $\dfrac{x^3 - 3x^2 + 4x - 12}{x^4 - 16} \cdot \dfrac{3x^2 + 5x - 2}{3x^2 - 10x + 3}$

46. $\dfrac{x^3 - 5x^2 + 9x - 45}{x^4 - 81} \cdot \dfrac{5x^2 + 18x + 9}{5x^2 - 22x - 15}$

Simplify each expression. Work inside parentheses first, and then divide out common factors.

47. $\left(1 - \dfrac{1}{2}\right)\left(1 - \dfrac{1}{3}\right)\left(1 - \dfrac{1}{4}\right)\left(1 - \dfrac{1}{5}\right)$

48. $\left(1 + \dfrac{1}{2}\right)\left(1 + \dfrac{1}{3}\right)\left(1 + \dfrac{1}{4}\right)\left(1 + \dfrac{1}{5}\right)$

The dots in the following problems represent factors not written that are in the same pattern as the surrounding factors. Simplify.

49. $(1 - \tfrac{1}{2})(1 - \tfrac{1}{3})(1 - \tfrac{1}{4}) \cdots (1 - \tfrac{1}{99})(1 - \tfrac{1}{100})$

50. $(1 - \tfrac{1}{3})(1 - \tfrac{1}{4})(1 - \tfrac{1}{5}) \cdots (1 - \tfrac{1}{98})(1 - \tfrac{1}{99})$

Applying the Concepts

51. The top of Mount Whitney, the highest point in California, is 14,494 feet above sea level. Give this height in miles to the nearest tenth of a mile.

52. The relationship between liters and cubic inches, both of which are measures of volume, is 0.0164 liters = 1 cubic inch. If a Ford Mustang has a motor with a displacement of 4.9 liters, what is the displacement in cubic inches? Round your answer to the nearest cubic inch.

53. The speed of sound is 1,088 feet per second. Convert the speed of sound to miles per hour. Round your answer to the nearest whole number.

54. A car travels 122 miles in 3 hours. Find the average speed of the car in feet per second. Round to the nearest whole number.

55. As we mentioned in Problem Set 6.1, the first Ferris wheel was built in 1893. It was a large wheel with a circumference of 785 feet. If one trip around the circumference of the wheel took 20 minutes, find the average speed of a rider in miles per hour. Round to the nearest hundredth.

56. In 1897, a large Ferris wheel was built in Vienna. Known as *The Great Wheel,* it has a circumference of 618 feet. If one trip around the wheel takes 15 minutes, find the average speed of a rider on this wheel in miles per hour. Round to the nearest hundredth.

57. A Ferris wheel called *Colossus* has a circumference of 518 feet. If a trip around the circumference of *Colossus* takes 40 seconds, find the average speed of a rider in miles per hour. Round to the nearest tenth.

58. A person riding a Ferris wheel travels once around the wheel, a distance of 188 feet, in 30 seconds. What is the average speed of the rider in miles per hour? Round to the nearest tenth.

59. Tina is training for a biathlon. As part of her training she runs an 8-mile course, 2 miles of which is on level ground and 6 miles of which is downhill. It takes her 20 minutes to run the level part of the course and 40 minutes to run the downhill part of the course. Find her average speed in miles per hour for each part of the course.

60. A jogger covers a distance of 3 miles in 24 minutes. Find the average speed of the jogger in miles per hour. Round to the nearest tenth.

Review Problems

The following problems review material we covered in Sections 1.9 and 4.3. Reviewing these problems will help you in the next section.

Add the following fractions. [1.9]

61. $\dfrac{1}{2} + \dfrac{5}{2}$

62. $\dfrac{2}{3} + \dfrac{8}{3}$

63. $2 + \dfrac{3}{4}$

64. $1 + \dfrac{4}{7}$

65. $\dfrac{1}{10} + \dfrac{3}{14}$

66. $\dfrac{1}{12} + \dfrac{11}{30}$

Simplify each term, then add. [4.3]

67. $\dfrac{10x^4}{2x^2} + \dfrac{12x^6}{3x^4}$

68. $\dfrac{32x^8}{8x^3} + \dfrac{27x^7}{3x^2}$

69. $\dfrac{12a^2b^5}{3ab^3} + \dfrac{14a^4b^7}{7a^3b^5}$

70. $\dfrac{16a^3b^2}{4ab} + \dfrac{25a^6b^5}{5a^4b^4}$

SECTION

6.3   **Addition and Subtraction of Rational Expressions**

In Chapter 1 we combined fractions having the same denominator by combining their numerators and putting the result over the common denominator. We use the same process to add two rational expressions with the same denominator.

▶ **EXAMPLES**

1. Add $\dfrac{5}{x} + \dfrac{3}{x}$.

Solution   Adding numerators, we have

$$\frac{5}{x} + \frac{3}{x} = \frac{8}{x}$$

2. Add $\dfrac{x}{x^2 - 9} + \dfrac{3}{x^2 - 9}$.

Solution   Since both expressions have the same denominator, we add numerators and reduce to lowest terms:

$$\frac{x}{x^2 - 9} + \frac{3}{x^2 - 9} = \frac{x + 3}{x^2 - 9}$$

$$\left. \begin{aligned} &= \frac{x + 3}{(x + 3)(x - 3)} \\ &= \frac{1}{x - 3} \end{aligned} \right\} \quad \begin{aligned} &\text{Reduce to lowest terms by} \\ &\text{factoring the denominator} \\ &\text{and then dividing out} \\ &\text{common factor } x + 3 \end{aligned} \quad ◀$$

Remember, it is the distributive property that allows us to add rational expressions by simply adding numerators. Because of this, we must begin all addition problems involving rational expressions by first making sure all the expressions have the same denominator.

DEFINITION   The **least common denominator** (LCD) for a set of denominators is the simplest quantity that is exactly divisible by all the denominators.

▶ **EXAMPLE 3**   Add $\dfrac{1}{10} + \dfrac{3}{14}$.

Solution

Step 1: Find the LCD for 10 and 14. To do so, we factor each denominator and build the LCD from the factors:

$$\left. \begin{aligned} 10 &= 2 \cdot 5 \\ 14 &= 2 \cdot 7 \end{aligned} \right\} \quad \text{LCD} = 2 \cdot 5 \cdot 7 = 70$$

We know the LCD is divisible by 10 because it contains the factors 2 and 5. It is also divisible by 14 because it contains the factors 2 and 7.

Step 2: Change to equivalent fractions that each have denominator 70. To accomplish this task, we multiply the numerator and denominator of each fraction by the factor of the LCD that is not also a factor of its denominator:

Original Fractions	Denominators in Factored Form	Multiply by Factor Needed to Obtain LCD	These Have the Same Value as the Original Fractions
$\dfrac{1}{10} =$	$\dfrac{1}{2 \cdot 5} =$	$\dfrac{1}{2 \cdot 5} \cdot \dfrac{7}{7} =$	$\dfrac{7}{70}$
$\dfrac{3}{14} =$	$\dfrac{3}{2 \cdot 7} =$	$\dfrac{3}{2 \cdot 7} \cdot \dfrac{5}{5} =$	$\dfrac{15}{70}$

The fraction $\frac{7}{70}$ has the same value as the fraction $\frac{1}{10}$. Likewise, the fractions $\frac{15}{70}$ and $\frac{3}{14}$ are equivalent; they have the same value.

Step 3: Add numerators and put the result over the LCD:

$$\frac{7}{70} + \frac{15}{70} = \frac{7 + 15}{70} = \frac{22}{70}$$

Step 4: Reduce to lowest terms:

$$\frac{22}{70} = \frac{11}{35} \qquad \text{Divide numerator and denominator by 2} \quad \blacktriangleleft$$

The main idea in adding fractions is to write each fraction again with the LCD for a denominator. Once we have done that, we simply add numerators. The same process can be used to add (or subtract) rational expressions, as the next example illustrates.

▶ **EXAMPLE 4** Subtract $\dfrac{3}{x} - \dfrac{1}{2}$.

Solution

Step 1: The LCD for x and 2 is $2x$. It is the smallest expression divisible by x and by 2.

Step 2: To change to equivalent expressions with the denominator $2x$, we multiply the first fraction by 2/2 and the second by x/x.

$$\frac{3}{x} \cdot \frac{2}{2} = \frac{6}{2x}$$

$$\frac{1}{2} \cdot \frac{x}{x} = \frac{x}{2x}$$

Step 3: Subtracting numerators of the rational expressions in step 2, we have

$$\frac{6}{2x} - \frac{x}{2x} = \frac{6-x}{2x}$$

Step 4: Since $6 - x$ and $2x$ do not have any factors in common, we cannot reduce any further. Here is the complete problem:

$$\frac{3}{x} - \frac{1}{2} = \frac{3}{x} \cdot \frac{2}{2} - \frac{1}{2} \cdot \frac{x}{x}$$

$$= \frac{6}{2x} - \frac{x}{2x}$$

$$= \frac{6-x}{2x} \qquad \blacktriangleleft$$

▶ **EXAMPLE 5** Add $\dfrac{5}{2x-6} + \dfrac{x}{x-3}$.

Solution If we factor $2x - 6$, we have $2x - 6 = 2(x - 3)$. We need only multiply the second rational expression in our problem by $\dfrac{2}{2}$ to have two expressions with the same denominator:

$$\frac{5}{2x-6} + \frac{x}{x-3} = \frac{5}{2(x-3)} + \frac{x}{x-3}$$

$$= \frac{5}{2(x-3)} + \frac{2}{2}\left(\frac{x}{x-3}\right)$$

$$= \frac{5}{2(x-3)} + \frac{2x}{2(x-3)}$$

$$= \frac{2x+5}{2(x-3)} \qquad \blacktriangleleft$$

▶ **EXAMPLE 6** Add $\dfrac{1}{x+4} + \dfrac{8}{x^2-16}$.

Solution After writing each denominator in factored form, we find that the least common denominator is $(x + 4)(x - 4)$. To change the first rational expression to an equivalent rational expression with the common denominator, we multiply its numerator and denominator by $x - 4$.

$$\frac{1}{x+4} + \frac{8}{x^2-16}$$

$$= \frac{1}{x+4} + \frac{8}{(x+4)(x-4)} \qquad \text{Factor each denominator}$$

$$= \frac{1}{x+4} \cdot \frac{x-4}{x-4} + \frac{8}{(x+4)(x-4)} \qquad \begin{array}{l}\text{Change to equivalent}\\\text{rational expressions}\end{array}$$

$$= \frac{x - 4}{(x + 4)(x - 4)} + \frac{8}{(x + 4)(x - 4)} \qquad \text{Simplify}$$

$$= \frac{x + 4}{(x + 4)(x - 4)} \qquad \text{Add numerators}$$

$$= \frac{1}{x - 4} \qquad \begin{array}{l} \text{Divide out common} \\ \text{factor } x + 4 \end{array}$$

Note that in the last step we reduced the rational expression to lowest terms by dividing out the common factor of $x + 4$. The last thing you should do in any addition or subtraction problem involving rational expressions is see if your answer can be reduced further. ◀

▶ **EXAMPLE 7**   Add $\dfrac{2}{x^2 + 5x + 6} + \dfrac{x}{x^2 - 9}$.

Solution

Step 1: We factor each denominator and build the LCD from the factors:

$$\left. \begin{array}{l} x^2 + 5x + 6 = (x + 2)(x + 3) \\ x^2 - 9 = (x + 3)(x - 3) \end{array} \right\} \quad \text{LCD} = (x + 2)(x + 3)(x - 3)$$

Step 2: Change to equivalent rational expressions:

$$\frac{2}{x^2 + 5x + 6} = \frac{2}{(x + 2)(x + 3)} \cdot \frac{(x - 3)}{(x - 3)} = \frac{2x - 6}{(x + 2)(x + 3)(x - 3)}$$

$$\frac{x}{x^2 - 9} = \frac{x}{(x + 3)(x - 3)} \cdot \frac{(x + 2)}{(x + 2)} = \frac{x^2 + 2x}{(x + 2)(x + 3)(x - 3)}$$

Step 3: Add numerators of the rational expressions produced in step 2:

$$\frac{2x - 6}{(x + 2)(x + 3)(x - 3)} + \frac{x^2 + 2x}{(x + 2)(x + 3)(x - 3)}$$

$$= \frac{x^2 + 4x - 6}{(x + 2)(x + 3)(x - 3)}$$

The numerator and denominator do not have any factors in common. ◀

▶ **EXAMPLE 8**   Subtract $\dfrac{x + 4}{2x + 10} - \dfrac{5}{x^2 - 25}$.

Solution   We begin by factoring each denominator:

$$\frac{x + 4}{2x + 10} - \frac{5}{x^2 - 25} = \frac{x + 4}{2(x + 5)} - \frac{5}{(x + 5)(x - 5)}$$

The LCD is $2(x + 5)(x - 5)$. Completing the problem, we have

$$= \frac{x + 4}{2(x + 5)} \cdot \frac{(x - 5)}{(x - 5)} + \frac{-5}{(x + 5)(x - 5)} \cdot \frac{2}{2}$$

$$= \frac{x^2 - x - 20}{2(x + 5)(x - 5)} + \frac{-10}{2(x + 5)(x - 5)}$$

$$= \frac{x^2 - x - 30}{2(x + 5)(x - 5)}$$

To see if this expression will reduce, we factor the numerator into $(x - 6)(x + 5)$.

$$= \frac{(x - 6)(x + 5)}{2(x + 5)(x - 5)}$$

$$= \frac{x - 6}{2(x - 5)}$$

◄

Notice in Example 8 that we replaced subtraction by addition of the opposite. There seems to be less chance for error when this is done on longer problems.

▶ **EXAMPLE 9**   Write an expression for the sum of a number and its reciprocal and then simplify that expression.

Solution   If we let $x =$ the number, then its reciprocal is $\frac{1}{x}$. To find the sum of the number and its reciprocal we add them:

$$x + \frac{1}{x}$$

The first term, x, can be thought of as having a denominator of 1. Since the denominators are 1 and x, the least common denominator is x. Here is the complete problem:

$$x + \frac{1}{x} = \frac{x}{1} + \frac{1}{x} \qquad \text{Write } x \text{ as } \frac{x}{1}$$

$$= \frac{x}{1} \cdot \frac{x}{x} + \frac{1}{x} \qquad \text{The LCD is } x$$

$$= \frac{x^2}{x} + \frac{1}{x}$$

$$= \frac{x^2 + 1}{x} \qquad \text{Add numerators}$$

◄

P R O B L E M   S E T   6 . 3

Find the following sums and differences.

1. $\dfrac{3}{x} + \dfrac{4}{x}$

2. $\dfrac{5}{x} + \dfrac{3}{x}$

3. $\dfrac{9}{a} - \dfrac{5}{a}$

4. $\dfrac{8}{a} - \dfrac{7}{a}$

5. $\dfrac{1}{x+1} + \dfrac{x}{x+1}$

6. $\dfrac{x}{x-3} - \dfrac{3}{x-3}$

7. $\dfrac{y^2}{y-1} - \dfrac{1}{y-1}$

8. $\dfrac{y^2}{y+3} - \dfrac{9}{y+3}$

9. $\dfrac{x^2}{x+2} + \dfrac{4x+4}{x+2}$

10. $\dfrac{x^2-6x}{x-3} + \dfrac{9}{x-3}$

11. $\dfrac{x^2}{x-2} - \dfrac{4x-4}{x-2}$

12. $\dfrac{x^2}{x-5} - \dfrac{10x-25}{x-5}$

13. $\dfrac{x+2}{x+6} - \dfrac{x-4}{x+6}$

14. $\dfrac{x+5}{x+2} - \dfrac{x+3}{x+2}$

15. $\dfrac{y}{2} - \dfrac{2}{y}$

16. $\dfrac{3}{y} + \dfrac{y}{3}$

17. $\dfrac{1}{2} + \dfrac{a}{3}$

18. $\dfrac{2}{3} + \dfrac{2a}{5}$

19. $\dfrac{x}{x+1} + \dfrac{3}{4}$

20. $\dfrac{x}{x-3} + \dfrac{1}{3}$

21. $\dfrac{x+1}{x-2} - \dfrac{4x+7}{5x-10}$

22. $\dfrac{3x+1}{2x-6} - \dfrac{x+2}{x-3}$

23. $\dfrac{4x-2}{3x+12} - \dfrac{x-2}{x+4}$

24. $\dfrac{6x+5}{5x-25} - \dfrac{x+2}{x-5}$

25. $\dfrac{6}{x(x-2)} + \dfrac{3}{x}$

26. $\dfrac{10}{x(x+5)} - \dfrac{2}{x}$

27. $\dfrac{4}{a} - \dfrac{12}{a^2+3a}$

28. $\dfrac{5}{a} + \dfrac{20}{a^2-4a}$

29. $\dfrac{2}{x+5} - \dfrac{10}{x^2-25}$

30. $\dfrac{6}{x^2-1} + \dfrac{3}{x+1}$

31. $\dfrac{x-4}{x-3} + \dfrac{6}{x^2-9}$

32. $\dfrac{x+1}{x-1} - \dfrac{4}{x^2-1}$

33. $\dfrac{a-4}{a-3} + \dfrac{5}{a^2-a-6}$

34. $\dfrac{a+2}{a+1} + \dfrac{7}{a^2-5a-6}$

35. $\dfrac{8}{x^2-16} - \dfrac{7}{x^2-x-12}$

36. $\dfrac{6}{x^2-9} - \dfrac{5}{x^2-x-6}$

37. $\dfrac{4y}{y^2+6y+5} - \dfrac{3y}{y^2+5y+4}$

38. $\dfrac{3y}{y^2+7y+10} - \dfrac{2y}{y^2+6y+8}$

39. $\dfrac{4x+1}{x^2+5x+4} - \dfrac{x+3}{x^2+4x+3}$

40. $\dfrac{2x-1}{x^2+x-6} - \dfrac{x+2}{x^2+5x+6}$

41. $\dfrac{1}{x} + \dfrac{x}{3x+9} - \dfrac{3}{x^2+3x}$

42. $\dfrac{1}{x} + \dfrac{x}{2x+4} - \dfrac{2}{x^2+2x}$

Add and subtract as indicated.

43. $1 + \dfrac{1}{x}$

44. $1 - \dfrac{1}{x}$

45. $1 - \dfrac{1}{x+1}$

46. $1 + \dfrac{1}{x+1}$

47. $1 + \dfrac{1}{x+2}$

48. $1 - \dfrac{1}{x+2}$

49. $1 - \dfrac{1}{x+3}$

50. $1 + \dfrac{1}{x+3}$

51. Write an expression for the sum of a number and twice its reciprocal. Then, simplify that expression. (*Hint:* If the reciprocal of a number is $\dfrac{1}{x}$, then twice that is $\dfrac{2}{x}$, not $\dfrac{1}{2x}$.)

52. Write an expression for the sum of a number and three times its reciprocal. Then, simplify that expression.

53. One number is twice another. Write an expression for the sum of their reciprocals. Then, simplify that expression. (*Hint:* The numbers are x and $2x$. Their reciprocals are $\dfrac{1}{x}$ and $\dfrac{1}{2x}$, respectively.)

54. One number is three times another. Write an expression for the sum of their reciprocals. Then, simplify that expression.

Review Problems

The following problems review material we covered in Sections 2.4 and 5.6. Reviewing these problems will help you understand the next section.

Solve each equation. [2.4]

55. $2x + 3(x - 3) = 6$
56. $4x - 2(x - 5) = 6$
57. $x - 3(x + 3) = x - 3$
58. $x - 4(x + 4) = x - 4$
59. $7 - 2(3x + 1) = 4x + 3$
60. $8 - 5(2x - 1) = 2x + 4$

Solve each quadratic equation. [5.6]

61. $x^2 + 5x + 6 = 0$ **62.** $x^2 - 5x + 6 = 0$
63. $x^2 - x = 6$ **64.** $x^2 + x = 6$
65. $x^2 - 5x = 0$ **66.** $x^2 - 6x = 0$

SECTION

6.4   Equations Involving Rational Expressions

The first step in solving an equation that contains one or more rational expressions is to find the LCD for all denominators in the equation. Once the LCD has been found, we multiply both sides of the equation by it. The resulting equation should be equivalent to the original one (unless we inadvertently multiplied by zero) and free from any denominators except the number 1.

▶ **EXAMPLE 1**   Solve $\dfrac{x}{3} + \dfrac{5}{2} = \dfrac{1}{2}$ for x.

Solution   The LCD for 3 and 2 is 6. If we multiply both sides by 6, we have

$$6\left(\frac{x}{3} + \frac{5}{2}\right) = 6\left(\frac{1}{2}\right) \qquad \text{Multiply both sides by 6}$$

$$6\left(\frac{x}{3}\right) + 6\left(\frac{5}{2}\right) = 6\left(\frac{1}{2}\right) \qquad \text{Distributive property}$$

$$2x + 15 = 3$$

$$2x = -12$$

$$x = -6$$

We can check our solution by replacing x with -6 in the original equation:

$$-\frac{6}{3} + \frac{5}{2} \overset{?}{=} \frac{1}{2}$$

$$\frac{1}{2} = \frac{1}{2}$$ ◀

Multiplying both sides of an equation containing fractions by the LCD clears the equation of all denominators, because the LCD has the property that all denominators will divide it evenly.

▶ **EXAMPLE 2**   Solve for x: $\dfrac{3}{x-1} = \dfrac{3}{5}$.

Solution   The LCD for $(x-1)$ and 5 is $5(x-1)$. Multiplying both sides by $5(x-1)$ we have

$$5(x-1) \cdot \frac{3}{x-1} = 5(x-1) \cdot \frac{3}{5}$$

$$5 \cdot 3 = (x-1) \cdot 3$$

$$15 = 3x - 3$$

$$18 = 3x$$

$$6 = x$$

If we substitute $x = 6$ into the original equation, we have

$$\frac{3}{6-1} \overset{?}{=} \frac{3}{5}$$

$$\frac{3}{5} = \frac{3}{5}$$

The solution set is $\{6\}$.   ◀

▶ **EXAMPLE 3**   Solve $1 - \dfrac{5}{x} = \dfrac{-6}{x^2}$.

Solution   The LCD is x^2. Multiplying both sides by x^2, we have

$$x^2\left(1 - \frac{5}{x}\right) = x^2\left(\frac{-6}{x^2}\right) \qquad \text{Multiply both sides by } x^2$$

$$x^2(1) - x^2\left(\frac{5}{x}\right) = x^2\left(\frac{-6}{x^2}\right) \qquad \begin{array}{l}\text{Apply distributive property} \\ \text{to the left side}\end{array}$$

$$x^2 - 5x = -6 \qquad \text{Simplify each side}$$

We have a quadratic equation, which we write in standard form, factor, and solve as we did in Section 5.6.

$$x^2 - 5x + 6 = 0 \qquad \text{Standard form}$$

$$(x - 2)(x - 3) = 0 \qquad \text{Factor}$$

$$x - 2 = 0 \quad \text{or} \quad x - 3 = 0 \qquad \text{Set factors equal to 0}$$

$$x = 2 \quad \text{or} \qquad x = 3$$

The two possible solutions are 2 and 3. Checking each in the original equation, we find they both give true statements. They are both solutions to the original equation.

Check $x = 2$	Check $x = 3$
$1 - \dfrac{5}{2} \stackrel{?}{=} \dfrac{-6}{4}$	$1 - \dfrac{5}{3} \stackrel{?}{=} \dfrac{-6}{9}$
$\dfrac{2}{2} - \dfrac{5}{2} \stackrel{?}{=} -\dfrac{3}{2}$	$\dfrac{3}{3} - \dfrac{5}{3} \stackrel{?}{=} -\dfrac{2}{3}$
$-\dfrac{3}{2} = -\dfrac{3}{2}$	$-\dfrac{2}{3} = -\dfrac{2}{3}$

◀

▶ **EXAMPLE 4**   Solve $\dfrac{x}{x^2 - 9} - \dfrac{3}{x - 3} = \dfrac{1}{x + 3}$.

Solution   The factors of $x^2 - 9$ are $(x + 3)(x - 3)$. The LCD, then, is $(x + 3)(x - 3)$:

$$\cancel{(x+3)(x-3)} \cdot \frac{x}{\cancel{(x+3)(x-3)}} + (x + 3)\cancel{(x-3)} \cdot \frac{-3}{\cancel{x-3}}$$

$$= \cancel{(x+3)}(x - 3) \cdot \frac{1}{\cancel{x+3}}$$

$$x + (x + 3)(-3) = (x - 3)1$$

$$x + (-3x) + (-9) = x - 3$$

$$-2x - 9 = x - 3$$

$$-3x = 6$$

$$x = -2$$

The solution is $x = -2$. It checks when replaced for x in the original equation. ◀

▶ **EXAMPLE 5**   Solve $\dfrac{x}{x - 3} + \dfrac{3}{2} = \dfrac{3}{x - 3}$.

Solution   We begin by multiplying each term on both sides of the equation by $2(x - 3)$.

$$2(x - 3) \cdot \frac{x}{x - 3} + 2(x - 3) \cdot \frac{3}{2} = 2(x - 3) \cdot \frac{3}{x - 3}$$

$$2x + (x - 3) \cdot 3 = 2 \cdot 3$$

$$2x + 3x - 9 = 6$$

$$5x - 9 = 6$$

$$5x = 15$$

$$x = 3$$

Our only possible solution is $x = 3$. If we substitute $x = 3$ into our original equation, we get

$$\frac{3}{3 - 3} + \frac{3}{2} \overset{?}{=} \frac{3}{3 - 3}$$

$$\frac{3}{0} + \frac{3}{2} = \frac{3}{0}$$

Two of the terms are undefined, so the equation is meaningless. What has happened is that we have multiplied both sides of the original equation by zero. The equation produced by doing this is not equivalent to our original equation. We must always check our solution when we multiply both sides of an equation by an expression containing the variable in order to make sure we have not multiplied both sides by zero.

Our original equation has no solutions. That is, there is no real number x such that

$$\frac{x}{x - 3} + \frac{3}{2} = \frac{3}{x - 3}$$

The solution set is Ø. ◀

▶ **EXAMPLE 6** Solve $\dfrac{a + 4}{a^2 + 5a} = \dfrac{-2}{a^2 - 25}$ for a.

Solution Factoring each denominator, we have

$$a^2 + 5a = a(a + 5)$$

$$a^2 - 25 = (a + 5)(a - 5)$$

The LCD is $a(a + 5)(a - 5)$. Multiplying both sides of the equation by the LCD gives us

$$a(a + 5)(a - 5) \cdot \frac{a + 4}{a(a + 5)} = \frac{-2}{(a + 5)(a - 5)} \cdot a(a + 5)(a - 5)$$

$$(a - 5)(a + 4) = -2a$$

$$a^2 - a - 20 = -2a$$

The result is a quadratic equation, which we write in standard form, factor, and solve:

$$a^2 + a - 20 = 0 \qquad \text{Add } 2a \text{ to both sides}$$
$$(a + 5)(a - 4) = 0 \qquad \text{Factor}$$
$$a + 5 = 0 \quad \text{or} \quad a - 4 = 0 \qquad \text{Set each factor to 0}$$
$$a = -5 \quad \text{or} \qquad a = 4$$

The two possible solutions are -5 and 4. There is no problem with the 4. It checks when substituted for a in the original equation. However, -5 is not a solution. Substituting -5 into the original equation gives

$$\frac{-5 + 4}{(-5)^2 + 5(-5)} \stackrel{?}{=} \frac{-2}{(-5)^2 - 25}$$

$$\frac{-1}{0} = \frac{-2}{0}$$

This indicates -5 is not a solution. The solution is 4. ◀

PROBLEM SET 6.4

Solve the following equations. Be sure to check each answer in the original equation if you multiply both sides by an expression that contains the variable.

1. $\dfrac{x}{3} + \dfrac{1}{2} = -\dfrac{1}{2}$

2. $\dfrac{x}{2} + \dfrac{4}{3} = -\dfrac{2}{3}$

3. $\dfrac{4}{a} = \dfrac{1}{5}$

4. $\dfrac{2}{3} = \dfrac{6}{a}$

5. $\dfrac{3}{x} + 1 = \dfrac{2}{x}$

6. $\dfrac{4}{x} + 3 = \dfrac{1}{x}$

7. $\dfrac{3}{a} - \dfrac{2}{a} = \dfrac{1}{5}$

8. $\dfrac{7}{a} + \dfrac{1}{a} = 2$

9. $\dfrac{3}{x} + 2 = \dfrac{1}{2}$

10. $\dfrac{5}{x} + 3 = \dfrac{4}{3}$

11. $\dfrac{1}{y} - \dfrac{1}{2} = -\dfrac{1}{4}$

12. $\dfrac{3}{y} - \dfrac{4}{5} = -\dfrac{1}{5}$

13. $1 - \dfrac{8}{x} = \dfrac{-15}{x^2}$

14. $1 - \dfrac{3}{x} = \dfrac{-2}{x^2}$

15. $\dfrac{x}{2} - \dfrac{4}{x} = -\dfrac{7}{2}$

16. $\dfrac{x}{2} - \dfrac{5}{x} = -\dfrac{3}{2}$

17. $\dfrac{x - 3}{2} + \dfrac{2x}{3} = \dfrac{5}{6}$

18. $\dfrac{x - 2}{3} + \dfrac{5x}{2} = 5$

19. $\dfrac{x + 1}{3} + \dfrac{x - 3}{4} = \dfrac{1}{6}$

20. $\dfrac{x + 2}{3} + \dfrac{x - 1}{5} = -\dfrac{3}{5}$

21. $\dfrac{6}{x + 2} = \dfrac{3}{5}$

22. $\dfrac{4}{x + 3} = \dfrac{1}{2}$

23. $\dfrac{3}{y - 2} = \dfrac{2}{y - 3}$

24. $\dfrac{5}{y + 1} = \dfrac{4}{y + 2}$

25. $\dfrac{x}{x - 2} + \dfrac{2}{3} = \dfrac{2}{x - 2}$

26. $\dfrac{x}{x-5} + \dfrac{1}{5} = \dfrac{5}{x-5}$

27. $\dfrac{x}{x-2} + \dfrac{3}{2} = \dfrac{9}{2(x-2)}$

28. $\dfrac{x}{x+1} + \dfrac{4}{5} = \dfrac{-14}{5(x+1)}$

29. $\dfrac{5}{x+2} + \dfrac{1}{x+3} = \dfrac{-1}{x^2+5x+6}$

30. $\dfrac{3}{x-1} + \dfrac{2}{x+3} = \dfrac{-3}{x^2+2x-3}$

31. $\dfrac{8}{x^2-4} + \dfrac{3}{x+2} = \dfrac{1}{x-2}$

32. $\dfrac{10}{x^2-25} - \dfrac{1}{x-5} = \dfrac{3}{x+5}$

33. $\dfrac{a}{2} + \dfrac{3}{a-3} = \dfrac{a}{a-3}$

34. $\dfrac{a}{2} + \dfrac{4}{a-4} = \dfrac{a}{a-4}$

35. $\dfrac{6}{y^2-4} = \dfrac{4}{y^2+2y}$

36. $\dfrac{2}{y^2-9} = \dfrac{5}{y^2-3y}$

37. $\dfrac{2}{a^2-9} = \dfrac{3}{a^2+a-12}$

38. $\dfrac{2}{a^2-1} = \dfrac{6}{a^2-2a-3}$

39. $\dfrac{3x}{x-5} - \dfrac{2x}{x+1} = \dfrac{-42}{x^2-4x-5}$

40. $\dfrac{4x}{x-4} - \dfrac{3x}{x-2} = \dfrac{-3}{x^2-6x+8}$

41. $\dfrac{2x}{x+2} = \dfrac{x}{x+3} - \dfrac{3}{x^2+5x+6}$

42. $\dfrac{3x}{x-4} = \dfrac{2x}{x-3} + \dfrac{6}{x^2-7x+12}$

Review Problems

The problems that follow review material we covered in Sections 2.6 and 5.7. Reviewing these problems will help you with the next section.

Solve each application. [2.6]

43. If twice the difference of a number and three were decreased by five, the result would be three. Find the number.

44. If three times the sum of a number and two were increased by six, the result would be 27. Find the number.

45. The length of a rectangle is five more than twice the width. The perimeter is 34 inches. Find the length and width.

46. The length of a rectangle is two more than three times the width. The perimeter is 44 feet. Find the length and width.

Solve each problem. Be sure to show the equation that describes the situation. [5.7]

47. The product of two consecutive even integers is 48. Find the two integers.

48. The product of two consecutive odd integers is 35. Find the two integers.

49. The hypotenuse (the longest side) of a right triangle is 10 inches, while the lengths of the two legs (the other two sides) are given by two consecutive even integers. Find the lengths of the two legs.

50. One leg of a right triangle is two more than twice the other. If the hypotenuse is 13 feet, find the lengths of the two legs.

SECTION

6.5 **Applications**

In this section we will solve some word problems whose equations involve rational expressions. Like the other word problems we have encountered, the more you work with them the easier they become.

▶ **EXAMPLE 1**   One number is twice another. The sum of their reciprocals is $\frac{9}{2}$. Find the two numbers.

Solution   Let x = the smaller number. The larger then must be $2x$. Their reciprocals are $1/x$ and $1/2x$, respectively. An equation that describes the situation is

$$\frac{1}{x} + \frac{1}{2x} = \frac{9}{2}$$

We can multiply both sides by the LCD $2x$ and then solve the resulting equation:

$$2x\left(\frac{1}{x}\right) + 2x\left(\frac{1}{2x}\right) = 2x\left(\frac{9}{2}\right)$$

$$2 + 1 = 9x$$

$$3 = 9x$$

$$x = \frac{3}{9} = \frac{1}{3}$$

The smaller number is $\frac{1}{3}$. The other number is twice as large, or $\frac{2}{3}$. If we add their reciprocals, we have

$$\frac{3}{1} + \frac{3}{2} = \frac{6}{2} + \frac{3}{2} = \frac{9}{2}$$

The solutions check with the original problem.   ◀

▶ **EXAMPLE 2**   A boat travels 30 miles up a river in the same amount of time it takes to travel 50 miles down the same river. If the current is 5 mph, what is the speed of the boat in still water?

Solution   The easiest way to work a problem like this is with a table. The top row of the table is labeled with d for distance, r for rate, and t for time. The left column of the table is labeled with the two trips: upstream and downstream. Here is what the table looks like:

	d	r	t
Upstream			
Downstream			

The next step is to read the problem over again and fill in as much of the table as we can with the information in the problem. The distance the boat travels upstream is 30 and the distance downstream is 50. Since we are asked for the speed of the boat in still water, we will let that be x. If the speed of the boat in still water is x, then its speed upstream (against the current) must be $x - 5$, and its speed downstream (with the current) must be $x + 5$. Putting these four quantities into the appropriate positions in the table we have

	d	r	t
Upstream	30	$x - 5$	
Downstream	50	$x + 5$	

The last positions in the table are filled in by using the equation $t = d/r$.

	d	r	t
Upstream	30	$x - 5$	$\dfrac{30}{x - 5}$
Downstream	50	$x + 5$	$\dfrac{50}{x + 5}$

Reading the problem again, we find that the time for the trip upstream is equal to the time for the trip downstream. Setting these two quantities equal to each other, we have our equation.

$$\text{Time (downstream)} = \text{time (upstream)}$$

$$\frac{50}{x + 5} = \frac{30}{x - 5}$$

The LCD is $(x + 5)(x - 5)$. We multiply both sides of the equation by the LCD to clear it of all denominators. Here is the solution:

$$(x + 5)(x - 5) \cdot \frac{50}{x + 5} = (x + 5)(x - 5) \cdot \frac{30}{x - 5}$$

$$50x - 250 = 30x + 150$$

$$20x = 400$$

$$x = 20$$

The speed of the boat in still water is 20 mph.   ◀

▶ **EXAMPLE 3**   Tina is training for a biathlon. To train for the bicycle portion, she rides her bike 15 miles up a hill and then 15 miles back down the same hill. The complete trip takes her 2 hours. If her downhill speed is 20 miles per hour faster than her uphill speed, how fast does she ride uphill?

Solution Again, we make a table. As in the previous example, we label the top row with distance, rate, and time. We label the left column with the two trips, uphill and downhill.

	d	r	t
Uphill			
Downhill			

Next, we fill in the table with as much information as we can from the problem. We know the distance traveled is 15 miles uphill and 15 miles downhill, which allows us to fill in the distance column. To fill in the rate column, we first note that she rides 20 miles per hour faster downhill than uphill. Therefore, if we let x equal her rate uphill, then her rate downhill is $x + 20$. Filling in the table with this information gives us

	d	r	t
Uphill	15	x	
Downhill	15	$x + 20$	

Since time is distance divided by rate, $t = d/r$, we can fill in the last column in the table.

	d	r	t
Uphill	15	x	$\dfrac{15}{x}$
Downhill	15	$x + 20$	$\dfrac{15}{x + 20}$

Rereading the problem, we find that the total time (the time riding uphill plus the time riding downhill) is two hours. We write our equation as follows:

$$\text{Time (uphill)} + \text{time (downhill)} = 2$$

$$\frac{15}{x} + \frac{15}{x + 20} = 2$$

We solve this equation for x by first finding the LCD and then multiplying each term in the equation by it in order to clear the equation of all denominators. Our LCD is $x(x + 20)$. Here is our solution:

$$x(x + 20)\,\frac{15}{x} + x(x + 20)\,\frac{15}{x + 20} = 2 \cdot [x(x + 20)]$$

$$15(x + 20) + 15x = 2x(x + 20)$$

$$15x + 300 + 15x = 2x^2 + 40x$$

$$0 = 2x^2 + 10x - 300$$

$$0 = x^2 + 5x - 150 \qquad \text{Divide both}$$

$$0 = (x + 15)(x - 10) \qquad \text{sides by 2}$$

$$x = -15 \quad \text{or} \quad x = 10$$

Since we cannot have a negative speed, our only solution is $x = 10$. Tina rides her bike at a rate of 10 miles per hour when going uphill. (Her downhill speed is $x + 20 = 30$ miles per hour.) ◀

▶ **EXAMPLE 4** An inlet pipe can fill a water tank in 10 hours, while an outlet pipe can empty the same tank in 15 hours. By mistake, both pipes are left open. How long will it take to fill the water tank with both pipes open?

Solution Let $x =$ amount of time to fill the tank with both pipes open.
One method of solving this type of problem is to think in terms of how much of the job is done by a pipe in 1 hour:

1. If the inlet pipe fills the tank in 10 hours, then in 1 hour the inlet pipe fills $\frac{1}{10}$ of the tank.
2. If the outlet pipe empties the tank in 15 hours, then in 1 hour the outlet pipe empties $\frac{1}{15}$ of the tank.
3. If it takes x hours to fill the tank with both pipes open, then in 1 hour the tank is $1/x$ full.

Here is how we set up the equation. *In one hour,*

$$\underset{\substack{\text{Amount of water let} \\ \text{in by inlet pipe}}}{\frac{1}{10}} \quad - \quad \underset{\substack{\text{Amount of water let} \\ \text{out by outlet pipe}}}{\frac{1}{15}} \quad = \quad \underset{\substack{\text{Total amount of} \\ \text{water in tank}}}{\frac{1}{x}}$$

The LCD for our equation is $30x$. We multiply both sides by the LCD and solve:

$$30x\left(\frac{1}{10}\right) - 30x\left(\frac{1}{15}\right) = 30x\left(\frac{1}{x}\right)$$

$$3x - 2x = 30$$

$$x = 30$$

It takes 30 hours with both pipes open to fill the tank. ◀

Note In solving a problem like the one in Example 4, we have to assume that the thing doing the work (whether it is a pipe, a person, or a machine) is working at a constant rate. That is, as much work gets done in the first hour as is done in the last hour and any other hour in between.

▶ **EXAMPLE 5** Graph the equation $y = \dfrac{1}{x}$.

Solution Since this is the first time we have graphed an equation of this form, we will make a table of values for x and y that satisfy the equation. Before we do, let's make some generalizations about the graph.

First, notice that, since y is equal to 1 divided by x, y will be positive when x is positive. (The quotient of two positive numbers is a positive number.) Likewise, when x is negative, y will be negative. In other words x and y will always have the same sign. Thus, our graph will appear in quadrants 1 and 3 only, because in those quadrants x and y have the same sign.

Next, notice that the expression $\frac{1}{x}$ will be undefined when x is 0, meaning that there is no value of y corresponding to $x = 0$. Because of this, the graph will not cross the y-axis. Further, the graph will not cross the x-axis either. If we try to find the x-intercept by letting $y = 0$, we have

$$0 = \frac{1}{x}$$

But there is no value of x to divide into 1 to obtain 0. Therefore, since there is no solution to this equation, our graph will not cross the x-axis.

To summarize, we can expect to find the graph in quadrants 1 and 3 only, and the graph will cross neither axis.

x	y
-3	$-\frac{1}{3}$
-2	$-\frac{1}{2}$
-1	-1
$-\frac{1}{2}$	-2
$-\frac{1}{3}$	-3
0	undefined
$\frac{1}{3}$	3
$\frac{1}{2}$	2
1	1
2	$\frac{1}{2}$
3	$\frac{1}{3}$

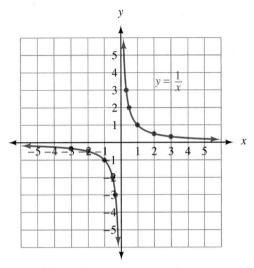

Figure 1

▶ **EXAMPLE 6**   Graph the equation $y = \dfrac{-6}{x}$.

Solution   Since y is -6 divided by x, when x is positive, y will be negative (a negative divided by a positive is negative), and when x is negative, y will be positive (a negative divided by a negative). Thus, the graph will appear in quadrants 2 and 4 only. As was the case in Example 5, the graph will not cross either axis.

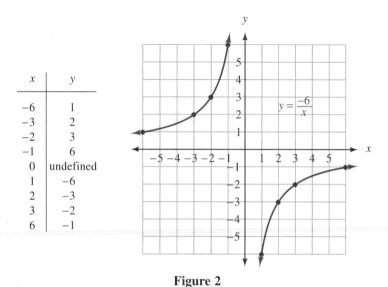

x	y
-6	1
-3	2
-2	3
-1	6
0	undefined
1	-6
2	-3
3	-2
6	-1

Figure 2   ◀

PROBLEM SET 6.5

Number Problems

1. One number is three times as large as another. The sum of their reciprocals is $\frac{16}{3}$. Find the two numbers.
2. If $\frac{3}{5}$ is added to twice the reciprocal of a number, the result is 1. Find the number.
3. The sum of a number and its reciprocal is $\frac{13}{6}$. Find the number.
4. The sum of a number and 10 times its reciprocal is 7. Find the number.
5. If a certain number is added to both the numerator and denominator of the fraction $\frac{7}{9}$, the result is $\frac{5}{7}$. Find the number.

6. The numerator of a certain fraction is two more than the denominator. If $\frac{1}{3}$ is added to the fraction, the result is 2. Find the fraction.
7. The sum of the reciprocals of two consecutive even integers is $\frac{5}{12}$. Find the integers.
8. The sum of the reciprocals of two consecutive integers is $\frac{7}{12}$. Find the two integers.

Motion Problems

9. A boat travels 26 miles up a river in the same amount of time it takes to travel 38 miles down the same river. If the current is 3 mph, what is the speed of the boat in still water?

	d	r	t
Upstream			
Downstream			

10. A boat can travel 9 miles up a river in the same amount of time it takes to travel 11 miles down the same river. If the current is 2 mph, what is the speed of the boat in still water?

	d	r	t
Upstream			
Downstream			

11. An airplane flying against the wind travels 140 miles in the same amount of time it would take the same plane to travel 160 miles with the wind. If the wind speed is a constant 20 mph, how fast would the plane travel in still air?

12. An airplane flying against the wind travels 500 miles in the same amount of time that it would take to travel 600 miles with the wind. If the speed of the wind is 50 mph, what is the speed of the plane in still air?

13. One plane can travel 20 mph faster than another. One of them goes 285 miles in the same time it takes the other to go 255 miles. What are their speeds?

14. One car travels 300 miles in the same amount of time it takes a second car traveling 5 mph slower than the first to go 275 miles. What are the speeds of the cars?

15. Tina, whom we mentioned in Example 3 of this section, is training for a biathlon. To train for the running portion of the race, she runs 8 miles each day, over the same course. The first 2 miles of the course is on level ground, while the last 6 miles is downhill. She runs 3 miles per hour slower on level ground than she runs downhill. If the complete course takes 1 hour, how fast does she run on the downhill part of the course?

16. Jerri is training for the same biathlon as Tina (Example 3 and Problem 15). To train for the bicycle portion of the race, she rides 24 miles out a straight road, then turns around and rides 24 miles back. The trip out is against the wind, while the trip back is with the wind. If she rides 10 miles per hour faster with the wind than she does against the wind, and the complete trip out and back takes 2 hours, how fast does she ride when she rides against the wind?

17. To train for the running of a triathlon, Jerri jogs 1 hour each day over the same 9-mile course. Five miles of the course is downhill while the other 4 miles is on level ground. Jerri figures that she runs 2 miles per hour faster downhill than she runs on level ground. Find the rate at which Jerri runs on level ground.

18. Travis paddles his kayak in the harbor at Morro Bay, California, where the incoming tide has caused a current in the water. From the point where he enters the water, he paddles 1 mile against the current, then turns around and paddles 1 mile back to where he started. His average speed when paddling with the current is 4 miles per hour faster than his speed against the current. If the complete trip (out and back) takes him 1.2 hours, find his average speed when he paddles against the current.

Work Problems

19. An inlet pipe can fill a pool in 12 hours, while an outlet pipe can empty it in 15 hours. If both pipes are left open, how long will it take to fill the pool?

20. A water tank can be filled in 20 hours by an inlet pipe and emptied in 25 hours by an outlet pipe. How long will it take to fill the tank if both pipes are left open?

21. A bathtub can be filled by the cold water faucet in 10 minutes and by the hot water faucet in 12 minutes. How long does it take to fill the tub if both faucets are open?

22. A water faucet can fill a sink in 6 minutes, while the drain can empty it in 4 minutes. If the sink is full, how long will it take to empty if both the faucet and the drain are open?

23. A sink can be filled by the cold water faucet in 3 minutes. The drain can empty a full sink in 4 minutes. If the sink is empty and both the cold water faucet and the drain are open, how long will it take the sink to overflow?

24. A bathtub can be filled by the cold water faucet in 9 minutes and by the hot water faucet in 10 minutes. The drain can empty the tub in 5 minutes. Can the tub be filled if both faucets and the drain are open?

Graph each of the following equations.

25. $y = \dfrac{2}{x}$

26. $y = \dfrac{-2}{x}$

27. $y = \dfrac{-4}{x}$

28. $y = \dfrac{4}{x}$

29. $y = \dfrac{8}{x}$

30. $y = \dfrac{-8}{x}$

31. Graph $y = \dfrac{3}{x}$ and $x + y = 4$ on the same coordinate system. At what points do the two graphs intersect?

32. Graph $y = \dfrac{4}{x}$ and $x - y = 3$ on the same coordinate system. At what points do the two graphs intersect?

Review Problems

The following problems review material we covered in Sections 3.8 and 3.9.

Solve each system of equations by the elimination method. [3.8]

33. $2x + y = 3$
$3x - y = 7$

34. $3x - y = -6$
$4x + y = -8$

35. $4x - 5y = 1$
$x - 2y = -2$

36. $6x - 4y = 2$
$2x + y = 10$

Solve by the substitution method. [3.9]

37. $5x + 2y = 7$
$y = 3x - 2$

38. $-7x - 5y = -1$
$y = x + 5$

39. $2x - 3y = 4$
$x = 2y + 1$

40. $4x - 5y = 2$
$x = 2y - 1$

SECTION

6.6   **Complex Fractions**

A **complex fraction** is a fraction or rational expression that contains other fractions in its numerator or denominator. Each of the following is a complex fraction:

$$\frac{\dfrac{1}{2}}{\dfrac{2}{3}} \qquad \frac{x + \dfrac{1}{y}}{y + \dfrac{1}{x}} \qquad \frac{\dfrac{a + 1}{a^2 - 9}}{\dfrac{2}{a + 3}}$$

We will begin this section by simplifying the first of these complex fractions. Before we do, though, let's agree on some vocabulary. So that we won't have to use phrases such as ''the numerator of the denominator,'' let's call the numerator of a complex fraction the *top* and the denominator of a complex fraction the *bottom*.

▶ **EXAMPLE 1**   Simplify $\dfrac{\frac{1}{2}}{\frac{2}{3}}$.

Solution   There are two methods we can use to solve this problem.

Method 1   We can multiply the top and bottom of this complex fraction by the LCD for both fractions. In this case, the LCD is 6:

$$\frac{\frac{1}{2}}{\frac{2}{3}} = \frac{6 \cdot \frac{1}{2}}{6 \cdot \frac{2}{3}} = \frac{3}{4}$$

Method 2   We can treat this as a division problem. Instead of dividing by $\frac{2}{3}$, we can multiply by its reciprocal $\frac{3}{2}$:

$$\frac{\frac{1}{2}}{\frac{2}{3}} = \frac{1}{2} \cdot \frac{3}{2} = \frac{3}{4}$$

Using either method, we obtain the same result.   ◀

▶ **EXAMPLE 2**   Simplify

$$\frac{\dfrac{2x^3}{y^2}}{\dfrac{4x}{y^5}}$$

Solution

Method 1   The LCD for each rational expression is y^5. Multiplying the top and bottom of the complex fraction by y^5, we have

$$\frac{\dfrac{2x^3}{y^2}}{\dfrac{4x}{y^5}} = \frac{y^5 \cdot \dfrac{2x^3}{y^2}}{y^5 \cdot \dfrac{4x}{y^5}} = \frac{2x^3y^3}{4x} = \frac{x^2y^3}{2}$$

Method 2   Instead of dividing by $4x/y^5$ we can multiply by its reciprocal, $y^5/4x$:

$$\frac{\dfrac{2x^3}{y^2}}{\dfrac{4x}{y^5}} = \frac{2x^3}{y^2} \cdot \frac{y^5}{4x} = \frac{x^2 y^3}{2}$$

Again the result is the same, whether we use Method 1 or Method 2. ◀

▶ **EXAMPLE 3**   Simplify

$$\frac{x + \dfrac{1}{y}}{y + \dfrac{1}{x}}$$

Solution   To apply Method 2 as we did in the first two examples, we would have to simplify the top and bottom separately to obtain a single rational expression for both before we could multiply by the reciprocal. It is much easier, in this case, to multiply the top and bottom by the LCD xy:

$$\frac{x + \dfrac{1}{y}}{y + \dfrac{1}{x}} = \frac{\mathbf{xy}\left(x + \dfrac{1}{y}\right)}{\mathbf{xy}\left(y + \dfrac{1}{x}\right)} \qquad \text{Multiply top and bottom by } \mathbf{xy}$$

$$= \frac{xy \cdot x + xy \cdot \dfrac{1}{y}}{xy \cdot y + xy \cdot \dfrac{1}{x}} \qquad \text{Distributive property}$$

$$= \frac{x^2 y + x}{xy^2 + y} \qquad \text{Simplify}$$

We can factor an x from $x^2 y + x$ and a y from $xy^2 + y$ and then reduce to lowest terms:

$$= \frac{x(xy + 1)}{y(xy + 1)}$$

$$= \frac{x}{y}$$ ◀

▶ **EXAMPLE 4**   Simplify

$$\frac{1 - \dfrac{4}{x^2}}{1 - \dfrac{1}{x} - \dfrac{6}{x^2}}$$

Solution The simplest way to simplify this complex fraction is to multiply the top and bottom by the LCD, x^2:

$$\frac{1 - \dfrac{4}{x^2}}{1 - \dfrac{1}{x} - \dfrac{6}{x^2}} = \frac{x^2\left(1 - \dfrac{4}{x^2}\right)}{x^2\left(1 - \dfrac{1}{x} - \dfrac{6}{x^2}\right)} \qquad \text{Multiply top and bottom by } x^2$$

$$= \frac{x^2 \cdot 1 - x^2 \cdot \dfrac{4}{x^2}}{x^2 \cdot 1 - x^2 \cdot \dfrac{1}{x} - x^2 \cdot \dfrac{6}{x^2}} \qquad \text{Distributive property}$$

$$= \frac{x^2 - 4}{x^2 - x - 6} \qquad \text{Simplify}$$

$$= \frac{(x - 2)(x + 2)}{(x - 3)(x + 2)} \qquad \text{Factor}$$

$$= \frac{x - 2}{x - 3} \qquad \text{Reduce} \qquad \blacktriangleleft$$

In the introduction to this chapter we mentioned a sequence of numbers that was closely related to the numbers in the Fibonacci sequence. In our next example we find the relationship between that sequence and the numbers in the Fibonacci sequence.

▶ **EXAMPLE 5** Simplify each term in the sequence below and then explain how this sequence is related to the Fibonacci sequence.

$$1 + \frac{1}{1 + 1}, \ 1 + \frac{1}{1 + \dfrac{1}{1 + 1}}, \ 1 + \frac{1}{1 + \dfrac{1}{1 + \dfrac{1}{1 + 1}}}, \ \cdots$$

Solution We can simplify our work somewhat if we notice that the first term $1 + \dfrac{1}{1 + 1}$ is the larger denominator in the second term, and that the second term is the largest denominator in the third term.

First term: $1 + \dfrac{1}{1 + 1} = 1 + \dfrac{1}{2} = \dfrac{2}{2} + \dfrac{1}{2} = \dfrac{3}{2}$

Second term: $1 + \dfrac{1}{1 + \dfrac{1}{1 + 1}} = 1 + \dfrac{1}{\dfrac{3}{2}} = 1 + \dfrac{2}{3} = \dfrac{3}{3} + \dfrac{2}{3} = \dfrac{5}{3}$

$$\text{Third term: } 1 + \cfrac{1}{1 + \cfrac{1}{1 + \cfrac{1}{1 + 1}}} = 1 + \cfrac{1}{\cfrac{5}{3}} = 1 + \frac{3}{5} = \frac{5}{5} + \frac{3}{5} = \frac{8}{5}$$

Here are the simplified numbers for the first three terms in our sequence:

$$\frac{3}{2}, \frac{5}{3}, \frac{8}{5}, \cdots$$

Recall the Fibonacci sequence: 1, 1, 2, 3, 5, 8, 13, 21, . . .

As you can see, each term in the sequence we have simplified is the ratio of two consecutive numbers in the Fibonacci sequence. If the pattern continues in this manner, the next number in our sequence will be $\frac{13}{8}$. ◀

P R O B L E M   S E T   6 . 6

Simplify each complex fraction.

1. $\cfrac{\dfrac{3}{4}}{\dfrac{1}{8}}$

2. $\cfrac{\dfrac{1}{3}}{\dfrac{5}{6}}$

3. $\cfrac{\dfrac{2}{3}}{\dfrac{3}{4}}$

4. $\cfrac{\dfrac{5}{1}}{\dfrac{1}{2}}$

5. $\cfrac{\dfrac{x^2}{y}}{\dfrac{x}{y^3}}$

6. $\cfrac{\dfrac{x^5}{y^3}}{\dfrac{x^2}{y^8}}$

7. $\cfrac{\dfrac{4x^3}{y^6}}{\dfrac{8x^2}{y^7}}$

8. $\cfrac{\dfrac{6x^4}{y}}{\dfrac{2x}{y^5}}$

9. $\cfrac{y + \dfrac{1}{x}}{x + \dfrac{1}{y}}$

10. $\cfrac{y - \dfrac{1}{x}}{x - \dfrac{1}{y}}$

11. $\cfrac{1 + \dfrac{1}{a}}{1 - \dfrac{1}{a}}$

12. $\cfrac{\dfrac{1}{a} - 1}{\dfrac{1}{a} + 1}$

13. $\cfrac{\dfrac{x + 1}{x^2 - 9}}{\dfrac{2}{x + 3}}$

14. $\cfrac{\dfrac{3}{x - 5}}{\dfrac{x + 1}{x^2 - 25}}$

15. $\cfrac{\dfrac{1}{a + 2}}{\dfrac{1}{a^2 - a - 6}}$

16. $\cfrac{\dfrac{1}{a^2 + 5a + 6}}{\dfrac{1}{a + 3}}$

17. $\cfrac{1 - \dfrac{9}{y^2}}{1 - \dfrac{1}{y} - \dfrac{6}{y^2}}$

18. $\cfrac{1 - \dfrac{4}{y^2}}{1 - \dfrac{2}{y} - \dfrac{8}{y^2}}$

19. $\cfrac{\dfrac{1}{y} + \dfrac{1}{x}}{\dfrac{1}{xy}}$

20. $\cfrac{\dfrac{1}{xy}}{\dfrac{1}{y} - \dfrac{1}{x}}$

21. $\dfrac{1 - \dfrac{1}{a^2}}{1 - \dfrac{1}{a}}$

22. $\dfrac{1 + \dfrac{1}{a}}{1 - \dfrac{1}{a^2}}$

23. $\dfrac{\dfrac{1}{10x} - \dfrac{y}{10x^2}}{\dfrac{1}{10} - \dfrac{y}{10x}}$

24. $\dfrac{\dfrac{1}{2x} + \dfrac{y}{2x^2}}{\dfrac{1}{4} + \dfrac{y}{4x}}$

25. $\dfrac{\dfrac{1}{a+1} + 2}{\dfrac{1}{a+1} + 3}$

26. $\dfrac{\dfrac{2}{a+1} + 3}{\dfrac{3}{a+1} + 4}$

Although the following problems do not contain complex fractions, they do involve more than one operation. Simplify inside the parentheses first, then multiply.

27. $\left(1 - \dfrac{1}{x}\right)\left(1 - \dfrac{1}{x+1}\right)\left(1 - \dfrac{1}{x+2}\right)$

28. $\left(1 + \dfrac{1}{x}\right)\left(1 + \dfrac{1}{x+1}\right)\left(1 + \dfrac{1}{x+2}\right)$

29. $\left(1 + \dfrac{1}{x+3}\right)\left(1 + \dfrac{1}{x+2}\right)\left(1 + \dfrac{1}{x+1}\right)$

30. $\left(1 - \dfrac{1}{x+3}\right)\left(1 - \dfrac{1}{x+2}\right)\left(1 - \dfrac{1}{x+1}\right)$

31. Simplify each term in the sequence below.

$$2 + \dfrac{1}{2+1}, 2 + \dfrac{1}{2 + \dfrac{1}{2+1}}, 2 + \dfrac{1}{2 + \dfrac{1}{2 + \dfrac{1}{2+1}}}, \cdots$$

32. Simplify each term in the sequence below.

$$2 + \dfrac{3}{2+3}, 2 + \dfrac{3}{2 + \dfrac{3}{2+3}}, 2 + \dfrac{3}{2 + \dfrac{3}{2 + \dfrac{3}{2+3}}}, \cdots$$

Review Problems

The following problems review material we covered in Section 2.8.

Solve each inequality.

33. $2x + 3 < 5$ **34.** $3x - 2 > 7$
35. $-3x \le 21$ **36.** $-5x \ge -10$
37. $-2x + 8 > -4$ **38.** $-4x - 1 < 11$
39. $4 - 2(x + 1) \ge -2$ **40.** $6 - 2(x + 3) \le -8$

Research Project 9

The first woman mentioned in the history of mathematics is Hypatia of Alexandria. Research the life of Hypatia and then write an essay that begins with a description of the time and place in which she lived and then goes on to give an indication of the type of person she was, her accomplishments in areas other than mathematics, and how she was viewed by her contemporaries.

6.7 Proportions

A **proportion** is two equal ratios. That is, if $\dfrac{a}{b}$ and $\dfrac{c}{d}$ are ratios, then

$$\frac{a}{b} = \frac{c}{d}$$

is a proportion.

Each of the four numbers in a proportion is called a **term** of the proportion. We number the terms as follows:

$$\text{First term} \to \frac{a}{b} = \frac{c}{d} \begin{matrix} \leftarrow \text{Third term} \\ \leftarrow \text{Fourth term} \end{matrix}$$

The first and fourth terms are called the **extremes,** and the second and third terms are called the **means:**

$$\text{Means} \quad \frac{a}{b} = \frac{c}{d} \quad \text{Extremes}$$

For example, in the proportion

$$\frac{3}{8} = \frac{12}{32}$$

the extremes are 3 and 32, and the means are 8 and 12.

Means-Extremes Property

If a, b, c, and d are real numbers with $b \neq 0$ and $d \neq 0$, then

$$\text{if} \quad \frac{a}{b} = \frac{c}{d}$$

$$\text{then} \quad ad = bc$$

In words: In any proportion, the product of the extremes is equal to the product of the means.

This property of proportions comes from the multiplication property of equality. We can use it to solve for a missing term in a proportion.

▶ **EXAMPLE 1** Solve the proportion $\dfrac{3}{x} = \dfrac{6}{7}$ for x.

Solution We could solve for x by using the method developed in Section 6.4, that is, multiplying both sides by the LCD $7x$. Instead, let's use our new means-extremes property:

$$\frac{3}{x} = \frac{6}{7} \qquad \text{Extremes are 3 and 7;}$$
$$\text{means are } x \text{ and 6}$$

$$21 = 6x \qquad \begin{matrix}\text{Product of extremes} = \\ \text{product of means}\end{matrix}$$

$$\frac{21}{6} = x \qquad \text{Divide both sides by 6}$$

$$x = \frac{7}{2} \qquad \text{Reduce to lowest terms} \qquad \blacktriangleleft$$

▶ **EXAMPLE 2** Solve for x: $\dfrac{x+1}{2} = \dfrac{3}{x}$.

Solution Again we want to point out that we could solve for x by using the method we used in Section 6.4. Using the means-extremes property is simply an alternative to the method developed in Section 6.4:

$$\frac{x+1}{2} = \frac{3}{x} \qquad \begin{array}{l} \text{Extremes are } x+1 \text{ and } x; \\ \text{means are 2 and 3} \end{array}$$

$$x^2 + x = 6 \qquad \begin{array}{l} \text{Product of extremes } = \\ \text{product of means} \end{array}$$

$$x^2 + x - 6 = 0 \qquad \begin{array}{l} \text{Standard form for a} \\ \text{quadratic equation} \end{array}$$

$$(x+3)(x-2) = 0 \qquad \text{Factor}$$

$$x + 3 = 0 \quad \text{or} \quad x - 2 = 0 \qquad \text{Set factors equal to 0}$$

$$x = -3 \quad \text{or} \quad x = 2$$

This time we have two solutions, -3 and 2. $\qquad \blacktriangleleft$

▶ **EXAMPLE 3** A manufacturer knows that during a production run, 8 out of every 100 parts produced by a certain machine will be defective. If the machine produces 1,450 parts, how many can be expected to be defective?

Solution The ratio of defective parts to total parts produced is $\frac{8}{100}$. If we let x represent the number of defective parts out of the total of 1,450 parts, then we can write this ratio again as $\dfrac{x}{1,450}$. This gives us a proportion to solve:

Defective parts in numerator $\dfrac{x}{1,450} = \dfrac{8}{100}$ Extremes are x and 100; means are 1,450 and 8

Total parts in denominator $\quad 100x = 11,600$ Product of extremes = product of means

$$x = 116$$

The manufacturer can expect 116 defective parts out of the total of 1,450 parts if the machine usually produces 8 defective parts for every 100 parts it produces. $\qquad \blacktriangleleft$

P R O B L E M   S E T   6 . 7

Solve each of the following proportions.

1. $\dfrac{x}{2} = \dfrac{6}{12}$

2. $\dfrac{x}{4} = \dfrac{6}{8}$

3. $\dfrac{2}{5} = \dfrac{4}{x}$

4. $\dfrac{3}{8} = \dfrac{9}{x}$

5. $\dfrac{10}{20} = \dfrac{20}{x}$

6. $\dfrac{15}{60} = \dfrac{60}{x}$

7. $\dfrac{a}{3} = \dfrac{5}{12}$

8. $\dfrac{a}{2} = \dfrac{7}{20}$

9. $\dfrac{2}{x} = \dfrac{6}{7}$

10. $\dfrac{4}{x} = \dfrac{6}{7}$

11. $\dfrac{x+1}{3} = \dfrac{4}{x}$

12. $\dfrac{x+1}{6} = \dfrac{7}{x}$

13. $\dfrac{x}{2} = \dfrac{8}{x}$

14. $\dfrac{x}{9} = \dfrac{4}{x}$

15. $\dfrac{4}{a+2} = \dfrac{a}{2}$

16. $\dfrac{3}{a+2} = \dfrac{a}{5}$

17. $\dfrac{1}{x} = \dfrac{x-5}{6}$

18. $\dfrac{1}{x} = \dfrac{x-6}{7}$

Applying the Concepts

19. A baseball player gets 6 hits in the first 18 games of the season. If he continues hitting at the same rate, how many hits will he get in the first 45 games?

20. A basketball player makes 8 of 12 free throws in the first game of the season. If she shoots with the same accuracy in the second game, how many of the 15 free throws she attempts will she make?

21. A solution contains 12 milliliters of alcohol and 16 milliliters of water. If another solution is to have the same concentration of alcohol in water but is to contain 28 milliliters of water, how much alcohol must it contain?

22. A solution contains 15 milliliters of HCl and 42 milliliters of water. If another solution is to have the same concentration of HCl in water but is to contain 140 milliliters of water, how much HCl must it contain?

23. If 100 grams of ice cream contain 13 grams of fat, how much fat is in 350 grams of ice cream?

24. A 6-ounce serving of grapefruit juice contains 159 grams of water. How many grams of water are in 20 ounces of grapefruit juice?

25. A map is drawn so that every 3.5 inches on the map corresponds to an actual distance of 100 miles. If the actual distance between two cities is 420 miles, how far apart are they on the map?

26. The scale on a map indicates that 1 inch on the map corresponds to an actual distance of 105 miles. Two cities are 4.5 inches apart on the map. What is the actual distance between the two cities?

27. A man drives his car 245 miles in 5 hours. At this rate, how far will he travel in 7 hours?

28. An airplane flies 1,380 miles in 3 hours. How far will it fly in 5 hours?

Review Problems

The following problems review material we covered in Sections 6.1, 6.2, and 6.3.

Reduce to lowest terms. [6.1]

29. $\dfrac{x^2 - x - 6}{x^2 - 9}$

30. $\dfrac{xy + 5x + 3y + 15}{x^2 + ax + 3x + 3a}$

Multiply and divide. [6.2]

31. $\dfrac{x^2 - 25}{x + 4} \cdot \dfrac{2x + 8}{x^2 - 9x + 20}$

32. $\dfrac{3x + 6}{x^2 + 4x + 3} \div \dfrac{x^2 + x - 2}{x^2 + 2x - 3}$

Add and subtract. [6.3]

33. $\dfrac{x}{x^2 - 16} + \dfrac{4}{x^2 - 16}$

34. $\dfrac{2}{x^2 - 1} - \dfrac{5}{x^2 + 3x - 4}$

CHAPTER 6 SUMMARY

Examples

1. We can reduce $\frac{6}{8}$ to lowest terms by dividing the numerator and denominator by their greatest common factor 2:

$$\frac{6}{8} = \frac{\cancel{2} \cdot 3}{\cancel{2} \cdot 4} = \frac{3}{4}$$

Rational Numbers [6.1]

Any number that can be put in the form $\frac{a}{b}$, where a and b are integers ($b \neq 0$), is called a **rational number.**

Multiplying or dividing the numerator and denominator of a rational number by the same nonzero number never changes the value of the rational number.

2. We reduce rational expressions to lowest terms by factoring the numerator and denominator and then dividing out any factors they have in common:

$$\frac{x-3}{x^2-9} = \frac{\cancel{x-3}}{(\cancel{x-3})(x+3)} = \frac{1}{x+3}$$

Rational Expressions [6.1]

Any expression of the form $\frac{P}{Q}$, where P and Q are polynomials ($Q \neq 0$), is a **rational expression.**

Multiplying or dividing the numerator and denominator of a rational expression by the same nonzero quantity always produces a rational expression equivalent to the original one.

3. $\dfrac{x-1}{x^2+2x-3} \cdot \dfrac{x^2-9}{x-2}$

$= \dfrac{\cancel{x-1}}{(x+3)(\cancel{x-1})} \cdot \dfrac{(x-3)(\cancel{x+3})}{x-2}$

$= \dfrac{x-3}{x-2}$

Multiplication [6.2]

To multiply two rational numbers or two rational expressions, multiply numerators, multiply denominators, and divide out any factors common to the numerator and denominator:

For rational numbers $\dfrac{a}{b}$ and $\dfrac{c}{d}$, $\quad \dfrac{a}{b} \cdot \dfrac{c}{d} = \dfrac{ac}{bd}$

For rational expressions $\dfrac{P}{Q}$ and $\dfrac{R}{S}$, $\quad \dfrac{P}{Q} \cdot \dfrac{R}{S} = \dfrac{PR}{QS}$

4. $\dfrac{2x}{x^2-25} \div \dfrac{4}{x-5}$

$= \dfrac{2x}{(\cancel{x-5})(x+5)} \cdot \dfrac{\cancel{x-5}}{4}$

$= \dfrac{x}{2(x+5)}$

Division [6.2]

To divide by a rational number or rational expression, simply multiply by its reciprocal:

For rational numbers $\dfrac{a}{b}$ and $\dfrac{c}{d}$, $\quad \dfrac{a}{b} \div \dfrac{c}{d} = \dfrac{a}{b} \cdot \dfrac{d}{c}$

For rational expressions $\dfrac{P}{Q}$ and $\dfrac{R}{S}$, $\quad \dfrac{P}{Q} \div \dfrac{R}{S} = \dfrac{P}{Q} \cdot \dfrac{S}{R}$

5. $\dfrac{3}{x-1} + \dfrac{x}{2}$

$= \dfrac{3}{x-1} \cdot \dfrac{2}{2} + \dfrac{x}{2} \cdot \dfrac{x-1}{x-1}$

$= \dfrac{6}{2(x-1)} + \dfrac{x^2 - x}{2(x-1)}$

$= \dfrac{x^2 - x + 6}{2(x-1)}$

Addition [6.3]
To add two rational numbers or rational expressions, find a common denominator, change each expression to an equivalent expression having the common denominator, then add numerators and reduce if possible:

For rational numbers $\dfrac{a}{c}$ and $\dfrac{b}{c}$, $\dfrac{a}{c} + \dfrac{b}{c} = \dfrac{a+b}{c}$

For rational expressions $\dfrac{P}{S}$ and $\dfrac{Q}{S}$, $\dfrac{P}{S} + \dfrac{Q}{S} = \dfrac{P+Q}{S}$

6. $\dfrac{x}{x^2 - 4} - \dfrac{2}{x^2 - 4}$

$= \dfrac{x - 2}{x^2 - 4}$

$= \dfrac{\cancel{x - 2}}{\cancel{(x - 2)}(x + 2)}$

$= \dfrac{1}{x + 2}$

Subtraction [6.3]
To subtract a rational number or rational expression, simply add its opposite:

For rational numbers $\dfrac{a}{c}$ and $\dfrac{b}{c}$, $\dfrac{a}{c} - \dfrac{b}{c} = \dfrac{a}{c} + \left(\dfrac{-b}{c}\right)$

For rational expressions $\dfrac{P}{S}$ and $\dfrac{Q}{S}$, $\dfrac{P}{S} - \dfrac{Q}{S} = \dfrac{P}{S} + \left(\dfrac{-Q}{S}\right)$

7. Solve $\dfrac{1}{2} + \dfrac{3}{x} = 5$.

$2x\left(\dfrac{1}{2}\right) + 2x\left(\dfrac{3}{x}\right) = 2x(5)$

$x + 6 = 10x$

$6 = 9x$

$x = \dfrac{2}{3}$

Equations [6.4]
To solve equations involving rational expressions, first find the least common denominator (LCD) for all denominators. Then multiply both sides by the LCD and solve as usual. Be sure to check all solutions in the original equation to be sure there are no undefined terms.

8. $\dfrac{1 - \dfrac{4}{x}}{x - \dfrac{16}{x}} = \dfrac{x\left(1 - \dfrac{4}{x}\right)}{x\left(x - \dfrac{16}{x}\right)}$

$= \dfrac{x - 4}{x^2 - 16}$

$= \dfrac{\cancel{x - 4}}{\cancel{(x - 4)}(x + 4)}$

$= \dfrac{1}{x + 4}$

Complex Fractions [6.6]
A rational expression that contains a fraction in its numerator or denominator is called a **complex fraction.** The most common method of simplifying a complex fraction is to multiply the top and bottom by the LCD for all denominators.

9. Solve for x: $\dfrac{3}{x} = \dfrac{5}{20}$

$3 \cdot 20 = 5 \cdot x$

$60 = 5x$

$x = 12$

Ratio and Proportion [6.1, 6.7]

The ratio of a to b is

$$\frac{a}{b}$$

Two equal ratios form a proportion. In the proportion

$$\frac{a}{b} = \frac{c}{d}$$

a and d are the *extremes,* and b and c are the *means.* In any proportion, the product of the extremes is equal to the product of the means.

COMMON MISTAKES

1. Trying to reduce by dividing by a quantity that is not a factor of both the top and bottom:

$$\frac{x^2 - \overset{2}{\cancel{4}x} + \cancel{4}}{x^2 - \underset{3}{\cancel{6}x} + \underset{2}{\cancel{8}}} \qquad \text{Mistake}$$

This makes no sense at all. The numerator and denominator must be factored completely before any factors common to the numerator and denominator can be recognized:

$$\frac{x^2 - 4x + 4}{x^2 - 6x + 8} = \frac{\cancel{(x - 2)}(x - 2)}{\cancel{(x - 2)}(x - 4)}$$

$$= \frac{x - 2}{x - 4}$$

2. Forgetting to check solutions to equations involving rational expressions. If you multiply both sides of an equation by an expression that contains the variable, you must check your solutions to be sure you have not multiplied by zero.

CHAPTER 6 REVIEW

The numbers in brackets refer to the sections of the text in which similar problems can be found.

Reduce to lowest terms. Also specify any restriction on the variable. [6.1]

1. $\dfrac{7}{14x - 28}$

2. $\dfrac{-6}{6x + 18}$

3. $\dfrac{a + 6}{a^2 - 36}$

4. $\dfrac{a - 9}{a^2 - 81}$

5. $\dfrac{8x - 4}{4x + 12}$

6. $\dfrac{5x + 15}{x + 3}$

Reduce to lowest terms. [6.1]

7. $\dfrac{x + 4}{x^2 + 8x + 16}$

8. $\dfrac{x^2 - 14x + 49}{x - 7}$

9. $\dfrac{3x^3 + 16x^2 - 12x}{2x^3 + 9x^2 - 18x}$

10. $\dfrac{8x^3 + 10x^2 - 3x}{20x^3 + 3x^2 - 2x}$

11. $\dfrac{x + 2}{x^4 - 16}$

12. $\dfrac{x^2 - 9}{x^4 - 81}$

13. $\dfrac{x^2 + 5x - 14}{x + 7}$

14. $\dfrac{x^2 + 5x - 24}{x + 8}$

15. $\dfrac{a^2 + 16a + 64}{a + 8}$

16. $\dfrac{a^2 - 12a + 36}{a - 6}$

17. $\dfrac{xy + bx + ay + ab}{xy + 5x + ay + 5a}$

18. $\dfrac{2xy + 10x + 3y + 15}{3xy + 15x + 2y + 10}$

Multiply or divide as indicated. [6.2]

19. $\dfrac{3x + 9}{x^2} \cdot \dfrac{x^3}{6x + 18}$

20. $\dfrac{6x - 12}{6x + 12} \cdot \dfrac{3x + 3}{12x - 24}$

21. $\dfrac{x^2 + 8x + 16}{x^2 + x - 12} \div \dfrac{x^2 - 16}{x^2 - x - 6}$

22. $\dfrac{x^2 + 3x}{x^2 + 4x + 4} \cdot \dfrac{x^2 - 5x - 14}{x^2 + 6x + 9}$

23. $(a^2 - 4a - 12)\left(\dfrac{a - 6}{a + 2}\right)$

24. $(a^2 + 5a - 24)\left(\dfrac{a + 8}{a - 3}\right)$

25. $\dfrac{3x^2 - 2x - 1}{x^2 + 6x + 8} \div \dfrac{3x^2 + 13x + 4}{x^2 + 8x + 16}$

26. $\dfrac{x^2 + 5x + 6}{x^2 - x - 6} \div \dfrac{x^2 + 6x + 9}{2x^2 - 5x - 3}$

Find the following sums and differences. [6.3]

27. $\dfrac{2x}{2x + 3} + \dfrac{3}{2x + 3}$

28. $\dfrac{3x}{3x - 4} - \dfrac{4}{3x - 4}$

29. $\dfrac{x^2}{x - 9} - \dfrac{18x - 81}{x - 9}$

30. $\dfrac{x^2}{x - 7} - \dfrac{14x - 49}{x - 7}$

31. $\dfrac{a + 4}{a + 8} - \dfrac{a - 9}{a + 8}$

32. $\dfrac{a - 3}{a - 7} - \dfrac{a + 10}{a - 7}$

Find the following sums and differences. [6.3]

33. $\dfrac{x}{x + 9} + \dfrac{5}{x}$

34. $\dfrac{x}{x - 3} - \dfrac{9}{x}$

35. $\dfrac{5}{4x + 20} + \dfrac{x}{x + 5}$

36. $\dfrac{6}{10x + 30} + \dfrac{x}{x + 3}$

37. $\dfrac{3}{x^2 - 36} - \dfrac{2}{x^2 - 4x - 12}$

38. $\dfrac{-1}{x^2 - 4} - \dfrac{-2}{x^2 - 4x - 12}$

39. $\dfrac{3a}{a^2 + 8a + 15} - \dfrac{2}{a + 5}$

40. $\dfrac{7a}{a^2 - 3a - 54} + \dfrac{5}{a - 9}$

Solve each equation. [6.4]

41. $\dfrac{3}{x} + \dfrac{1}{2} = \dfrac{5}{x}$

42. $\dfrac{5}{x} - \dfrac{1}{3} = \dfrac{3}{x}$

43. $\dfrac{a}{a - 3} = \dfrac{3}{2}$

44. $\dfrac{a}{a + 4} = \dfrac{7}{3}$

45. $1 - \dfrac{7}{x} = \dfrac{-6}{x^2}$

46. $1 + \dfrac{12}{x} = \dfrac{-35}{x^2}$

47. $\dfrac{3}{x + 6} - \dfrac{1}{x - 2} = \dfrac{-8}{x^2 + 4x - 12}$

48. $\dfrac{4}{x - 5} - \dfrac{3}{x + 2} = \dfrac{28}{x^2 - 3x - 10}$

49. $\dfrac{2}{y^2 - 16} = \dfrac{10}{y^2 + 4y}$

50. $\dfrac{4}{y^2 - 25} = \dfrac{24}{y^2 - 5y}$

Solve. [6.5]

51. The sum of a number and seven times its reciprocal is $\frac{16}{3}$. Find the number.

52. The sum of a number and five times its reciprocal is 6. Find the number.

53. A boat travels 48 miles up a river in the same amount of time it takes to travel 72 miles down the same river. If the current is 3 mph, what is the speed of the boat in still water?

54. A boat travels 54 miles up a river in the same amount of time it takes to travel 66 miles down the same river. If the current is 2 mph, what is the speed of the boat in still water?

55. An inlet pipe can fill a pool in 21 hours, while an outlet pipe can empty it in 28 hours. If both pipes are left open, how long will it take to fill the pool?

56. A bathtub can be filled by the cold water faucet in 6 hours and by the hot water faucet in 12 hours. How long does it take to fill the tub if both faucets are open?

Graph each equation. [6.5]

57. $y = \dfrac{4}{x}$

58. $y = \dfrac{-1}{x}$

Simplify each complex fraction. [6.6]

59. $\dfrac{\dfrac{x-2}{x^2+6x+8}}{\dfrac{4}{x+4}}$

60. $\dfrac{1-\dfrac{9}{y^2}}{1+\dfrac{4}{y}-\dfrac{21}{y^2}}$

61. $\dfrac{1-\dfrac{16}{y^2}}{1-\dfrac{4}{y}-\dfrac{32}{y^2}}$

62. $\dfrac{\dfrac{1}{a-2}+4}{\dfrac{1}{a-2}+1}$

Write each ratio as a fraction in lowest terms. [6.1]

63. 40 to 100

64. $\dfrac{10}{3}$ to $\dfrac{5}{2}$

65. If there are 60 seconds in 1 minute, what is the ratio of 40 seconds to 3 minutes?

66. If there are 12 inches in 1 foot, what is the ratio of 10 inches to 5 feet?

Solve each proportion. [6.7]

67. $\dfrac{x}{9} = \dfrac{4}{3}$

68. $\dfrac{4}{x} = \dfrac{7}{3}$

69. $\dfrac{a}{3} = \dfrac{12}{a}$

70. $\dfrac{4}{a} = \dfrac{a}{16}$

71. $\dfrac{8}{x-2} = \dfrac{x}{6}$

72. $\dfrac{x}{3} = \dfrac{6}{x-3}$

73. A man drives his car 124 miles in 4 hours. At this rate, how far will he travel in 9 hours?

74. An airplane flies 1,860 miles in 5 hours. How far will it fly in 7 hours?

CHAPTER 6 TEST

Reduce to lowest terms. [6.1]

1. $\dfrac{x^2-16}{x^2-8x+16}$

2. $\dfrac{10a+20}{5a^2+20a+20}$

3. $\dfrac{xy+7x+5y+35}{x^2+ax+5x+5a}$

Multiply or divide as indicated. [6.2]

4. $\dfrac{3x-12}{4} \cdot \dfrac{8}{2x-8}$

5. $\dfrac{x^2-49}{x+1} \div \dfrac{x+7}{x^2-1}$

6. $\dfrac{x^2 - 3x - 10}{x^2 - 8x + 15} \div \dfrac{3x^2 + 2x - 8}{x^2 + x - 12}$

7. $(x^2 - 9)\left(\dfrac{x + 2}{x + 3}\right)$

Add or subtract as indicated. [6.3]

8. $\dfrac{3}{x - 2} - \dfrac{6}{x - 2}$

9. $\dfrac{x}{x^2 - 9} + \dfrac{4}{4x - 12}$

10. $\dfrac{2x}{x^2 - 1} + \dfrac{x}{x^2 - 3x + 2}$

Solve the following equations. [6.4]

11. $\dfrac{7}{5} = \dfrac{x + 2}{3}$

12. $\dfrac{10}{x + 4} = \dfrac{6}{x} - \dfrac{4}{x}$

13. $\dfrac{3}{x - 2} - \dfrac{4}{x + 1} = \dfrac{5}{x^2 - x - 2}$

Solve the following problems. [6.5]

14. A boat travels 26 miles up a river in the same amount of time it takes to travel 34 miles down the same river. If the current is 2 mph, what is the speed of the boat in still water?

15. An inlet pipe can fill a pool in 15 hours, while an outlet pipe can empty it in 12 hours. If the pool is full and both pipes are open, how long will it take to empty?

16. Graph $y = \dfrac{8}{x}$. [6.5]

Solve the following problems involving ratio and proportion. [6.1, 6.7]

17. A solution of alcohol and water contains 27 milliliters of alcohol and 54 milliliters of water. What is the ratio of alcohol to water and the ratio of alcohol to total volume?

18. A manufacturer knows that during a production run 8 out of every 100 parts produced by a certain machine will be defective. If the machine produces 1,650 parts, how many can be expected to be defective?

Simplify each complex fraction. [6.6]

19. $\dfrac{1 + \dfrac{1}{x}}{1 - \dfrac{1}{x}}$

20. $\dfrac{1 - \dfrac{16}{x^2}}{1 - \dfrac{2}{x} - \dfrac{8}{x^2}}$

7

ROOTS AND RADICALS

INTRODUCTION

The diagram shown in Figure 1 is called the **spiral of roots.** It is constructed using the Pythagorean Theorem, which we introduced in Chapter 5. The spiral of roots gives us a way to visualize positive square roots, one of the topics we will study in detail in this chapter.

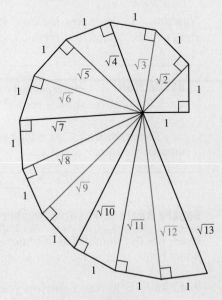

Figure 1

OVERVIEW

Chapter 7 is concerned with operations on and simplification of radicals. Most of the radical expressions we will work with involve square roots. Occasionally we will encounter higher roots. Finding the square root of a number is the reverse of raising a number to the second power.

Since much of what we will do in this chapter is the reverse of what we did in Chapter 4, the more familiar we are with the properties of exponents, the better equipped we will be to handle this chapter.

The main ideas necessary for success in Chapter 7 are

1. properties of exponents
2. operations on polynomials

SECTION

7.1   # Definitions and Common Roots

In Chapter 4 we developed notation (exponents) that would take us from a number to its square. If we wanted the square of 5, we wrote $5^2 = 25$. In this section we will use another type of notation that will take us in the reverse direction—from the square of a number back to the number itself.

In general we are interested in going from a number, say, 49, back to the number we squared to get 49. Since the square of 7 is 49, we say 7 is a square root of 49. The notation we use looks like this:

$$\sqrt{49}$$

Notation   In the expression $\sqrt{49}$, 49 is called the **radicand**, $\sqrt{}$ is the **radical sign,** and the complete expression, $\sqrt{49}$, is called the **radical.**

DEFINITION   If x represents any positive real number, then the expression $\sqrt{x}$ is the **positive square root** of x. It is the *positive* number we square to get x.

The expression $-\sqrt{x}$ is the **negative square root** of x. It is the *negative* number we square to get x.

Square Roots of Positive Numbers

Every positive number has two square roots, one positive and the other negative. Some books refer to the positive square root of a number as the principal root.

▶ **EXAMPLE 1**   The positive square root of 25 is 5 and can be written $\sqrt{25} = 5$. The negative square root of 25 is -5 and can be written $-\sqrt{25} = -5$. ◀

If we want to consider the negative square root of a number, we must put a negative sign in front of the radical. It is a common mistake to think of $\sqrt{25}$ as meaning either 5 or -5. The expression $\sqrt{25}$ means the *positive* square root of 25, which is 5. If we want the negative square root, we write $-\sqrt{25}$ to begin with.

▶ **EXAMPLES** Find the following roots.

2. $\sqrt{49} = 7$ 7 is the positive number we square to get 49
3. $-\sqrt{49} = -7$ -7 is the negative number we square to get 49
4. $\sqrt{121} = 11$ 11 is the positive number we square to get 121 ◀

▶ **EXAMPLE 5** The positive square root of 17 is written $\sqrt{17}$. The negative square root of 17 is written $-\sqrt{17}$. ◀

We have no other exact representation for the two roots in Example 5. Since 17 itself is not a perfect square (the square of an integer), its two square roots, $\sqrt{17}$ and $-\sqrt{17}$, are irrational numbers. They have a place on the real number line but cannot be written as the ratio of two integers. The square roots of any number that is not itself a perfect square are irrational numbers.

▶ **EXAMPLE 6**

Number	Positive Square Root	Negative Square Root	Roots Are
9	3	-3	Rational numbers
36	6	-6	Rational numbers
7	$\sqrt{7}$	$-\sqrt{7}$	Irrational numbers
22	$\sqrt{22}$	$-\sqrt{22}$	Irrational numbers
100	10	-10	Rational numbers

◀

Square Root of Zero

The number 0 is the only real number with one square root. It is also its own square root:

$$\sqrt{0} = 0$$

Square Roots of Negative Numbers

Negative numbers have square roots, but their square roots are not real numbers. They do not have a place on the real number line. We will consider square roots of negative numbers later in the book.

▶ **EXAMPLE 7** The expression $\sqrt{-4}$ does not represent a real number, since there is no real number we can square and end up with -4. The same is true of square roots of any negative number. ◀

Other Roots

There are many other roots of numbers besides square roots, although square roots seem to be the most commonly used. The cube root of a number is the number we cube (raise to the third power) to get the original number. The cube root of 8 is 2, since $2^3 = 8$. The cube root of 27 is 3, since $3^3 = 27$. The notation for cube roots looks like this:

The 3 is called the index $\longrightarrow$ $\sqrt[3]{8} = 2$
$\longrightarrow$ $\sqrt[3]{27} = 3$

We can go as high as we want with roots. The fourth root of 16 is 2 because $2^4 = 16$. We can write this in symbols as $\sqrt[4]{16} = 2$.

Here is a list of the most common roots. They are the roots that will come up most often in the remainder of the book and should be memorized.

Square Roots		Cube Roots	Fourth Roots
$\sqrt{1} = 1$	$\sqrt{49} = 7$	$\sqrt[3]{1} = 1$	$\sqrt[4]{1} = 1$
$\sqrt{4} = 2$	$\sqrt{64} = 8$	$\sqrt[3]{8} = 2$	$\sqrt[4]{16} = 2$
$\sqrt{9} = 3$	$\sqrt{81} = 9$	$\sqrt[3]{27} = 3$	$\sqrt[4]{81} = 3$
$\sqrt{16} = 4$	$\sqrt{100} = 10$	$\sqrt[3]{64} = 4$	$\sqrt[4]{256} = 4$
$\sqrt{25} = 5$	$\sqrt{121} = 11$	$\sqrt[3]{125} = 5$	$\sqrt[4]{625} = 5$
$\sqrt{36} = 6$	$\sqrt{144} = 12$		

With even roots—square roots, fourth roots, sixth roots, and so on—we cannot have negative numbers *under* the radical sign. With odd roots, negative numbers under the radical sign do not cause problems.

▶ **EXAMPLES** Find the following roots, if possible.

8. $\sqrt[3]{-8} = -2$ Because $(-2)^3 = -8$

9. $\sqrt[3]{-27} = -3$ Because $(-3)^3 = -27$

10. $\sqrt{-4}$ Not a real number since there is no real number whose square is -4

11. $-\sqrt{4} = -2$ Because -2 is the negative number we square to get 4

12. $\sqrt[4]{-16}$ Not a real number since there is no real number that can be raised to the fourth power to obtain -16

13. $-\sqrt[4]{16} = -2$ Because -2 is the negative number we raise to the fourth power to get 16 ◀

Variables under the Radical Sign

In this chapter, unless we say otherwise, we will assume that all variables that appear under a radical sign represent positive numbers. That way we can simplify expressions involving radicals that contain variables. Here are some examples.

▶ **EXAMPLE 14** Simplify $\sqrt{49x^2}$.

Solution We are looking for the expression we square to get $49x^2$. Since the square of 7 is 49 and the square of x is x^2, we can square $7x$ and get $49x^2$:

$$\sqrt{49x^2} = 7x \qquad \text{Because } (7x)^2 = 49x^2 \qquad ◀$$

▶ **EXAMPLE 15** Simplify $\sqrt{16a^2b^2}$.

Solution We want an expression whose square is $16a^2b^2$. That expression is $4ab$:

$$\sqrt{16a^2b^2} = 4ab \qquad \text{Because } (4ab)^2 = 16a^2b^2 \qquad ◀$$

▶ **EXAMPLE 16** Simplify $\sqrt[3]{125a^3}$.

Solution We are looking for the expression we cube to get $125a^3$. That expression is $5a$:

$$\sqrt[3]{125a^3} = 5a \qquad \text{Because } (5a)^3 = 125a^3 \qquad ◀$$

▶ **EXAMPLE 17** Simplify $\sqrt{x^6}$.

Solution The number we square to obtain x^6 is x^3.

$$\sqrt{x^6} = x^3 \qquad \text{Because } (x^3)^2 = x^6 \qquad ◀$$

▶ **EXAMPLE 18** Simplify $\sqrt[4]{16a^8b^4}$.

Solution The number we raise to the fourth power to obtain $16a^8b^4$ is $2a^2b$.

$$\sqrt[4]{16a^8b^4} = 2a^2b \qquad \text{Because } (2a^2b)^4 = 16a^8b^4 \qquad ◀$$

The next example involves expressions that are similar to many of the expressions we will encounter when we solve quadratic equations in Chapter 8.

▶ **EXAMPLE 19** Simplify $\dfrac{2 + \sqrt{25}}{3}$ and $\dfrac{2 - \sqrt{25}}{3}$.

Solution We begin each simplification by writing $\sqrt{25}$ as 5.

$$\frac{2 + \sqrt{25}}{3} = \frac{2 + 5}{3} = \frac{7}{3}$$

$$\frac{2 - \sqrt{25}}{3} = \frac{2 - 5}{3} = \frac{-3}{3} = -1 \qquad ◀$$

There are many application problems involving radicals that require decimal approximations. When we need a decimal approximation to a square root, we can use a calculator.

▶ **EXAMPLE 20**   If you invest P dollars in an account and after two years the account has A dollars in it, then the annual rate of return, r, on the money you originally invested is given by the formula

$$r = \frac{\sqrt{A} - \sqrt{P}}{\sqrt{P}}$$

Suppose you pay $65 for a coin collection and find that the same coins sell for $84 two years later. Find the annual rate of return on your investment.

Solution   Substituting $A = 84$ and $P = 65$ in the formula, we have

$$r = \frac{\sqrt{84} - \sqrt{65}}{\sqrt{65}}$$

Using a calculator, we find that $\sqrt{84} \approx 9.165$ and $\sqrt{65} = 8.062$. Using these numbers in our formula gives us

$$r = \frac{9.165 - 8.062}{8.062}$$

$$= \frac{1.103}{8.062}$$

$$= 0.137 \text{ or } 13.7\%$$

To earn as much as this in a savings account that compounds interest once a year, you have to find an account that paid 13.7% in annual interest.   ◀

The Pythagorean Theorem Again

Now that we have had some experience working with square roots, we can rewrite the Pythagorean Theorem using a square root. In Figure 1, if triangle ABC is a right triangle with $C = 90°$, then the length of the longest side is the *square root* of the sum of the squares of the other two sides.

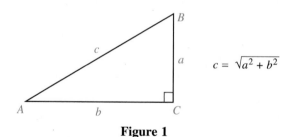

Figure 1

▶ **EXAMPLE 21** A tent pole is 8 feet in length and makes an angle of 90° with the ground. One end of a rope is attached to the top of the pole, while the other end of the rope is anchored to the ground 6 feet from the bottom of the pole. Find the length of the rope.

Solution The diagram in Figure 2 is a visual representation of the situation. To find the length of the rope we apply the Pythagorean Theorem.

$$x = \sqrt{6^2 + 8^2}$$
$$= \sqrt{36 + 64}$$
$$= \sqrt{100}$$
$$= 10 \text{ feet}$$

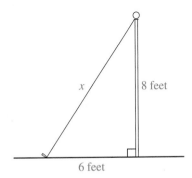

Figure 2

PROBLEM SET 7.1

Find the following roots. If the root does not exist as a real number, write ''not a real number.''

1. $\sqrt{9}$

2. $\sqrt{16}$

3. $-\sqrt{9}$

4. $-\sqrt{16}$

5. $\sqrt{-25}$

6. $\sqrt{-36}$

7. $-\sqrt{144}$

8. $\sqrt{256}$

9. $\sqrt{625}$

10. $-\sqrt{625}$

11. $\sqrt{-49}$

12. $\sqrt{-169}$

13. $-\sqrt{64}$

14. $-\sqrt{25}$

15. $-\sqrt{100}$

16. $\sqrt{121}$

17. $\sqrt{1,225}$

18. $-\sqrt{1,681}$

19. $\sqrt[4]{1}$

20. $-\sqrt[4]{81}$

21. $\sqrt[3]{-8}$

22. $\sqrt[3]{125}$

23. $-\sqrt[3]{125}$

24. $-\sqrt[3]{-8}$

25. $\sqrt[3]{-1}$

26. $-\sqrt[3]{-1}$

27. $\sqrt[3]{-27}$

28. $-\sqrt[3]{27}$

29. $-\sqrt[4]{16}$

30. $\sqrt[4]{-16}$

Assume all variables are positive and find the following roots.

31. $\sqrt{x^2}$

32. $\sqrt{a^2}$

33. $\sqrt{9x^2}$

34. $\sqrt{25x^2}$

35. $\sqrt{x^2y^2}$

36. $\sqrt{a^2b^2}$

37. $\sqrt{(a + b)^2}$

38. $\sqrt{(x + y)^2}$

39. $\sqrt{49x^2y^2}$

40. $\sqrt{81x^2y^2}$

41. $\sqrt[3]{x^3}$

42. $\sqrt[3]{a^3}$

43. $\sqrt[3]{8x^3}$

44. $\sqrt[3]{27x^3}$

45. $\sqrt{x^4}$

46. $\sqrt{x^6}$

47. $\sqrt{36a^6}$

48. $\sqrt{64a^4}$

49. $\sqrt{25a^8b^4}$

50. $\sqrt{16a^4b^8}$

51. $\sqrt[3]{x^6}$

52. $\sqrt[3]{x^9}$

53. $\sqrt[3]{27a^{12}}$

54. $\sqrt[3]{8a^6}$

55. $\sqrt[4]{x^8}$

56. $\sqrt[4]{x^{12}}$

Simplify each expression.

57. $\sqrt{9} + \sqrt{16}$

58. $\sqrt{64} + \sqrt{36}$

59. $\sqrt{9 + 16}$

60. $\sqrt{64 + 36}$

61. $\sqrt{144} + \sqrt{25}$

62. $\sqrt{25} - \sqrt{16}$

63. $\sqrt{144 + 25}$

64. $\sqrt{25 - 16}$

Simplify each pair of expressions below.

65. $\dfrac{5 + \sqrt{49}}{2}$ and $\dfrac{5 - \sqrt{49}}{2}$

66. $\dfrac{3 + \sqrt{25}}{4}$ and $\dfrac{3 - \sqrt{25}}{4}$

67. $\dfrac{2 + \sqrt{16}}{2}$ and $\dfrac{2 - \sqrt{16}}{2}$

68. $\dfrac{3 + \sqrt{9}}{2}$ and $\dfrac{3 - \sqrt{9}}{2}$

69. We know that the trinomial $x^2 + 6x + 9$ is the square of the binomial $x + 3$. That is,

$$x^2 + 6x + 9 = (x + 3)^2$$

Use this fact to find $\sqrt{x^2 + 6x + 9}$.

70. Use the fact that $x^2 + 10x + 25 = (x + 5)^2$ to find $\sqrt{x^2 + 10x + 25}$.

71. Replace x with 4 in the expression $\sqrt{x^2 + 9}$ and in the expression $x + 3$. Simplify each result.

72. Replace x with 8 in the expression $\sqrt{x^2 + 36}$ and in the expression $x + 6$. Simplify both results.

Applying the Concepts

In the next four problems you will need the following formula. The formula gives the annual rate of return, r, on an investment of P dollars that grows to A dollars over a two-year period of time:

$$r = \frac{\sqrt{A} - \sqrt{P}}{\sqrt{P}}$$

73. If you invest $50 in silver coins and two years later you find the coins are worth $65, what is the annual rate of return on your investment?

74. If you pay $35 for some rare stamps and two years later find that the same stamps cost $45, what is the annual rate of return on your investment?

75. Suppose you invest $500 in the stock market and two years later your stocks are worth $600. Find the annual rate of return on your stocks.

76. Suppose you invest $700 in the stock market and two years later your stocks are worth $800. Find the annual rate of return on your stocks.

Find x in each right triangle below.

77.

78.

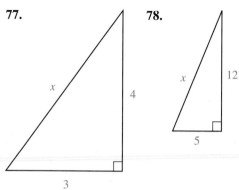

79.

80.

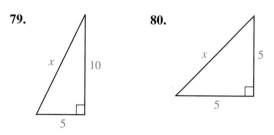

81. One end of a wire is attached to the top of a 24-foot pole, while the other end of the wire is anchored to the ground 18 feet from the bottom of the pole. If the pole makes an angle of 90° with the ground, find the length of the wire.

82. The screen on a television set is in the shape of a rectangle. If the length is 20 inches and the width is 12 inches, how many inches is it from one corner of the screen to the opposite corner?

Review Problems

The following problems review material we covered in Section 6.1.

Reduce each expression to lowest terms.

83. $\dfrac{x^2 - 16}{x + 4}$

84. $\dfrac{x - 5}{x^2 - 25}$

85. $\dfrac{10a + 20}{5a^2 - 20}$

86. $\dfrac{8a - 16}{4a^2 - 16}$

87. $\dfrac{2x^2 - 5x - 3}{x^2 - 3x}$

88. $\dfrac{x^2 - 5x}{3x^2 - 13x - 10}$

89. $\dfrac{xy + 3x + 2y + 6}{xy + 3x + ay + 3a}$

90. $\dfrac{xy + 5x + 4y + 20}{x^2 + bx + 4x + 4b}$

Research Project 10

Although Pythagoras preceded William Shakespeare by twenty-one centuries, the philosophy of the Pythagoreans is mentioned in Shakespeare's *The Merchant of Venice*. Here is a quote from that play:

> Thou almost mak'st me waver in my faith,
>
> To hold opinion with Pythagoras,
>
> That souls of animals infuse themselves
>
> Into the trunks of men.

What part of the philosophy of the Pythagoreans was Shakespeare referring to with this quote? What present-day religions share a similar belief?

SECTION
7.2   Properties of Radicals

In this section we will consider the first part of what is called **simplified form** for radical expressions. A **radical expression** is any expression containing a radical, whether it is a square root, a cube root, or a higher root. Simplified form for a radical expression is the form that is easiest to work with. The first step in putting a radical expression in simplified form is to take as much out from under the radical sign as possible. To do this, we must first develop two properties of radicals in general.

Consider the following two problems:

$$\sqrt{9 \cdot 16} = \sqrt{144} = 12$$
$$\sqrt{9} \cdot \sqrt{16} = 3 \cdot 4 = 12$$

Since the answers to both are equal, the original problems must also be equal. That is, $\sqrt{9 \cdot 16} = \sqrt{9} \cdot \sqrt{16}$. We can generalize this property as follows.

> **Property 1 for Radicals**
> If x and y represent nonnegative real numbers, then it is always true that
> $$\sqrt{xy} = \sqrt{x}\,\sqrt{y}$$
> *In words:* The square root of a product is the product of the square roots.

We can use this property to simplify radical expressions.

▶ **EXAMPLE 1**   Simplify $\sqrt{20}$.

Solution   To simplify $\sqrt{20}$ we want to take as much out from under the radical sign as possible. We begin by looking for the largest perfect square that is a factor of 20. The largest perfect square that divides 20 is 4, so we write 20 as $4 \cdot 5$:

$$\sqrt{20} = \sqrt{4 \cdot 5}$$

Next we apply the first property of radicals and write

$$\sqrt{4 \cdot 5} = \sqrt{4}\,\sqrt{5}$$

And since $\sqrt{4} = 2$, we have

$$\sqrt{4}\,\sqrt{5} = 2\sqrt{5}$$

The expression $2\sqrt{5}$ is the simplified form of $\sqrt{20}$, since we have taken as much out from under the radical sign as possible.   ◀

Note   Working a problem like the one in Example 1 depends on recognizing the largest perfect square that divides (is a factor of) the number under the radical sign. The set of perfect squares is the set

$$\{1, 4, 9, 16, 25, 36, \ldots \}$$

To simplify an expression like $\sqrt{20}$, we must first find the largest number in this set that is a factor of the number under the radical sign.

▶ **EXAMPLE 2**   Simplify $\sqrt{75}$.

Solution   Since 25 is the largest perfect square that divides 75, we have

$$\begin{aligned}
\sqrt{75} &= \sqrt{25 \cdot 3} & &\text{Factor 75 into } 25 \cdot 3 \\
&= \sqrt{25}\,\sqrt{3} & &\text{Property 1 for radicals} \\
&= 5\sqrt{3} & &\sqrt{25} = 5
\end{aligned}$$

The expression $5\sqrt{3}$ is the simplified form for $\sqrt{75}$, since we have taken as much out from under the radical sign as possible.   ◀

The next two examples involve square roots of expressions that contain variables. Remember, we are assuming that all variables that appear under a radical sign represent positive numbers.

▶ **EXAMPLE 3**   Simplify $\sqrt{25x^3}$.

Solution   The largest perfect square that is a factor of $25x^3$ is $25x^2$. We write $25x^3$ as $25x^2 \cdot x$ and apply Property 1:

$$\sqrt{25x^3} = \sqrt{25x^2 \cdot x} \qquad \text{Factor } 25x^3 \text{ into } 25x^2 \cdot x$$
$$= \sqrt{25x^2}\sqrt{x} \qquad \text{Property 1 for radicals}$$
$$= 5x\sqrt{x} \qquad \sqrt{25x^2} = 5x$$

◀

▶ **EXAMPLE 4** Simplify $\sqrt{18y^4}$.

Solution The largest perfect square that is a factor of $18y^4$ is $9y^4$. We write $18y^4$ as $9y^4 \cdot 2$ and apply Property 1:

$$\sqrt{18y^4} = \sqrt{9y^4 \cdot 2} \qquad \text{Factor } 18y^4 \text{ into } 9y^4 \cdot 2$$
$$= \sqrt{9y^4}\sqrt{2} \qquad \text{Property 1 for radicals}$$
$$= 3y^2\sqrt{2} \qquad \sqrt{9y^4} = 3y^2$$

◀

▶ **EXAMPLE 5** Simplify $3\sqrt{32}$.

Solution We want to get as much out from under $\sqrt{32}$ as possible. Since 16 is the largest perfect square that divides 32, we have

$$3\sqrt{32} = 3\sqrt{16 \cdot 2} \qquad \text{Factor 32 into } 16 \cdot 2$$
$$= 3\sqrt{16}\sqrt{2} \qquad \text{Property 1 for radicals}$$
$$= 3 \cdot 4\sqrt{2} \qquad \sqrt{16} = 4$$
$$= 12\sqrt{2} \qquad 3 \cdot 4 = 12$$

◀

Although we have stated Property 1 for radicals in terms of square roots only, it holds for higher roots as well. If we were to state Property 1 again for cube roots, it would look like this:

$$\sqrt[3]{xy} = \sqrt[3]{x}\sqrt[3]{y}$$

▶ **EXAMPLE 6** Simplify $\sqrt[3]{24x^3}$.

Solution Since we are simplifying a cube root, we look for the largest perfect cube that is a factor of $24x^3$. Since 8 is a perfect cube, the largest perfect cube that is a factor of $24x^3$ is $8x^3$:

$$\sqrt[3]{24x^3} = \sqrt[3]{8x^3 \cdot 3} \qquad \text{Factor } 24x^3 \text{ into } 8x^3 \cdot 3$$
$$= \sqrt[3]{8x^3}\sqrt[3]{3} \qquad \text{Property 1 for radicals}$$
$$= 2x\sqrt[3]{3} \qquad \sqrt[3]{8x^3} = 2x$$

◀

The second property of radicals has to do with division. The property becomes apparent when we consider the following two problems:

$$\sqrt{\frac{64}{16}} = \sqrt{4} = 2 \qquad\qquad \frac{\sqrt{64}}{\sqrt{16}} = \frac{8}{4} = 2$$

Since the answers in each case are equal, the original problems must be also:

$$\sqrt{\frac{64}{16}} = \frac{\sqrt{64}}{\sqrt{16}}$$

Here is the property in general.

Property 2 for Radicals

If x and y both represent nonnegative real numbers and $y \neq 0$, then it is always true that

$$\sqrt{\frac{x}{y}} = \frac{\sqrt{x}}{\sqrt{y}}$$

In words: The square root of a quotient is the quotient of the square roots.

Although we have stated Property 2 for square roots only, it holds for higher roots as well.

We can use Property 2 for radicals in much the same way as we used Property 1 to simplify radical expressions.

▶ **EXAMPLE 7**   Simplify $\sqrt{\dfrac{49}{81}}$.

Solution   We begin by applying Property 2 for radicals to separate the fraction into two separate radicals. Then, we simplify each radical separately:

$$\sqrt{\frac{49}{81}} = \frac{\sqrt{49}}{\sqrt{81}} \qquad \text{Property 2 for radicals}$$

$$= \frac{7}{9} \qquad \sqrt{49} = 7 \text{ and } \sqrt{81} = 9 \qquad \blacktriangleleft$$

▶ **EXAMPLE 8**   Simplify $\sqrt[4]{\dfrac{81}{16}}$.

Solution   Remember, although Property 2 has been stated in terms of square roots, it holds for higher roots as well. Proceeding as we did in Example 7, we have

$$\sqrt[4]{\frac{81}{16}} = \frac{\sqrt[4]{81}}{\sqrt[4]{16}} \qquad \text{Property 2}$$

$$= \frac{3}{2} \qquad \sqrt[4]{81} = 3 \text{ and } \sqrt[4]{16} = 2 \qquad \blacktriangleleft$$

▶ **EXAMPLE 9** Simplify $\sqrt{\dfrac{50}{49}}$.

Solution Applying Property 2 for radicals and then simplifying each resulting radical separately, we have

$$\sqrt{\frac{50}{49}} = \frac{\sqrt{50}}{\sqrt{49}} \qquad \text{Property 2 for radicals}$$

$$= \frac{\sqrt{25 \cdot 2}}{7} \qquad \begin{array}{l} \text{Factor } 50 = 25 \cdot 2 \\ \sqrt{49} = 7 \end{array}$$

$$= \frac{\sqrt{25}\sqrt{2}}{7} \qquad \text{Property 1 for radicals}$$

$$= \frac{5\sqrt{2}}{7} \qquad \sqrt{25} = 5 \qquad \blacktriangleleft$$

▶ **EXAMPLE 10** Simplify $\sqrt{\dfrac{12x^2}{25}}$.

Solution Proceeding as we have in the previous three examples, we use Property 2 for radicals to separate the numerator and denominator into two separate radicals. Then, we simplify each radical separately:

$$\sqrt{\frac{12x^2}{25}} = \frac{\sqrt{12x^2}}{\sqrt{25}} \qquad \text{Property 2 for radicals}$$

$$= \frac{\sqrt{4x^2 \cdot 3}}{5} \qquad \begin{array}{l} \text{Factor } 12x^2 = 4x^2 \cdot 3 \\ \sqrt{25} = 5 \end{array}$$

$$= \frac{\sqrt{4x^2}\sqrt{3}}{5} \qquad \text{Property 1 for radicals}$$

$$= \frac{2x\sqrt{3}}{5} \qquad \sqrt{4x^2} = 2x \qquad \blacktriangleleft$$

▶ **EXAMPLE 11** Simplify $\sqrt{\dfrac{50x^3y^2}{49}}$.

Solution We begin by taking the square roots of $50x^3y^2$ and 49 separately and then writing $\sqrt{49}$ as 7:

$$\sqrt{\frac{50x^3y^2}{49}} = \frac{\sqrt{50x^3y^2}}{\sqrt{49}} \qquad \text{Property 2 for radicals}$$

$$= \frac{\sqrt{50x^3y^2}}{7} \qquad \sqrt{49} = 7$$

To simplify the numerator of this last expression, we determine that the largest perfect square that is a factor of $50x^3y^2$ is $25x^2y^2$. Continuing, we have

$$= \frac{\sqrt{25x^2y^2 \cdot 2x}}{7} \qquad \begin{array}{l} \text{Factor } 50x^3y^2 \text{ into} \\ \qquad 25x^2y^2 \cdot 2x \end{array}$$

$$= \frac{\sqrt{25x^2y^2}\,\sqrt{2x}}{7} \qquad \text{Property 1}$$

$$= \frac{5xy\sqrt{2x}}{7} \qquad \sqrt{25x^2y^2} = 5xy$$

◀

PROBLEM SET 7.2

Use Property 1 for radicals to simplify the following radical expressions as much as possible. Assume that all variables represent positive numbers.

1. $\sqrt{8}$ **2.** $\sqrt{18}$

3. $\sqrt{12}$ **4.** $\sqrt{27}$

5. $\sqrt[3]{24}$ **6.** $\sqrt[3]{54}$

7. $\sqrt{50x^2}$ **8.** $\sqrt{32x^2}$

9. $\sqrt{45a^2b^2}$ **10.** $\sqrt{128a^2b^2}$

11. $\sqrt[3]{54x^3}$ **12.** $\sqrt[3]{128x^3}$

13. $\sqrt{32x^4}$ **14.** $\sqrt{48x^4}$

15. $5\sqrt{80}$ **16.** $3\sqrt{125}$

17. $\frac{1}{2}\sqrt{28x^3}$ **18.** $\frac{2}{3}\sqrt{54x^3}$

19. $x\sqrt[3]{8x^4}$ **20.** $x\sqrt[3]{8x^5}$

21. $2a\sqrt[3]{27a^5}$ **22.** $3a\sqrt[3]{27a^4}$

23. $\frac{4}{3}\sqrt{45a^3}$ **24.** $\frac{3}{5}\sqrt{300a^3}$

25. $3\sqrt{50xy^2}$ **26.** $4\sqrt{18xy^2}$

27. $7\sqrt{12x^2y}$ **28.** $6\sqrt{20x^2y}$

Use Property 2 for radicals to simplify each of the following. Assume all variables represent positive numbers.

29. $\sqrt{\dfrac{16}{25}}$ **30.** $\sqrt{\dfrac{81}{64}}$

31. $\sqrt{\dfrac{4}{9}}$ **32.** $\sqrt{\dfrac{49}{16}}$

33. $\sqrt[3]{\dfrac{8}{27}}$ **34.** $\sqrt[3]{\dfrac{64}{27}}$

35. $\sqrt[4]{\dfrac{16}{81}}$ **36.** $\sqrt[4]{\dfrac{81}{16}}$

37. $\sqrt{\dfrac{100x^2}{25}}$ **38.** $\sqrt{\dfrac{100x^2}{4}}$

39. $\sqrt{\dfrac{81a^2b^2}{9}}$ **40.** $\sqrt{\dfrac{64a^2b^2}{16}}$

41. $\sqrt[3]{\dfrac{27x^3}{8y^3}}$ **42.** $\sqrt[3]{\dfrac{125x^3}{64y^3}}$

Use combinations of Properties 1 and 2 for radicals to simplify the following problems as much as possible. Assume all variables represent positive numbers.

43. $\sqrt{\dfrac{50}{9}}$ **44.** $\sqrt{\dfrac{32}{49}}$

45. $\sqrt{\dfrac{75}{25}}$ **46.** $\sqrt{\dfrac{300}{4}}$

47. $\sqrt{\dfrac{128}{49}}$

48. $\sqrt{\dfrac{32}{64}}$

49. $\sqrt{\dfrac{288x}{25}}$

50. $\sqrt{\dfrac{28y}{81}}$

51. $\sqrt{\dfrac{54a^2}{25}}$

52. $\sqrt{\dfrac{243a^2}{49}}$

53. $\dfrac{3\sqrt{50}}{2}$

54. $\dfrac{5\sqrt{48}}{3}$

55. $\dfrac{7\sqrt{28y^2}}{3}$

56. $\dfrac{9\sqrt{243x^2}}{2}$

57. $\dfrac{5\sqrt{72a^2b^2}}{\sqrt{36}}$

58. $\dfrac{2\sqrt{27a^2b^2}}{\sqrt{9}}$

59. $\dfrac{6\sqrt{8x^2y}}{\sqrt{4}}$

60. $\dfrac{5\sqrt{32xy^2}}{\sqrt{25}}$

61. $\dfrac{8\sqrt{12x^2y^3}}{\sqrt{100}}$

62. $\dfrac{6\sqrt{18x^3y^2}}{\sqrt{81}}$

Applying the Concepts

63. Use a calculator to find decimal approximations for $\sqrt{12}$ and $2\sqrt{3}$. Round your answers to three places past the decimal point.

64. Use a calculator to find decimal approximations for $\sqrt{50}$ and $5\sqrt{2}$. Round your answers to three places past the decimal point.

65. The formula that gives the number of seconds, t, it takes for an object to reach the ground when dropped from the top of a building h feet high is

$$t = \sqrt{\dfrac{h}{16}}$$

If a rock is dropped from the top of a building 25 feet high, how long will it take for the rock to hit the ground?

66. Using the formula given in Problem 65, how long will it take for an object dropped from a building 100 feet high to reach the ground?

Review Problems

The following problems review material we covered in Section 6.2.

Multiply or divide as indicated.

67. $\dfrac{8x}{x^2 - 5x} \cdot \dfrac{x^2 - 25}{4x^2 + 4x}$

68. $\dfrac{x^2 + 4x}{4x^2} \cdot \dfrac{7x}{x^2 + 5x}$

69. $\dfrac{x^2 + 3x - 4}{3x^2 + 7x - 20} \div \dfrac{x^2 - 2x + 1}{3x^2 - 2x - 5}$

70. $\dfrac{x^2 - 16}{2x^2 + 7x + 3} \div \dfrac{2x^2 - 5x - 12}{4x^2 + 8x + 3}$

71. $(x^2 - 36)\left(\dfrac{x + 3}{x - 6}\right)$

72. $(x^2 - 49)\left(\dfrac{x + 5}{x + 7}\right)$

Simplified Form for Radicals

A radical expression is in simplified form if it has three special characteristics. Radical expressions that are in simplified form are generally easier to work with.

DEFINITION A radical expression is in **simplified form** if

1. there are no perfect squares larger than 1 that are factors of the quantity under the square root sign, no perfect cubes that are factors of the quantity under the cube root sign, and so on. We want as little as possible under the radical sign.
2. there are no fractions under the radical sign.
3. there are no radicals in the denominator.

A radical expression that has these three characteristics is said to be in simplified form. As we will see, simplified form is not always the least complicated expression. In many cases, the simplified expression looks more complicated than the original expression. The important thing about simplified form for radicals is that simplified expressions are easier to work with.

The tools we will use to put radical expressions into simplified form are the properties of radicals. We list the properties again for clarity.

Properties of Radicals

If a and b represent any two nonnegative real numbers, then it is always true that

1. $\sqrt{a}\sqrt{b} = \sqrt{a \cdot b}$

2. $\dfrac{\sqrt{a}}{\sqrt{b}} = \sqrt{\dfrac{a}{b}}$ $\qquad b \neq 0$

3. $\sqrt{a}\sqrt{a} = (\sqrt{a})^2 = a$ $\qquad$ This property comes directly from the definition of radicals

---

The following examples illustrate how we put a radical expression into simplified form using the three properties of radicals. Although the properties are stated for square roots only, they hold for all roots. [Property 3 written for cube roots would be $\sqrt[3]{a}\sqrt[3]{a}\sqrt[3]{a} = (\sqrt[3]{a})^3 = a$.]

▶ **EXAMPLE 1** Put $\sqrt{\dfrac{1}{2}}$ into simplified form.

Solution The expression $\sqrt{\dfrac{1}{2}}$ is not in simplified form, because there is a fraction under the radical sign. We can change this by applying Property 2 for radicals:

$$\sqrt{\frac{1}{2}} = \frac{\sqrt{1}}{\sqrt{2}} \qquad \text{Property 2 for radicals}$$

$$= \frac{1}{\sqrt{2}} \qquad \sqrt{1} = 1$$

The expression $\dfrac{1}{\sqrt{2}}$ is not in simplified form, because there is a radical sign

in the denominator. If we multiply the numerator and denominator of $\dfrac{1}{\sqrt{2}}$ by

$\sqrt{2}$, the denominator becomes $\sqrt{2} \cdot \sqrt{2} = 2$:

$$\frac{1}{\sqrt{2}} = \frac{1}{\sqrt{2}} \cdot \frac{\sqrt{2}}{\sqrt{2}} \qquad \text{Multiply numerator and}$$
$$\text{denominator by } \sqrt{2}$$

$$= \frac{\sqrt{2}}{2} \qquad \begin{aligned} 1 \cdot \sqrt{2} &= \sqrt{2} \\ \sqrt{2} \cdot \sqrt{2} &= \sqrt{4} = 2 \end{aligned} \qquad \blacktriangleleft$$

If we check the expression $\dfrac{\sqrt{2}}{2}$ against our definition of simplified form

for radicals, we find that all three rules hold. There are no perfect squares larger

than 1 that are factors of 2. There are no fractions under the radical sign. No

radicals appear in the denominator. The expression $\dfrac{\sqrt{2}}{2}$, therefore, must be in

simplified form.

▶ **EXAMPLE 2** Write $\sqrt{\tfrac{2}{3}}$ in simplified form.

Solution We proceed as we did in Example 1:

$$\sqrt{\frac{2}{3}} = \frac{\sqrt{2}}{\sqrt{3}} \qquad \text{Use Property 2 to separate radicals}$$

$$= \frac{\sqrt{2}}{\sqrt{3}} \cdot \frac{\sqrt{3}}{\sqrt{3}} \qquad \text{Multiply by } \frac{\sqrt{3}}{\sqrt{3}} \text{ to remove}$$
$$\text{the radical from the denominator}$$

$$= \frac{\sqrt{6}}{3} \qquad \begin{aligned} \sqrt{2} \cdot \sqrt{3} &= \sqrt{6} \\ \sqrt{3} \cdot \sqrt{3} &= \sqrt{9} = 3 \end{aligned} \qquad \blacktriangleleft$$

▶ **EXAMPLE 3** Put the expression $\dfrac{6\sqrt{20}}{2\sqrt{5}}$ into simplified form.

Solution Although there are many ways to begin this problem, we notice
that 20 is divisible by 5. Using Property 2 for radicals as the first step, we can
quickly put the expression into simplified form:

$$\frac{6\sqrt{20}}{2\sqrt{5}} = \frac{6}{2}\sqrt{\frac{20}{5}} \qquad \text{Property 2 for radicals}$$

$$= 3\sqrt{4} \qquad \frac{20}{5} = 4$$

$$= 3 \cdot 2 \qquad \sqrt{4} = 2$$

$$= 6 \qquad \blacktriangleleft$$

Remember, when working with square roots of variable quantities, we will always assume that the variables represent positive numbers. That way we can say that $\sqrt{x^2} = x$. Our next example uses this fact.

▶ **EXAMPLE 4** Simplify $\sqrt{\dfrac{4x^3y^2}{3}}$.

Solution We begin by separating the numerator and denominator and then taking the perfect squares out of the numerator:

$$\sqrt{\frac{4x^3y^2}{3}} = \frac{\sqrt{4x^3y^2}}{\sqrt{3}} \qquad \text{Property 2 for radicals}$$

$$= \frac{\sqrt{4x^2y^2}\sqrt{x}}{\sqrt{3}} \qquad \text{Property 1 for radicals}$$

$$= \frac{2xy\sqrt{x}}{\sqrt{3}} \qquad \sqrt{4x^2y^2} = 2xy$$

The only thing keeping our expression from being in simplified form is the $\sqrt{3}$ in the denominator. We can take care of this by multiplying the numerator and denominator by $\sqrt{3}$:

$$\frac{2xy\sqrt{x}}{\sqrt{3}} = \frac{2xy\sqrt{x}}{\sqrt{3}} \cdot \frac{\sqrt{3}}{\sqrt{3}} \qquad \begin{array}{l}\text{Multiply numerator and} \\ \text{denominator by } \sqrt{3}\end{array}$$

$$= \frac{2xy\sqrt{3x}}{3} \qquad \sqrt{3} \cdot \sqrt{3} = \sqrt{9} = 3 \qquad \blacktriangleleft$$

Although the final expression may look as complicated as the original expression, it is in simplified form. The last step is called **rationalizing the denominator.** We have taken the radical out of the denominator and replaced it with a rational number.

▶ **EXAMPLE 5** Simplify $\sqrt[3]{\dfrac{2}{3}}$.

Solution We can apply Property 2 first to separate the cube roots:

$$\sqrt[3]{\frac{2}{3}} = \frac{\sqrt[3]{2}}{\sqrt[3]{3}}$$

To write this expression in simplified form, we must remove the radical from the denominator. Since the radical is a cube root, we will need to multiply it by an expression that will give us a perfect cube under that cube root sign. We can accomplish this by multiplying the numerator and denominator by $\sqrt[3]{9}$. Here is what it looks like:

$$\frac{\sqrt[3]{2}}{\sqrt[3]{3}} = \frac{\sqrt[3]{2}}{\sqrt[3]{3}} \cdot \frac{\sqrt[3]{9}}{\sqrt[3]{9}}$$

$$= \frac{\sqrt[3]{18}}{\sqrt[3]{27}} \qquad \begin{array}{l} \sqrt[3]{2} \cdot \sqrt[3]{9} = \sqrt[3]{18} \\ \sqrt[3]{3} \cdot \sqrt[3]{9} = \sqrt[3]{27} \end{array}$$

$$= \frac{\sqrt[3]{18}}{3} \qquad \sqrt[3]{27} = 3$$

To see why multiplying numerator and denominator by $\sqrt[3]{9}$ works in this example, you must first convince yourself that multiplying numerator and denominator by $\sqrt[3]{3}$ would not have worked. ◄

▶ **EXAMPLE 6** Simplify $\sqrt[3]{\frac{1}{4}}$.

Solution We begin by separating the numerator and denominator:

$$\sqrt[3]{\frac{1}{4}} = \frac{\sqrt[3]{1}}{\sqrt[3]{4}} \qquad \text{Property 2 for radicals}$$

$$= \frac{1}{\sqrt[3]{4}} \qquad \sqrt[3]{1} = 1$$

To rationalize the denominator, we need to have a perfect cube under the cube root sign. If we multiply numerator and denominator by $\sqrt[3]{2}$, we will have $\sqrt[3]{4} \cdot \sqrt[3]{2} = \sqrt[3]{8}$ in the denominator:

$$\frac{1}{\sqrt[3]{4}} = \frac{1}{\sqrt[3]{4}} \cdot \frac{\sqrt[3]{2}}{\sqrt[3]{2}} \qquad \begin{array}{l} \text{Multiply numerator and} \\ \text{denominator by } \sqrt[3]{2} \end{array}$$

$$= \frac{\sqrt[3]{2}}{\sqrt[3]{8}} \qquad \sqrt[3]{4} \cdot \sqrt[3]{2} = \sqrt[3]{8}$$

$$= \frac{\sqrt[3]{2}}{2} \qquad \sqrt[3]{8} = 2$$

The final expression has no radical sign in the denominator and therefore is in simplified form. ◄

The Spiral of Roots

In order to visualize the square roots of the positive integers we can construct the *spiral of roots,* which we mentioned in the introduction to this chapter. To begin, we draw two line segments, each of length 1, at right angles to each other. Then we use the Pythagorean Theorem to find the length of the diagonal. Figure 1 illustrates:

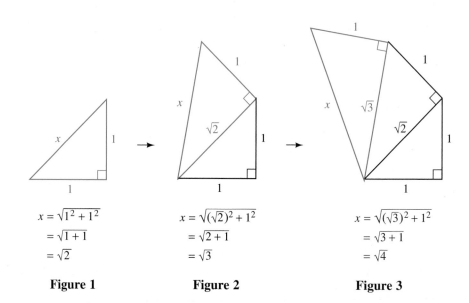

$$x = \sqrt{1^2 + 1^2}$$
$$= \sqrt{1 + 1}$$
$$= \sqrt{2}$$

Figure 1

$$x = \sqrt{(\sqrt{2})^2 + 1^2}$$
$$= \sqrt{2 + 1}$$
$$= \sqrt{3}$$

Figure 2

$$x = \sqrt{(\sqrt{3})^2 + 1^2}$$
$$= \sqrt{3 + 1}$$
$$= \sqrt{4}$$

Figure 3

Next, we construct a second triangle by connecting a line segment of length 1 to the end of the first diagonal so that the angle formed is a right angle. We find the length of the second diagonal using the Pythagorean Theorem. Figure 2 illustrates this procedure. Continuing to draw new triangles by connecting line segments of length 1 to the end of each new diagonal, so that the angle formed is a right angle, the spiral of roots begins to appear.

PROBLEM SET 7.3

Put each of the following radical expressions into simplified form. Assume all variables represent positive numbers.

1. $\sqrt{\dfrac{1}{2}}$

2. $\sqrt{\dfrac{1}{5}}$

3. $\sqrt{\dfrac{1}{3}}$

4. $\sqrt{\dfrac{1}{6}}$

5. $\sqrt{\dfrac{2}{5}}$

6. $\sqrt{\dfrac{3}{7}}$

7. $\sqrt{\dfrac{3}{2}}$

8. $\sqrt{\dfrac{5}{3}}$

9. $\sqrt{\dfrac{20}{3}}$

10. $\sqrt{\dfrac{32}{5}}$

11. $\sqrt{\dfrac{45}{6}}$

12. $\sqrt{\dfrac{48}{7}}$

13. $\sqrt{\dfrac{20}{5}}$

14. $\sqrt{\dfrac{12}{3}}$

15. $\dfrac{\sqrt{21}}{\sqrt{3}}$

16. $\dfrac{\sqrt{21}}{\sqrt{7}}$

17. $\dfrac{\sqrt{35}}{\sqrt{7}}$

18. $\dfrac{\sqrt{35}}{\sqrt{5}}$

19. $\dfrac{10\sqrt{15}}{5\sqrt{3}}$

20. $\dfrac{4\sqrt{12}}{8\sqrt{3}}$

21. $\dfrac{6\sqrt{21}}{3\sqrt{7}}$

22. $\dfrac{8\sqrt{50}}{16\sqrt{2}}$

23. $\dfrac{6\sqrt{35}}{12\sqrt{5}}$

24. $\dfrac{8\sqrt{35}}{16\sqrt{7}}$

25. $\sqrt{\dfrac{4x^2y^2}{2}}$

26. $\sqrt{\dfrac{9x^2y^2}{3}}$

27. $\sqrt{\dfrac{5x^2y}{3}}$

28. $\sqrt{\dfrac{7x^2y}{5}}$

29. $\sqrt{\dfrac{16a^4}{5}}$

30. $\sqrt{\dfrac{25a^4}{7}}$

31. $\sqrt{\dfrac{72a^5}{5}}$

32. $\sqrt{\dfrac{12a^5}{5}}$

33. $\sqrt{\dfrac{20x^2y^3}{3}}$

34. $\sqrt{\dfrac{27x^2y^3}{2}}$

35. $\dfrac{2\sqrt{20x^2y^3}}{3}$

36. $\dfrac{5\sqrt{27x^3y^2}}{2}$

37. $\dfrac{6\sqrt{54a^2b^3}}{5}$

38. $\dfrac{7\sqrt{75a^3b^2}}{6}$

39. $\dfrac{3\sqrt{72x^4}}{\sqrt{2x}}$

40. $\dfrac{2\sqrt{45x^4}}{\sqrt{5x}}$

41. $\sqrt[3]{\dfrac{1}{2}}$

42. $\sqrt[3]{\dfrac{1}{4}}$

43. $\sqrt[3]{\dfrac{1}{9}}$

44. $\sqrt[3]{\dfrac{1}{3}}$

45. $\sqrt[3]{\dfrac{3}{2}}$

46. $\sqrt[3]{\dfrac{7}{9}}$

47. $\sqrt[3]{\dfrac{3}{4}}$

48. $\sqrt[3]{\dfrac{5}{3}}$

Applying the Concepts

49. A calculator gives the four-digit decimal approximation of $\sqrt{2}$ as 1.414. Use this number to find approximations for the expressions $\dfrac{1}{\sqrt{2}}$ and $\dfrac{\sqrt{2}}{2}$. Round your answers to three places past the decimal point.

50. Find the decimal approximation for $\sqrt{3}$ and use it to find decimal approximations for $\dfrac{1}{\sqrt{3}}$ and $\dfrac{\sqrt{3}}{3}$. Round your answers to three places past the decimal point.

51. The higher you are above the surface of the earth, the farther you can see. The formula that gives the approximate distance in miles, d, that you can see from a height of h feet above the surface of the earth is

$$d = \sqrt{\dfrac{3h}{2}}$$

How far can you see from a window that is 24 feet above the ground? (This is assuming that your view is unobstructed.)

52. Using the formula given in Problem 51, how far can you see from the window of an airplane flying at 6,000 feet?

53. Construct your own spiral of roots by using a ruler. Draw the first triangle by using two 1-inch lines. The first diagonal will have a length of $\sqrt{2}$ inches. Each new triangle will be formed by drawing a 1-inch line segment at the end of the previous diagonal so that the angle formed is 90°. Continue until you see $\sqrt{6}$.

54. Construct a spiral of roots by using line segments of length 2 inches. The length of the first diagonal will be $2\sqrt{2}$ inches. The length of the second diagonal will be $2\sqrt{3}$ inches.

55. Simplify the terms in the sequence below. The result will be a sequence that gives the lengths of the diagonals in the spiral of roots you constructed in Problem 53.

$$\sqrt{1^2 + 1},\ \sqrt{(\sqrt{2})^2 + 1},\ \sqrt{(\sqrt{3})^2 + 1},\ \dots$$

56. Simplify the terms in the following sequence. The result will be a sequence that gives the lengths of the diagonals in the spiral of roots you constructed in Problem 54.

$$\sqrt{2^2 + 4}, \sqrt{(2\sqrt{2})^2 + 4}, \sqrt{(2\sqrt{3})^2 + 4}, \ldots$$

Review Problems

The following problems review material we covered in Sections 2.1 and 6.3. Reviewing the problems from Section 2.1 will help you understand the next section.

Use the distributive property to combine the following. [2.1]

57. $3x + 7x$ **58.** $3x - 7x$

59. $15x + 8x$ **60.** $15x - 8x$

61. $7a - 3a + 6a$ **62.** $25a + 3a - a$

Add or subtract as indicated. [6.3]

63. $\dfrac{x^2}{x+5} + \dfrac{10x+25}{x+5}$ **64.** $\dfrac{x^2}{x-3} - \dfrac{9}{x-3}$

65. $\dfrac{a}{3} + \dfrac{2}{5}$ **66.** $\dfrac{4}{a} + \dfrac{2}{3}$

67. $\dfrac{6}{a^2-9} - \dfrac{5}{a^2-a-6}$

68. $\dfrac{4a}{a^2+6a+5} - \dfrac{3a}{a^2+5a+4}$

SECTION 7.4   Addition and Subtraction of Radical Expressions

To add two or more radical expressions we apply the distributive property. Adding radical expressions is very similar to adding similar terms of polynomials.

▶ **EXAMPLE 1** Combine terms in the expression $3\sqrt{5} - 7\sqrt{5}$.

Solution The two terms $3\sqrt{5}$ and $7\sqrt{5}$ each have $\sqrt{5}$ in common. Since $3\sqrt{5}$ means 3 times $\sqrt{5}$, or $3 \cdot \sqrt{5}$, we apply the distributive property:

$$3\sqrt{5} - 7\sqrt{5} = (3-7)\sqrt{5} \quad \text{Distributive property}$$
$$= -4\sqrt{5} \quad 3 - 7 = -4 \quad ◀$$

Since we use the distributive property to add radical expressions, each expression must contain exactly the same radical.

▶ **EXAMPLE 2** Combine terms in the expression $7\sqrt{2} - 3\sqrt{2} + 6\sqrt{2}$.

Solution

$$7\sqrt{2} - 3\sqrt{2} + 6\sqrt{2} = (7-3+6)\sqrt{2} \quad \text{Distributive property}$$
$$= 10\sqrt{2} \quad \text{Addition} \quad ◀$$

In Examples 1 and 2 each term was a radical expression in simplified form. If one or more terms are not in simplified form, we must put them into simplified form and then combine terms, if possible. It is not always possible to combine terms containing radicals. Occasionally two or more of the terms will not have a radical in common. If there is a possibility of combining terms, it will always become apparent when each term is in simplified form.

Rule

To combine two or more radical expressions, put each expression in simplified form and then apply the distributive property, if possible.

▶ **EXAMPLE 3** Combine terms in the expression $3\sqrt{50} + 2\sqrt{32}$.

Solution We begin by putting each term into simplified form:

$$3\sqrt{50} + 2\sqrt{32} = 3\sqrt{25}\sqrt{2} + 2\sqrt{16}\sqrt{2} \quad \text{Property 1 for radicals}$$
$$= 3 \cdot 5\sqrt{2} + 2 \cdot 4\sqrt{2} \quad \sqrt{25} = 5 \text{ and } \sqrt{16} = 4$$
$$= 15\sqrt{2} + 8\sqrt{2} \quad \text{Multiplication}$$

Applying the distributive property to the last line, we have

$$15\sqrt{2} + 8\sqrt{2} = (15 + 8)\sqrt{2} \quad \text{Distributive property}$$
$$= 23\sqrt{2} \quad 15 + 8 = 23 \quad \blacktriangleleft$$

▶ **EXAMPLE 4** Combine terms in the expression $5\sqrt{75} + \sqrt{27} - \sqrt{3}$.

Solution

$$5\sqrt{75} + \sqrt{27} - \sqrt{3}$$
$$= 5\sqrt{25}\sqrt{3} + \sqrt{9}\sqrt{3} - \sqrt{3} \quad \text{Property 1 for radicals}$$
$$= 5 \cdot 5\sqrt{3} + 3\sqrt{3} - \sqrt{3} \quad \sqrt{25} = 5 \text{ and } \sqrt{9} = 3$$
$$= 25\sqrt{3} + 3\sqrt{3} - \sqrt{3} \quad 5 \cdot 5 = 25$$
$$= (25 + 3 - 1)\sqrt{3} \quad \text{Distributive property}$$
$$= 27\sqrt{3} \quad \text{Addition} \quad \blacktriangleleft$$

The most time-consuming part of combining most radical expressions is simplifying each term in the expression. Once this has been done, applying the distributive property is simple and fast.

▶ **EXAMPLE 5** Simplify $a\sqrt{12} + 5\sqrt{3a^2}$.

Solution We must assume that a represents a positive number. Then we simplify each term in the expression by putting it in simplified form for radicals:

$$\begin{aligned}
a\sqrt{12} + 5\sqrt{3a^2} &= a\sqrt{4}\sqrt{3} + 5\sqrt{a^2}\sqrt{3} &&\text{Property 1 for radicals} \\
&= a \cdot 2\sqrt{3} + 5 \cdot a\sqrt{3} &&\sqrt{4} = 2 \text{ and } \sqrt{a^2} = a \\
&= 2a\sqrt{3} + 5a\sqrt{3} &&\text{Commutative property} \\
&= (2a + 5a)\sqrt{3} &&\text{Distributive property} \\
&= 7a\sqrt{3} &&\text{Addition}
\end{aligned}$$

▶ **EXAMPLE 6** Combine terms in the expression

$$\sqrt{20x^3} - 3x\sqrt{45x} + 10\sqrt{25x^2}$$

(Assume x is a positive real number.)

Solution

$$\begin{aligned}
\sqrt{20x^3} &- 3x\sqrt{45x} + 10\sqrt{25x^2} \\
&= \sqrt{4x^2}\sqrt{5x} - 3x\sqrt{9}\sqrt{5x} + 10\sqrt{25x^2} \\
&= 2x\sqrt{5x} - 3x \cdot 3\sqrt{5x} + 10 \cdot 5x \\
&= 2x\sqrt{5x} - 9x\sqrt{5x} + 50x
\end{aligned}$$

Each term is now in simplified form. The best we can do next is to combine the first two terms. The last term does not have the common radical $\sqrt{5x}$.

$$\begin{aligned}
2x\sqrt{5x} - 9x\sqrt{5x} + 50x &= (2x - 9x)\sqrt{5x} + 50x \\
&= -7x\sqrt{5x} + 50x
\end{aligned}$$

We have, in any case, succeeded in reducing the number of terms in our original problem.

Our next example involves an expression that is similar to many of the expressions we will find when we solve quadratic equations in Chapter 8.

▶ **EXAMPLE 7** Simplify $\dfrac{6 + \sqrt{12}}{4}$.

Solution We begin by writing $\sqrt{12}$ as $2\sqrt{3}$. (Remember, $\sqrt{12} = \sqrt{4 \cdot 3} = \sqrt{4}\sqrt{3} = 2\sqrt{3}$.)

$$\frac{6 + \sqrt{12}}{4} = \frac{6 + 2\sqrt{3}}{4}$$

Next, we factor 2 from the numerator and denominator and then reduce to lowest terms.

$$\begin{aligned}
\frac{6 + 2\sqrt{3}}{4} &= \frac{\cancel{2}(3 + \sqrt{3})}{\cancel{2} \cdot 2} \\
&= \frac{3 + \sqrt{3}}{2}
\end{aligned}$$

P R O B L E M   S E T   7 . 4

In each of the following problems, simplify each term, if necessary, and then use the distributive property to combine terms, if possible.

1. $3\sqrt{2} + 4\sqrt{2}$

2. $7\sqrt{3} + 2\sqrt{3}$

3. $9\sqrt{5} - 7\sqrt{5}$

4. $6\sqrt{7} - 10\sqrt{7}$

5. $\sqrt{3} + 6\sqrt{3}$

6. $\sqrt{2} + 10\sqrt{2}$

7. $\frac{5}{8}\sqrt{5} - \frac{3}{7}\sqrt{5}$

8. $\frac{5}{6}\sqrt{11} - \frac{7}{9}\sqrt{11}$

9. $14\sqrt{13} - \sqrt{13}$

10. $-2\sqrt{6} - 9\sqrt{6}$

11. $-3\sqrt{10} + 9\sqrt{10}$

12. $11\sqrt{11} + \sqrt{11}$

13. $5\sqrt{5} + \sqrt{5}$

14. $\sqrt{6} - 10\sqrt{6}$

15. $\sqrt{8} + 2\sqrt{2}$

16. $\sqrt{20} + 3\sqrt{5}$

17. $3\sqrt{3} - \sqrt{27}$

18. $4\sqrt{5} - \sqrt{80}$

19. $5\sqrt{12} - 10\sqrt{48}$

20. $3\sqrt{300} - 5\sqrt{27}$

21. $-\sqrt{75} - \sqrt{3}$

22. $5\sqrt{20} + 8\sqrt{80}$

23. $\frac{1}{5}\sqrt{75} - \frac{1}{2}\sqrt{12}$

24. $\frac{1}{2}\sqrt{24} + \frac{1}{5}\sqrt{150}$

25. $\frac{3}{4}\sqrt{8} + \frac{3}{10}\sqrt{75}$

26. $\frac{5}{6}\sqrt{54} - \frac{3}{4}\sqrt{24}$

27. $\sqrt{27} - 2\sqrt{12} + \sqrt{3}$

28. $\sqrt{20} + 3\sqrt{45} - \sqrt{5}$

29. $\frac{5}{6}\sqrt{72} - \frac{3}{8}\sqrt{8} + \frac{3}{10}\sqrt{50}$

30. $\frac{3}{4}\sqrt{24} - \frac{5}{6}\sqrt{54} - \frac{7}{10}\sqrt{150}$

31. $5\sqrt{7} + 2\sqrt{28} - 4\sqrt{63}$

32. $3\sqrt{3} - 5\sqrt{27} + 8\sqrt{75}$

33. $6\sqrt{48} - 2\sqrt{12} + 5\sqrt{27}$

34. $5\sqrt{50} + 8\sqrt{12} - \sqrt{32}$

35. $6\sqrt{48} - \sqrt{72} - 3\sqrt{300}$

36. $7\sqrt{44} - 8\sqrt{99} + \sqrt{176}$

All variables in the following problems represent positive real numbers. Simplify each term and combine, if possible.

37. $\sqrt{x^3} + x\sqrt{x}$

38. $2\sqrt{x} - 2\sqrt{4x}$

39. $5\sqrt{3a^2} - a\sqrt{3}$

40. $6a\sqrt{a} + 7\sqrt{a^3}$

41. $5\sqrt{8x^3} + x\sqrt{50x}$

42. $2\sqrt{27x^2} - x\sqrt{48}$

43. $3\sqrt{75x^3y} - 2x\sqrt{3xy}$

44. $9\sqrt{24x^3y^2} - 5x\sqrt{54xy^2}$

45. $\sqrt{20ab^2} - b\sqrt{45a}$

46. $4\sqrt{a^3b^2} - 5a\sqrt{ab^2}$

47. $9\sqrt{18x^3} - 2x\sqrt{48x}$

48. $8\sqrt{72x^2} - x\sqrt{8}$

49. $7\sqrt{50x^2y} + 8x\sqrt{8y} - 7\sqrt{32x^2y}$

50. $6\sqrt{44x^3y^3} - 8x\sqrt{99xy^3} - 6y\sqrt{176x^3y}$

Simplify each expression.

51. $\dfrac{8 - \sqrt{24}}{6}$

52. $\dfrac{8 + \sqrt{48}}{8}$

53. $\dfrac{6 + \sqrt{8}}{2}$

54. $\dfrac{4 - \sqrt{12}}{2}$

55. $\dfrac{-10 + \sqrt{50}}{10}$

56. $\dfrac{-12 + \sqrt{20}}{6}$

57. Use a calculator to find a decimal approximation of the expression $\sqrt{5} + \sqrt{3}$. Is the answer equal to the decimal approximation of $\sqrt{8}$?

58. Find a decimal approximation of the expression $\sqrt{5} - \sqrt{3}$. Is the answer equal to the decimal approximation of $\sqrt{2}$?

59. The statement below is false. Correct the right side to make the statement true.

$$4\sqrt{3} + 5\sqrt{3} = 9\sqrt{6}$$

60. The statement below is false. Correct the right side to make the statement true.

$$7\sqrt{5} - 3\sqrt{5} = 4\sqrt{25}$$

Review Problems

The following problems review material we covered in Sections 4.6 and 6.4. Reviewing the problems from Section 4.6 will help you in the next section.

Multiply. [4.6]

61. $(3x + y)^2$

62. $(2x - 3y)^2$

63. $(3x - 4y)(3x + 4y)$

64. $(7x + 2y)(7x - 2y)$

Solve each equation. [6.4]

65. $\dfrac{x}{3} - \dfrac{1}{2} = \dfrac{5}{2}$

66. $\dfrac{3}{x} + \dfrac{1}{5} = \dfrac{4}{5}$

69. $\dfrac{a}{a-4} - \dfrac{a}{2} = \dfrac{4}{a-4}$

67. $1 - \dfrac{5}{x} = \dfrac{-6}{x^2}$

68. $1 - \dfrac{1}{x} = \dfrac{6}{x^2}$

70. $\dfrac{a}{a-3} - \dfrac{a}{2} = \dfrac{3}{a-3}$

SECTION

7.5 # Multiplication and Division of Radicals

In this section we will look at multiplication and division of expressions that contain radicals. As you will see, multiplication of expressions that contain radicals is very similar to multiplication of polynomials. The division problems in this section are just an extension of the work we did previously when we rationalized denominators.

▶ **EXAMPLE 1**   Multiply $(3\sqrt{5})(2\sqrt{7})$.

Solution   We can rearrange the order and grouping of the numbers in this product by applying the commutative and associative properties. Following that, we apply Property 1 for radicals and multiply:

$$(3\sqrt{5})(2\sqrt{7}) = (3 \cdot 2)(\sqrt{5}\sqrt{7}) \qquad \text{Commutative and}$$
$$\text{associative properties}$$
$$= (3 \cdot 2)(\sqrt{5 \cdot 7}) \qquad \text{Property 1 for radicals}$$
$$= 6\sqrt{35} \qquad \text{Multiplication}$$

In actual practice, it is not necessary to show either of the first two steps, although you may want to show them on the first few problems you work, just to be sure you understand them. ◀

▶ **EXAMPLE 2**   Multiply $\sqrt{5}(\sqrt{2} + \sqrt{5})$.

Solution

$$\sqrt{5}(\sqrt{2} + \sqrt{5}) = \sqrt{5} \cdot \sqrt{2} + \sqrt{5} \cdot \sqrt{5} \qquad \text{Distributive property}$$
$$= \sqrt{10} + 5 \qquad \text{Multiplication}$$ ◀

▶ **EXAMPLE 3**   Multiply $3\sqrt{2}(2\sqrt{5} + 5\sqrt{3})$.

Solution

$$3\sqrt{2}(2\sqrt{5} + 5\sqrt{3})$$
$$= 3\sqrt{2} \cdot 2\sqrt{5} + 3\sqrt{2} \cdot 5\sqrt{3} \qquad \text{Distributive property}$$
$$= 3 \cdot 2 \cdot \sqrt{2}\sqrt{5} + 3 \cdot 5\sqrt{2}\sqrt{3} \qquad \text{Commutative property}$$
$$= 6\sqrt{10} + 15\sqrt{6}$$

Each item in the last line is in simplified form, so the problem is complete. ◀

▶ **EXAMPLE 4** Multiply $(\sqrt{5} + 2)(\sqrt{5} + 7)$.

Solution We multiply using the FOIL method we used to multiply binomials:

$$(\sqrt{5} + 2)(\sqrt{5} + 7) = \underset{F}{\sqrt{5} \cdot \sqrt{5}} + \underset{O}{7\sqrt{5}} + \underset{I}{2\sqrt{5}} + \underset{L}{14}$$
$$= 5 + 9\sqrt{5} + 14$$
$$= 19 + 9\sqrt{5}$$

We must be careful not to try to simplify further by adding 19 and 9. We can add only radical expressions that have a common radical part; 19 and $9\sqrt{5}$ are not similar. ◀

▶ **EXAMPLE 5** Multiply $(\sqrt{x} + 3)(\sqrt{x} - 7)$.

Solution Remember, we are assuming that any variables that appear under a radical represent positive numbers.

$$(\sqrt{x} + 3)(\sqrt{x} - 7) = \underset{F}{\sqrt{x}\sqrt{x}} - \underset{O}{7\sqrt{x}} + \underset{I}{3\sqrt{x}} - \underset{L}{21}$$
$$= x - 4\sqrt{x} - 21 \qquad \blacktriangleleft$$

▶ **EXAMPLE 6** Expand and simplify $(\sqrt{3} - 2)^2$.

Solution Multiplying $\sqrt{3} - 2$ times itself, we have
$$(\sqrt{3} - 2)^2 = (\sqrt{3} - 2)(\sqrt{3} - 2)$$
$$= \sqrt{3}\sqrt{3} - 2\sqrt{3} - 2\sqrt{3} + 4$$
$$= 3 - 4\sqrt{3} + 4$$
$$= 7 - 4\sqrt{3} \qquad \blacktriangleleft$$

▶ **EXAMPLE 7** Multiply $(\sqrt{5} + \sqrt{2})(\sqrt{5} - \sqrt{2})$.

Solution We can apply the formula $(x + y)(x - y) = x^2 - y^2$ to obtain

$$(\sqrt{5} + \sqrt{2})(\sqrt{5} - \sqrt{2}) = (\sqrt{5})^2 - (\sqrt{2})^2$$
$$= 5 - 2$$
$$= 3$$

We could also have multiplied the two expressions using the FOIL method. If we were to do so, the work would look like this:

$$(\sqrt{5} + \sqrt{2})(\sqrt{5} - \sqrt{2})$$
$$= \underset{\text{F}}{\sqrt{5}\sqrt{5}} - \underset{\text{O}}{\sqrt{2}\sqrt{5}} + \underset{\text{I}}{\sqrt{2}\sqrt{5}} - \underset{\text{L}}{\sqrt{2}\sqrt{2}}$$
$$= 5 - \sqrt{10} + \sqrt{10} - 2$$
$$= 5 - 2$$
$$= 3$$

In either case, the product is 3. Also, the expressions $\sqrt{5} + \sqrt{2}$ and $\sqrt{5} - \sqrt{2}$ are called **conjugates** of each other.   ◀

▶ **EXAMPLE 8** Multiply $(\sqrt{a} + \sqrt{b})(\sqrt{a} - \sqrt{b})$.

Solution We can apply the formula $(x + y)(x - y) = x^2 - y^2$ to obtain

$$(\sqrt{a} + \sqrt{b})(\sqrt{a} - \sqrt{b}) = (\sqrt{a})^2 - (\sqrt{b})^2$$
$$= a - b$$ ◀

▶ **EXAMPLE 9** Rationalize the denominator in the expression

$$\frac{\sqrt{3}}{\sqrt{3} - \sqrt{2}}$$

Solution To remove the two radicals in the denominator, we must multiply both the numerator and denominator by $\sqrt{3} + \sqrt{2}$. That way, when we multiply $\sqrt{3} - \sqrt{2}$ and $\sqrt{3} + \sqrt{2}$ we will obtain the difference of two squares in the denominator:

$$\frac{\sqrt{3}}{\sqrt{3} - \sqrt{2}} = \frac{\sqrt{3}}{(\sqrt{3} - \sqrt{2})} \cdot \frac{(\sqrt{3} + \sqrt{2})}{(\sqrt{3} + \sqrt{2})}$$
$$= \frac{\sqrt{3}\sqrt{3} + \sqrt{3}\sqrt{2}}{(\sqrt{3})^2 - (\sqrt{2})^2}$$
$$= \frac{3 + \sqrt{6}}{3 - 2}$$
$$= \frac{3 + \sqrt{6}}{1}$$
$$= 3 + \sqrt{6}$$ ◀

▶ **EXAMPLE 10**   Rationalize the denominator in the expression $\dfrac{2}{5 - \sqrt{3}}$.

Solution   We use the same procedure as in Example 9. Multiply the numerator and denominator by the conjugate of the denominator, which is $5 + \sqrt{3}$:

$$\left(\frac{2}{5 - \sqrt{3}}\right)\left(\frac{\mathbf{5 + \sqrt{3}}}{\mathbf{5 + \sqrt{3}}}\right) = \frac{10 + 2\sqrt{3}}{5^2 - (\sqrt{3})^2}$$

$$= \frac{10 + 2\sqrt{3}}{25 - 3}$$

$$= \frac{10 + 2\sqrt{3}}{22}$$

The numerator and denominator of this last expression have a factor of 2 in common. We can reduce to lowest terms by dividing out the common factor 2. Continuing, we have

$$= \frac{\cancel{2}(5 + \sqrt{3})}{\cancel{2} \cdot 11}$$

$$= \frac{5 + \sqrt{3}}{11}$$

The final expression is in simplified form.   ◀

▶ **EXAMPLE 11**   Rationalize the denominator in the expression

$$\frac{\sqrt{2} + \sqrt{3}}{\sqrt{2} - \sqrt{3}}$$

Solution   We remove the two radicals in the denominator by multiplying both the numerator and denominator by the conjugate of $\sqrt{2} - \sqrt{3}$, which is $\sqrt{2} + \sqrt{3}$:

$$\frac{\sqrt{2} + \sqrt{3}}{\sqrt{2} - \sqrt{3}} = \left(\frac{\sqrt{2} + \sqrt{3}}{\sqrt{2} - \sqrt{3}}\right)\frac{(\mathbf{\sqrt{2} + \sqrt{3}})}{(\mathbf{\sqrt{2} + \sqrt{3}})}$$

$$= \frac{\sqrt{2}\sqrt{2} + \sqrt{2}\sqrt{3} + \sqrt{3}\sqrt{2} + \sqrt{3}\sqrt{3}}{(\sqrt{2})^2 - (\sqrt{3})^2}$$

$$= \frac{2 + \sqrt{6} + \sqrt{6} + 3}{2 - 3}$$

$$= \frac{5 + 2\sqrt{6}}{-1}$$

$$= -(5 + 2\sqrt{6}) \quad \text{or} \quad -5 - 2\sqrt{6} \qquad ◀$$

P R O B L E M   S E T   7 . 5

Perform the following multiplications. All answers should be in simplified form for radical expressions.

1. $\sqrt{3}\sqrt{2}$ **2.** $\sqrt{5}\sqrt{6}$

3. $\sqrt{6}\sqrt{2}$ **4.** $\sqrt{6}\sqrt{3}$

5. $(2\sqrt{3})(5\sqrt{7})$ **6.** $(3\sqrt{2})(4\sqrt{5})$

7. $(4\sqrt{3})(2\sqrt{6})$ **8.** $(7\sqrt{6})(3\sqrt{2})$

9. $\sqrt{2}(\sqrt{3}-1)$ **10.** $\sqrt{3}(\sqrt{5}+2)$

11. $\sqrt{2}(\sqrt{3}+\sqrt{2})$ **12.** $\sqrt{5}(\sqrt{7}-\sqrt{5})$

13. $\sqrt{3}(2\sqrt{2}+\sqrt{3})$ **14.** $\sqrt{11}(3\sqrt{2}-\sqrt{11})$

15. $\sqrt{3}(2\sqrt{3}-\sqrt{5})$ **16.** $\sqrt{7}(\sqrt{14}-\sqrt{7})$

17. $2\sqrt{3}(\sqrt{2}+\sqrt{5})$ **18.** $3\sqrt{2}(\sqrt{3}+\sqrt{2})$

19. $(\sqrt{2}+1)^2$ **20.** $(\sqrt{5}-4)^2$

21. $(\sqrt{x}+3)^2$ **22.** $(\sqrt{x}-4)^2$

23. $(5-\sqrt{2})^2$ **24.** $(2+\sqrt{5})^2$

25. $\left(\sqrt{a}-\dfrac{1}{2}\right)^2$ **26.** $\left(\sqrt{a}+\dfrac{1}{2}\right)^2$

27. $(3+\sqrt{7})^2$ **28.** $(3-\sqrt{2})^2$

29. $(\sqrt{5}+3)(\sqrt{5}+2)$ **30.** $(\sqrt{7}+4)(\sqrt{7}-5)$

31. $(\sqrt{2}-5)(\sqrt{2}+6)$ **32.** $(\sqrt{3}+8)(\sqrt{3}-2)$

33. $\left(\sqrt{3}+\dfrac{1}{2}\right)\left(\sqrt{2}+\dfrac{1}{3}\right)$

34. $\left(\sqrt{5}-\dfrac{1}{4}\right)\left(\sqrt{3}+\dfrac{1}{5}\right)$

35. $(\sqrt{x}+6)(\sqrt{x}-6)$ **36.** $(\sqrt{x}+7)(\sqrt{x}-7)$

37. $\left(\sqrt{a}+\dfrac{1}{3}\right)\left(\sqrt{a}+\dfrac{2}{3}\right)$

38. $\left(\sqrt{a}+\dfrac{1}{4}\right)\left(\sqrt{a}+\dfrac{3}{4}\right)$

39. $(\sqrt{5}-2)(\sqrt{5}+2)$ **40.** $(\sqrt{6}-3)(\sqrt{6}+3)$

41. $(2\sqrt{7}+3)(3\sqrt{7}-4)$

42. $(3\sqrt{5}+1)(4\sqrt{5}+3)$

43. $(2\sqrt{x}+4)(3\sqrt{x}+2)$

44. $(3\sqrt{x}+5)(4\sqrt{x}+2)$

45. $(7\sqrt{a}+2\sqrt{b})(7\sqrt{a}-2\sqrt{b})$

46. $(3\sqrt{a}-2\sqrt{b})(3\sqrt{a}+2\sqrt{b})$

Rationalize the denominator. All answers should be expressed in simplified form.

47. $\dfrac{\sqrt{3}}{\sqrt{5}-\sqrt{2}}$ **48.** $\dfrac{\sqrt{2}}{\sqrt{6}+\sqrt{3}}$

49. $\dfrac{\sqrt{5}}{\sqrt{5}+\sqrt{2}}$ **50.** $\dfrac{\sqrt{7}}{\sqrt{7}-\sqrt{2}}$

51. $\dfrac{8}{3-\sqrt{5}}$ **52.** $\dfrac{10}{5+\sqrt{5}}$

53. $\dfrac{\sqrt{3}+\sqrt{2}}{\sqrt{3}-\sqrt{2}}$ **54.** $\dfrac{\sqrt{5}-\sqrt{2}}{\sqrt{5}+\sqrt{2}}$

55. $\dfrac{\sqrt{7}-\sqrt{3}}{\sqrt{7}+\sqrt{3}}$ **56.** $\dfrac{\sqrt{11}-\sqrt{6}}{\sqrt{11}+\sqrt{6}}$

57. $\dfrac{\sqrt{x}+2}{\sqrt{x}-2}$ **58.** $\dfrac{\sqrt{x}-3}{\sqrt{x}+3}$

59. $\dfrac{\sqrt{5}-\sqrt{2}}{\sqrt{5}+\sqrt{3}}$ **60.** $\dfrac{\sqrt{7}-\sqrt{3}}{\sqrt{5}+\sqrt{2}}$

61. The statement below is incorrect. Correct the right side to make the statement true.

$$2(3\sqrt{5}) = 6\sqrt{15}$$

62. The statement below is incorrect. Correct the right side to make the statement true.

$$5(2\sqrt{6}) = 10\sqrt{30}$$

63. The statement below is incorrect. Correct the right side to make the statement true.

$$(\sqrt{3}+7)^2 = 3 + 49$$

64. The statement below is incorrect. Correct the right side to make the statement true.

$$(\sqrt{5}+\sqrt{2})^2 = 5 + 2$$

Review Problems

The following problems review material we covered in Sections 5.6 and 6.7. Reviewing the problems from Section 5.6 will help you understand the next section.

Solve each equation. [5.6]

65. $x^2 + 5x - 6 = 0$ **66.** $x^2 + 5x + 6 = 0$

67. $x^2 - 3x = 0$ **68.** $x^2 + 5x = 0$

Solve each proportion. [6.7]

69. $\dfrac{x}{3} = \dfrac{27}{x}$ **70.** $\dfrac{x}{2} = \dfrac{8}{x}$

71. $\dfrac{x}{5} = \dfrac{3}{x + 2}$ **72.** $\dfrac{x}{2} = \dfrac{4}{x + 2}$

73. Suppose you drive your car 375 miles in 15 hours. At this rate, how far will you drive in 20 hours?

74. If 2.5 inches on a map correspond to an actual distance of 100 miles, then how many miles apart are two cities that are 3.5 inches apart on the map?

SECTION

7.6

Equations Involving Radicals

In order to solve equations that contain one or more radical expressions we need an additional property. From our work with exponents we know that if two quantities are equal, then so are the squares of those quantities. That is, for real numbers a and b

$$\text{if} \qquad a = b$$

$$\text{then} \qquad a^2 = b^2$$

The only problem with squaring both sides of an equation is that occasionally we will change a false statement into a true statement. Let's take the false statement $3 = -3$ as an example.

$$3 = -3 \qquad \text{A false statement}$$

$$(3)^2 = (-3)^2 \qquad \text{Square both sides}$$

$$9 = 9 \qquad \text{A true statement}$$

We can avoid this problem by always checking our solutions if, at any time during the process of solving an equation, we have squared both sides of the equation. Here is how the property is stated.

> **Squaring Property of Equality**
> We can square both sides of an equation any time it is convenient to do so, as long as we check all solutions in the original equation.

▶ **EXAMPLE 1**   Solve for x: $\sqrt{x + 1} = 7$.

Solution   In order to solve this equation by our usual methods, we must first eliminate the radical sign. We can accomplish this by squaring both sides of the equation:

$$\sqrt{x + 1} = 7$$

$$(\sqrt{x + 1})^2 = 7^2 \qquad \text{Square both sides}$$

$$x + 1 = 49$$

$$x = 48$$

To check our solution, we substitute $x = 48$ into the original equation:

$$\sqrt{48 + 1} \overset{?}{=} 7$$

$$\sqrt{49} \overset{?}{=} 7$$

$$7 = 7 \qquad \text{A true statement}$$

The solution checks. ◀

▶ **EXAMPLE 2**   Solve for x: $\sqrt{2x - 3} = -9$.

Solution   We square both sides and proceed as in Example 1:

$$\sqrt{2x - 3} = -9$$

$$(\sqrt{2x - 3})^2 = (-9)^2 \qquad \text{Square both sides}$$

$$2x - 3 = 81$$

$$2x = 84$$

$$x = 42$$

Checking our solution in the original equation, we have

$$\sqrt{2(42) - 3} \overset{?}{=} -9$$

$$\sqrt{84 - 3} \overset{?}{=} -9$$

$$\sqrt{81} \overset{?}{=} -9$$

$$9 = -9 \qquad \text{A false statement}$$

Our solution does not check, because we end up with a false statement. ◀

Squaring both sides of the equation has produced what is called an *extraneous solution*. This happens occasionally when we use the squaring property of equality. We can always eliminate extraneous solutions by checking each solution in the original equation.

Note   As you can see, when we check $x = 42$ in the original equation in Example 2, we find that it is not a solution to the equation. Actually, it was apparent from the beginning that the equation had no solution. That is, no matter what x is, the equation

$$\sqrt{2x - 3} = -9$$

can never be true, because the left side is a positive number (or zero) for any value of x, while the right side is always negative.

▶ **EXAMPLE 3** Solve for a: $\sqrt{3a - 2} + 3 = 5$.

Solution Before we can square both sides to eliminate the radical, we must isolate the radical on the left side of the equation. To do so, we add -3 to both sides:

$$\sqrt{3a - 2} + 3 = 5$$
$$\sqrt{3a - 2} = 2 \qquad \text{Add } -3 \text{ to both sides}$$
$$(\sqrt{3a - 2})^2 = 2^2 \qquad \text{Square both sides}$$
$$3a - 2 = 4$$
$$3a = 6$$
$$a = 2$$

Checking $a = 2$ in the original equation, we have

$$\sqrt{3 \cdot 2 - 2} + 3 \stackrel{?}{=} 5$$
$$\sqrt{4} + 3 \stackrel{?}{=} 5$$
$$5 = 5 \qquad \text{A true statement}$$

◀

▶ **EXAMPLE 4** Solve for x: $\sqrt{x + 15} = x + 3$.

Solution We begin by squaring both sides:

$$(\sqrt{x + 15})^2 = (x + 3)^2 \qquad \text{Square both sides}$$
$$x + 15 = x^2 + 6x + 9$$

We have a quadratic equation. We put it into standard form by adding $-x$ and -15 to both sides. Then we factor and solve as usual:

$$0 = x^2 + 5x - 6 \qquad \text{Standard form}$$
$$0 = (x + 6)(x - 1) \qquad \text{Factor}$$
$$x + 6 = 0 \quad \text{or} \quad x - 1 = 0 \qquad \text{Set factors equal to 0}$$
$$x = -6 \quad \text{or} \qquad x = 1$$

We check each solution in the original equation:

Check -6	Check 1
$\sqrt{-6 + 15} \stackrel{?}{=} -6 + 3$	$\sqrt{1 + 15} \stackrel{?}{=} 1 + 3$
$\sqrt{9} \stackrel{?}{=} -3$	$\sqrt{16} \stackrel{?}{=} 4$
$3 = -3$	$4 = 4$
A false statement	A true statement

Since $x = -6$ does not check in the original equation, it cannot be a solution. The only solution is $x = 1$.

◀

▶ **EXAMPLE 5**   Graph $y = \sqrt{x}$ and $y = \sqrt[3]{x}$.

Solution   The graphs are shown in Figures 1 and 2. Notice that the graph of $y = \sqrt{x}$ appears in the first quadrant only, because in the equation $y = \sqrt{x}$, x and y cannot be negative.

The graph of $y = \sqrt[3]{x}$ appears in quadrants 1 and 3, since the cube root of a positive number is also a positive number and the cube root of a negative number is a negative number. That is, when x is positive, y will be positive, and when x is negative, y will be negative.

The graphs of both equations will contain the origin, since $y = 0$ when $x = 0$ in both equations.

x	y
-4	undefined
-1	undefined
0	0
1	1
4	2
9	3
16	4

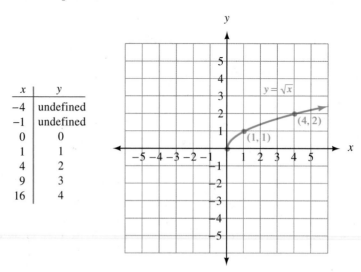

Figure 1

x	y
-27	-3
-8	-2
-1	-1
0	0
1	1
8	2
27	3

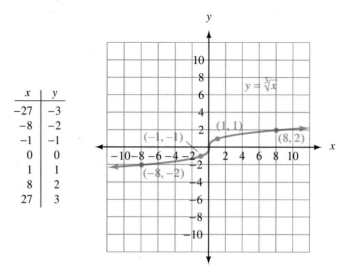

Figure 2

PROBLEM SET 7.6

Solve each equation by applying the squaring property of equality. Be sure to check all solutions in the original equation.

1. $\sqrt{x + 1} = 2$

2. $\sqrt{x - 3} = 4$

3. $\sqrt{x + 5} = 7$

4. $\sqrt{x + 8} = 5$

5. $\sqrt{x - 9} = -6$

6. $\sqrt{x + 10} = -3$

7. $\sqrt{x - 5} = -4$

8. $\sqrt{x + 7} = -5$

9. $\sqrt{x - 8} = 0$

10. $\sqrt{x - 9} = 0$

11. $\sqrt{2x + 1} = 3$

12. $\sqrt{2x - 5} = 7$

13. $\sqrt{2x - 3} = -5$

14. $\sqrt{3x - 8} = -4$

15. $\sqrt{3x + 6} = 2$

16. $\sqrt{5x - 1} = 5$

17. $2\sqrt{x} = 10$

18. $3\sqrt{x} = 9$

19. $3\sqrt{a} = 6$

20. $2\sqrt{a} = 12$

21. $\sqrt{3x + 4} - 3 = 2$

22. $\sqrt{2x - 1} + 2 = 5$

23. $\sqrt{5y - 4} - 2 = 4$

24. $\sqrt{3y + 1} + 7 = 2$

25. $\sqrt{2x + 1} + 5 = 2$

26. $\sqrt{6x - 8} - 1 = 3$

27. $\sqrt{x + 3} = x - 3$

28. $\sqrt{x - 3} = x - 3$

29. $\sqrt{a + 2} = a + 2$

30. $\sqrt{a + 10} = a - 2$

31. $\sqrt{2x + 9} = x + 5$

32. $\sqrt{x + 6} = x + 4$

33. $\sqrt{y - 4} = y - 6$

34. $\sqrt{2y + 13} = y + 7$

35. $\sqrt{5x + 1} = x + 1$

36. $\sqrt{6x + 1} = x - 1$

To solve each of the following equations you will square each side of the equation two times. Begin by squaring both sides. Then, simplify the result as much as possible. Since your result still contains a square root, square both sides again to eliminate it.

37. $\sqrt{x + 3} = \sqrt{x} + 3$ *(Hint:* $(\sqrt{x} + 3)^2$ *is not* $x + 9$.)

38. $\sqrt{x + 2} = \sqrt{x} + 2$

39. $\sqrt{x + 8} = \sqrt{x} + 2$

40. $\sqrt{x + 15} = \sqrt{x} + 3$

Applying the Concepts

41. The sum of a number and two is equal to the positive square root of eight times the number. Find the number.

42. The sum of twice a number and one is equal to three times the positive square root of the number. Find the number.

43. The difference of a number and three is equal to twice the positive square root of the number. Find the number.

44. The difference of a number and two is equal to the positive square root of the number. Find the number.

45. The number of seconds, T, it takes the pendulum of a grandfather clock to swing through one complete cycle is given by the formula

$$T = \frac{11}{7} \sqrt{\frac{L}{2}}$$

where L is the length, in feet, of the pendulum. Find how long the pendulum must be for one complete cycle to take 2 seconds by substituting 2 for T in the formula and then solving for L.

46. How long must the pendulum on a grandfather clock be if one complete cycle is to take 1 second? (Use formula in Problem 45.)

Graph each equation. Begin by making a table using the given values of x. (In some cases you must use a calculator to approximate a square root.)

47. $y = \sqrt{x}$ $x = 0, 1, 2, 3, 4$

48. $y = \sqrt[3]{x}$ $x = -8, -4, -1, 0, 4, 8$

49. $y = 2\sqrt{x}$ $x = 0, 1, 4, 9$

50. $y = 3\sqrt[3]{x}$ $x = -8, -1, 0, 1, 8$

51. $y = \sqrt{x} + 2$ $x = 0, 1, 2, 4, 9$

52. $y = \sqrt[3]{x} + 3$ $x = -8, -1, 0, 1, 8$

Review Problems

The problems that follow review material we covered in Sections 6.5 and 6.6.

Simplify each complex fraction. [6.6]

53. $\dfrac{\dfrac{2}{5}}{\dfrac{4}{15}}$

54. $\dfrac{\dfrac{x^4}{y^2}}{\dfrac{x^3}{y^5}}$

55. $\dfrac{1 - \dfrac{1}{a}}{1 + \dfrac{1}{a}}$

56. $\dfrac{1 - \dfrac{16}{x^2}}{1 - \dfrac{2}{x} - \dfrac{24}{x^2}}$

Solve each application. [6.5]

57. The sum of a number and its reciprocal is $\frac{10}{3}$. Find the number.

58. One number is twice another. The sum of their reciprocals is $\frac{1}{2}$. Find the two numbers.

59. Suppose a person can jog four miles an hour faster than she can walk. If she jogs 12 miles in the same amount of time it takes her to walk 6 miles, how fast can she walk?

60. A boat can travel 6 miles up a river in the same amount of time it takes to travel 14 miles down the same river. If the speed of the current is 2 miles per hour, what is the speed of the boat in still water?

61. A hot water faucet can fill a sink in 4 minutes, while the cold water faucet can fill the sink in 3 minutes. If both faucets are open, how long will it take to fill the sink?

62. A hose can fill a swimming pool in 6 hours, while the drain can empty the pool in 18 hours. If the hose is on and the drain is open, how long will it take to fill the pool?

CHAPTER 7 SUMMARY

Examples

1. The two square roots of 9 are 3 and -3:

$$\sqrt{9} = 3 \quad \text{and} \quad -\sqrt{9} = -3$$

Roots [7.1]

Every positive real number x has two square roots, one positive and one negative. The positive square root is written $\sqrt{x}$. The negative square root of x is written $-\sqrt{x}$. In both cases the square root of x is a number we square to get x.

The cube root of x is written $\sqrt[3]{x}$ and is the number we cube to get x.

2. Index

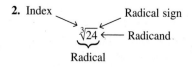

Radical

Notation [7.1]

In the expression $\sqrt[3]{8}$, 8 is called the *radicand,* 3 is the *index,* $\sqrt{\ }$ is called the *radical sign,* and the whole expression $\sqrt[3]{8}$ is called the *radical.*

3. (a) $\sqrt{3} \cdot \sqrt{2} = \sqrt{3 \cdot 2} = \sqrt{6}$

Properties of Radicals [7.2]

If a and b represent nonnegative real numbers, then

(b) $\dfrac{\sqrt{12}}{\sqrt{3}} = \sqrt{\dfrac{12}{3}} = \sqrt{4} = 2$

1. $\sqrt{a}\sqrt{b} = \sqrt{ab}$ The product of the square roots is the square root of the product

2. $\dfrac{\sqrt{a}}{\sqrt{b}} = \sqrt{\dfrac{a}{b}} \quad (b \neq 0)$ The quotient of the square roots is the square root of the quotient

(c) $\sqrt{5} \cdot \sqrt{5} = (\sqrt{5})^2 = 5$

3. $\sqrt{a} \cdot \sqrt{a} = (\sqrt{a})^2 = a$ This property shows that squaring and square roots are inverse operations

4. Simplify $\sqrt{20}$ and $\sqrt{\frac{2}{3}}$.

$$\sqrt{20} = \sqrt{4 \cdot 5} = \sqrt{4}\sqrt{5} = 2\sqrt{5}$$

$$\sqrt{\frac{2}{3}} = \frac{\sqrt{2}}{\sqrt{3}} = \frac{\sqrt{2}}{\sqrt{3}} \cdot \frac{\sqrt{3}}{\sqrt{3}} = \frac{\sqrt{6}}{3}$$

Simplified Form for Radicals [7.3]

A radical expression is in *simplified form* if:

1. There are no perfect squares that are factors of the quantity under the square root sign, no perfect cubes that are factors of the quantity under the cube root sign, and so on. We want as little as possible under the radical sign.
2. There are no fractions under the radical sign.
3. There are no radicals in the denominator.

5. (a) $5\sqrt{7} + 3\sqrt{7} = 8\sqrt{7}$
  (b) $2\sqrt{18} - 3\sqrt{50}$
    $= 2 \cdot 3\sqrt{2} - 3 \cdot 5\sqrt{2}$
    $= 6\sqrt{2} - 15\sqrt{2}$
    $= -9\sqrt{2}$

Addition and Subtraction of Radical Expressions [7.4]

We add and subtract radical expressions by using the distributive property to combine terms that have the same radical parts. If the radicals are not in simplified form, we begin by writing them in simplified form and then combining similar terms, if possible.

6. (a) $\sqrt{3}(\sqrt{5} - \sqrt{3}) = \sqrt{15} - 3$
  (b) $(\sqrt{7} + 3)(\sqrt{7} - 5)$
    $= 7 - 5\sqrt{7} + 3\sqrt{7} - 15$
    $= -8 - 2\sqrt{7}$

Multiplication of Radical Expressions [7.5]

We multiply radical expressions by applying the distributive property or the FOIL method.

7. $\dfrac{7}{\sqrt{5} - \sqrt{3}}$

$= \dfrac{7}{\sqrt{5} - \sqrt{3}} \cdot \dfrac{\sqrt{5} + \sqrt{3}}{\sqrt{5} + \sqrt{3}}$

$= \dfrac{7\sqrt{5} + 7\sqrt{3}}{2}$

Division of Radical Expressions [7.5]

To divide by an expression like $\sqrt{5} - \sqrt{3}$, we multiply the numerator and denominator by its *conjugate*, $\sqrt{5} + \sqrt{3}$. This process is also called *rationalizing the denominator.*

8. Solve $\sqrt{x - 3} = 2$.

$$(\sqrt{x - 3})^2 = 2^2$$

$$x - 3 = 4$$

$$x = 7$$

The solution checks in the original equation.

Squaring Property of Equality [7.6]

We are free to square both sides of an equation whenever it is convenient, as long as we check all solutions in the original equation. We must check solutions because squaring both sides of an equation occasionally produces extraneous solutions.

COMMON MISTAKES

1. A very common mistake with radicals is to think of $\sqrt{25}$ as representing both the positive and negative square roots of 25. The notation $\sqrt{25}$ stands for the *positive* square root of 25. If we want the negative square root of 25, we write $-\sqrt{25}$.

2. The most common mistake when working with radicals is to try to apply a property similar to Property 1 for radicals involving addition instead of multiplication. Here is an example:

$$\sqrt{16 + 9} = \sqrt{16} + \sqrt{9} \qquad \text{Mistake}$$

Although this example looks like it may be true, it isn't. If we carry it out further, the mistake becomes obvious:

$$\sqrt{16 + 9} \overset{?}{=} \sqrt{16} + \sqrt{9}$$

$$\sqrt{25} \overset{?}{=} 4 + 3$$

$$5 = 7 \qquad \text{False}$$

3. It is a mistake to try to simplify expressions like $2 + 3\sqrt{7}$. The 2 and 3 cannot be combined, because the terms they appear in are not similar. Therefore, $2 + 3\sqrt{7} \neq 5\sqrt{7}$. The expression $2 + 3\sqrt{7}$ cannot be simplified further.

CHAPTER 7 REVIEW

The numbers in brackets refer to the sections of the text in which similar problems can be found.

Find the following roots. Assume all variables are positive. [7.1]

1. $\sqrt{25}$ **2.** $\sqrt{81}$

3. $\sqrt{169}$ **4.** $\sqrt{400}$

5. $\sqrt[3]{-1}$ **6.** $-\sqrt[3]{-64}$

7. $\sqrt[4]{625}$ **8.** $-\sqrt[4]{81}$

9. $\sqrt{100x^2y^4}$ **10.** $\sqrt{121x^4y^2}$

11. $\sqrt[3]{8a^3}$ **12.** $\sqrt[3]{-27a^3}$

Simplify. Assume all variables represent positive numbers. [7.2]

13. $\sqrt{24}$ **14.** $\sqrt{44}$

15. $\sqrt{60x^2}$ **16.** $\sqrt{80x^2}$

17. $\sqrt{90x^3y^4}$ **18.** $\sqrt{120x^4y^3}$

19. $-\sqrt{32}$ **20.** $-\sqrt{72}$

21. $3\sqrt{20x^3y}$ **22.** $4\sqrt{28x^2y^3}$

Simplify. Assume all variables represent positive numbers. [7.2]

23. $\sqrt{\dfrac{3}{49}}$ **24.** $\sqrt{\dfrac{5}{64}}$

25. $\sqrt{\dfrac{8}{81}}$ **26.** $\sqrt{\dfrac{18}{121}}$

27. $\sqrt{\dfrac{49}{64}}$ **28.** $\sqrt{\dfrac{121}{144}}$

29. $\sqrt{\dfrac{49a^2b^2}{16}}$ **30.** $\sqrt{\dfrac{100a^2b^2}{81}}$

31. $\sqrt{\dfrac{80}{49}}$ **32.** $\sqrt{\dfrac{200}{81}}$

33. $\sqrt{\dfrac{40a^2}{121}}$ **34.** $\sqrt{\dfrac{90a^2}{169}}$

35. $\dfrac{5\sqrt{84}}{7}$ **36.** $\dfrac{3\sqrt{96}}{5}$

37. $\dfrac{3\sqrt{120a^2b^2}}{\sqrt{25}}$ **38.** $\dfrac{8\sqrt{150a^2b^2}}{\sqrt{81}}$

39. $\dfrac{-5\sqrt{20x^3y^2}}{\sqrt{144}}$ **40.** $\dfrac{-9\sqrt{45x^2y^3}}{\sqrt{36}}$

Write in simplified form. Assume all variables represent positive numbers. [7.3]

41. $\dfrac{2}{\sqrt{7}}$ **42.** $\dfrac{5}{\sqrt{3}}$

43. $\sqrt{\dfrac{32}{5}}$ **44.** $\sqrt{\dfrac{40}{3}}$

45. $\sqrt{\dfrac{5}{48}}$ **46.** $\sqrt{\dfrac{7}{50}}$

47. $\dfrac{-3\sqrt{60}}{\sqrt{5}}$ **48.** $\dfrac{-5\sqrt{80}}{\sqrt{3}}$

49. $\sqrt{\dfrac{32ab^2}{3}}$

50. $\sqrt{\dfrac{40a^2b}{7}}$

51. $\sqrt[3]{\dfrac{3}{4}}$

52. $\sqrt[3]{\dfrac{2}{3}}$

Write in simplified form. Assume all variables represent positive numbers. [7.5]

53. $\dfrac{3}{\sqrt{3}-4}$

54. $\dfrac{-2}{\sqrt{5}+1}$

55. $\dfrac{2}{3+\sqrt{7}}$

56. $\dfrac{6}{2-\sqrt{5}}$

57. $\dfrac{3}{\sqrt{5}-\sqrt{2}}$

58. $\dfrac{5}{\sqrt{7}+\sqrt{3}}$

59. $\dfrac{\sqrt{5}}{\sqrt{3}-\sqrt{5}}$

60. $\dfrac{\sqrt{2}}{\sqrt{5}+\sqrt{2}}$

61. $\dfrac{\sqrt{5}-\sqrt{2}}{\sqrt{5}+\sqrt{2}}$

62. $\dfrac{\sqrt{7}+\sqrt{3}}{\sqrt{7}-\sqrt{3}}$

63. $\dfrac{\sqrt{x}+3}{\sqrt{x}-3}$

64. $\dfrac{\sqrt{x}-2}{\sqrt{x}+2}$

Combine the following expressions. [7.4]

65. $3\sqrt{5}-7\sqrt{5}$

66. $-5\sqrt{11}+9\sqrt{11}$

67. $3\sqrt{27}-5\sqrt{48}$

68. $5\sqrt{200}+9\sqrt{50}$

69. $-2\sqrt{45}-5\sqrt{80}+2\sqrt{20}$

70. $5\sqrt{12}+3\sqrt{48}-2\sqrt{300}$

71. $3\sqrt{50x^2}-x\sqrt{200}$

72. $5\sqrt{63x^2}-x\sqrt{28}$

73. $\sqrt{40a^3b^2}-a\sqrt{90ab^2}$

74. $\sqrt{99a^2b^2}-5ab\sqrt{44}$

Multiply. Write all answers in simplified form. [7.5]

75. $\sqrt{3}(\sqrt{3}+3)$

76. $\sqrt{5}(\sqrt{5}-3)$

77. $4\sqrt{2}(\sqrt{3}+\sqrt{5})$

78. $3\sqrt{7}(\sqrt{5}-\sqrt{2})$

79. $(\sqrt{x}+7)(\sqrt{x}-7)$

80. $(\sqrt{x}+4)(\sqrt{x}-4)$

81. $(2\sqrt{5}-4)(\sqrt{5}+3)$

82. $(5\sqrt{7}-8)(\sqrt{7}+1)$

83. $(\sqrt{x}+5)^2$

84. $(\sqrt{x}-2)^2$

Solve each equation. [7.6]

85. $\sqrt{x-3}=3$

86. $\sqrt{x+4}=5$

87. $\sqrt{3x-5}=4$

88. $\sqrt{2x+3}=5$

89. $5\sqrt{a}=20$

90. $7\sqrt{a}=63$

91. $\sqrt{3x-7}+6=2$

92. $\sqrt{2x+1}+10=8$

93. $\sqrt{7x+1}=x+1$

94. $\sqrt{6x-2}=3x-5$

Find x in each right triangle below. [7.1]

95.

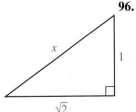

96.

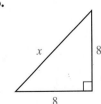

Graph each equation. [7.6]

97. $y=4\sqrt[3]{x}$

98. $y=3\sqrt{x}$

99. $y=\sqrt{x}+3$

100. $y=\sqrt[3]{x}+2$

CHAPTER 7 TEST

Find the following roots. [7.1]

1. $\sqrt{16}$

2. $-\sqrt{36}$

3. The square roots of 49

4. $\sqrt[3]{27}$

5. $\sqrt[3]{-8}$

6. $-\sqrt[4]{81}$

Put the following expressions into simplified form. [7.2, 7.3]

7. $\sqrt{75}$

8. $\sqrt{32}$

9. $\sqrt{\dfrac{2}{3}}$

10. $\dfrac{1}{\sqrt[3]{4}}$

11. $3\sqrt{50x^2}$

12. $\sqrt{\dfrac{12x^2y^3}{5}}$

Combine. [7.4]

13. $5\sqrt{12} - 2\sqrt{27}$

14. $2x\sqrt{18} + 5\sqrt{2x^2}$

Multiply. [7.5]

15. $\sqrt{3}(\sqrt{5} - 2)$

16. $(\sqrt{5} + 7)(\sqrt{5} - 8)$

17. $(\sqrt{x} + 6)(\sqrt{x} - 6)$

18. $(\sqrt{5} - \sqrt{3})^2$

Divide. (Rationalize the denominator.) [7.5]

19. $\dfrac{\sqrt{7} - \sqrt{3}}{\sqrt{7} + \sqrt{3}}$

20. $\dfrac{\sqrt{x}}{\sqrt{x} + 5}$

Solve the following equations. [7.6]

21. $\sqrt{2x + 1} + 2 = 7$ **22.** $\sqrt{3x + 1} + 6 = 2$

23. $\sqrt{2x - 3} = x - 3$

24. The difference of a number and four is equal to three times the positive square root of the number. Find the number.

25. Find x in the right triangle below. [7.1]

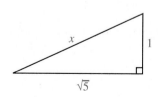

26. Graph $y = \sqrt{x} - 2$. [7.6]

MORE QUADRATIC EQUATIONS

INTRODUCTION

Now that we have become familiar with square roots, we can expand on some of the topics we covered earlier in the book. For instance, the two triangles in Figure 1 are special triangles that are used frequently in trigonometry.

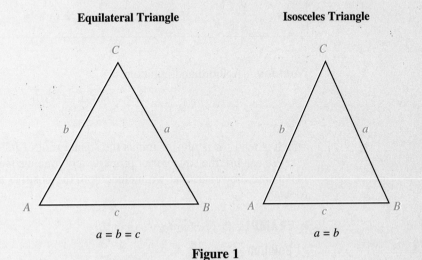

Equilateral Triangle

$a = b = c$

Isosceles Triangle

$a = b$

Figure 1

We will use the equations that we learn to solve in this chapter along with the Pythagorean Theorem to solve some problems involving these new triangles.

OVERVIEW

In this chapter we will develop a method of solving quadratic equations that can be used on all quadratic equations, whether or not they are factorable. We will then apply this method to the general quadratic equation $ax^2 + bx + c = 0$ and arrive at what is known as the quadratic formula. We will end the chapter by considering the graphs of second-degree equations.

To be successful in this chapter you should have a working knowledge of

1. square roots
2. binomial squares
3. graphing by use of tables

SECTION

8.1 **More Quadratic Equations**

Consider the equation $x^2 = 9$. Inspection shows that there are two solutions: $x = 3$ and $x = -3$, the two square roots of 9. Since every positive real number has two square roots, we can write the following property.

Square Root Property for Equations
For all positive real numbers b,

$$\text{If } a^2 = b, \text{ then } a = \sqrt{b} \text{ or } a = -\sqrt{b}$$

Notation   A shorthand notation for

$$a = \sqrt{b} \text{ or } a = -\sqrt{b}$$

is

$$a = \pm\sqrt{b}$$

which is read "a is plus or minus the square root of b."

We can use the square root property any time we feel it is helpful. We must make sure, however, that we include both the positive and the negative square roots.

▶ **EXAMPLE 1**   Solve for x: $x^2 = 7$.

Solution

$$x^2 = 7$$

$$x = \pm\sqrt{7} \qquad \text{Square root property}$$

The two solutions are $\sqrt{7}$ and $-\sqrt{7}$.   ◀

Note This method of solving quadratic equations is sometimes called **extraction of roots.** The key to solving equations using this method is to isolate the squared term on one side of the equation.

▶ **EXAMPLE 2** Solve for y: $3y^2 = 60$.

Solution We begin by dividing both sides by 3 (which is the same as multiplying both sides by $\frac{1}{3}$).

$$3y^2 = 60$$
$$y^2 = 20 \qquad \text{Divide each side by 3}$$
$$y = \pm\sqrt{20} \qquad \text{Square root property}$$
$$y = \pm 2\sqrt{5} \qquad \sqrt{20} = \sqrt{4 \cdot 5} = \sqrt{4}\,\sqrt{5} = 2\sqrt{5}$$

Our two solutions are $2\sqrt{5}$ and $-2\sqrt{5}$. Each of them will yield a true statement when used in place of the variable in the original equation $3y^2 = 60$. ◀

▶ **EXAMPLE 3** Solve for a: $(a + 3)^2 = 16$.

Solution We proceed as we did in the last example. The square root of $(a + 3)^2$ is $(a + 3)$:

$$(a + 3)^2 = 16$$
$$a + 3 = \pm 4$$

At this point we add -3 to both sides to get

$$a = -3 \pm 4$$

which we can write as

$$a = -3 + 4 \quad \text{or} \quad a = -3 - 4$$
$$a = 1 \qquad\quad \text{or} \quad a = -7$$

Our solutions are 1 and -7. ◀

▶ **EXAMPLE 4** Solve for x: $(3x - 2)^2 = 25$.

Solution

$$(3x - 2)^2 = 25$$
$$3x - 2 = \pm 5$$

Adding 2 to both sides, we have

$$3x = 2 \pm 5$$

Dividing both sides by 3 gives us

$$x = \frac{2 \pm 5}{3}$$

We separate this last equation into two separate statements:

$$x = \frac{2 + 5}{3} \quad \text{or} \quad x = \frac{2 - 5}{3}$$

$$x = \frac{7}{3} \quad \text{or} \quad x = \frac{-3}{3} = -1$$

◀

▶ **EXAMPLE 5**   Solve for y: $(4y - 5)^2 = 6$.

Solution

$$(4y - 5)^2 = 6$$

$$4y - 5 = \pm\sqrt{6}$$

$$4y = 5 \pm \sqrt{6} \qquad \text{Add 5 to both sides}$$

$$y = \frac{5 \pm \sqrt{6}}{4} \qquad \text{Divide both sides by 4}$$

Since $\sqrt{6}$ is irrational, we cannot simplify the expression further. The solution set is $\left\{ \dfrac{5 + \sqrt{6}}{4}, \dfrac{5 - \sqrt{6}}{4} \right\}$.

◀

▶ **EXAMPLE 6**   Solve for x: $(2x + 6)^2 = 8$.

Solution

$$(2x + 6)^2 = 8$$

$$2x + 6 = \pm\sqrt{8}$$

$$2x + 6 = \pm 2\sqrt{2} \qquad \sqrt{8} = \sqrt{4 \cdot 2} = 2\sqrt{2}$$

$$2x = -6 \pm 2\sqrt{2} \qquad \text{Add } -6 \text{ to both sides}$$

$$x = \frac{-6 \pm 2\sqrt{2}}{2} \qquad \text{Divide each side by 2}$$

We can reduce this last expression to lowest terms by factoring a 2 from each term in the numerator and then dividing that 2 by the 2 in the denominator. This is equivalent to dividing each term in the numerator by the 2 in the denominator. Here is what it looks like:

$$x = \frac{\cancel{2}(-3 \pm \sqrt{2})}{\cancel{2}} \qquad \text{Factor a 2 from each term in numerator}$$

$$x = -3 \pm \sqrt{2} \qquad \text{Divide numerator and denominator by 2}$$

The two solutions are $-3 + \sqrt{2}$ and $-3 - \sqrt{2}$.

We can check our two solutions in the original equation. Let's check our first solution, $-3 + \sqrt{2}$.

When $x = -3 + \sqrt{2}$

the equation

$$(2x + 6)^2 = 8$$

becomes

$$[2(-3 + \sqrt{2}) + 6]^2 \stackrel{?}{=} 8$$

$$(-6 + 2\sqrt{2} + 6)^2 \stackrel{?}{=} 8$$

$$(2\sqrt{2})^2 \stackrel{?}{=} 8$$

$$4 \cdot 2 \stackrel{?}{=} 8$$

$$8 = 8 \qquad \text{A true statement}$$

The second solution, $-3 - \sqrt{2}$, checks also.   ◀

▶ **EXAMPLE 7** If an object is dropped from a height of h feet, the amount of time in seconds it will take for the object to reach the ground (neglecting the resistance of air) is given by the formula

$$h = 16t^2$$

Solve this formula for t.

Solution To solve for t, we apply the square root property:

$$h = 16t^2 \qquad \text{Original formula}$$

$$\pm\sqrt{h} = 4t \qquad \text{Square root property}$$

$$\pm\frac{\sqrt{h}}{4} = t \qquad \text{Divide each side by 4}$$

Since t represents the time it takes for the object to fall h feet, t will never be negative. Therefore, the formula that gives t in terms of h is

$$t = \frac{\sqrt{h}}{4}$$ ◀

Whenever we are solving an application problem like this one and we obtain a result that includes the $\pm$ sign, we must ask ourselves if the result can actually be negative. If it cannot be, we delete the negative result and use only the positive result.

Facts from Geometry: Special Triangles

The two special triangles we mentioned in the introduction to this chapter are shown below. An **equilateral triangle** (Figure 1) is a triangle with three sides of equal length. If all three sides in a triangle have the same length, then the three interior angles in the triangle must also be equal. Since the sum of the interior angles in a triangle is always 180°, each of the three interior angles in any equilateral triangle must be 60°.

An **isosceles triangle** (Figure 2) is a triangle with two sides of equal length. Angles A and B in the isosceles triangle in Figure 2 are called the **base angles;** they are the angles opposite the two equal sides. In every isosceles triangle, the base angles are equal.

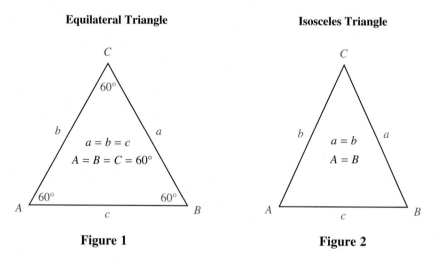

Equilateral Triangle

Isosceles Triangle

Figure 1

Figure 2

▶ **EXAMPLE 8** The lengths of all three sides of an equilateral triangle are 8 centimeters. Find the height of the triangle.

Solution Because the three sides of an equilateral triangle are equal, the height always divides the base into two equal line segments. Figure 3 illustrates this fact. We find the height by applying the Pythagorean Theorem.

$$8^2 = x^2 + 4^2$$
$$64 = x^2 + 16$$
$$x^2 = 64 - 16$$
$$x^2 = 48$$
$$x = \sqrt{48}$$
$$= \sqrt{16 \cdot 3}$$
$$= 4\sqrt{3} \text{ cm}$$

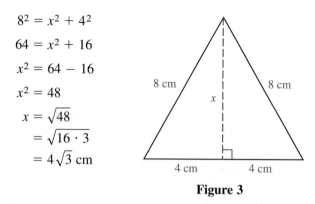

Figure 3 ◀

Note A calculator gives a decimal approximation of 6.9 cm for the height of the triangle in Example 8.

▶ **EXAMPLE 9**   The base of an isosceles triangle is 10 feet, while the length of the two equal sides is 7 feet, as shown in Figure 4. Find the height of the triangle.

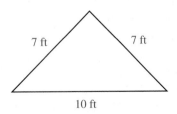

10 ft

Figure 4

Solution   Because two sides in every isosceles triangle are equal, the height that is drawn to the base always divides the base into two equal line segments, as shown in Figure 5. This fact allows us to form a right triangle in which two sides are known. We apply the Pythagorean Theorem and solve for x.

$$7^2 = x^2 + 5^2$$

$$49 = x^2 + 25$$

$$x^2 = 49 - 25$$

$$x^2 = 24$$

$$x = \sqrt{24}$$

$$= 2\sqrt{6} \text{ ft} = 4.9 \text{ ft to the nearest tenth}$$

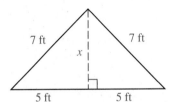

Figure 5 ◀

P R O B L E M   S E T   8 . 1

Solve each of the following equations using the methods learned in this section.

1. $x^2 = 9$ **2.** $x^2 = 16$

3. $a^2 = 25$ **4.** $a^2 = 36$

5. $y^2 = 8$ **6.** $y^2 = 75$

7. $2x^2 = 100$ **8.** $2x^2 = 54$

9. $3a^2 = 54$ **10.** $2a^2 = 64$

11. $(x + 2)^2 = 4$ **12.** $(x - 3)^2 = 16$

13. $(x + 1)^2 = 25$ **14.** $(x + 3)^2 = 64$

15. $(a - 5)^2 = 75$ **16.** $(a - 4)^2 = 32$

17. $(y + 1)^2 = 50$ **18.** $(y - 5)^2 = 27$

19. $(2x + 1)^2 = 25$ **20.** $(3x - 2)^2 = 16$

21. $(4a - 5)^2 = 36$ **22.** $(2a + 6)^2 = 64$

23. $(3y - 1)^2 = 12$ **24.** $(5y - 4)^2 = 12$

25. $(6x + 2)^2 = 27$ **26.** $(8x - 1)^2 = 20$

27. $(3x - 9)^2 = 27$ **28.** $(2x + 8)^2 = 32$

29. $(3x + 6)^2 = 45$ **30.** $(5x - 10)^2 = 75$

31. $(2y - 4)^2 = 8$ **32.** $(4y - 6)^2 = 48$

33. $\left(x - \dfrac{2}{3}\right)^2 = \dfrac{25}{9}$ **34.** $\left(x - \dfrac{3}{4}\right)^2 = \dfrac{49}{16}$

35. $\left(x + \dfrac{1}{2}\right)^2 = \dfrac{7}{4}$ **36.** $\left(x + \dfrac{1}{3}\right)^2 = \dfrac{5}{9}$

37. $\left(a - \dfrac{4}{5}\right)^2 = \dfrac{12}{25}$ **38.** $\left(a - \dfrac{3}{7}\right)^2 = \dfrac{18}{49}$

Since $a^2 + 2ab + b^2$ can be written as $(a + b)^2$, each of the following equations can be solved using our square root method. The first step is to write the trinomial on the left side of the equal sign as the square of a binomial. Solve each equation.

39. $x^2 + 10x + 25 = 7$
40. $x^2 + 6x + 9 = 11$ **41.** $x^2 - 2x + 1 = 9$
42. $x^2 + 8x + 16 = 25$
43. $x^2 + 12x + 36 = 8$
44. $x^2 - 4x + 4 = 12$
45. Check the solution $x = -1 + 5\sqrt{2}$ in the equation $(x + 1)^2 = 50$.
46. Check the solution $x = -8 + 2\sqrt{6}$ in the equation $(x + 8)^2 = 24$.

Applying the Concepts

47. The square of the sum of a number and 3 is 16. Find the number. (There are two solutions.)
48. The square of the sum of twice a number and 3 is 25. Find the number. (There are two solutions.)
49. If you invest $100 in an account with interest rate r compounded annually, the amount of money, A, in the account after two years is given by the formula

$$A = 100(1 + r)^2$$

Solve this formula for r.

50. If you invest P dollars in an account with interest rate r compounded annually, the amount of money, A, in the account after two years is given by the formula

$$A = P(1 + r)^2$$

Solve this formula for r.

51. The lengths of all three sides of an equilateral triangle are 10 feet. Find the height of the triangle.
52. The lengths of all three sides of an equilateral triangle are 12 meters. Find the height of the triangle.
53. The front of a tent forms an equilateral triangle with sides of 6 feet. Can a person 5 feet 8 inches tall stand up inside the tent?

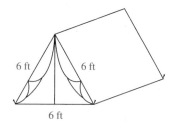

54. The front of a tent forms an equilateral triangle. The tent must be constructed so that a person 6 feet tall can stand up inside. Find the length of the three sides of the front of the tent.

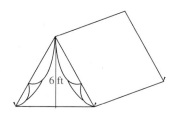

55. The base of an isosceles triangle is 8 feet, while the length of the two equal sides is 5 feet. Find the height of the triangle.
56. An isosceles triangle has a base 8 feet long. If the length of the two equal sides is 6 feet, find the height of the triangle.

Review Problems

The following problems review material we covered in Sections 4.6, 5.4, and 7.1. Reviewing the problems from Sections 4.6 and 5.4 will help you understand the material in the next section.

Multiply. [4.6]

57. $(x + 3)^2$ **58.** $(x - 3)^2$
59. $(x - 5)^2$ **60.** $(x + 5)^2$

Factor. [5.4]

61. $x^2 - 12x + 36$ **62.** $x^2 + 12x + 36$
63. $x^2 + 4x + 4$ **64.** $x^2 - 4x + 4$

Find the following roots. [7.1]

65. $\sqrt[3]{8}$ **66.** $\sqrt[3]{27}$
67. $\sqrt[4]{16}$ **68.** $\sqrt[4]{81}$

Completing the Square

In this section we will develop a method of solving quadratic equations that works whether or not the equation can be factored. Since we will be working with the individual terms of trinomials, we need some new definitions so that we can keep our vocabulary straight.

DEFINITION   In the trinomial $2x^2 + 3x + 4$, the first term, $2x^2$, is called the **quadratic term;** the middle term, $3x$, is called the **linear term;** and the last term, 4, is called the **constant term.**

Now consider the following list of perfect square trinomials and their corresponding binomial squares.

$$x^2 +\ \ 6x +\ \ 9 = (x + 3)^2$$
$$x^2 -\ \ 8x + 16 = (x - 4)^2$$
$$x^2 - 10x + 25 = (x - 5)^2$$
$$x^2 + 12x + 36 = (x + 6)^2$$

In each case the coefficient of x^2 is 1. A more important observation, however, stems from the relationship between the linear terms (middle terms) and the constant terms (last terms). Notice that the constant in each case is the square of half the coefficient of x in the linear term. That is,

1. for the first trinomial, $x^2 + 6x + 9$, the last term, 9, is the square of half the coefficient of the middle term: $9 = (\frac{6}{2})^2$.
2. for the second trinomial, $x^2 - 8x + 16$, $16 = [\frac{1}{2}(-8)]^2$.
3. for the trinomial $x^2 - 10x + 25$, it also holds: $25 = [\frac{1}{2}(-10)]^2$.

In summary, then, for every perfect square trinomial in which the coefficient of x^2 is 1, the last term is always the square of half the coefficient of the linear term. We can use this fact to build our own perfect square trinomials.

▶ **EXAMPLES**   Write the correct last term to each of the following expressions so each becomes a perfect square trinomial.

1. $x^2 - 2x$

   Solution   The coefficient of the linear term is -2. If we take the square of half of -2, we get 1. Adding the 1 as the last term, we have the perfect square trinomial:

   $$x^2 - 2x + 1 = (x - 1)^2$$

2. $x^2 + 18x$

Solution   Half of 18 is 9, the square of which is 81. If we add 81 at the end, we have

$$x^2 + 18x + 81 = (x + 9)^2$$

3. $x^2 + 3x$

Solution   Half of 3 is $\frac{3}{2}$, the square of which is $\frac{9}{4}$.

$$x^2 + 3x + \frac{9}{4} = \left(x + \frac{3}{2} \right)^2$$

◄

We can use this procedure, along with the method developed in Section 8.1, to solve some quadratic equations.

► **EXAMPLE 4**   Solve $x^2 - 6x + 5 = 0$ by completing the square.

Solution   We begin by adding -5 to both sides of the equation. We want just $x^2 - 6x$ on the left side so that we can add on our own last term to get a perfect square trinomial:

$$x^2 - 6x + 5 = 0$$

$$x^2 - 6x \quad\; = -5 \qquad \text{Add } -5 \text{ to both sides}$$

Now we can add 9 to both sides and the left side will be a perfect square:

$$x^2 - 6x + \mathbf{9} = -5 + \mathbf{9}$$

$$(x - 3)^2 = 4$$

The last line is in the form of the equations we solved in Section 8.1.

$$x - 3 = \pm 2$$

$$x = 3 \pm 2 \qquad \text{Add 3 to both sides}$$

$$x = 3 + 2 \quad \text{or} \quad x = 3 - 2$$

$$x = 5 \qquad \text{or} \quad x = 1$$

The two solutions are 5 and 1.

◄

The preceding method of solution is called **completing the square.**

Note   The equation in Example 4 can be solved quickly by factoring:

$$x^2 - 6x + 5 = 0$$

$$(x - 5)(x - 1) = 0$$

$$x - 5 = 0 \quad \text{or} \quad x - 1 = 0$$

$$x = 5 \quad \text{or} \qquad x = 1$$

The reason we didn't solve it by factoring is that we want to practice completing the square on some simple equations.

▶ **EXAMPLE 5** Solve by completing the square: $2x^2 + 16x - 18 = 0$.

Solution We begin by moving the constant term to the other side:

$$2x^2 + 16x - 18 = 0$$

$$2x^2 + 16x \quad\quad = 18 \quad\quad \text{Add 18 to both sides}$$

In order to complete the square, we must be sure the coefficient of x^2 is 1. To accomplish this, we divide both sides by 2:

$$\frac{2x^2}{2} + \frac{16x}{2} = \frac{18}{2}$$

$$x^2 + 8x = 9$$

We now complete the square by adding the square of half the coefficient of the linear term to both sides:

$$x^2 + 8x + \mathbf{16} = 9 + \mathbf{16} \quad\quad \text{Add } \mathbf{16} \text{ to both sides}$$

$$(x + 4)^2 = 25$$

$$x + 4 = \pm 5 \quad\quad \text{Square root property}$$

$$x = -4 \pm 5 \quad\quad \text{Add } -4 \text{ to both sides}$$

$$x = -4 + 5 \quad \text{or} \quad x = -4 - 5$$

$$x = 1 \quad\quad\quad \text{or} \quad x = -9$$

The solution set arrived at by completing the square is $\{1, -9\}$. ◀

We will now summarize the preceding examples by listing the steps involved in solving quadratic equations by completing the square.

Strategy for Solving a Quadratic Equation by Completing the Square

Step 1: Put the equation in the form $ax^2 + bx = c$. This usually involves moving only the constant term to the opposite side.

Step 2: Make sure the coefficient of the squared term is 1. If it is not 1, simply divide both sides by whatever it is.

Step 3: Add the square of half the coefficient of the linear term to both sides of the equation.

Step 4: Write the left-hand side of the equation as a binomial square and solve, using the methods developed in Section 8.1.

Here is one final example.

▶ **EXAMPLE 6** Solve for y: $3y^2 - 9y + 3 = 0$.

Solution

$$3y^2 - 9y + 3 = 0$$

$$3y^2 - 9y \qquad = -3 \qquad \text{Add } -3 \text{ to both sides}$$

$$y^2 - 3y \qquad = -1 \qquad \text{Divide by 3}$$

$$y^2 - 3y + \frac{9}{4} = -1 + \frac{9}{4} \qquad \text{Complete the square}$$

$$\left(y - \frac{3}{2}\right)^2 = \frac{5}{4} \qquad\qquad -1 + \frac{9}{4} = -\frac{4}{4} + \frac{9}{4} = \frac{5}{4}$$

$$y - \frac{3}{2} = \pm\frac{\sqrt{5}}{2} \qquad \text{Square root property}$$

$$y = \frac{3}{2} \pm \frac{\sqrt{5}}{2} \qquad \text{Add } \frac{3}{2} \text{ to both sides}$$

$$y = \frac{3}{2} + \frac{\sqrt{5}}{2} \quad \text{or} \quad y = \frac{3}{2} - \frac{\sqrt{5}}{2}$$

$$y = \frac{3 + \sqrt{5}}{2} \quad \text{or} \quad y = \frac{3 - \sqrt{5}}{2}$$

The solutions are $\dfrac{3 + \sqrt{5}}{2}$ and $\dfrac{3 - \sqrt{5}}{2}$, which can be written in a shorter form as

$$\frac{3 \pm \sqrt{5}}{2}$$

Note   We can use a calculator to get decimal approximations to these solutions. If we use the approximation $\sqrt{5} \approx 2.236$, then

$$\frac{3 + \sqrt{5}}{2} \approx \frac{3 + 2.236}{2}$$

$$= \frac{5.236}{2} = 2.618$$

$$\frac{3 - \sqrt{5}}{2} \approx \frac{3 - 2.236}{2}$$

$$= \frac{0.764}{2} = 0.382$$

◀

P R O B L E M   S E T   8 . 2

Give the correct last term for each of the following expressions to ensure that the resulting trinomial is a perfect square trinomial.

1. $x^2 + 6x$
2. $x^2 - 10x$
3. $x^2 + 2x$
4. $x^2 + 14x$
5. $y^2 - 8y$
6. $y^2 + 12y$
7. $y^2 - 2y$
8. $y^2 - 6y$
9. $x^2 + 16x$
10. $x^2 - 4x$
11. $a^2 - 3a$
12. $a^2 + 5a$
13. $x^2 - 7x$
14. $x^2 - 9x$
15. $y^2 + y$

16. $y^2 - y$
17. $x^2 - \dfrac{3}{2}x$

18. $x^2 + \dfrac{2}{3}x$

Solve each of the following equations by completing the square. Follow the steps given at the end of this section.

19. $x^2 + 4x = 12$
20. $x^2 - 2x = 8$
21. $x^2 - 6x = 16$
22. $x^2 + 12x = -27$
23. $a^2 + 2a = 3$
24. $a^2 - 8a = -7$
25. $x^2 - 10x = 0$
26. $x^2 + 4x = 0$
27. $y^2 + 2y - 15 = 0$
28. $y^2 - 10y - 11 = 0$
29. $x^2 + 4x - 3 = 0$
30. $x^2 + 6x + 5 = 0$
31. $x^2 - 4x = 4$
32. $x^2 + 4x = -1$
33. $a^2 = 7a + 8$
34. $a^2 = 3a + 1$
35. $4x^2 + 8x - 4 = 0$
36. $3x^2 + 12x + 6 = 0$
37. $2x^2 + 2x - 4 = 0$
38. $4x^2 + 4x - 3 = 0$
39. $4x^2 + 8x + 1 = 0$
40. $3x^2 + 6x + 2 = 0$
41. $2x^2 - 2x = 1$
42. $3x^2 - 3x = 1$
43. $4a^2 - 4a + 1 = 0$
44. $2a^2 + 4a + 1 = 0$
45. $3y^2 - 9y = 2$
46. $5y^2 - 10y = 4$

47. The two solutions to Problem 29 are $-2 + \sqrt{7}$ and $-2 - \sqrt{7}$. A calculator gives the decimal approximation of $\sqrt{7}$ as 2.646. Use this approximation for $\sqrt{7}$ to find decimal approximations of $-2 + \sqrt{7}$ and $-2 - \sqrt{7}$.

48. The solutions to Problem 41 are $\dfrac{1 + \sqrt{3}}{2}$ and

$\dfrac{1 - \sqrt{3}}{2}$. Use the decimal approximation $\sqrt{3} = 1.732$ to find decimal approximations of these two numbers.

49. One of the solutions to the equation in Problem 35 is $-1 + \sqrt{2}$. Check this solution in the original equation.

50. Check the solution $x = -1 - \sqrt{2}$ from Problem 35 in the original equation $4x^2 + 8x - 4 = 0$.

51. Find the sum of the two solutions to the equation in Problem 29. (Add $-2 + \sqrt{7}$ and $-2 - \sqrt{7}$.)

52. Find the product of the two solutions to the equation in Problem 29.

Completing the Square, Visually

At the end of Problem Set 4.6 we gave some visual representations for $(a + b)^2$ and $(x + 5)^2$. The diagram below can be used to show why we add 16 to $x^2 + 8x$ in order to complete the square.

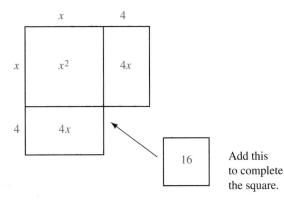

Draw a similar diagram to visualize completing the square on each expression below.

53. $x^2 + 6x$
54. $x^2 + 10x$
55. $x^2 + 2x$
56. $x^2 + 4x$

Review Problems

The following problems review material we covered in Sections 2.1 and 7.2. Reviewing the problems from Section 2.1 will help you with the next section.

Find the value of each expression if $a = 2$, $b = 4$, and $c = -3$. [2.1]

57. $2a$

58. b^2

59. $4ac$

60. $b^2 - 4ac$

61. $\sqrt{b^2 - 4ac}$

62. $-b + \sqrt{b^2 - 4ac}$

Put each expression in simplified form for radicals. [7.2]

63. $\sqrt{12}$

64. $\sqrt{50x^2}$

65. $\sqrt{20x^2y^3}$

66. $3\sqrt{48x^4}$

67. $\sqrt{\dfrac{81}{25}}$

68. $\dfrac{6\sqrt{8x^2y}}{\sqrt{9}}$

SECTION
8.3

The Quadratic Formula

In this section we will derive the quadratic formula. It is one formula that you will use in almost all types of mathematics. We will first state the formula as a theorem and then prove it. The proof is based on the method of completing the square developed in the last section.

> **The Quadratic Theorem**
>
> For any quadratic equation in the form $ax^2 + bx + c = 0$, where a, b, and c are real numbers and $a \neq 0$, the two solutions are
>
> $$x = \frac{-b + \sqrt{b^2 - 4ac}}{2a} \quad \text{and} \quad x = \frac{-b - \sqrt{b^2 - 4ac}}{2a}$$

PROOF We will prove the theorem by completing the square on

$$ax^2 + bx + c = 0$$

Adding $-c$ to both sides, we have

$$ax^2 + bx = -c$$

To make the coefficient of x^2 one, we divide both sides by a:

$$\frac{ax^2}{a} + \frac{bx}{a} = -\frac{c}{a}$$

$$x^2 + \frac{b}{a}x = -\frac{c}{a}$$

Now, to complete the square, we add the square of half of $\frac{b}{a}$ to both sides:

$$x^2 + \frac{b}{a}x + \left(\frac{b}{2a}\right)^2 = -\frac{c}{a} + \left(\frac{b}{2a}\right)^2 \qquad \frac{1}{2} \text{ of } \frac{b}{a} \text{ is } \frac{b}{2a}$$

Let's simplify the right side separately:

$$-\frac{c}{a} + \left(\frac{b}{2a}\right)^2 = -\frac{c}{a} + \frac{b^2}{4a^2}$$

The least common denominator is $4a^2$. We multiply the numerator and denominator of $-\dfrac{c}{a}$ by $4a$ to give it the common denominator. Then we combine numerators:

$$\frac{4a}{4a}\left(-\frac{c}{a}\right) + \frac{b^2}{4a^2} = -\frac{4ac}{4a^2} + \frac{b^2}{4a^2}$$

$$= \frac{-4ac + b^2}{4a^2}$$

$$= \frac{b^2 - 4ac}{4a^2}$$

Now back to the equation. We use our simplified expression for the right side:

$$x^2 + \frac{b}{a}x + \left(\frac{b}{2a}\right)^2 = \frac{b^2 - 4ac}{4a^2}$$

$$\left(x + \frac{b}{2a}\right)^2 = \frac{b^2 - 4ac}{4a^2}$$

Applying the square root property, we have

$$x + \frac{b}{2a} = \pm\frac{\sqrt{b^2 - 4ac}}{2a}$$

$$x = \frac{-b}{2a} \pm \frac{\sqrt{b^2 - 4ac}}{2a} \qquad \text{Add } \frac{-b}{2a} \text{ to both sides}$$

$$x = \frac{-b \pm \sqrt{b^2 - 4ac}}{2a}$$

Our proof is now complete. What we have is this: if our equation is in the form $ax^2 + bx + c = 0$ (standard form), then the solution is given by:

$$x = \frac{-b \pm \sqrt{b^2 - 4ac}}{2a}$$

This formula is called the **quadratic** formula. You will see it many times if you continue taking math classes. By the time you are finished with this section and the problems in the problem set, you should have it memorized.

▶ **EXAMPLE 1** Solve $x^2 - 5x - 6 = 0$ by using the quadratic formula.

Solution To use the quadratic formula, we must make sure the equation is in standard form; identify a, b, and c; substitute them into the formula, and work out the arithmetic.

For the equation $x^2 - 5x - 6 = 0$, $a = 1$, $b = -5$, and $c = -6$:

$$x = \frac{-b \pm \sqrt{b^2 - 4ac}}{2a} = \frac{-(-5) \pm \sqrt{(-5)^2 - 4(1)(-6)}}{2(1)}$$

$$= \frac{5 \pm \sqrt{49}}{2}$$

$$= \frac{5 \pm 7}{2}$$

$$x = \frac{5 + 7}{2} \quad \text{or} \quad x = \frac{5 - 7}{2}$$

$$x = \frac{12}{2} \qquad\qquad x = -\frac{2}{2}$$

$$x = 6 \qquad\qquad x = -1$$

The two solutions are 6 and -1. ◀

Note Whenever the solutions to our quadratic equations turn out to be rational numbers, as in Example 1, it means the original equation could have been solved by factoring. (We didn't solve the equation in Example 1 by factoring because we were trying to get some practice with the quadratic formula.)

▶ **EXAMPLE 2** Solve for x: $2x^2 = -4x + 3$.

Solution Before we can identify a, b, and c, we must write the equation in standard form. To do so we add $4x$ and -3 to each side of the equation:

$$2x^2 = -4x + 3$$

$$2x^2 + 4x - 3 = 0 \qquad \text{Add } 4x \text{ and } -3 \text{ to each side}$$

Now that the equation is in standard form, we see that $a = 2$, $b = 4$, and $c = -3$. Using the quadratic formula we have

$$x = \frac{-b \pm \sqrt{b^2 - 4ac}}{2a}$$

$$= \frac{-4 \pm \sqrt{4^2 - 4(2)(-3)}}{2(2)}$$

$$= \frac{-4 \pm \sqrt{40}}{4}$$

$$= \frac{-4 \pm 2\sqrt{10}}{4}$$

We can reduce this last expression to lowest terms by factoring 2 from the numerator and denominator and then dividing it out:

$$x = \frac{\cancel{2}(-2 \pm \sqrt{10})}{\cancel{2} \cdot 2}$$

$$= \frac{-2 \pm \sqrt{10}}{2}$$

Our two solutions are $\dfrac{-2 + \sqrt{10}}{2}$ and $\dfrac{-2 - \sqrt{10}}{2}$. ◀

▶ **EXAMPLE 3** Solve for x: $(x - 2)(x + 3) = 5$.

Solution We must put the equation into standard form before we can use the formula:

$$(x - 2)(x + 3) = 5$$

$$x^2 + x - 6 = 5 \qquad \text{Multiply out the left side}$$

$$x^2 + x - 11 = 0 \qquad \text{Add } -5 \text{ to each side}$$

Now $a = 1$, $b = 1$, and $c = -11$; therefore

$$x = \frac{-1 \pm \sqrt{1^2 - 4(1)(-11)}}{2(1)}$$

$$= \frac{-1 \pm \sqrt{45}}{2}$$

$$= \frac{-1 \pm 3\sqrt{5}}{2}$$

The solution set is $\left\{ \dfrac{-1 + 3\sqrt{5}}{2}, \dfrac{-1 - 3\sqrt{5}}{2} \right\}$. ◀

▶ **EXAMPLE 4** Solve $x^2 - 6x = -7$.

Solution We begin by writing the equation in standard form:

$$x^2 - 6x = -7$$

$$x^2 - 6x + 7 = 0 \qquad \text{Add 7 to each side}$$

Using $a = 1$, $b = -6$, and $c = 7$ in the quadratic formula

$$x = \frac{-b \pm \sqrt{b^2 - 4ac}}{2a}$$

we have

$$x = \frac{-(-6) \pm \sqrt{(-6)^2 - 4(1)(7)}}{2(1)}$$

$$= \frac{6 \pm \sqrt{36 - 28}}{2}$$

$$= \frac{6 \pm \sqrt{8}}{2}$$

$$= \frac{6 \pm 2\sqrt{2}}{2}$$

The two terms in the numerator have a 2 in common. We reduce to lowest terms by factoring the 2 from the numerator and then dividing numerator and denominator by 2:

$$= \frac{\cancel{2}(3 \pm \sqrt{2})}{\cancel{2}}$$

$$= 3 \pm \sqrt{2}$$

The two solutions are $3 + \sqrt{2}$ and $3 - \sqrt{2}$. This time, let's check our solutions in the original equation $x^2 - 6x = -7$.
Checking $x = 3 + \sqrt{2}$, we have

$$(3 + \sqrt{2})^2 - 6(3 + \sqrt{2}) \stackrel{?}{=} -7$$

$$9 + 6\sqrt{2} + 2 - 18 - 6\sqrt{2} \stackrel{?}{=} -7 \qquad \text{Multiply}$$

$$11 - 18 + 6\sqrt{2} - 6\sqrt{2} \stackrel{?}{=} -7 \qquad \text{Add 9 and 2}$$

$$-7 + 0 \stackrel{?}{=} -7 \qquad \text{Subtraction}$$

$$-7 = -7 \qquad \text{A true statement}$$

Checking $x = 3 - \sqrt{2}$, we have

$$(3 - \sqrt{2})^2 - 6(3 - \sqrt{2}) \stackrel{?}{=} -7$$

$$9 - 6\sqrt{2} + 2 - 18 + 6\sqrt{2} \stackrel{?}{=} -7 \qquad \text{Multiply}$$

$$11 - 18 - 6\sqrt{2} + 6\sqrt{2} \stackrel{?}{=} -7 \qquad \text{Add 9 and 2}$$

$$-7 + 0 \stackrel{?}{=} -7 \qquad \text{Subtraction}$$

$$-7 = -7 \qquad \text{A true statement}$$

As you can see, both solutions yield true statements when used in place of the variable in the original equation. ◀

P R O B L E M   S E T   8 . 3

Solve the following equations by using the quadratic formula.

1. $x^2 + 3x + 2 = 0$ **2.** $x^2 - 5x + 6 = 0$
3. $x^2 + 5x + 6 = 0$ **4.** $x^2 - 7x - 8 = 0$
5. $x^2 + 6x + 9 = 0$
6. $x^2 - 10x + 25 = 0$
7. $x^2 + 6x + 7 = 0$ **8.** $x^2 - 4x - 1 = 0$
9. $2x^2 + 5x + 3 = 0$
10. $2x^2 + 3x - 20 = 0$
11. $4x^2 + 8x + 1 = 0$ **12.** $3x^2 + 6x + 2 = 0$
13. $x^2 - 2x + 1 = 0$ **14.** $x^2 + 2x - 3 = 0$
15. $x^2 - 5x = 7$ **16.** $2x^2 - 6x = 8$
17. $6x^2 - x - 2 = 0$ **18.** $6x^2 + 5x - 4 = 0$
19. $(x - 2)(x + 1) = 3$ **20.** $(x - 8)(x + 7) = 5$
21. $(2x - 3)(x + 2) = 1$
22. $(4x - 5)(x - 3) = 6$
23. $2x^2 - 3x = 5$ **24.** $3x^2 - 4x = 5$
25. $2x^2 = -6x + 7$ **26.** $5x^2 = -6x + 3$
27. $3x^2 = -4x + 2$ **28.** $3x^2 = 4x + 2$
29. $2x^2 - 5 = 2x$ **30.** $5x^2 + 1 = -10x$
31. Solve the equation $2x^3 + 3x^2 - 4x = 0$ by first factoring out the common factor, x, and then using the quadratic formula. There are three solutions to this equation.
32. Solve the equation $5y^3 - 10y^2 + 4y = 0$ by first factoring out the common factor, y, and then using the quadratic formula.
33. To apply the quadratic formula to the equation $3x^2 - 4x = 0$, you have to notice that $c = 0$. Solve the equation using the quadratic formula.

34. Solve the equation $9x^2 - 16 = 0$ using the quadratic formula. (Notice $b = 0$.)
35. Solve the following equation by first multiplying both sides by the LCD and then applying the quadratic formula to the result.

$$\frac{1}{2}x^2 - \frac{1}{2}x - \frac{1}{6} = 0$$

36. Solve the following equation by first multiplying both sides by the LCD and then applying the quadratic formula to the result.

$$\frac{1}{2}y^2 - y - \frac{3}{2} = 0$$

Review Problems

The following problems review the material we covered in Section 7.5. Reviewing these problems will help you with the material in the next section.

Multiply.

37. $(2\sqrt{3})(3\sqrt{5})$ **38.** $\sqrt{7}(\sqrt{7} + 4)$
39. $(\sqrt{6} + 2)(\sqrt{6} - 5)$ **40.** $(\sqrt{x} + 4)^2$
41. $(\sqrt{7} - \sqrt{2})(\sqrt{7} + \sqrt{2})$
42. $(\sqrt{a} + \sqrt{b})(\sqrt{a} - \sqrt{b})$

Rationalize the denominator.

43. $\dfrac{2}{3 + \sqrt{5}}$ **44.** $\dfrac{2 + \sqrt{3}}{2 - \sqrt{3}}$

SECTION

8.4   # Complex Numbers

In order to solve quadratic equations such as $x^2 = -4$, we need to introduce a new set of numbers. If we try to solve $x^2 = -4$ using real numbers, we always get no solution. There is no real number whose square is -4.

The new set of numbers is called **complex numbers** and is based on the following definition.

> **DEFINITION** **The number i** is a number such that $i = \sqrt{-1}$.

The first thing we notice about this definition is that i is not a real number. There are no real numbers that represent the square root of -1. The other observation we make about i is $i^2 = -1$. If $i = \sqrt{-1}$, then, squaring both sides, we must have $i^2 = -1$. The most common power of i is i^2. Whenever we see i^2, we can write it as -1. We are now ready for a definition of complex numbers.

> **DEFINITION** A **complex number** is any number that can be put in the form $a + bi$, where a and b are real numbers and $i = \sqrt{-1}$.

Note The form $a + bi$ is called **standard form** for complex numbers. The definition above indicates that if a number can be written in the form $a + bi$, then it is a complex number.

EXAMPLE The following are complex numbers:

$$3 + 4i \qquad \frac{1}{2} - 6i \qquad 8 + i\sqrt{2} \qquad \frac{3}{4} - 2i\sqrt{5}$$

EXAMPLE The number $4i$ is a complex number because $4i = 0 + 4i$.

EXAMPLE The number 8 is a complex number because $8 = 8 + 0i$.

From the last example we can see that the real numbers are a subset of the complex numbers because any real number x can be written as $x + 0i$.

Addition and Subtraction of Complex Numbers

We add and subtract complex numbers according to the same procedure we used to add and subtract polynomials: we combine similar terms.

▶ **EXAMPLE 1** Combine $(3 + 4i) + (2 - 6i)$.

Solution

$$\begin{aligned}
(3 + 4i) + (2 - 6i) &= (3 + 2) + (4i - 6i) && \text{Commutative and} \\
&&& \text{associative properties} \\
&= 5 + (-2i) && \text{Combine similar terms} \\
&= 5 - 2i
\end{aligned}$$

◀

▶ **EXAMPLE 2**   Combine $(2 - 5i) - (3 + 7i) + (2 - i)$.

Solution

$$(2 - 5i) - (3 + 7i) + (2 - i) = 2 - 5i - 3 - 7i + 2 - i$$
$$= (2 - 3 + 2) + (-5i - 7i - i)$$
$$= 1 - 13i$$ ◀

Multiplication of Complex Numbers

Multiplication of complex numbers is very similar to multiplication of polynomials. We can simplify many answers by using the fact that $i^2 = -1$.

▶ **EXAMPLE 3**   Multiply $4i(3 + 5i)$.

Solution

$$4i(3 + 5i) = 4i(3) + 4i(5i) \qquad \text{Distributive property}$$
$$= 12i + 20i^2 \qquad \text{Multiplication}$$
$$= 12i + 20(-1) \qquad i^2 = -1$$
$$= -20 + 12i$$ ◀

▶ **EXAMPLE 4**   Multiply $(3 + 2i)(4 - 3i)$.

Solution

$$(3 + 2i)(4 - 3i)$$
$$= 3 \cdot 4 + 3(-3i) + 2i(4) + 2i(-3i) \qquad \text{FOIL method}$$
$$= 12 - 9i + 8i - 6i^2$$
$$= 12 - 9i + 8i - 6(-1) \qquad\qquad i^2 = -1$$
$$= (12 + 6) + (-9i + 8i)$$
$$= 18 - i$$ ◀

Division of Complex Numbers

We divide complex numbers by applying the same process we used to rationalize denominators.

▶ **EXAMPLE 5**   Divide $\dfrac{2i}{3 - 4i}$.

Solution   We multiply numerator and denominator by the conjugate of the denominator, which is $3 + 4i$:

$$\left(\frac{2i}{3-4i}\right)\left(\frac{3+4i}{3+4i}\right) = \frac{6i + 8i^2}{9 - 16i^2}$$

$$= \frac{6i + 8(-1)}{9 - 16(-1)}$$

$$= \frac{-8 + 6i}{25}$$ ◄

▶ **EXAMPLE 6** Divide $\dfrac{2+i}{5+2i}$.

Solution The conjugate of the denominator is $5 - 2i$:

$$\left(\frac{2+i}{5+2i}\right)\left(\frac{5-2i}{5-2i}\right) = \frac{10 - 4i + 5i - 2i^2}{25 - 4i^2}$$

$$= \frac{10 - 4i + 5i - 2(-1)}{25 - 4(-1)} \qquad i^2 = -1$$

$$= \frac{12 + i}{29}$$

If we write our answer in standard form for complex numbers, it will look like this:

$$\frac{12}{29} + \frac{1}{29}i$$ ◄

PROBLEM SET 8.4

Combine the following complex numbers.

1. $(3 - 2i) + 3i$
2. $(5 - 4i) - 8i$
3. $(6 + 2i) - 10i$
4. $(8 - 10i) + 7i$
5. $(11 + 9i) - 9i$
6. $(12 + 2i) + 6i$
7. $(3 + 2i) + (6 - i)$
8. $(4 + 8i) - (7 + i)$
9. $(5 + 7i) - (6 + 8i)$
10. $(11 + 6i) - (3 + 6i)$
11. $(9 - i) + (2 - i)$
12. $(8 + 3i) - (8 - 3i)$
13. $(6 + i) - 4i - (2 - i)$
14. $(3 + 2i) - 5i - (5 + 4i)$
15. $(6 - 11i) + 3i + (2 + i)$
16. $(3 + 4i) - (5 + 7i) - (6 - i)$
17. $(2 + 3i) - (6 - 2i) + (3 - i)$
18. $(8 + 9i) + (5 - 6i) - (4 - 3i)$

Multiply the following complex numbers.

19. $3(2 - i)$
20. $4(5 + 3i)$
21. $2i(8 - 7i)$
22. $-3i(2 + 5i)$
23. $(2 + i)(4 - i)$
24. $(6 + 3i)(4 + 3i)$
25. $(2 + i)(3 - 5i)$
26. $(4 - i)(2 - i)$
27. $(3 + 5i)(3 - 5i)$
28. $(8 + 6i)(8 - 6i)$
29. $(2 + i)(2 - i)$
30. $(3 + i)(3 - i)$

Divide the following complex numbers.

31. $\dfrac{2}{3 - 2i}$
32. $\dfrac{3}{5 + 6i}$
33. $\dfrac{-3i}{2 + 3i}$
34. $\dfrac{4i}{3 + i}$

35. $\dfrac{6i}{3 - i}$

36. $\dfrac{-7i}{5 - 4i}$

37. $\dfrac{2 + i}{2 - i}$

38. $\dfrac{3 + 2i}{3 - 2i}$

39. $\dfrac{4 + 5i}{3 - 6i}$

40. $\dfrac{-2 + i}{5 + 6i}$

41. Use the FOIL method to multiply $(x + 3i)(x - 3i)$.

42. Use the FOIL method to multiply $(x + 5i)(x - 5i)$.

43. The opposite of i is $-i$. The reciprocal of i is $\dfrac{1}{i}$.

Multiply the numerator and denominator of $\dfrac{1}{i}$ by i and simplify the result to see that the opposite of i and the reciprocal of i are the same number.

44. If $i^2 = -1$, what are i^3 and i^4? (*Hint:* $i^3 = i^2 \cdot i$.)

Review Problems

The following problems review material we covered in Sections 8.1 and 7.3. Reviewing the problems from Section 8.1 will help you with the next section.

Solve each equation by taking the square root of each side. [8.1]

45. $(x - 3)^2 = 25$

46. $(x - 2)^2 = 9$

47. $(2x - 6)^2 = 16$

48. $(2x + 1)^2 = 49$

49. $(x + 3)^2 = 12$

50. $(x + 3)^2 = 8$

Put each expression into simplified form for radicals. [7.3]

51. $\sqrt{\dfrac{1}{2}}$

52. $\sqrt{\dfrac{5}{6}}$

53. $\sqrt{\dfrac{8x^2y^3}{3}}$

54. $\sqrt{\dfrac{45xy^4}{7}}$

55. $\sqrt[3]{\dfrac{1}{4}}$

56. $\sqrt[3]{\dfrac{2}{3}}$

SECTION

8.5

Complex Solutions to Quadratic Equations

The quadratic formula tells us that solutions to equations of the form $ax^2 + bx + c = 0$ are always

$$x = \frac{-b \pm \sqrt{b^2 - 4ac}}{2a}$$

The part of the quadratic formula under the radical sign is called the **discriminant:**

$$\text{Discriminant} = b^2 - 4ac$$

When the discriminant is negative, we have to deal with the square root of a negative number. We handle square roots of negative numbers by using the definition $i = \sqrt{-1}$. To illustrate, suppose we want to simplify an expression that contains $\sqrt{-9}$, which is not a real number. We begin by writing $\sqrt{-9}$ as $\sqrt{9(-1)}$. Then, we write this expression as the product of two separate radicals: $\sqrt{9}\,\sqrt{-1}$. Applying the definition $i = \sqrt{-1}$ to this last expression, we have

$$\sqrt{9}\sqrt{-1} = 3i$$

As you may recall from the previous section, the number $3i$ is called a complex number. Here are some further examples.

▶ **EXAMPLE 1** Write the following radicals as complex numbers.

(a) $\sqrt{-4} = \sqrt{4(-1)} = \sqrt{4}\sqrt{-1} = 2i$

(b) $\sqrt{-36} = \sqrt{36(-1)} = \sqrt{36}\sqrt{-1} = 6i$

(c) $\sqrt{-7} = \sqrt{7(-1)} = \sqrt{7}\sqrt{-1} = i\sqrt{7}$

(d) $\sqrt{-75} = \sqrt{75(-1)} = \sqrt{75}\sqrt{-1} = 5i\sqrt{3}$ ◀

In parts (c) and (d) of Example 1, we wrote i before the radical because it is less confusing that way. If we put i after the radical, it is sometimes mistaken for being under the radical.

Let's see how complex numbers relate to quadratic equations by looking at some examples of quadratic equations whose solutions are complex numbers.

▶ **EXAMPLE 2** Solve for x: $(x + 2)^2 = -9$.

Solution We can solve this equation by expanding the left side, putting the results into standard form, and then applying the quadratic formula. It is faster, however, simply to apply the square root property.

$$(x + 2)^2 = -9$$
$$x + 2 = \pm\sqrt{-9} \qquad \text{Square root property}$$
$$x + 2 = \pm 3i \qquad \sqrt{-9} = \sqrt{9}\sqrt{-1} = 3i$$
$$x = -2 \pm 3i \qquad \text{Add } -2 \text{ to both sides}$$

The solution set contains two complex solutions. Notice that the two solutions are conjugates.

The solution set is $\{-2 + 3i, -2 - 3i\}$.   ◀

▶ **EXAMPLE 3** Solve for x: $\dfrac{1}{10}x^2 - \dfrac{1}{5}x = -\dfrac{1}{2}$.

Solution It will be easier to apply the quadratic formula if we clear the equation of fractions. Multiplying both sides of the equation by the LCD 10 gives us

$$x^2 - 2x = -5$$

Next we add 5 to both sides to put the equation into standard form:

$$x^2 - 2x + 5 = 0 \qquad \text{Add 5 to both sides}$$

Applying the quadratic formula with $a = 1$, $b = -2$, and $c = 5$, we have

$$x = \frac{-(-2) \pm \sqrt{(-2)^2 - 4(1)(5)}}{2(1)}$$

$$= \frac{2 \pm \sqrt{-16}}{2}$$

$$= \frac{2 \pm 4i}{2}$$

Dividing the numerator and denominator by 2, we have the two solutions:

$$x = 1 \pm 2i$$

The two solutions are $1 + 2i$ and $1 - 2i$.   ◀

▶ **EXAMPLE 4**   Solve $(2x - 3)(2x - 1) = -4$.

Solution   We multiply the binomials on the left side and then add 4 to each side to write the equation in standard form. From there we identify a, b, and c and apply the quadratic formula.

$$(2x - 3)(2x - 1) = -4$$

$$4x^2 - 8x + 3 = -4 \qquad \text{Multiply binomials on left side}$$

$$4x^2 - 8x + 7 = 0 \qquad \text{Add 4 to each side}$$

Placing $a = 4$, $b = -8$, and $c = 7$ in the quadratic formula we have

$$x = \frac{-(-8) \pm \sqrt{(-8)^2 - 4(4)(7)}}{2(4)}$$

$$= \frac{8 \pm \sqrt{64 - 112}}{8}$$

$$= \frac{8 \pm \sqrt{-48}}{8}$$

$$= \frac{8 \pm 4i\sqrt{3}}{8} \qquad \sqrt{-48} = i\sqrt{48} = i\sqrt{16}\sqrt{3} = 4i\sqrt{3}$$

To reduce this last expression to lowest terms we factor a 4 from the numerator and then divide numerator and denominator by 4:

$$= \frac{\cancel{4}(2 \pm i\sqrt{3})}{\cancel{4} \cdot 2}$$

$$= \frac{2 \pm i\sqrt{3}}{2}$$

   ◀

Note   It would be a mistake to try to reduce this last expression further. Sometimes first-year algebra students will try to divide the 2 in the denominator into the 2 in the numerator, which is a mistake. Remember, when we reduce to lowest terms we do so by dividing the numerator and denominator by any factors they have in common. In this case, 2 is not a factor of the numerator. This expression is in lowest terms.

This completes our work with solving quadratic equations. We can solve any quadratic equation. Factoring is probably still the fastest method of solution, but again, factoring works only if the equation is factorable. Applying the quadratic formula always produces solutions, whether the equation is factorable or not.

PROBLEM SET 8.5

Write the following radicals as complex numbers.

1. $\sqrt{-16}$

2. $\sqrt{-25}$

3. $\sqrt{-49}$

4. $\sqrt{-81}$

5. $\sqrt{-6}$

6. $\sqrt{-10}$

7. $\sqrt{-11}$

8. $\sqrt{-19}$

9. $\sqrt{-32}$

10. $\sqrt{-288}$

11. $\sqrt{-50}$

12. $\sqrt{-45}$

13. $\sqrt{-8}$

14. $\sqrt{-24}$

15. $\sqrt{-48}$

16. $\sqrt{-27}$

Solve the following quadratic equations. Use whatever method seems to fit the situation or is convenient for you.

17. $x^2 = 2x - 2$

18. $x^2 = 4x - 5$

19. $x^2 - 4x = -4$

20. $x^2 - 4x = 4$

21. $2x^2 + 5x = 12$

22. $2x^2 + 30 = 16x$

23. $(x - 2)^2 = -4$

24. $(x - 5)^2 = -25$

25. $\left(x + \dfrac{1}{2}\right)^2 = -\dfrac{9}{4}$

26. $\left(x - \dfrac{1}{4}\right)^2 = -\dfrac{1}{2}$

27. $\left(x - \dfrac{1}{2}\right)^2 = -\dfrac{27}{36}$

28. $\left(x + \dfrac{1}{2}\right)^2 = -\dfrac{32}{64}$

29. $x^2 + x + 1 = 0$

30. $x^2 - 3x + 4 = 0$

31. $x^2 - 5x + 6 = 0$

32. $x^2 + 2x + 2 = 0$

33. $\dfrac{1}{2}x^2 + \dfrac{1}{3}x + \dfrac{1}{6} = 0$

34. $\dfrac{1}{5}x^2 + \dfrac{1}{20}x + \dfrac{1}{4} = 0$

35. $\dfrac{1}{3}x^2 = -\dfrac{1}{2}x + \dfrac{1}{3}$

36. $\dfrac{1}{2}x^2 = -\dfrac{1}{3}x + \dfrac{1}{6}$

37. $(x + 2)(x - 3) = 5$

38. $(x - 1)(x + 1) = 6$

39. $(x - 5)(x - 3) = -10$

40. $(x - 2)(x - 4) = -5$

41. $(2x - 2)(x - 3) = 9$

42. $(x - 1)(2x + 6) = 9$

43. Is $x = 2 + 2i$ a solution to the equation $x^2 - 4x + 8 = 0$?

44. Is $x = 5 + 3i$ a solution to the equation $x^2 - 10x + 34 = 0$?

45. If one solution to a quadratic equation is $3 + 7i$, what do you think the other solution is?

46. If one solution to a quadratic equation is $4 - 2i$, what do you think the other solution is?

Review Problems

The following problems review material we covered in Sections 3.2, 3.3, and 7.4. Reviewing the problems from Sections 3.2 and 3.3 will help you understand the next section.

Graph each line. [3.2, 3.3]

47. $y = x - 2$

48. $y = -x - 2$

49. $2x + 4y = 8$

50. $2x - 4y = 8$

Write each term in simplified form for radicals. Then combine similar terms. [7.4]

51. $3\sqrt{50} + 2\sqrt{32}$

52. $4\sqrt{18} + \sqrt{32} - \sqrt{2}$

53. $\sqrt{24} - \sqrt{54} - \sqrt{150}$

54. $\sqrt{72} - \sqrt{8} + \sqrt{50}$

55. $2\sqrt{27x^2} - x\sqrt{48}$

56. $5\sqrt{8x^3} + x\sqrt{50x}$

SECTION

8.6

Graphing Parabolas

In this section we will graph equations of the form $y = ax^2 + bx + c$ and equations that can be put into this form. The graphs of this type of equation all have similar shapes.

We will begin by reviewing how we graphed the simplest quadratic equation, $y = x^2$, in Chapter 4. To get the idea of the shape of this graph, we need to find some ordered pairs that are solutions. We can do this by setting up the following table:

x	$y = x^2$	y

We can choose any convenient numbers for x and then use the equation $y = x^2$ to find the corresponding values for y. Let's use the values $-3, -2, -1, 0, 1, 2,$ and 3 for x and find corresponding values for y. Here is how the table looks when we let x have these values:

x	$y = x^2$	y
-3	$y = (-3)^2 = 9$	9
-2	$y = (-2)^2 = 4$	4
-1	$y = (-1)^2 = 1$	1
0	$y = 0^2 = 0$	0
1	$y = 1^2 = 1$	1
2	$y = 2^2 = 4$	4
3	$y = 3^2 = 9$	9

The table gives us the solutions $(-3, 9), (-2, 4), (-1, 1), (0, 0), (1, 1), (2, 4),$ and $(3, 9)$ for the equation $y = x^2$. We plot each of the points on a rectangular coordinate system and draw a smooth curve between them, as shown in Figure 1.

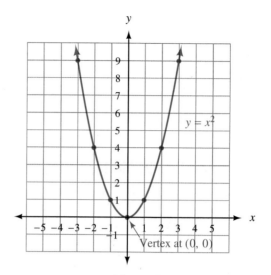

Figure 1

This graph is called a **parabola**. All equations of the form $y = ax^2 + bx + c$ $(a \neq 0)$ produce parabolas when graphed.

Note that the point $(0, 0)$ is called the **vertex** of the parabola in Figure 1. It is the lowest point on the graph. Although all the parabolas in this section will open up, there are some parabolas that open downward. For those parabolas, the vertex is the highest point on the graph.

▶ **EXAMPLE 1** Graph the equation $y = x^2 - 3$.

Solution We begin by making a table using convenient values for x:

x	$y = x^2 - 3$	y
-2	$y = (-2)^2 - 3 = 4 - 3 = 1$	1
-1	$y = (-1)^2 - 3 = 1 - 3 = -2$	-2
0	$y = 0^2 - 3 = -3$	-3
1	$y = 1^2 - 3 = 1 - 3 = -2$	-2
2	$y = 2^2 - 3 = 4 - 3 = 1$	1

The table gives us the ordered pairs $(-2, 1)$, $(-1, -2)$, $(0, -3)$, $(1, -2)$, and $(2, 1)$ as solutions to $y = x^2 - 3$. The graph is shown in Figure 2.

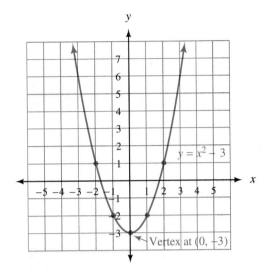

$y = x^2 - 3$

Vertex at $(0, -3)$

Figure 2

◀

▶ **EXAMPLE 2**   Graph $y = (x - 2)^2$.

Solution

x	$y = (x - 2)^2$	y
-1	$y = (-1 - 2)^2 = (-3)^2 = 9$	9
0	$y = (0 - 2)^2 = (-2)^2 = 4$	4
1	$y = (1 - 2)^2 = (-1)^2 = 1$	1
2	$y = (2 - 2)^2 = 0^2 = 0$	0

We can continue the table if we feel more solutions will make the graph clearer.

3	$y = (3 - 2)^2 = 1^2 = 1$	1
4	$y = (4 - 2)^2 = 2^2 = 4$	4
5	$y = (5 - 2)^2 = 3^2 = 9$	9

Putting the results of the table onto a coordinate system, we have the graph in Figure 3.

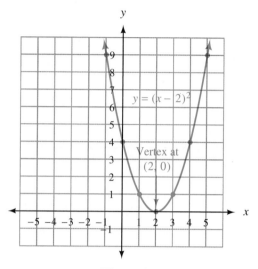

Figure 3

When graphing parabolas, it is sometimes easier to obtain the correct graph if you have some idea what the graph will look like before you start making your table.

If you look over the first two examples closely, you will see that the graphs in Figures 2 and 3 have the same shape as the graph of $y = x^2$. The difference is in the position of the vertex. For example, the graph of $y = x^2 - 3$, as shown in Figure 2, looks like the graph of $y = x^2$ with its vertex moved down three units vertically. Similarly, the graph of $y = (x - 2)^2$, shown in Figure 3, looks like the graph of $y = x^2$ with its vertex moved two units to the right.

Without showing the tables necessary to graph them, Figures 4–7 (below and next page) are four more parabolas and their corresponding equations. Can

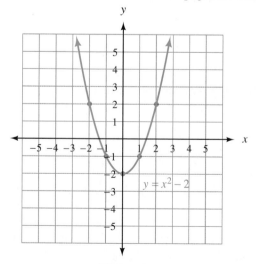

Figure 4

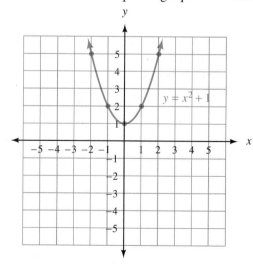

Figure 5

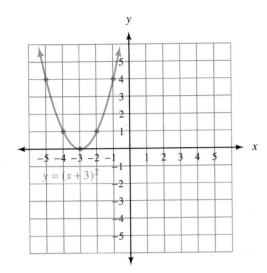

Figure 6

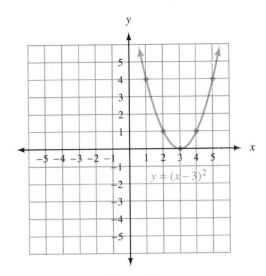

Figure 7

you see the correlation between the numbers in the equations and the position of the graphs on the coordinate systems?

For our next example, we will graph a parabola in which the vertex has been moved both vertically and horizontally from the origin.

▶ **EXAMPLE 3**   Graph $y = (x + 1)^2 - 3$.

Solution

x	$y = (x + 1)^2 - 3$	y
-4	$y = (-4 + 1)^2 - 3 = 9 - 3$	6
-3	$y = (-3 + 1)^2 - 3 = 4 - 3$	1
-1	$y = (-1 + 1)^2 - 3 = 0 - 3$	-3
1	$y = (1 + 1)^2 - 3 = 4 - 3$	1
2	$y = (2 + 1)^2 - 3 = 9 - 3$	6

Graphing the results of the table, we have Figure 8.

Note that the vertex is at the point $(-1, -3)$. As was the case with the parabolas we graphed previously, the number inside the parentheses, in this case 1, moves the vertex away from the origin in a horizontal direction. Likewise, the number outside the parentheses, in this case -3, moves the vertex away from the origin in a vertical direction.

We summarize the information shown in the examples and graphs above with the following statement.

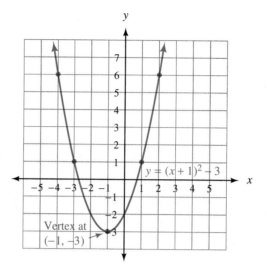

$y = (x + 1)^2 - 3$

Vertex at $(-1, -3)$

Figure 8

Graphing Parabolas

The graph of each equation below is a parabola with the same basic shape as the graph of $y = x^2$.

Equation	Vertex	To obtain the graph, move the graph of $y = x^2$
$y = x^2 + k$	$(0, k)$	k units vertically
$y = (x - h)^2$	$(h, 0)$	h units horizontally
$y = (x - h)^2 + k$	(h, k)	h units horizontally and k units vertically

---

Our last example for this section shows what we do if the equation of the parabola we want to graph is not in the form $y = (x - h)^2 + k$.

▶ **EXAMPLE 4** Graph $y = x^2 - 6x + 5$.

Solution We could graph this equation by making a table of values of x and y as we have previously. However, there is an easier way! If we can write the equation in the form

$$y = (x - h)^2 + k$$

then, to graph it, we simply place the vertex at (h, k) and draw a graph from there that has the same shape as $y = x^2$.

To write our equation in the form given above, we simply complete the square on the first two terms on the right side of the equation, $x^2 - 6x$. We do so by adding 9 to and subtracting 9 from the right side of the equation. This amounts to adding 0 to the equation, so we know we haven't changed its solutions. This is what it looks like:

$$y = (x^2 - 6x \quad\quad) + 5$$

$$y = (x^2 - 6x + \mathbf{9}) + 5 - \mathbf{9}$$

$$y = (x - 3)^2 - 4$$

This last equation has the form we are looking for. Our graph will have the same shape as the graph of $y = x^2$. The vertex of our graph is at $(3, -4)$. The graph is shown in Figure 9.

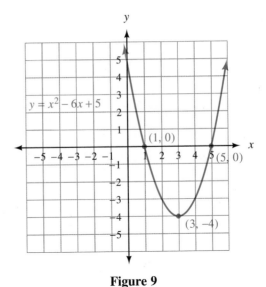

Figure 9

One final note: We can check to see that the x-intercepts in Figure 9 are correct by finding the x-intercepts from our original equation. Remember, the x-intercepts occur when y is 0. That is, to find the x-intercepts, we let $y = 0$ and solve for x:

$$0 = x^2 - 6x + 5$$

$$0 = (x - 5)(x - 1)$$

$$x = 5 \quad \text{or} \quad x = 1$$

P R O B L E M   S E T   8 . 6

Graph each of the following equations.

1. $y = x^2 - 4$ **2.** $y = x^2 + 2$
3. $y = x^2 + 5$ **4.** $y = x^2 - 2$
5. $y = (x + 2)^2$ **6.** $y = (x + 5)^2$
7. $y = (x - 3)^2$ **8.** $y = (x - 2)^2$
9. $y = (x - 5)^2$ **10.** $y = (x + 3)^2$
11. $y = (x + 1)^2 - 2$ **12.** $y = (x - 1)^2 + 2$
13. $y = (x + 2)^2 - 3$ **14.** $y = (x - 2)^2 + 3$
15. $y = (x - 3)^2 + 2$ **16.** $y = (x + 4)^2 + 1$

Graph each of the following equations. Begin by completing the square on the first two terms as we did in Example 4 of this section.

17. $y = x^2 + 6x + 5$ **18.** $y = x^2 - 8x + 12$
19. $y = x^2 - 2x - 3$ **20.** $y = x^2 + 2x - 3$

The following equations have graphs that are also parabolas. However, the graphs of these equations will open downward; the vertex of each will be the highest point on the graph. Graph each equation by first making a table of ordered pairs using the given values of x.

21. $y = 4 - x^2$ $x = -3, -2, -1, 0, 1, 2, 3$
22. $y = 3 - x^2$ $x = -3, -2, -1, 0, 1, 2, 3$

23. $y = -1 - x^2$ $x = -2, -1, 0, 1, 2$
24. $y = -2 - x^2$ $x = -2, -1, 0, 1, 2$
25. Graph the line $y = x + 2$ and the parabola $y = x^2$ on the same coordinate system. Name the points where the two graphs intersect.
26. Graph the line $y = x$ and the parabola $y = x^2 - 2$ on the same coordinate system. Name the points where the two graphs intersect.
27. Graph the parabola $y = 2x^2$ and the parabola $y = \frac{1}{2}x^2$ on the same coordinate system.
28. Graph the parabola $y = 3x^2$ and the parabola $y = \frac{1}{3}x^2$ on the same coordinate system.

Review Problems

The following problems review material we covered in Section 7.6.

Solve each equation.

29. $\sqrt{x + 5} = 4$ **30.** $\sqrt{x - 3} = 2$
31. $\sqrt{2x + 3} = -3$ **32.** $\sqrt{3x - 3} = -5$
33. $\sqrt{3a + 2} - 3 = 5$ **34.** $\sqrt{2a - 4} + 3 = 7$
35. $\sqrt{x + 10} = x - 2$ **36.** $\sqrt{6x + 1} = x - 1$

C H A P T E R   8   S U M M A R Y

Examples

1. $(x - 3)^2 = 25$

$x - 3 = \pm 5$

$x = 3 \pm 5$

$x = -2$ or $x = 8$

Solving Quadratic Equations of the Form $(ax + b)^2 = c$ [8.1]
We can solve equations of the form $(ax + b)^2 = c$ by applying the square root property for equations to write

$$ax + b = \pm\sqrt{c}$$

2. $x^2 - 6x + 2 = 0$

$x^2 - 6x \quad = -2$

$x^2 - 6x + 9 = -2 + 9$

$(x - 3)^2 = 7$

$x - 3 = \pm\sqrt{7}$

$x = 3 \pm \sqrt{7}$

Strategy for Completing the Square [8.2]

To complete the square on a quadratic equation as a method of solution we use the following steps:

Step 1: Move the constant term to one side and the variable terms to the other. Then, divide each side by the coefficient of x^2 if it is other than one.

Step 2: Take the square of half the coefficient of the linear term and add it to both sides of the equation.

Step 3: Write the left side as a binomial square and then apply the square root property.

Step 4: Solve the resulting equation.

3. If $2x^2 + 3x - 4 = 0$

then $x = \dfrac{-3 \pm \sqrt{9 - 4(2)(-4)}}{2(2)}$

$= \dfrac{-3 \pm \sqrt{41}}{4}$

The Quadratic Formula [8.3]

Any equation that is in the form $ax^2 + bx + c = 0$, where $a \neq 0$, has as its solutions

$$x = \frac{-b \pm \sqrt{b^2 - 4ac}}{2a}$$

The expression under the square root sign, $b^2 - 4ac$, is known as the discriminant. When the discriminant is negative, the solutions are complex numbers.

4. The numbers 5, $3i$, $2 + 4i$, and $7 - i$ are all complex numbers.

Complex Numbers [8.4]

Any number that can be put in the form $a + bi$, where $i = \sqrt{-1}$, is called a complex number.

5. $(3 + 4i) + (6 - 7i)$

$= (3 + 6) + (4i - 7i)$

$= 9 - 3i$

Addition and Subtraction of Complex Numbers [8.4]

We add (or subtract) complex numbers by using the same procedure we used to add (or subtract) polynomials: we combine similar terms.

6. $(2 + 3i)(3 - i)$

$= 6 - 2i + 9i - 3i^2$

$= 6 + 7i + 3$

$= 9 + 7i$

Multiplication of Complex Numbers [8.4]

We multiply complex numbers in the same way we multiply binomials. The result, however, can be simplified further by substituting -1 for i^2 whenever it appears.

7. $\dfrac{3}{2 + 5i}$

$= \dfrac{3}{2 + 5i} \cdot \dfrac{2 - 5i}{2 - 5i}$

$= \dfrac{6 - 15i}{29}$

Division of Complex Numbers [8.4]

Division with complex numbers is accomplished with the method for rationalizing the denominator that we developed while working with radical expressions. If the denominator has the form $a + bi$, we multiply both the numerator and the denominator by its conjugate, $a - bi$.

8. Graph $y = x^2 - 2$.

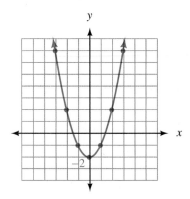

Graphing Parabolas [8.6]

The graph of each equation below is a parabola with the same basic shape as the graph of $y = x^2$.

Equation	Vertex	To obtain the graph, move the graph of $y = x^2$
$y = x^2 + k$	$(0, k)$	k units vertically
$y = (x - h)^2$	$(h, 0)$	h units horizontally
$y = (x - h)^2 + k$	(h, k)	h units horizontally and k units vertically

COMMON MISTAKES

1. The most common mistake when working with complex numbers is to say $i = -1$. It does not; i is the *square root* of -1, not -1 itself.

2. The most common mistake when working with the quadratic formula is to try to identify the constants a, b, and c before putting the equation into standard form.

CHAPTER 8 REVIEW

The numbers in brackets refer to the sections of the text in which similar problems can be found.

Solve each quadratic equation. [8.1, 8.5]

1. $a^2 = 32$
2. $a^2 = 60$
3. $2x^2 = 32$
4. $3x^2 = 27$
5. $(x + 3)^2 = 36$
6. $(x - 2)^2 = 81$
7. $(3x + 2)^2 = 16$
8. $(2x - 3)^2 = 49$
9. $(2x + 5)^2 = 32$
10. $(3x - 4)^2 = 27$
11. $\left(x - \dfrac{2}{3}\right)^2 = -\dfrac{25}{9}$
12. $\left(x + \dfrac{1}{2}\right)^2 = -\dfrac{9}{2}$

Solve by completing the square. [8.2]

13. $x^2 + 8x = 4$
14. $x^2 - 8x = 4$
15. $x^2 - 4x - 7 = 0$
16. $x^2 + 4x + 3 = 0$
17. $a^2 = 9a + 3$
18. $a^2 = 5a + 6$
19. $2x^2 + 4x - 6 = 0$
20. $3x^2 - 6x - 2 = 0$

Solve by using the quadratic formula. [8.3, 8.5]

21. $x^2 + 7x + 12 = 0$
22. $x^2 - 8x + 16 = 0$
23. $x^2 + 5x + 7 = 0$
24. $x^2 + 4x - 2 = 0$
25. $2x^2 = -8x + 5$
26. $3x^2 = -3x - 4$
27. $\dfrac{1}{5}x^2 - \dfrac{1}{2}x = \dfrac{3}{10}$
28. $\dfrac{1}{2}x^2 - \dfrac{1}{3}x = -1$
29. $(2x + 1)(2x - 3) = -6$
30. $(2x + 1)(2x - 5) = -4$

Add and subtract the following complex numbers. [8.4]

31. $(4 - 3i) + 5i$
32. $(2 - 5i) - 3i$
33. $(5 + 6i) + (5 - i)$
34. $(2 + 5i) + (3 - 7i)$
35. $(3 - 2i) - (3 - i)$
36. $(5 - 7i) - (6 - 2i)$
37. $(3 + i) - 5i - (4 - i)$

38. $(3 + 3i) - 7i - (2 + 2i)$
39. $(2 - 3i) - (5 - 2i) + (5 - i)$
40. $(3 + 5i) + (4 - 3i) - (6 - i)$

Multiply the following complex numbers. [8.4]

41. $2(3 - i)$ **42.** $-6(4 + 2i)$
43. $4i(6 - 5i)$ **44.** $-3i(7 + 6i)$
45. $(2 - i)(3 + i)$ **46.** $(3 - 4i)(5 + i)$
47. $(3 - i)^2$ **48.** $(3 + 5i)^2$
49. $(4 + i)(4 - i)$ **50.** $(3 + 2i)(3 - 2i)$

Divide the following complex numbers. [8.4]

51. $\dfrac{i}{3 + i}$ **52.** $\dfrac{i}{2 - i}$

53. $\dfrac{5}{2 + 5i}$ **54.** $\dfrac{3}{3 - 4i}$

55. $\dfrac{2i}{4 - i}$ **56.** $\dfrac{-3i}{3 - 2i}$

57. $\dfrac{3 + i}{3 - i}$ **58.** $\dfrac{4 - 5i}{4 + 5i}$

59. $\dfrac{2 + 3i}{3 - 5i}$ **60.** $\dfrac{3 - 2i}{5 - i}$

Write the following radicals as complex numbers. [8.5]

61. $\sqrt{-36}$ **62.** $\sqrt{-144}$
63. $\sqrt{-17}$ **64.** $\sqrt{-31}$
65. $\sqrt{-40}$ **66.** $\sqrt{-72}$
67. $\sqrt{-200}$ **68.** $\sqrt{-242}$

Graph each of the following equations. [8.6]

69. $y = x^2 + 2$ **70.** $y = x^2 - 2$
71. $y = (x - 2)^2$ **72.** $y = (x + 2)^2$
73. $y = (x + 3)^2 - 2$ **74.** $y = (x - 3)^2 - 2$
75. $y = x^2 + 4x + 7$ **76.** $y = x^2 - 4x + 7$

CHAPTER 8 TEST

Solve the following quadratic equations. [8.1, 8.3, 8.5]

1. $x^2 - 7x - 8 = 0$ **2.** $(x - 3)^2 = 12$

3. $\left(x - \dfrac{5}{2}\right)^2 = -\dfrac{75}{4}$ **4.** $\dfrac{1}{3}x^2 = \dfrac{1}{2}x - \dfrac{5}{6}$

5. $3x^2 = -2x + 1$ **6.** $(x + 2)(x - 1) = 6$
7. $9x^2 + 12x + 4 = 0$
8. Solve $x^2 - 6x - 6 = 0$ by completing the square. [8.2]

Write as complex numbers. [8.5]

9. $\sqrt{-9}$ **10.** $\sqrt{-121}$

11. $\sqrt{-72}$ **12.** $\sqrt{-18}$

Work the following problems involving complex numbers. [8.4]

13. $(3i + 1) + (2 + 5i)$
14. $(6 - 2i) - (7 - 4i)$
15. $(2 + i)(2 - i)$ **16.** $(3 + 2i)(1 + i)$

17. $\dfrac{i}{3 - i}$ **18.** $\dfrac{2 + i}{2 - i}$

Graph the following equations. [8.6]

19. $y = x^2 - 4$ **20.** $y = (x - 4)^2$
21. $y = (x + 3)^2 - 4$ **22.** $y = x^2 - 6x + 11$

9

ADDITIONAL TOPICS

INTRODUCTION AND OVERVIEW

In this chapter we will extend some of the topics we have already covered and preview some other topics that are common to intermediate algebra. The sections in this chapter do not all fit together in a smooth progression. Many of them can be studied individually and taken up in any order.

The review problems in this chapter cover the first eight chapters of the book. If you are getting ready to take a final exam in this course, doing the review problems at the end of each section will show you which of the first eight chapters you need to work on most.

9.1 Compound Inequalities

The **union** of two sets A and B is the set of all elements that are in A or in B. The word "or" is the key word in the definition. The **intersection** of two sets A and B is the set of elements contained in both A and B. The key word in this definition is the word "and." We can put the words "and" and "or" together with our methods of graphing inequalities to find the solution sets for compound inequalities.

> **DEFINITION** A **compound inequality** is two or more inequalities connected by the word "and" or "or."

▶ **EXAMPLE 1** Graph the solution set for the compound inequality

$$x < -1 \quad \text{or} \quad x \ge 3$$

Solution   Graphing each inequality separately, we have

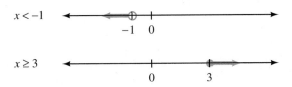

Since the two inequalities are connected by "or," we want to graph their union: that is, we graph all points that are on either the first graph or the second graph. Essentially, we put the two graphs together on the same number line.

$$x < -1 \quad \text{or} \quad x \geq 3$$

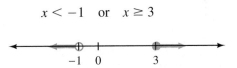

▶ **EXAMPLE 2**   Graph the solution set for the compound inequality

$$x > -2 \quad \text{and} \quad x < 3$$

Solution   Graphing each inequality separately, we have

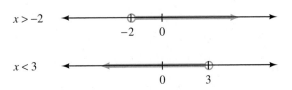

Since the two inequalities are connected by the word "and," we will graph their intersection, which consists of all points that are common to both graphs: that is, we graph the region where the two graphs overlap.

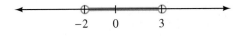

▶ **EXAMPLE 3**   Solve and graph the solution set for

$$2x - 1 \geq 3 \quad \text{and} \quad -3x > -12$$

Solution   Solving the two inequalities separately, we have

$$2x - 1 \geq 3 \qquad \text{and} \qquad -3x > -12$$

$$2x \geq 4$$

$$x \geq 2 \qquad \text{and} \quad -\tfrac{1}{3}(-3x) < \tfrac{1}{3}(-12)$$

$$x \leq 4$$

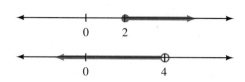

Since the word "and" connects the two graphs, we will graph their intersection—the points they have in common:

Notation Sometimes compound inequalities that use the word "and" can be written in a shorter form. For example, the compound inequality $-2 < x$ and $x < 3$ can be written as $-2 < x < 3$. The word "and" does not appear when an inequality is written in this form. It is implied. The solution set for $-2 < x$ and $x < 3$ is

It is all the numbers between -2 and 3 on the number line. It seems reasonable, then, that this graph should be the graph of

$$-2 < x < 3$$

In both the graph and the inequality, x is said to be between -2 and 3.

▶ **EXAMPLE 4** Solve and graph $-3 \leq 2x - 1 \leq 9$.

Solution To solve for x we must add 1 to the center expression and then divide the result by 2. Whatever we do to the center expression, we must also do to the two expressions on the ends. In this way, we can be sure we are producing equivalent inequalities. The solution set will not be affected.

$$-3 \leq 2x - 1 \leq 9$$

$$-2 \leq 2x \leq 10 \qquad \text{Add 1 to each expression}$$

$$-1 \leq x \leq 5 \qquad \text{Multiply each expression by } \tfrac{1}{2}$$

▶ **EXAMPLE 5**   Solve and graph $-5 < 3x + 1 < 10$.

Solution

$$-5 < 3x + 1 < 10$$
$$-6 < \quad 3x \quad < 9 \qquad \text{Add } -1 \text{ to each expression}$$
$$-2 < \quad x \quad < 3 \qquad \text{Multiply through by } \tfrac{1}{3}$$

The solution set is all real numbers between -2 and 3. We must keep in mind that the notation $-2 < x < 3$ is equivalent to the expression $-2 < x$ and $x < 3$. This notation always implies the word "and"; it never implies the word "or." ◀

▶ **EXAMPLE 6**   The length of a rectangle is 3 feet longer than the width. If the perimeter must be between 14 feet and 30 feet, find all possible values for the width.

Solution   If $x =$ the width of the rectangle, then the length is $x + 3$. Since the perimeter is the sum of twice the width and twice the length, we can write it as $2x + 2(x + 3)$. This last expression must be between 14 feet and 30 feet. The inequality that describes the situation is

$$14 < 2x + 2(x + 3) < 30 \qquad \text{The perimeter is between 14 and 30}$$
$$14 < \quad 2x + 2x + 6 \quad < 30 \qquad \text{Multiply 2 and } (x + 3)$$
$$14 < \quad\quad 4x + 6 \quad\quad < 30 \qquad \text{Add } 2x \text{ and } 2x$$
$$8 < \quad\quad\quad 4x \quad\quad\quad < 24 \qquad \text{Add } -6 \text{ to each expression}$$
$$2 < \quad\quad\quad x \quad\quad\quad < 6 \qquad \text{Multiply each expression by } \tfrac{1}{4}$$

The width is between 2 feet and 6 feet. ◀

P R O B L E M   S E T   9 . 1

Graph the following compound inequalities.

1. $x < -1$ or $x > 5$
2. $x \leq -2$ or $x \geq -1$
3. $x < -3$ or $x \geq 0$
4. $x < 5$ and $x > 1$
5. $x \leq 6$ and $x > -1$
6. $x \leq 7$ and $x > 0$
7. $x > 2$ and $x < 4$
8. $x < 2$ or $x > 4$
9. $x \geq -2$ and $x \leq 4$
10. $x \leq 2$ or $x \geq 4$
11. $x < 5$ and $x > -1$
12. $x > 5$ or $x < -1$
13. $-1 < x < 3$
14. $-1 \leq x \leq 3$
15. $-3 < x \leq -2$
16. $-5 \leq x \leq 0$

Solve the following. Graph the solution set in each case.

17. $3x - 1 < 5$ or $5x - 5 > 10$
18. $x + 1 < -3$ or $x - 2 > 6$
19. $x - 2 > -5$ and $x + 7 < 13$
20. $3x + 2 \leq 11$ and $2x + 2 \geq 0$
21. $11x < 22$ or $12x > 36$
22. $-5x < 25$ and $-2x \geq -12$
23. $3x - 5 < 10$ and $2x + 1 > -5$

24. $5x + 8 < -7$ or $3x - 8 > 10$
25. $2x - 3 < 8$ and $3x + 1 > -10$
26. $11x - 8 > 3$ or $12x + 7 < -5$
27. $2x - 1 < 3$ and $3x - 2 > 1$
28. $3x + 9 < 7$ or $2x - 7 > 11$

Solve and graph each of the following.

29. $-1 \leq x - 5 \leq 2$ **30.** $0 \leq x + 2 \leq 3$
31. $-4 \leq 2x \leq 6$ **32.** $-5 < 5x < 10$
33. $-3 < 2x + 1 < 5$ **34.** $-7 \leq 2x - 3 \leq 7$
35. $0 \leq 3x + 2 \leq 7$ **36.** $2 \leq 5x - 3 \leq 12$
37. $-7 < 2x + 3 < 11$
38. $-5 < 6x - 2 < 8$ **39.** $-1 \leq 4x + 5 \leq 9$
40. $-8 \leq 7x - 1 \leq 13$
41. The sum of a number and 5 is between 10 and 20. Find the number.
42. The difference of a number and 2 is between 6 and 14. Find the number.
43. The difference of twice a number and 3 is between 5 and 7. Find the number.
44. The sum of twice a number and 5 is between 7 and 13. Find the number.
45. The length of a rectangle is 4 inches longer than the width. If the perimeter is between 20 inches and 30 inches, find all possible values for the width.
46. The length of a rectangle is 6 feet longer than the width. If the perimeter is between 24 feet and 36 feet, find all possible values for the width.

Review Problems

The problems that follow review the material we covered in Chapter 1.

Simplify each expression.

47. $-|-5|$ **48.** $\left(-\dfrac{2}{3}\right)^3$

49. $-3 - 4(-2)$ **50.** $2^4 + 3^3 \div 9 - 4^2$
51. $5(3 - 8) - 6(2 - 5)$
52. $7 - 3(2 - 6)$
53. $5 - 2[-3(5 - 7) - 8]$
54. $\dfrac{5 + 3(7 - 2)}{2(-3) - 4}$

55. Find the difference of -3 and -9.
56. If you add -4 to the product of -3 and 5, what number results?
57. Apply the distributive property to $\frac{1}{2}(4x - 6)$.
58. Use the associative property to simplify $-6(\frac{1}{3}x)$.

For the set $\left\{-3, -\frac{4}{5}, 0, \frac{5}{8}, 2, \sqrt{5}\right\}$, which numbers are

59. Integers   **60.** Rational numbers

The problems that follow review material we covered in Chapter 2.

61. Simplify the expression $7 - 3(2x - 4) - 8$.
62. Find the value of $x^2 - 2xy + y^2$ when $x = 3$ and $y = -4$.

Solve each equation.

63. $-\dfrac{3}{2}x = 12$ **64.** $2x - 4 = 5x + 2$

65. $8 - 2(x + 7) = 2$
66. $3(2x - 5) - (2x - 4) = 6 - (4x + 5)$
67. Solve the formula $P = 2l + 2w$ for w.

Solve each inequality and graph the solution.

68. $-4x < 20$ **69.** $3 - 2x > 5$
70. $3 - 4(x - 2) \geq -5x + 6$
71. Solve the formula $3x - 2y \leq 12$ for y.
72. What number is 12% of 2,000?
73. The length of a rectangle is 5 inches more than three times the width. If the perimeter is 26 inches, find the length and width.

SECTION

9.2   **Linear Inequalities in Two Variables**

A **linear inequality in two variables** is any expression that can be put in the form

$$ax + by < c$$

where a, b, and c are real numbers (a and b not both 0). The inequality symbol can be any one of the following four: $<$, $\leq$, $>$, $\geq$.

Some examples of linear inequalities are

$$2x + 3y < 6 \qquad y \geq 2x + 1 \qquad x - y \leq 0$$

Although not all of these inequalities have the form $ax + by < c$, each one can be put in that form.

The solution set for a linear inequality is a section of the coordinate plane. The boundary for the section is found by replacing the inequality symbol with an equal sign and graphing the resulting equation. The boundary is included in the solution set (and represented with a solid line) if the inequality symbol used originally is $\leq$ or $\geq$. The boundary is not included (and is represented with a broken line) if the original symbol is $<$ or $>$.

Let's look at some examples.

▶ **EXAMPLE 1**   Graph the solution set for $x + y \leq 4$.

Solution   The boundary for the graph is the graph of $x + y = 4$; we graph $x + y = 4$ by using the method described in Chapter 3. The boundary is included in the solution set because the inequality symbol is $\leq$.

The graph of the boundary is shown in Figure 1.

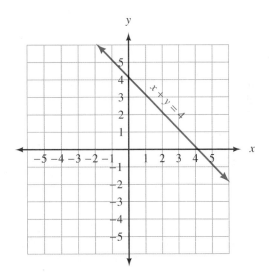

Figure 1

The boundary separates the coordinate plane into two sections, or regions—the region above the boundary and the region below the boundary. The solution set for $x + y \leq 4$ is one of these two regions along with the boundary. To find the correct region, we simply choose any convenient point that is *not* on the boundary. We then substitute the coordinates of the point into the original inequality $x + y \leq 4$. If the point we choose satisfies the inequality, then it is a member of the solution set, and we can assume that all points on the same side of the boundary as the chosen point are also in the solution set. If the coordinates of our point do not satisfy the original inequality, then the solution set lies on the other side of the boundary.

In this example, a convenient point not on the boundary is the origin. Substituting $(0, 0)$ into $x + y \leq 4$ gives us

$$0 + 0 \overset{?}{\leq} 4$$

$$0 \leq 4 \qquad \text{A true statement}$$

Since the origin is a solution to the inequality $x + y \leq 4$, and the origin is below the boundary, all other points below the boundary are also solutions. The graph of $x + y \leq 4$ is shown in Figure 2.

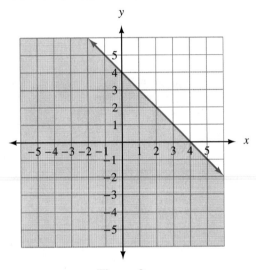

Figure 2

The region above the boundary is described by the inequality $x + y > 4$.
◄

Here is a list of steps to follow when graphing the solution set for linear inequalities in two variables.

Strategy for Graphing the Solution Set for Linear Inequalities in Two Variables

Step 1: Replace the inequality symbol with an equal sign. The resulting equation represents the boundary for the solution set.

> *Step 2:* Graph the boundary found in step 1, using a *solid line* if the boundary is included in the solution set (that is, if the original inequality symbol was either ≤ or ≥). Use a *broken line* to graph the boundary if it is *not* included in the solution set. (It is not included if the original inequality was either < or >.)
>
> *Step 3:* Choose any convenient point not on the boundary and substitute the coordinates into the *original* inequality. If the resulting statement is *true*, the graph lies on the *same* side of the boundary as the chosen point. If the resulting statement is *false*, the solution set lies on the *opposite* side of the boundary.

▶ **EXAMPLE 2**   Graph the solution set for $y < 2x - 3$.

Solution   The boundary is the graph of $y = 2x - 3$. The boundary is not included since the original inequality symbol is <. Therefore we use a broken line to represent the boundary, as shown in Figure 3.

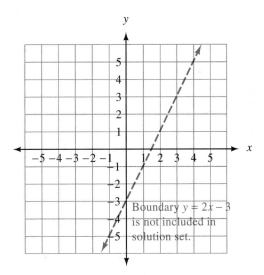

Figure 3

A convenient test point is again the origin. Using $(0, 0)$ in $y < 2x - 3$, we have

$$0 \overset{?}{<} 2(0) - 3$$

$$0 < -3 \qquad \text{A false statement}$$

Since our test point gives us a false statement and it lies above the boundary, the solution set must lie on the other side of the boundary, as shown in Figure 4.

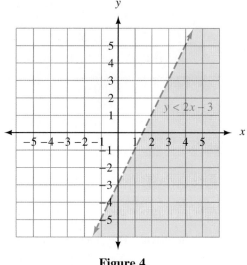

Figure 4 ◄

▶ **EXAMPLE 3** Graph the inequality $2x + 3y \leq 6$.

Solution We begin by graphing the boundary $2x + 3y = 6$. The boundary is included in the solution because the inequality symbol is $\leq$.

If we use $(0, 0)$ as our test point, we see that it yields a true statement when its coordinates are substituted into $2x + 3y \leq 6$. The graph, therefore, lies below the boundary, as shown in Figure 5.

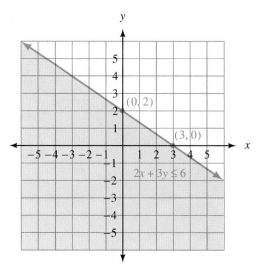

Figure 5

The ordered pair $(0, 0)$ is a solution to $2x + 3y \leq 6$; all points on the same side of the boundary as $(0, 0)$ must also be solutions to the inequality $2x + 3y \leq 6$.   ◀

▶ **EXAMPLE 4**   Graph the solution set for $x \leq 5$.

Solution   The boundary is $x = 5$, which is a vertical line. All points to the left have x-coordinates less than 5 and all points to the right have x-coordinates greater than 5, as shown in Figure 6.

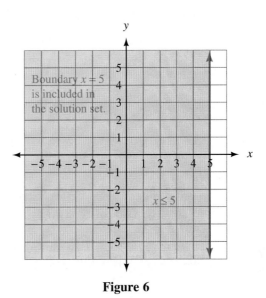

Figure 6   ◀

P R O B L E M   S E T   9 . 2

Graph the following linear inequalities.

1. $2x - 3y < 6$

2. $3x + 2y \geq 6$

3. $x - 2y \leq 4$

4. $2x + y > 4$

5. $x - y \leq 2$

6. $x - y \leq 1$

7. $3x - 4y \geq 12$

8. $4x + 3y < 12$

9. $5x - y \leq 5$

10. $4x + y > 4$

11. $2x + 6y \leq 12$

12. $x - 5y > 5$

13. $x \geq 1$

14. $x < 5$

15. $x \geq -3$

16. $y \leq -4$

17. $y < 2$

18. $3x - y > 1$

19. $2x + y > 3$

20. $5x + 2y < 2$

21. $y \leq 3x - 1$

22. $y \geq 3x + 2$

23. $y \geq x - 5$

24. $y > x + 3$

25. $y \leq -\dfrac{1}{2}x + 2$

26. $y < \dfrac{1}{3}x + 3$

27. $y < -x + 4$

28. $y \geq -x - 3$

Review Problems

The problems that follow review material we covered in Chapter 3.

29. Fill in each ordered pair so that it is a solution to $y = \frac{1}{2}x + 3$.

$$(-2, \ ) \quad (0, \ ) \quad (2, \ )$$

30. Graph the line $y = \frac{1}{2}x + 3$.
31. Graph the line $x = -2$.
32. Find the intercepts and slope for $2x - 3y = 6$. Then, graph the line.
33. Find the slope of the line through $(-1, 2)$ and $(5, -4)$.
34. Find the equation of the line through $(-1, 2)$ and $(5, -4)$.
35. Find the equation of the line with slope -3 that contains the point $(-1, 4)$.
36. Solve this system by graphing.

$$x + y = 4$$
$$x - y = 2$$

Solve each system by the elimination method.

37. $3x + 2y = 1$
    $2x + \ y = 3$

38. $2x + 3y = -1$
    $3x + 5y = -2$

Solve each system by the substitution method.

39. $x + y = 20$
    $y = 5x + 2$

40. $2x - 6y = 2$
    $y = 3x + 1$

41. A total of $1200 is invested in two accounts. One of the accounts pays 8% interest annually and the other pays 10% interest annually. If the total amount of interest earned from both accounts for the year is $104, how much is invested in each account?

42. Amy has $1.85 in dimes and quarters. If she has a total of 11 coins, how many of each coin does she have?

SECTION

9.3   Factoring the Sum and Difference of Two Cubes

In Chapter 5 we factored a variety of polynomials. Among the polynomials we factored were polynomials that were the difference of two squares. The formula we used to factor the difference of two squares looks like this:

$$a^2 - b^2 = (a + b)(a - b)$$

If we ran across a binomial that had the form of the difference of two squares, we factored it by applying this formula. For example, to factor $x^2 - 25$, we simply notice that it can be written in the form $x^2 - 5^2$, which looks like the difference of two squares. According to the formula above, this binomial factors into $(x + 5)(x - 5)$.

In this section we want to use two new formulas that will allow us to factor the sum and difference of two cubes. For example, we want to factor the binomial $x^3 - 8$, which is the difference of two cubes. (To see that it is the difference of two cubes, notice that it can be written as $x^3 - 2^3$.) We also want to factor $y^3 + 27$, which is the sum of two cubes. (To see this, notice that $y^3 + 27$ can be written as $y^3 + 3^3$.)

The formulas that allow us to factor the sum of two cubes and the difference of two cubes are not as simple as the formula for factoring the difference of two squares. Here is what they look like:

$$a^3 + b^3 = (a + b)(a^2 - ab + b^2)$$

$$a^3 - b^3 = (a - b)(a^2 + ab + b^2)$$

Let's begin our work with these two formulas by showing that they are true. To do so, we multiply out the right side of each formula.

▶ **EXAMPLE 1**   Verify the two formulas.

Solution   We verify the formulas by multiplying the right sides and comparing the results with the left sides:

$$
\begin{array}{r}
a^2 - ab + b^2 \\
a \quad + b \\
\hline
a^3 - a^2b + ab^2 \\
a^2b - ab^2 + b^3 \\
\hline
a^3 \qquad\qquad + b^3
\end{array}
$$

The first formula is correct.

$$
\begin{array}{r}
a^2 + ab + b^2 \\
a \quad - b \\
\hline
a^3 + a^2b + ab^2 \\
- a^2b - ab^2 - b^3 \\
\hline
a^3 \qquad\qquad - b^3
\end{array}
$$

The second formula is correct.   ◀

Here are some examples that use the formulas for factoring the sum and difference of two cubes.

▶ **EXAMPLE 2**   Factor $x^3 - 8$.

Solution   Since the two terms are perfect cubes, we write them as such and apply the formula:

$$
\begin{aligned}
x^3 - 8 &= x^3 - 2^3 \\
&= (x - 2)(x^2 + 2x + 2^2) \\
&= (x - 2)(x^2 + 2x + 4)
\end{aligned}
$$

◀

▶ **EXAMPLE 3**   Factor $y^3 + 27$.

Solution   Proceeding as we did in Example 2, we first write 27 as 3^3. Then, we apply the formula for factoring the sum of two cubes, which is $a^3 + b^3 = (a + b)(a^2 - ab + b^2)$:

$$
\begin{aligned}
y^3 + 27 &= y^3 + 3^3 \\
&= (y + 3)(y^2 - 3y + 3^2) \\
&= (y + 3)(y^2 - 3y + 9)
\end{aligned}
$$

◀

▶ **EXAMPLE 4** Factor $8x^3 - y^3$.

Solution Since we can write $8x^3$ as $(2x)^3$, we have a binomial that is the difference of two cubes:

$$8x^3 - y^3 = (2x)^3 - y^3$$
$$= (2x - y)[(2x)^2 + (2x)y + y^2]$$
$$= (2x - y)(4x^2 + 2xy + y^2) \qquad ◀$$

If you are wondering if the second factor, $4x^2 + 2xy + y^2$, can be factored further, it cannot. If you think that it can be factored further, write down the factors you think it has and multiply them. Doing so will convince you that it does not factor.

▶ **EXAMPLE 5** Factor $27x^3 + 125y^3$.

Solution Since $27x^3$ can be written as $(3x)^3$ and $125y^3$ can be written as $(5y)^3$, we have a binomial that is the sum of two cubes:

$$27x^3 + 125y^3 = (3x)^3 + (5y)^3$$
$$= (3x + 5y)[(3x)^2 - (3x)(5y) + (5y)^2]$$
$$= (3x + 5y)(9x^2 - 15xy + 25y^2)$$

Again, the second factor, $9x^2 - 15xy + 25y^2$, cannot be factored further. ◀

P R O B L E M   S E T   9 . 3

Multiply.

1. $(x + 4)(x - 4)$ **2.** $(x + 7)(x - 7)$
3. $(2x + 9)(2x - 9)$ **4.** $(3x + 1)(3x - 1)$
5. $(x + 2)(x^2 - 2x + 4)$
6. $(x - 2)(x^2 + 2x + 4)$
7. $(x - 1)(x^2 + x + 1)$
8. $(x + 1)(x^2 - x + 1)$
9. $(x + 4)(x^2 - 4x + 16)$
10. $(x - 4)(x^2 + 4x + 16)$
11. $(2x - 3)(4x^2 + 6x + 9)$
12. $(2x + 3)(4x^2 - 6x + 9)$
13. $(x + y)(x^2 - xy + y^2)$
14. $(x - y)(x^2 + xy + y^2)$

Factor each of the following.

15. $x^3 - 2^3$ **16.** $x^3 + 2^3$
17. $x^3 + 7^3$ **18.** $x^3 - 7^3$
19. $(3x)^3 - (5y)^3$ **20.** $(5x)^3 - (3y)^3$
21. $x^3 + 8$ **22.** $x^3 - 27$

23. $x^3 - 125$ **24.** $x^3 - 64$
25. $x^3 + 1$ **26.** $x^3 - 1$
27. $8x^3 + y^3$ **28.** $x^3 + 8y^3$
29. $27x^3 - 8y^3$ **30.** $8x^3 - 27y^3$
31. $125x^3 + 8$ **32.** $8x^3 + 125$
33. $x^3 + 64y^3$ **34.** $x^3 - 64y^3$

Review Problems

The problems that follow review material we covered in Chapter 4.

Simplify each expression. (Write all answers with positive exponents only.)

35. $(5x^3)^2(2x^6)^3$ **36.** 2^{-3}

37. $\dfrac{x^4}{x^{-3}}$ **38.** $\dfrac{(20x^2y^3)(5x^4y)}{(2xy^5)(10x^2y^3)}$

39. $(2 \times 10^{-4})(4 \times 10^5)$

40. $\dfrac{9 \times 10^{-3}}{3 \times 10^{-2}}$

41. $20ab^2 - 16ab^2 + 6ab^2$

42. Subtract $6x^2 - 5x - 7$ from $9x^2 + 3x - 2$.

Graph each equation.

43. $y = \dfrac{1}{2}x^2$ **44.** $y = x^2 - 4$

Multiply.

45. $2x^2(3x^2 + 3x - 1)$ **46.** $(2x + 3)(5x - 2)$
47. $(3y - 5)^2$
48. $(a - 4)(a^2 + 4a + 16)$
49. $(2a^2 + 7)(2a^2 - 7)$
50. Divide $15x^{10} - 10x^8 + 25x^6$ by $5x^6$.

51. Divide, using long division: $\dfrac{x^2 - 2x + 6}{x - 4}$

The problems that follow review material we covered in Chapter 5.

52. Factor out the greatest common factor:
$15a^3b^3 - 20a^2b - 35ab^2$.

53. Factor by grouping: $3ax - 2a + 15x - 10$.

Factor completely.

54. $x^2 - 4x - 12$
55. $4x^2 - 20xy + 25y^2$
56. $x^4 - 16$ **57.** $2x^2 + xy - 21y^2$
58. $5x^3 - 25x^2 - 30x$

Solve each equation.

59. $x^2 - 9x + 18 = 0$ **60.** $x^2 - 6x = 0$
61. $8x^2 = -2x + 15$ **62.** $x(x + 2) = 80$
63. The product of two consecutive even integers is four more than twice their sum. Find the two integers.
64. The hypotenuse of a right triangle is 15 inches. One of the legs is 3 inches more than the other. Find the lengths of the two legs.

SECTION

9.4   # Variation

Two variables are said to **vary directly** if one is a constant multiple of the other. For instance, y varies directly as x if $y = Kx$, where K is a constant. The constant K is called the **constant of variation.** The following table gives the relation between direct variation statements and their equivalent algebraic equations.

Statement	Equation (K = constant of variation)
y varies directly as x	$y = Kx$
y varies directly as the square of x	$y = Kx^2$
s varies directly as the square root of t	$s = K\sqrt{t}$
r varies directly as the cube of s	$r = Ks^3$

Any time we run across a statement similar to those in the table, we can immediately write an equivalent expression involving variables and a constant of variation K.

▶ **EXAMPLE 1**   Suppose y varies directly as x. When y is 15, x is 3. Find y when x is 4.

Solution   From the first sentence we can write the relationship between x and y as

$$y = Kx$$

We now use the second sentence to find the value of K. Since y is 15 when x is 3, we have

$$15 = K(3) \quad \text{or} \quad K = 5$$

Now we can rewrite the relationship between x and y more specifically as

$$y = 5x$$

To find the value of y when x is 4 we simply substitute $x = 4$ into our last equation. Substituting

$$x = 4$$

into

$$y = 5x$$

we have

$$y = 5(4)$$

$$y = 20$$ ◀

▶ **EXAMPLE 2**   Suppose y varies directly as the square of x. When x is 4, y is 32. Find x when y is 50.

Solution   The first sentence gives us

$$y = Kx^2$$

Since y is 32 when x is 4, we have

$$32 = K(4)^2$$

$$32 = 16K$$

$$K = 2$$

The equation now becomes

$$y = 2x^2$$

When y is 50, we have

$$50 = 2x^2$$

$$25 = x^2$$

$$x = \pm 5$$

There are two possible solutions, $x = 5$ or $x = -5$.   ◀

▶ **EXAMPLE 3.** The cost of a certain kind of candy varies directly with the weight of the candy. If 12 ounces of the candy cost $1.68, how much will 16 ounces cost?

Solution   Let x = the number of ounces of candy and y = the cost of the candy. Then $y = Kx$. Since y is 1.68 when x is 12, we have

$$1.68 = K \cdot 12$$

$$K = \frac{1.68}{12}$$

$$= 0.14$$

The equation must be

$$y = 0.14x$$

When x is 16, we have

$$y = 0.14(16)$$

$$= 2.24$$

The cost of 16 ounces of candy is $2.24.   ◀

Inverse Variation

Two variables are said to **vary inversely** if one is a constant multiple of the reciprocal of the other. For example, y varies inversely as x if $y = \dfrac{K}{x}$, where K is a real number constant. Again, K is called the constant of variation. The table that follows gives some examples of inverse variation statements and their associated algebraic equations.

Statement	Equation (K = constant of variation)
y varies inversely as x	$y = \dfrac{K}{x}$
y varies inversely as the square of x	$y = \dfrac{K}{x^2}$
F varies inversely as the square root of t	$F = \dfrac{K}{\sqrt{t}}$
r varies inversely as the cube of s	$r = \dfrac{K}{s^3}$

Every inverse variation statement has an associated inverse variation equation.

▶ **EXAMPLE 4**   Suppose y varies inversely as x. When y is 4, x is 5. Find y when x is 10.

Solution   The first sentence gives us the relationship between x and y:

$$y = \frac{K}{x}$$

We use the second sentence to find the value of the constant K:

$$4 = \frac{K}{5}$$

or

$$K = 20$$

We can now write the relationship between x and y more specifically as

$$y = \frac{20}{x}$$

We use this equation to find the value of y when x is 10. Substituting

$$x = 10$$

into

$$y = \frac{20}{x}$$

we have

$$y = \frac{20}{10}$$

$$y = 2$$ ◀

▶ **EXAMPLE 5**   The intensity (I) of light from a source varies inversely as the square of the distance (d) from the source. Ten feet away from the source the intensity is 200 candlepower. What is the intensity 5 feet from the source?

Solution

$$I = \frac{K}{d^2}$$

Since $I = 200$ when $d = 10$, we have

$$200 = \frac{K}{10^2}$$

$$200 = \frac{K}{100}$$

$$K = 20{,}000$$

The equation becomes

$$I = \frac{20{,}000}{d^2}$$

When $d = 5$, we have

$$I = \frac{20{,}000}{5^2}$$

$$= \frac{20{,}000}{25}$$

$$= 800 \text{ candlepower} \qquad \blacktriangleleft$$

P R O B L E M   S E T   9 . 4

For each of the following problems, y varies directly as x.

1. If $y = 10$ when $x = 5$, find y when x is 4.
2. If $y = 20$ when $x = 4$, find y when x is 11.
3. If $y = 39$ when $x = 3$, find y when x is 10.
4. If $y = -18$ when $x = 6$, find y when x is 3.
5. If $y = -24$ when $x = 4$, find x when y is -30.
6. If $y = 30$ when $x = -15$, find x when y is 8.
7. If $y = -7$ when $x = -1$, find x when y is -21.
8. If $y = 30$ when $x = 4$, find y when x is 7.

For each of the following problems, y varies directly as the square of x.

9. If $y = 75$ when $x = 5$, find y when x is 1.
10. If $y = -72$ when $x = 6$, find y when x is 3.
11. If $y = 48$ when $x = 4$, find y when x is 9.
12. If $y = 27$ when $x = 3$, find x when y is 75.

For each of the following problems, y varies inversely with x.

13. If $y = 5$ when $x = 2$, find y when x is 5.
14. If $y = 2$ when $x = 10$, find y when x is 4.
15. If $y = 2$ when $x = 1$, find y when x is 4.
16. If $y = 4$ when $x = 3$, find y when x is 6.

17. If $y = 5$ when $x = 3$, find x when y is 15.
18. If $y = 12$ when $x = 10$, find x when y is 60.
19. If $y = 10$ when $x = 10$, find x when y is 20.
20. If $y = 15$ when $x = 2$, find x when y is 6.

For each of the following problems, y varies inversely as the square of x.

21. If $y = 4$ when $x = 5$, find y when x is 2.
22. If $y = 5$ when $x = 2$, find y when x is 6.
23. If $y = 4$ when $x = 3$, find y when x is 2.
24. If $y = 9$ when $x = 4$, find y when x is 3.

Solve the following problems.

25. The tension t in a spring varies directly with the distance d the spring is stretched. If the tension is 42 pounds when the spring is stretched 2 inches, find the tension when the spring is stretched twice as far.

26. The time t it takes to fill a bucket varies directly with the volume g of the bucket. If it takes 1 minute to fill a 4-gallon bucket, how long will it take to fill a 6-gallon bucket?

27. The power P in an electric circuit varies directly with the square of the current I. If $P = 30$ when $I = 2$, find P when $I = 7$.

28. The resistance R in an electric circuit varies directly with the voltage V. If $R = 20$ when $V = 120$, find R when $V = 240$.

29. The amount of money M a woman makes per week varies directly with the number of hours h she works per week. If she works 20 hours and earns \$157, how much does she make if she works 30 hours?

30. The volume V of a gas varies directly as the temperature T. If $V = 3$ when $T = 150$, find V when T is 200.

31. The weight F of a body varies inversely with the square of the distance d between the body and the center of the earth. If a man weighs 150 pounds 4,000 miles from the center of the earth, how much will he weigh at a distance of 5,000 miles from the center of the earth?

32. The intensity I of a light source varies inversely with the square of the distance d from the source. Four feet from the source, the intensity is 9 footcandles. What is the intensity 3 feet from the source?

33. The current I in an electric circuit varies inversely with the resistance R. If a current of 30 amps is produced by a resistance of 2 ohms, what current will be produced by a resistance of 5 ohms?

34. The pressure exerted by a gas on the container in which it is held varies inversely with the volume of the container. A pressure of 40 pounds per square inch is exerted on a container of volume 2 cubic feet. What is the pressure on a container whose volume is 8 cubic feet?

Review Problems

The problems that follow review material we covered in Chapter 6.

35. Reduce to lowest terms: $\dfrac{x^2 - x - 6}{x^2 - 9}$.

36. Divide using long division: $\dfrac{x^2 - 2x - 8}{x - 4}$.

Perform the indicated operations.

37. $\dfrac{x^2 - 25}{x + 4} \cdot \dfrac{2x + 8}{x^2 - 9x + 20}$

38. $\dfrac{3x + 6}{x^2 + 4x + 3} \div \dfrac{x^2 + x - 2}{x^2 + 2x - 3}$

39. $\dfrac{x}{x^2 - 16} + \dfrac{4}{x^2 - 16}$

40. $\dfrac{2}{x^2 - 1} - \dfrac{5}{x^2 + 3x - 4}$

41. $\dfrac{1 - \dfrac{25}{x^2}}{1 - \dfrac{8}{x} + \dfrac{15}{x^2}}$

Solve each equation.

42. $\dfrac{x}{2} - \dfrac{5}{x} = -\dfrac{3}{2}$

43. $\dfrac{x}{x^2 - 9} - \dfrac{3}{x - 3} = \dfrac{1}{x + 3}$

44. A boat travels 30 miles up a river in the same amount of time it takes to travel 50 miles down the same river. If the current is 5 mph, what is the speed of the boat in still water?

45. A pool can be filled by an inlet pipe in 8 hours. The drain will empty the pool in 12 hours. How long will it take to fill the pool if both the inlet pipe and the drain are open?

46. If 30 liters of a certain solution contains 2 liters of alcohol, how much alcohol is in 45 liters of the same solution?

SECTION

9.5 Functions and Function Notation

The idea of a function is very common in mathematics. Functions are studied in almost all branches of mathematics. They are the result of the process of working with quantities that vary with one another. That is, a change in one quantity brings about a change in another. The idea of a function is also a very simple concept mathematically.

DEFINITION A **function** is any set of ordered pairs in which no two different ordered pairs have the same first coordinate. The **domain** of a function is the set of all first coordinates. The **range** of a function is the set of all second coordinates.

EXAMPLE The set $\{(2, 3), (1, 5), (6, 3), (-3, 4)\}$ is a function because it is a set of ordered pairs and no two different ordered pairs have the same first coordinates. The domain of this function is the set $\{2, 1, 6, -3\}$. The range is the set $\{3, 5, 4\}$.

In this example we listed all the ordered pairs of the function. Usually the ordered pairs are given in the form of a rule instead of a list.

EXAMPLE The set $\{(x \ y)|y = 2x + 1; x = 1, 2, 3, 4\}$ is an example of a function. The domain is the set $\{1, 2, 3, 4\}$. The range is the set $\{3, 5, 7, 9\}$, since these are the values of y that correspond to the x values 1, 2, 3, 4.

In this example we specified the function by giving the rule for obtaining ordered pairs, $y = 2x + 1$, and then listing the values that the variable x could assume, $x = 1, 2, 3, 4$. Most of the time, when a function is given by a rule, the domain is not specified. When this happens, we assume the domain includes all replacements for x except those that make the function undefined. For example, the domain of the function

$$y = \frac{3}{x}$$

is all real numbers except $x = 0$, because when x is 0 the expression $\frac{3}{0}$ is undefined. There is no value of y that corresponds to $x = 0$.

▶ **EXAMPLE 1** Specify the domain for $y = \dfrac{3}{x - 2}$.

Solution We can replace x with any real number except 2, because $x = 2$ makes the denominator zero. The domain is

$$\text{Domain} = \{x | x \text{ is real}, x \neq 2\}$$ ◀

▶ **EXAMPLE 2** Specify the domain for $y = \dfrac{3}{x^2 - x - 6}$.

Solution We are interested in eliminating any values of x that will make the denominator zero:

$$x^2 - x - 6 = 0$$
$$(x - 3)(x + 2) = 0$$
$$x = 3 \quad \text{or} \quad x = -2$$

Since $x = 3$ and $x = -2$ both make the denominator zero, we cannot use them as replacements for x:

$$\text{Domain} = \{x \mid x \text{ is real}, x \neq 3, x \neq -2\} \qquad \blacktriangleleft$$

Function Notation

The notation $f(x)$ is read "f of x" and is defined to be the value of the function f at x. It is not to be read or interpreted as f times x. The equations $y = 2x + 3$ and $f(x) = 2x + 3$ are essentially the same; that is, $y = f(x)$.

Suppose we are given the equation $y = 2x + 3$ and want to find the value of y when x is 4. We would have to say, "If $x = 4$, then $y = 2(4) + 3 = 11$." Using the new notation, $f(x) = 2x + 3$, we have $f(4) = 2(4) + 3 = 11$. The following table gives more examples of this concept.

Function Notation	y in Terms of x
$f(x) = 2x + 3$	$y = 2x + 3$
$f(1) = 2(1) + 3 = 5$	If $x = 1$, then $y = 2(1) + 3 = 5$
$f(0) = 2(0) + 3 = 3$	If $x = 0$, then $y = 2(0) + 3 = 3$
$f(-5) = 2(-5) + 3 = -7$	If $x = -5$, then $y = 2(-5) + 3 = -7$

Here are some additional examples of the use of function notation.

▶ **EXAMPLE 3**   If $f(x) = x^2 - 4$, find $f(0), f(-1), f(3), f(5), f(a)$.

Solution

$$f(x) = x^2 - 4$$
$$f(0) = 0^2 - 4 = -4$$
$$f(-1) = (-1)^2 - 4 = -3$$
$$f(3) = 3^2 - 4 = 5$$
$$f(5) = 5^2 - 4 = 21$$
$$f(a) = a^2 - 4 \qquad \blacktriangleleft$$

▶ **EXAMPLE 4**   If $f(x) = 2x + 5$ and $g(x) = 2x^2 + 1$, find $f(2), g(-3), g(a)$, $f(z)$.

Solution

$$f(2) = 2(2) + 5 = 4 + 5 = 9$$
$$g(-3) = 2(-3)^2 + 1 = 2(9) + 1 = 19$$
$$g(a) = 2a^2 + 1$$
$$f(z) = 2z + 5 \qquad \blacktriangleleft$$

PROBLEM SET 9.5

Specify the domain and range of the following functions.

1. $\{(1, 2), (3, 4), (5, 6)\}$
2. $\{(-1, 1), (-2, 2), (-3, 3)\}$
3. $\{(1, a), (3, b), (4, c)\}$
4. $\{(a, -1), (b, c), (d, -2)\}$
5. $y = x, \quad x = 1, 2, 3, 4$
6. $y = 2x, \quad x = 2, 4, 6, 8$
7. $y = 3x - 1, \quad x = 1, 3, 5, 7, 9$
8. $y = 2x - 3, \quad x = 2, 4, 6, 8, 10$
9. $2x + 3y = 6, \quad x = 0, 2, 4$
10. $3x - 2y = 6, \quad y = 0, -2, -3$
11. $2x - y = 4, y = 2, 4, 6, 8, 10$
12. $3x - 5y = 15, \quad x = 0, 5, 15$

Specify any restrictions on the domain of the following functions.

13. $y = \dfrac{1}{x - 2}$

14. $y = \dfrac{2}{x - 1}$

15. $y = \dfrac{3}{x + 3}$

16. $y = \dfrac{5}{x + 5}$

17. $y = \dfrac{4}{(x + 1)(x + 2)}$

18. $y = \dfrac{(x + 5)}{(x - 3)(x + 2)}$

19. $y = \dfrac{x}{(x + 5)(x - 1)}$

20. $y = \dfrac{2x}{(x - 7)(x + 6)}$

21. $y = \dfrac{3}{x^2 - 9}$

22. $y = \dfrac{2}{x^2 - 25}$

23. $y = \dfrac{3}{x^2 - 5x + 6}$

24. $y = \dfrac{2x}{x^2 + 4x + 4}$

25. $y = \dfrac{x}{2x^2 - 5x - 3}$

26. $y = \dfrac{3x}{3x^2 + x - 2}$

27. If $f(x) = 2x$, find $f(0), f(1), f(2)$.
28. If $f(x) = -3x$, find $f(-2), f(0), f(2)$.
29. If $f(x) = x - 5$, find $f(2), f(0), f(-1), f(a)$.
30. If $f(x) = x + 4$, find $f(-3), f(-2), f(0), f(b)$.
31. If $f(x) = 2x - 3$, find $f(1), f(2), f(\frac{1}{2}), f(z)$
32. If $f(x) = 5x - 6$, find $f(-3), f(-5), f(\frac{2}{5}), f(c)$.
33. If $f(x) = x^2 + 1$, find $f(2), f(3), f(4), f(a)$.
34. If $f(x) = 2x^2 - 3$, find $f(0), f(-3), f(2), f(a + 1)$.
35. If $f(x) = 1/x$, find $f(-1), f(-2), f(3), f(\frac{1}{2}), f(\frac{4}{3})$.
36. If $f(x) = \sqrt{x}$, find $f(4), f(a), f(16), f(b)$.
37. If $f(x) = x^2 + 2x$, find $f(3), f(-2), f(a), f(a - 2)$.
38. If $f(x) = 2x^2 + 3x$, find $f(5), f(-3), f(b), f(a - b)$.
39. If $f(x) = x^3 - 1$, find $f(1), f(2), f(-3), f(a)$.
40. If $f(x) = x^3 + 4$, find $f(0), f(-3), f(-2), f(a + 1)$.

Review Problems

The problems that follow review material we covered in Chapter 7.

Find each root.

41. $\sqrt{49}$

42. $\sqrt[3]{-8}$

Write in simplified form for radicals.

43. $\sqrt{50}$

44. $2\sqrt{18x^2 y^3}$

45. $\sqrt{\dfrac{2}{5}}$ **46.** $\sqrt[3]{\dfrac{1}{2}}$

Perform the indicated operations.

47. $3\sqrt{12} + 5\sqrt{27}$ **48.** $\sqrt{3}(\sqrt{3} - 4)$

49. $(\sqrt{6} + 2)(\sqrt{6} - 5)$ **50.** $(\sqrt{x} + 3)^2$

Rationalize the denominator.

51. $\dfrac{8}{\sqrt{5} - \sqrt{3}}$ **52.** $\dfrac{\sqrt{5} - \sqrt{2}}{\sqrt{5} + \sqrt{2}}$

Solve for x.

53. $\sqrt{2x - 5} = 3$ **54.** $\sqrt{x + 15} = x + 3$

SECTION 9.6   Fractional Exponents

Consider the expression $3^{1/2}$ and the expression $\sqrt{3}$. The second expression is the positive square root of 3. We have never encountered the first expression before. Assuming the properties of exponents apply to fractional exponents in the same way they apply to integer exponents, we can take a look at what happens when we square both of the preceding quantities:

$$\begin{array}{cc} (3^{1/2})^2 & (\sqrt{3})^2 \\ 3^{(1/2)2} & \sqrt{3} \cdot \sqrt{3} \\ 3^1 & \sqrt{9} \\ 3 & 3 \end{array}$$

The result is the same in both cases. The result allows us to define fractional exponents.

DEFINITION   If x represents a nonnegative real number, then

$$\sqrt{x} = x^{1/2}$$

$$\sqrt[3]{x} = x^{1/3}$$

$$\sqrt[4]{x} = x^{1/4}$$

and, in general, $\sqrt[n]{x} = x^{1/n}$

We can use this definition in many ways. We can change any expression involving radicals to an expression that involves only exponents. We can avoid radical notation altogether if we choose.

▶ **EXAMPLES**   Write each of the following as a radical and then simplify.

1. $25^{1/2}$

Solution   The exponent $\frac{1}{2}$ indicates that we are to find the square root of 25:

$$25^{1/2} = \sqrt{25} = 5$$

2. $9^{1/2}$

Solution Again, an exponent of $\frac{1}{2}$ indicates that we are to find the square root of 9:

$$9^{1/2} = \sqrt{9} = 3$$

3. $8^{1/3}$

Solution The exponent is $\frac{1}{3}$, indicating that we are to find the cube root of 8:

$$8^{1/3} = \sqrt[3]{8} = 2$$

4. $81^{1/4}$

Solution An exponent of $\frac{1}{4}$ indicates that we are to find the fourth root:

$$81^{1/4} = \sqrt[4]{81} = 3 \qquad \blacktriangleleft$$

Consider the expression $9^{3/2}$. If we assume that all our properties of exponents hold for fractional exponents, we can rewrite this expression as a power raised to another power:

$$9^{3/2} = (9^{1/2})^3$$

Since $9^{1/2}$ is equivalent to $\sqrt{9}$, we can simplify further:

$$(9^{1/2})^3 = (\sqrt{9})^3 = 3^3 = 27$$

We can summarize this discussion with the following definition.

DEFINITION If x represents a nonnegative real number, and n and m are integers,

$$x^{m/n} = \sqrt[n]{x^m} = (\sqrt[n]{x})^m$$

We can think of the exponent m/n as consisting of two parts. The numerator m is the power, and the denominator n is the root.

▶ **EXAMPLE 5** Simplify $8^{2/3}$.

Solution Using the preceding definition, we have

$$
\begin{aligned}
8^{2/3} &= (8^{1/3})^2 && \text{Separate exponents} \\
&= (\sqrt[3]{8})^2 && \text{Write as cube root} \\
&= 2^2 && \sqrt[3]{8} = 2 \\
&= 4 && 2^2 = 4 \qquad \blacktriangleleft
\end{aligned}
$$

▶ **EXAMPLE 6** Simplify $27^{4/3}$.

Solution

$$27^{4/3} = (27^{1/3})^4 \qquad \text{Separate exponents}$$
$$= (\sqrt[3]{27})^4 \qquad \text{Write as cube root}$$
$$= 3^4 \qquad\qquad \sqrt[3]{27} = 3$$
$$= 81 \qquad\qquad 3^4 = 81 \qquad\qquad \blacktriangleleft$$

We can apply the properties of exponents to expressions that involve fractional exponents.

▶ **EXAMPLES**   Use the properties of exponents to simplify each expression. Assume that all variables represent positive numbers.

7. $x^{1/3} \cdot x^{2/3}$

Solution   To multiply with the same base, we add exponents. The property we use is $a^r a^s = a^{r+s}$:

$$x^{1/3} \cdot x^{2/3} = x^{1/3 + 2/3} = x^{3/3} = x$$

8. $(x^{1/4})^8$

Solution   To raise a power to another power, we multiply exponents. The property we use is $(a^r)^s = a^{rs}$:

$$(x^{1/4})^8 = x^{(1/4)8} = x^2$$

9. $\dfrac{a^{5/6}}{a^{4/6}}$

Solution   To divide with the same base, we subtract the exponent in the denominator from the exponent in the numerator:

$$\frac{a^{5/6}}{a^{4/6}} = a^{5/6 - 4/6} = a^{1/6}$$

10. $(8y^{12})^{1/3}$

Solution   Distributing the exponent $\frac{1}{3}$ across the product, we have

$$(8y^{12})^{1/3} = 8^{1/3}(y^{12})^{1/3} = 2y^4 \qquad\qquad \blacktriangleleft$$

P R O B L E M   S E T   9 . 6

Change each of the following to an expression involving roots and then simplify.

1. $4^{1/2}$	**2.** $9^{1/2}$	**11.** $81^{1/2}$	**12.** $144^{1/2}$
3. $16^{1/2}$	**4.** $25^{1/2}$	**13.** $8^{2/3}$	**14.** $25^{3/2}$
5. $27^{1/3}$	**6.** $8^{1/3}$	**15.** $125^{2/3}$	**16.** $36^{3/2}$
7. $125^{1/3}$	**8.** $16^{1/4}$	**17.** $16^{3/4}$	**18.** $9^{3/2}$
9. $81^{1/4}$	**10.** $36^{1/2}$	**19.** $16^{3/2}$	**20.** $8^{5/3}$

21. $4^{3/2}$

22. $4^{5/2}$

23. $(-8)^{2/3}$

24. $(-27)^{2/3}$

25. $(-32)^{1/5}$

26. $(-32)^{3/5}$

27. $4^{1/2} + 9^{1/2}$

28. $16^{1/2} + 25^{1/2}$

29. $16^{3/4} + 27^{2/3}$

30. $49^{1/2} + 64^{1/2}$

31. $4^{1/2} \cdot 27^{1/3}$

32. $8^{1/3} \cdot 25^{1/2}$

Use the properties of exponents to simplify each of the following expressions. Assume that all variables represent positive numbers.

33. $x^{1/4} \cdot x^{3/4}$

34. $x^{1/8} \cdot x^{3/8}$

35. $(x^{2/3})^3$

36. $(x^{1/4})^{12}$

37. $\dfrac{a^{3/5}}{a^{1/5}}$

38. $\dfrac{a^{5/7}}{a^{3/7}}$

39. $(27y^6)^{1/3}$

40. $(81y^8)^{1/4}$

41. $(9a^4b^2)^{1/2}$

42. $(25a^8b^4)^{1/2}$

43. $\dfrac{x^{3/5} \cdot x^{4/5}}{x^{2/5}}$

44. $\dfrac{x^{4/7} \cdot x^{6/7}}{x^{3/7}}$

Recall that negative exponents indicate reciprocals. That is, $a^{-r} = \dfrac{1}{a^r}$. Use this property of exponents, along with what you have learned in this section, to simplify the following expressions.

45. $25^{-1/2}$

46. $8^{-1/3}$

47. $8^{-2/3}$

48. $81^{-3/4}$

49. $27^{-2/3}$

50. $16^{-3/4}$

Review Problems

The problems that follow review material we covered in Chapter 8.

Solve each equation. Use any convenient method.

51. $(2x + 1)^2 = 25$

52. $x^2 - 2x - 8 = 0$

53. $3x^2 = 4x + 2$

54. $x^2 - 5x - 5 = 0$

55. $(2x - 3)(2x - 1) = -4$

56. Solve by completing the square:
$x^2 + 4x - 3 = 0$

Perform the indicated operations.

57. $(4 + 3i) - (2 - 5i)$

58. $2i(1 - 3i)$

59. $(4 + 5i)(2 - 3i)$

60. $\dfrac{8}{2 - 3i}$

CHAPTER 9 SUMMARY

Examples

1. $x < -3$ or $x > 1$

$-2 \le x \le 3$

Compound Inequalities [9.1]

Two inequalities connected by the word ''and'' or ''or'' form a **compound inequality.** If the connecting word is ''or,'' we graph all points that are on either graph. If the connecting word is ''and,'' we graph only those points that are common to both graphs. The inequality $-2 \le x \le 3$ is equivalent to the compound inequality $-2 \le x$ and $x \le 3$.

2. Graph $x - y > 3$.

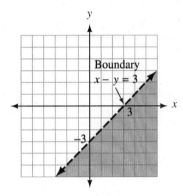

Strategy for Graphing a Linear Inequality in Two Variables [9.2]

Step 1: Replace the inequality symbol with an equal sign. The resulting equation represents the boundary for the solution set.

Step 2: Graph the boundary found in step 1, using a *solid line* if the original inequality symbol was either $\leq$ or $\geq$. Use a *broken line* otherwise.

Step 3: Choose any convenient point not on the boundary and substitute the coordinates into the *original* inequality. If the resulting statement is *true,* the graph lies on the *same* side of the boundary as the chosen point. If the resulting statement is *false,* the solution set lies on the *opposite* side of the boundary.

3. $x^3 - 8 = x^3 - 2^3$
$\qquad = (x - 2)(x^2 + 2x + 4)$

$x^3 + 125 = x^3 + 5^3$
$\qquad = (x + 5)(x^2 - 5x + 25)$

Factoring the Sum and Difference of Two Cubes [9.3]

We factor the sum of two cubes and the difference of two cubes according to the following two formulas:

$$a^3 + b^3 = (a + b)(a^2 - ab + b^2)$$

$$a^3 - b^3 = (a - b)(a^2 + ab + b^2)$$

4. If y varies directly with the square of x, then

$$y = Kx^2$$

Direct Variation [9.4]

The variable y is said to *vary directly* with the variable x if $y = Kx$, where K is a real number.

5. If y varies inversely with the cube of x, then

$$y = \frac{K}{x^3}$$

Inverse Variation [9.4]

The variable y is said to *vary inversely* with the variable x if $y = K/x$, where K is a real number.

6. If $\quad f(x) = 3x - 5$
then $f(0) = 3(0) - 5 = -5$
$\qquad f(2) = 3(2) - 5 = 1$
$\qquad f(3) = 3(3) - 5 = 4$

Functions [9.5]

A *function* is any set of ordered pairs in which no two different ordered pairs have the same first coordinates. The *domain* of the function is the set of all first coordinates. The *range* is the set of all second coordinates.

The notation $f(x)$ is read "f of x" and is defined to be the value of the function f at x.

7. $8^{2/3} = (\sqrt[3]{8})^2 = 2^2 = 4$
$9^{3/2} = (\sqrt{9})^3 = 3^3 = 27$

Fractional Exponents [9.6]

Fractional exponents are used to specify roots and avoid radical notation. For any nonnegative number a, we have

$$a^{m/n} = \sqrt[n]{a^m}$$

where m is the power and n represents the root. If n is an odd number, then a can be negative.

CHAPTER 9 REVIEW

The numbers in brackets refer to the sections of the text in which similar problems can be found.

Graph the solution to each of the following compound inequalities. [9.1]

1. $x < -2$ or $x > 5$ **2.** $x < -5$ or $x > 2$
3. $x > -2$ and $x < 5$ **4.** $x > -5$ and $x < 2$
5. $-5x \geq 25$ or $2x - 3 \geq 9$
6. $-3x \geq -6$ or $2x - 7 \geq 7$
7. $-1 < 2x + 1 < 9$ **8.** $-3 < 2x + 3 < 5$
9. $-1 \leq 3x + 5 \leq 8$
10. $-5 \leq 4x + 3 \leq 11$

Graph the following linear inequalities. [9.2]

11. $x - y < 3$ **12.** $2x + 3y \geq 6$
13. $x \geq -3$ **14.** $y < 2$
15. $y \leq -2x + 3$ **16.** $y > 3x - 4$

Factor. [9.3]

17. $a^3 - 7^3$ **18.** $a^3 + 6^3$
19. $x^3 - 8$ **20.** $x^3 + 8$
21. $27x^3 + y^3$ **22.** $27x^3 - y^3$
23. $64x^3 + 125$ **24.** $8x^3 - 27$

Work the following problems involving variation. [9.4]

25. y varies directly as x. If $y = -20$ when $x = 4$, find y when $x = 7$.
26. y varies directly as the square of x. If $y = 45$ when $x = 3$, find y when $x = 5$.
27. y varies inversely with x. If $y = 3$ when $x = 2$, find y when $x = 12$.
28. y varies inversely with the square root of x. If $y = 8$ when $x = 36$, find y when $x = 144$.
29. The volume V of a gas varies directly as the temperature T. If $V = 5$ when $T = 200$, find V when T is 360.
30. The current I in an electric circuit varies inversely with the resistance R. If a current of 50

amps is produced by a resistance of 6 ohms, what current will be produced by a resistance of 10 ohms?

Specify the domain and range of the following functions. [9.5]

31. $\{(2, 1), (-3, 5), (4, -2)\}$
32. $\{(-3, -2), (1, -4), (-5, -2)\}$
33. $y = 3x - 1$, $x = 2, 4, 6$
34. $y = 2x + 3$, $y = -1, 1, 3$

Specify any restrictions on the domain of the following functions. [9.5]

35. $y = \dfrac{1}{(x + 3)(x + 7)}$ **36.** $y = \dfrac{1}{(x - 2)(x + 5)}$

37. $y = \dfrac{5}{x^2 - 8x + 15}$ **38.** $y = \dfrac{2x}{x^2 - 3x - 28}$

Work the following problems involving function notation. [9.5]

39. If $f(x) = -2x$, find $f(-3), f(0), f(1)$.
40. If $f(x) = -2x - 1$, find $f(-2), f(1), f(3)$.
41. If $f(x) = 2x^2 + 1$, find $f(-2), f(2), f(a)$.
42. If $f(x) = -3x^2 + 2x + 1$, find $f(-1), f(3), f(c)$.
43. If $f(x) = \dfrac{1}{x}$, find $f(-2), f(4), f\left(\dfrac{1}{2}\right), f\left(\dfrac{2}{3}\right)$.
44. If $f(x) = \sqrt{x}$, find $f(36), f(81), f(b)$.
45. If $f(x) = x^3 + 1$, find $f(-2), f(2), f(a)$.
46. If $f(x) = x^3 - 3$, find $f(-1), f(3), f(b)$.

Simplify each expression as much as possible. [9.6]

47. $36^{1/2}$ **48.** $100^{1/2}$
49. $27^{2/3}$ **50.** $8^{4/3}$
51. $(-8)^{2/3}$ **52.** $(-32)^{2/5}$
53. $4^{5/2} + 27^{2/3}$ **54.** $(-8)^{1/3} + (8)^{2/3}$
55. $(16y^{20})^{1/4}$ **56.** $(27y^{15})^{1/3}$
57. $27^{-1/3}$ **58.** $32^{-2/5}$

CHAPTER 9 TEST

Graph each of the following inequalities. [9.1]

1. $x < 2$ or $x > 3$ **2.** $x \geq -2$ and $x \leq 3$

Solve each inequality and graph the solution. [9.1]

3. $3 - 4x \geq -5$ or $2x \geq 10$

4. $-7 < 2x - 1 < 9$

Graph each linear inequality in two variables. [9.2]

5. $y < -x + 4$ **6.** $3x - 4y \geq 12$

Factor each of the following. [9.3]

7. $x^3 - 7^3$ **8.** $x^3 + 27$

9. $(2x)^3 + (5y)^3$ **10.** $8x^3 - 27y^3$

11. Suppose y varies directly with the square of x. If x is 3 when y is 36, find y when x is 5. [9.4]

12. If y varies inversely with x, and x is 3 when y is 6, find y when x is 9. [9.4]

13. State the domain and range for the function $\{(1, 2), (3, 4), (5, 6)\}$. [9.5]

14. State the domain for the function $y = \dfrac{5}{x^2 - 25}$. [9.5]

If $f(x) = 3x^2 - 1$ and $g(x) = 5x + 2$, find the following. [9.5]

15. $f(3)$ **16.** $f(-2)$

17. $g(0)$ **18.** $g(-5)$

Simplify each of the following expressions. [9.6]

19. $25^{1/2}$ **20.** $8^{2/3}$

21. $x^{3/4}x^{1/4}$ **22.** $(27x^6y^9)^{1/3}$

ANSWERS TO ODD-NUMBERED PROBLEMS, CHAPTER REVIEWS, AND CHAPTER TESTS

CHAPTER 1

Problem Set 1.1

1. $x + 5 = 14$ **3.** $5y < 30$ **5.** $3y \le y + 6$ **7.** $\dfrac{x}{3} = x + 2$ **9.** 9 **11.** 49 **13.** 8 **15.** 64

17. 16 **19.** 100 **21.** 121 **23.** 11 **25.** 16 **27.** 17 **29.** 42 **31.** 30 **33.** 30
35. 24 **37.** 80 **39.** 27 **41.** 35 **43.** 13 **45.** 4 **47.** 37 **49.** 37 **51.** 16 **53.** 16
55. 81 **57.** 41 **59.** 345 **61.** 2,345 **63.** 2 **65.** 148 **67.** 36 **69.** 36 **71.** 58
73. 62 **75.** $2(10 + 4)$ **77.** $3(50) - 14$ **79.** 10 **81.** 420 **83.** Approximately 224 **85.** 95g
87. 5 **89.** 10 **91.** 25 **93.** 10

Problem Set 1.2

1–7.

9. $\frac{18}{24}$ **11.** $\frac{12}{24}$ **13.** $\frac{15}{24}$ **15.** $\frac{36}{60}$ **17.** $\frac{22}{60}$ **19.** $-10, \frac{1}{10}, 10$ **21.** $-\frac{3}{4}, \frac{4}{3}, \frac{3}{4}$ **23.** $-\frac{11}{2}, \frac{2}{11}, \frac{11}{2}$

25. $3, -\frac{1}{3}, 3$ **27.** $\frac{2}{5}, -\frac{5}{2}, \frac{2}{5}$ **29.** $-x, \dfrac{1}{x}$, the distance between x and 0 on the number line **31.** $<$

33. $>$ **35.** $>$ **37.** $>$ **39.** $<$ **41.** $<$ **43.** 6 **45.** 22 **47.** 3 **49.** 7 **51.** 3

53. $\frac{8}{15}$ **55.** $\frac{3}{2}$ **57.** $\frac{5}{4}$ **59.** 1 **61.** 1 **63.** 1 **65.** $\frac{9}{16}$ **67.** $\frac{8}{27}$ **69.** $\dfrac{1}{10,000}$ **71.** $\frac{1}{9}$

73. $\frac{1}{25}$ **75.** 4 inches; 1 inch² **77.** 4.5 inches; 1.125 inch² **79.** 10.25 centimeters; 5 centimeters²
81. $-8, -2$ **83.** $-64°; -54°$ **85.** $-100; -105$ **87.** Area = 93.5 inches²; perimeter = 39 inches
89. 1,387 calories **91.** 654 more calories

Problem Set 1.3

1. $3 + 5 = 8, 3 + (-5) = -2, -3 + 5 = 2, -3 + (-5) = -8$
3. $15 + 20 = 35, 15 + (-20) = -5, -15 + 20 = 5, -15 + (-20) = -35$ **5.** 3 **7.** -7 **9.** -14
11. -3 **13.** -25 **15.** -12 **17.** -19 **19.** -25 **21.** -8 **23.** -4 **25.** 6 **27.** 6

29. 8   **31.** −4   **33.** −14   **35.** −17   **37.** 4   **39.** 3   **41.** 15   **43.** −8   **45.** 12
47. 23, 28   **49.** 30, 35   **51.** 0, −5   **53.** −12, −18   **55.** −4, −8   **57.** Yes   **59.** 5 + 9 = 14
61. [−7 + (−5)] + 4 = −8   **63.** [−2 + (−3)] + 10 = 5   **65.** 3   **67.** −3   **69.** −12 + 4
71. 10 + (−6) + (−8) = −4   **73.** −30 + 40 = 10   **75.** 6.50, 6.75, 7.00, 7.25, 7.50, 7.75, 8.00; yes

Problem Set 1.4

1. −3   **3.** −6   **5.** 0   **7.** −10   **9.** −16   **11.** −12   **13.** −7   **15.** 35   **17.** 0   **19.** −4
21. 4   **23.** −24   **25.** −28   **27.** 25   **29.** 4   **31.** 7   **33.** 17   **35.** 8   **37.** 4   **39.** 18
41. 10   **43.** 17   **45.** 1   **47.** 1   **49.** 27   **51.** −26   **53.** −2   **55.** 68   **57.** −7 − 4 = −11
59. 12 − (−8) = 20   **61.** −5 − (−7) = 2   **63.** [4 + (−5)] − 17 = −18   **65.** 8 − 5 = 3
67. −8 − 5 = −13   **69.** 8 − (−5) = 13   **71.** 10   **73.** −2   **75.** 1,500 − 730
77. −35 + 15 − 20 = −40   **79.** 98 − 65 − 53 = −$20   **81.** 4,500, 3,950, 3,400, 2,850, 2,300; yes
83. 769 feet   **85.** 439 feet   **87.** 2 seconds   **89.** 81 pounds more   **91.** First   **93.** 35°   **95.** 60°

Problem Set 1.5

1. Commutative   **3.** Multiplicative inverse   **5.** Commutative   **7.** Distributive
9. Commutative, associative   **11.** Commutative, associative   **13.** Commutative
15. Commutative, associative   **17.** Commutative   **19.** Additive inverse   **21.** $3x + 6$
23. $9a + 9b$ **25.** 0   **27.** 0   **29.** 10   **31.** $(4 + 2) + x = 6 + x$ **33.** $x + (2 + 7) = x + 9$
35. $(3 \cdot 5)x = 15x$ **37.** $(9 \cdot 6)y = 54y$ **39.** $(\frac{1}{2} \cdot 3)a = \frac{3}{2}a$ **41.** $(\frac{1}{3} \cdot 3)x = x$ **43.** $(\frac{1}{2} \cdot 2)y = y$
45. $(\frac{3}{4} \cdot \frac{4}{3})x = x$ **47.** $(\frac{6}{5} \cdot \frac{5}{6})a = a$ **49.** $8(x) + 8(2) = 8x + 16$ **51.** $8(x) − 8(2) = 8x − 16$
53. $4(y) + 4(1) = 4y + 4$ **55.** $3(6x) + 3(5) = 18x + 15$ **57.** $2(3a) + 2(7) = 6a + 14$
59. $9(6y) − 9(8) = 54y − 72$ **61.** $\frac{1}{2}(3x) − \frac{1}{2}(6) = \frac{3}{2}x − 3$ **63.** $\frac{1}{3}(3x) + \frac{1}{3}(6) = x + 2$ **65.** $3x + 3y$
67. $8a − 8b$ **69.** $12x + 18y$ **71.** $12a − 8b$ **73.** $3x + 2y$ **75.** $4a + 25$ **77.** $6x + 12$
79. $14x + 38$ **81.** No   **83.** No, not commutative   **85.** 8 ÷ 4 ≠ 4 ÷ 8
87. 12(2,400 − 480) = $23,040; 12(2,400) − 12(480) = $23,040   **89.** $P = 2(l + w)$

Problem Set 1.6

1. −42   **3.** −21   **5.** −16   **7.** 3   **9.** 121   **11.** 6   **13.** −60   **15.** 24   **17.** 49
19. −27   **21.** 6   **23.** 10   **25.** 9   **27.** 45   **29.** 14   **31.** −2   **33.** 216   **35.** −2
37. −18   **39.** 29   **41.** 38   **43.** −5   **45.** 37   **47.** 80   **49.** $−\frac{10}{21}$ **51.** −4   **53.** 1
55. $\frac{9}{16}$ **57.** $−\frac{8}{27}$ **59.** $−8x$ **61.** $42x$ **63.** x **65.** x **67.** $−4a − 8$ **69.** $−\frac{3}{2}x + 3$
71. $−6x + 8$ **73.** $−15x − 30$ **75.** −25   **77.** $2(−4x) = −8x$ **79.** −26   **81.** 8   **83.** −80
85. $\frac{1}{8}$ **87.** 1   **89.** −24   **91.** $60   **93.** 1° F   **95.** 500, 1,000, 2,000, 4,000, 8,000, 16,000; yes

Problem Set 1.7

1. −2   **3.** −3   **5.** $−\frac{1}{3}$ **7.** 3   **9.** $\frac{1}{7}$ **11.** 0   **13.** 9   **15.** −15   **17.** −36   **19.** $−\frac{1}{4}$
21. $\frac{16}{15}$ **23.** $\frac{4}{3}$ **25.** $−\frac{8}{13}$ **27.** −1   **29.** 1   **31.** $\frac{3}{5}$ **33.** $−\frac{5}{3}$ **35.** −2   **37.** −3
39. Undefined   **41.** Undefined   **43.** 5   **45.** $−\frac{7}{3}$ **47.** −1   **49.** −7   **51.** $\frac{15}{17}$ **53.** $−\frac{32}{17}$
55. $\frac{1}{3}$ **57.** 1   **59.** 1   **61.** −2   **63.** $\frac{9}{7}$ **65.** $\frac{16}{11}$ **67.** −1   **69.** 1   **71.** −7   **73.** −1
75. 3   **77.** −10   **79.** −3   **81.** −8   **83.** $350   **85.** Drops 3.5° F each hour

Problem Set 1.8

1. $0, 1$ **3.** $-3, -2.5, 0, 1, \frac{3}{2}$ **5.** All   **7.** $-10, -8, -2, 9$ **9.** π **11.** T   **13.** F   **15.** F   **17.** T
19. Composite, $2^4 \cdot 3$ **21.** Prime   **23.** Composite, $3 \cdot 11 \cdot 31$ **25.** $2^4 \cdot 3^2$ **27.** $2 \cdot 19$ **29.** $3 \cdot 5 \cdot 7$
31. $2^2 \cdot 3^2 \cdot 5$ **33.** $5 \cdot 7 \cdot 11$ **35.** 11^2 **37.** $2^2 \cdot 3 \cdot 5 \cdot 7$ **39.** $2^2 \cdot 5 \cdot 31$ **41.** $\frac{7}{11}$ **43.** $\frac{5}{7}$
45. $\frac{11}{13}$ **47.** $\frac{14}{15}$ **49.** $\frac{5}{9}$ **51.** $\frac{5}{8}$ **53.** $6^3 = (2 \cdot 3)^3 = 2^3 \cdot 3^3$ **55.** $9^4 \cdot 16^2 = (3^2)^4(2^4)^2 = 2^8 \cdot 3^8$
57. $3 \cdot 8 + 3 \cdot 7 + 3 \cdot 5 = 24 + 21 + 15 = 60 = 2^2 \cdot 3 \cdot 5$ **59.** Irrational numbers   **61.** $8, 21, 34$

Problem Set 1.9

1. $\frac{2}{3}$ **3.** $-\frac{1}{4}$ **5.** $\frac{1}{2}$ **7.** $\frac{x-1}{3}$ **9.** $\frac{3}{2}$ **11.** $\frac{x+6}{2}$ **13.** $-\frac{3}{5}$ **15.** $\frac{10}{a}$ **17.** $\frac{7}{8}$ **19.** $\frac{1}{10}$

21. $\frac{7}{9}$ **23.** $\frac{7}{3}$ **25.** $\frac{1}{4}$ **27.** $\frac{7}{6}$ **29.** $\frac{19}{24}$ **31.** $\frac{13}{60}$ **33.** $\frac{29}{35}$ **35.** $\frac{949}{1,260}$ **37.** $\frac{13}{420}$ **39.** $\frac{41}{24}$ **41.** $\frac{5}{4}$
43. $\frac{160}{63}$ **45.** $\frac{5}{8}$ **47.** $-\frac{2}{3}$ **49.** $\frac{7}{3}$ **51.** $\frac{1}{125}$

Chapter 1 Review

1. $-7 + (-10) = -17$ **2.** $(-7 + 4) + 5 = 2$ **3.** $(-3 + 12) + 5 = 14$ **4.** $4 - 9 = -5$
5. $9 - (-3) = 12$ **6.** $-7 - (-9) = 2$ **7.** $(-3)(-7) - 6 = 15$ **8.** $5(-6) + 10 = -20$
9. $2[(-8)(3x)] = -48x$ **10.** $\frac{-25}{-5} = 5$ **11.** $\frac{-40}{8} - 7 = -12$ **12.** $\frac{-45}{15} + 9 = 6$
13–18.

19. 12 **20.** 3 **21.** $\frac{4}{5}$ **22.** $\frac{7}{10}$ **23.** 1.8 **24.** -10 **25.** $-6, \frac{1}{6}$ **26.** $-\frac{3}{10}, \frac{10}{3}$ **27.** $9, -\frac{1}{9}$
28. $\frac{12}{5}, -\frac{5}{12}$ **29.** $\frac{6}{35}$ **30.** 1 **31.** -5 **32.** -3 **33.** $-\frac{5}{4}$ **34.** 1 **35.** 3 **36.** -45
37. -38 **38.** -6 **39.** 0 **40.** -6 **41.** -22 **42.** -24 **43.** -25 **44.** -16 **45.** 1
46. 0 **47.** -3 **48.** -3 **49.** 22 **50.** 29 **51.** -4 **52.** -12 **53.** -20 **54.** -9
55. -15 **56.** -11 **57.** 14 **58.** 14 **59.** 12 **60.** 18 **61.** 28 **62.** 20 **63.** -30
64. 63 **65.** -12 **66.** 45 **67.** -24 **68.** 24 **69.** 12 **70.** 30 **71.** -4 **72.** -4
73. $-\frac{1}{4}$ **74.** 9 **75.** $-\frac{2}{3}$ **76.** $-\frac{3}{2}$ **77.** 23 **78.** 24 **79.** 47 **80.** 105 **81.** -3
82. 83 **83.** 7 **84.** 12 **85.** 32 **86.** 53 **87.** 2 **88.** -7 **89.** 30 **90.** 36 **91.** -3
92. -28 **93.** -98 **94.** -75 **95.** 70 **96.** 100 **97.** 2 **98.** 9 **99.** $\frac{17}{2}$ **100.** -3
101. Undefined   **102.** Undefined   **103.** 8 **104.** 4 **105.** 3 **106.** 2 **107.** Associative
108. Multiplicative identity   **109.** Commutative   **110.** Additive inverse   **111.** Multiplicative inverse
112. Additive identity   **113.** Commutative, associative   **114.** Distributive   **115.** $12 + x$ **116.** $28a$
117. x **118.** y **119.** $14x + 21$ **120.** $6a - 12$ **121.** $\frac{5}{2}x - 3$ **122.** $-\frac{3}{2}x + 3$
123. $-\frac{1}{3}, 0, 5, -4.5, \frac{2}{5}, -3$ **124.** $0, 5$ **125.** $\sqrt{7}, \pi$ **126.** $0, 5, -3$ **127.** $2 \cdot 3^2 \cdot 5$ **128.** $2^3 \cdot 3 \cdot 5$
129. $2^3 \cdot 3 \cdot 5 \cdot 7$ **130.** 2^{10} **131.** $\frac{173}{210}$ **132.** $\frac{109}{420}$ **133.** -2 **134.** 810 **135.** 8 **136.** 12
137. -1 **138.** $\frac{1}{16}$

Chapter 1 Test

1. $x + 3 = 8$ **2.** $5y = 15$ **3.** 40 **4.** 16 **5.** Opposite 4, reciprocal $-\frac{1}{4}$, absolute value 4
6. Opposite $-\frac{3}{4}$, reciprocal $\frac{4}{3}$, absolute value $\frac{3}{4}$ **7.** -4 **8.** -13 **9.** -12 **10.** 0 **11.** c   **12.** e
13. d   **14.** a   **15.** -21 **16.** 64 **17.** -2 **18.** $-\frac{8}{27}$ **19.** 4 **20.** 204 **21.** 25
22. 52 **23.** 2 **24.** 2 **25.** $8 + 2x$ **26.** $10x$ **27.** $6x + 10$ **28.** $-2x + 1$ **29.** $1, -8$
30. $1, 1.5, \frac{3}{4}, -8$ **31.** $\sqrt{2}$ **32.** All of them   **33.** $2^4 \cdot 37$ **34.** $2^2 \cdot 5 \cdot 67$ **35.** $\frac{25}{42}$ **36.** $\frac{287}{180}$
37. $8 + (-3) = 5$ **38.** $-24 - 2 = -26$ **39.** $-5(-4) = 20$ **40.** $\frac{-24}{-2} = 12$ **41.** 12 **42.** $\frac{1}{2}$

CHAPTER 2

Problem Set 2.1

1. $-3x$ **3.** $-a$ **5.** $12x$ **7.** $6a$ **9.** $6x - 3$ **11.** $7a + 5$ **13.** $5x - 5$ **15.** $4a + 2$
17. $-9x - 2$ **19.** $12a + 3$ **21.** $10x - 1$ **23.** $21y + 6$ **25.** $-6x + 8$ **27.** $-2a + 3$
29. $-4x + 26$ **31.** $4y - 16$ **33.** $-6x - 1$ **35.** $2x - 12$ **37.** $10a + 33$ **39.** $2x - 3$
41. $7y - 39$ **43.** $-19x - 14$ **45.** 5 **47.** -9 **49.** 4 **51.** 4 **53.** -37 **55.** -41
57. 64 **59.** 64 **61.** 144 **63.** 144 **65.** 3 **67.** 0 **69.** 15 **71.** 6 **73.** $5, 7, 9, 11$
75. $1, 8, 27, 64$ **77.** $2, 5, 10, 17$ **79.** $1, 4, 7, 10, \ldots$ an arithmetic sequence
81. $0, 1, 4, 9, \ldots$ a sequence of squares **83.** $x + 5; 3$ **85.** $x - 5; -7$ **87.** $2(x + 10); 16$
89. $\dfrac{10}{x}; -5$ **91.** $[3x + (-2)] - 5; -13$ **93.** $-\frac{7}{2}$ **95.** $\frac{51}{40}$

Problem Set 2.2

1. 11 **3.** 4 **5.** $-\frac{3}{4}$ **7.** -5.8 **9.** -17 **11.** $-\frac{1}{8}$ **13.** -4 **15.** -3.6 **17.** 1
19. $-\frac{7}{45}$ **21.** 3 **23.** $\frac{11}{8}$ **25.** 21 **27.** 7 **29.** 3.5 **31.** 22 **33.** -6 **35.** 0 **37.** -2
39. -16 **41.** -3 **43.** 10 **45.** -12 **47.** -1 **49.** 4 **51.** 2 **53.** -5 **55.** -1 **57.** -3
59. 8 **61.** -8 **63.** 2 **65.** 11 **67.** $18x$ **69.** x **71.** y **73.** x **75.** a

Problem Set 2.3

1. 2 **3.** 4 **5.** $-\frac{1}{2}$ **7.** -2 **9.** 3 **11.** 4 **13.** 0 **15.** 0 **17.** 6 **19.** -50 **21.** $\frac{3}{2}$
23. 12 **25.** -3 **27.** 32 **29.** -8 **31.** $\frac{1}{2}$ **33.** 4 **35.** 8 **37.** -4 **39.** 4 **41.** -15
43. $-\frac{1}{2}$ **45.** 3 **47.** 1 **49.** $\frac{1}{4}$ **51.** -3 **53.** 3 **55.** 2 **57.** $-\frac{3}{2}$ **59.** $-\frac{3}{2}$ **61.** 1
63. 1 **65.** -2 **67.** -2 **69.** $10x - 43$ **71.** $-3x - 13$ **73.** $-6y + 4$ **75.** $-5x + 7$

Problem Set 2.4

1. 3 **3.** -2 **5.** -1 **7.** 2 **9.** -4 **11.** -2 **13.** 0 **15.** 1 **17.** $\frac{1}{2}$ **19.** 7 **21.** 8
23. $-\frac{1}{3}$ **25.** $\frac{3}{4}$ **27.** 75 **29.** 2 **31.** 6 **33.** 8 **35.** 0 **37.** $\frac{3}{7}$ **39.** 1 **41.** $\frac{3}{2}$ **43.** 4
45. 1 **47.** $6x - 10$ **49.** $\frac{3}{2}x + 3$ **51.** $-x + 2$

Problem Set 2.5

1. 100 feet **3.** 0 **5.** 2 **7.** 15 **9.** 10 **11.** 4 **13.** 2 **15.** $l = \dfrac{A}{w}$ **17.** $r = \dfrac{d}{t}$ **19.** $h = \dfrac{V}{lw}$

21. $P = \dfrac{nRT}{V}$ **23.** $a = P - b - c$ **25.** $x = 3y - 1$ **27.** $y = 3x + 6$ **29.** $y = -\frac{2}{3}x + 2$

31. $y = -2x + 4$ **33.** $y = \frac{5}{2}x - \frac{3}{2}$ **35.** $w = \dfrac{P - 2l}{2}$ **37.** $v = \dfrac{h - 16t^2}{t}$ **39.** $h = \dfrac{A - \pi r^2}{2\pi r}$

41. $y = -\frac{3}{2}x + 3$ **43.** $y = \frac{3}{7}x - 3$ **45.** $y = 2x + 8$ **47.** $60°; 150°$ **49.** $45°; 135°$ **51.** 10
53. 240 **55.** 25% **57.** 35% **59.** 64 **61.** $2,000$ **63.** $100°$ C; yes **65.** $20°$ C; yes
67. $C = \frac{5}{9}(F - 32)$ **69.** 60% **71.** 26.5% **73.** 7 meters **75.** $\frac{3}{2}$ or 1.5 inches **77.** 132 feet
79. $\frac{2}{9}$ centimeters **81.** The sum of 4 and 1 is 5 **83.** The difference of 6 and 2 is 4 **85.** $2(6 + 3)$
87. $2 \cdot 5 + 3 = 13$

Problem Set 2.6

Along with the answers to the odd-numbered problems in this problem set we are including the equations used to solve each problem. Be sure that you try the problems on your own before looking here to see what the correct equations are.

1. $x + 5 = 13$; 8   **3.** $2x + 4 = 14$; 5   **5.** $5(x + 7) = 30$; -1

7. The two numbers are x and $x + 2$; $x + (x + 2) = 8$; 3 and 5

9. The two numbers are x and $3x - 4$; $x + (3x - 4) + 5 = 25$; 6 and 14

11. Barney's age is x, Fred's age is $x + 4$; $(x - 5) + (x - 1) = 48$; Barney is 27, Fred 31

13. Lacy's age is x, Jack's age is $2x$; $(x + 3) + (2x + 3) = 54$; Lacy is 16, Jack 32

15. Let $x =$ Nancy's age now; then $x - 28 =$ Erik's age now; $x + 14 + (x - 28) + 14 = 100$; Nancy is 50

17. The width is x, the length is $x + 5$; $2x + 2(x + 5) = 34$; length $= 11$ inches; width $= 6$ inches

19. The length of a side is x; $4x = 48$; 12 meters

21. The width is x, the length is $2x - 3$; $2x + 2(2x - 3) = 54$; length $= 17$ inches; width $= 10$ inches

23. If the number of nickels is x, then the number of dimes is $x + 9$; $5x + 10(x + 9) = 210$; 8 nickels, 17 dimes

25. If you have x dimes and $2x$ quarters, then $10x + 25(2x) = 900$; 15 dimes, 30 quarters

27. If she has x nickels, then she has $x + 3$ dimes and $x + 5$ quarters; $5x + 10(x + 3) + 25(x + 5) = 435$; 7 nickels, 10 dimes, 12 quarters

29. 4 is less than 10   **31.** 9 is greater than or equal to -5 **33.** $<$ **35.** $<$ **37.** 2   **39.** 12

Problem Set 2.7

1. If $x =$ the amount of money invested at 8%, then $x + 2{,}000$ is the amount invested at 9%; $0.08x + 0.09(x + 2{,}000) = 860$; $4,000 invested at 8%, $6,000 invested at 9%

3. If $x =$ the amount invested at 10%, then $x + 500$ is the amount invested at 12%; $0.10x + 0.12(x + 500) = 214$; $700 invested at 10%, $1,200 invested at 12%

5. Let $x =$ the amount invested at 8%, $2x =$ the amount invested at 9%, and $3x =$ the amount invested at 10%; $0.08x + 0.09(2x) + 0.10(3x) = 280$; $500 at 8%, $1,000 at 9%, $1,500 at 10%

7. 45°, 45°, 90°   **9.** 22.5°, 45°, 112.5°   **11.** 53°, 90°

13. Let $x =$ the number of minutes after the first minute; $41 + 32x = 521$; $x = 15$, so the call was 16 minutes

15. Let $x =$ the number of hours past 35 hours; $35(12) + 18x = 492$; $x = 4$, so she worked 39 hours that week

17. Let $x =$ the number of children's tickets; then $2x =$ the number of adult tickets; $4.5x + 6(2x) = 115.5$; $x = 7$; she sold 7 children's tickets and 14 adult tickets

19. 1,849 square feet   **21.** Jeff   **23.** $10.38   **25.** Yes   **27.** $54   **29.** Yes   **31.** -8 **33.** $\frac{3}{4}$

35. 25   **37.** 0

Problem Set 2.8

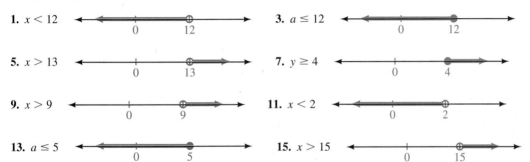

1. $x < 12$

3. $a \le 12$

5. $x > 13$

7. $y \ge 4$

9. $x > 9$

11. $x < 2$

13. $a \le 5$

15. $x > 15$

17. $x < -3$

19. $x \le 6$

21. $x \ge -50$

23. $y < -6$

25. $x < 6$

27. $y \ge -5$

29. $x < 3$

31. $x \le 18$

33. $a < -20$

35. $y < 25$

37. $a \le 3$

39. $x \ge \frac{15}{2}$

41. $x < -1$

43. $y \ge -2$

45. $x < -1$

47. $m \le -6$

49. $x \le -5$

51. $y < -\frac{3}{2}x + 3$ **53.** $y < \frac{2}{5}x - 2$ **55.** $y \le \frac{3}{7}x + 3$ **57.** $y \le \frac{1}{2}x + 1$ **59.** $2x + 6 < 10; x < 2$
61. $4x > x - 8; x > -\frac{8}{3}$ **63.** $2(x + 5) \le 12; x \le 1$ **65.** $3x - 5 < x + 7; x < 6$
67. $2x + 2(3x) \ge 48; x \ge 6$; the width is at least 6 meters
69. $x + (x + 2) + (x + 4) > 24; x > 6$; the shortest side is even and greater than 6 inches   **71.** b   **73.** a
75. b and c   **77.** 0, 2   **79.** $-3, -\frac{5}{3}, 0, 2, \frac{15}{2}$ **81.** $\frac{5}{33}$

Chapter 2 Review

1. $-3x$ **2.** $-3x$ **3.** $-2x - 3$ **4.** $-4x - 4$ **5.** $4a - 7$ **6.** $5a - 5$ **7.** $-2a + 3$
8. $9a + 15$ **9.** $-6y$ **10.** $-10a + 29$ **11.** $-6x + 1$ **12.** $-8a - 3$ **13.** 19   **14.** 21
15. -11 **16.** 13   **17.** -18 **18.** -21 **19.** -13 **20.** 5   **21.** 8   **22.** 3   **23.** 6
24. 6   **25.** -8 **26.** -7 **27.** $\frac{15}{14}$ **28.** $\frac{19}{12}$ **29.** 2   **30.** 4   **31.** -5 **32.** -8
33. -5 **34.** -6 **35.** 0   **36.** 0   **37.** 12   **38.** 10   **39.** -8 **40.** -21 **41.** -1
42. $\frac{5}{2}$ **43.** 1   **44.** 2   **45.** 5   **46.** -2 **47.** $\frac{5}{2}$ **48.** $-\frac{5}{2}$ **49.** 0   **50.** 0   **51.** 10
52. 7   **53.** 2   **54.** 4   **55.** 2   **56.** -2 **57.** $-\frac{8}{3}$ **58.** $\frac{11}{4}$ **59.** 0   **60.** 0   **61.** -5
62. -6 **63.** -4 **64.** -3 **65.** 0   **66.** -4 **67.** -8 **68.** 4   **69.** $y = \frac{2}{5}x - 2$

70. $y = \frac{5}{2}x - 5$ **71.** $h = \dfrac{V}{\pi r^2}$ **72.** $w = \dfrac{P - 2l}{2}$ **73.** 206.4   **74.** 9%   **75.** 11   **76.** 6

77. Tom is 15; Bob is 19   **78.** Tracey is 10; Debra is 16   **79.** 25 meters, 5 meters
80. 24 meters, 6 meters   **81.** 10 nickels, 5 dimes   **82.** 8 dimes, 7 quarters
83. $500 at 9%; $800 at 10%   **84.** $700 at 8%; $1,100 at 9%   **85.** $x > -2$ **86.** $x > -3$ **87.** $x < 2$
88. $x < 4$ **89.** $a \ge 6$ **90.** $a \ge 6$ **91.** $a < -15$ **92.** $a < -24$

93.    number line with open circle at -8, shaded right to 0

94. number line with closed circle at 9, shaded left to 0

95. number line with open circle at -5, shaded right

96. number line with closed circle at 4, shaded left

97. number line with open circle at 7, shaded right

98. number line with closed circle at 12, shaded left

99. number line with closed circle at -3, shaded left

100. number line with closed circle at -3, shaded right

Chapter 2 Test

1. $-4x + 5$ **2.** $3a - 4$ **3.** $-3y - 12$ **4.** $11x + 28$ **5.** 22 **6.** 25 **7.** 6 **8.** $\frac{4}{3}$ **9.** 2
10. $-\frac{1}{2}$ **11.** -3 **12.** 70 **13.** -3 **14.** -1 **15.** 5.7 **16.** $2,000$ **17.** 3 **18.** $\frac{28}{3}$ inches

19. $y = -\frac{2}{5}x + 4$ **20.** $v = \dfrac{h - x - 16t^2}{t}$ **21.** Rick is 20; Dave is 40

22. Width is 10 inches; length is 20 inches   **23.** 8 quarters, 15 dimes   **24.** \$800 at 7%; \$1,400 at 9%
25. $x < 1$ number line with open circle at 1, shaded left through 0 **26.** $a < -4$  number line with open circle at -4, shaded left

27. $x \leq -3$ number line with closed circle at -3, shaded left   **28.** $m \geq -2$ number line with closed circle at -2, shaded right

CHAPTER 3

Problem Set 3.1

1.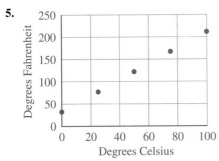

3.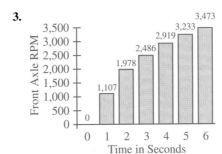

5.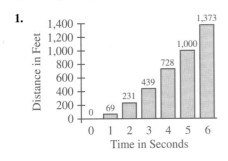

7. Yes

9.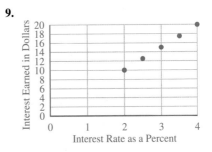

11. Yes
13–29.

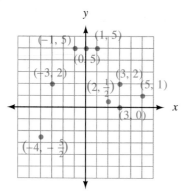

Odd-Numbered
Problems

31. $(-4, 4)$ **33.** $(-4, 2)$ **35.** $(-3, 0)$ **37.** $(2, -2)$ **39.** $(-5, -5)$

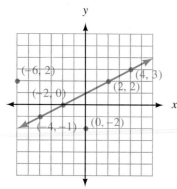

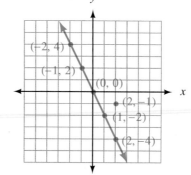

41. Yes
43. No

45. Yes
47. No

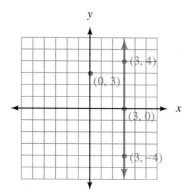

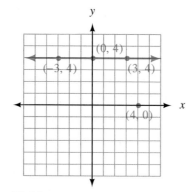

49. Yes
51. No

53. No
55. No

57.

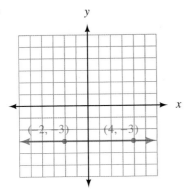

$(-2, -3)$ $(4, -3)$

Every point on this line has a y-coordinate of -3.

59. Along the y-axis   **61.** 31   **63.** $\frac{1}{27}$ **65.** -5 **67.** 567   **69.** 15

Problem Set 3.2

1. $(0, 6), (3, 0), (6, -6)$ **3.** $(0, 3), (4, 0), (-4, 6)$ **5.** $(1, 1), (\frac{3}{4}, 0), (5, 17)$ **7.** $(2, 13), (1, 6), (0, -1)$
9. $(-5, 4), (-5, -3), (-5, 0)$

11.

x	y
1	3
-3	-9
4	12
6	18

13.

x	y
0	0
$-\frac{1}{2}$	-2
-3	-12
3	12

15.

x	y
2	3
3	2
5	0
9	-4

17.

x	y
2	0
3	2
1	-2
-3	-10

19.

x	y
0	-1
-1	-7
-3	-19
$\frac{3}{2}$	8

21. $(0, -2)$ **23.** $(1, 5), (0, -2), (-2, -16)$ **25.** $(1, 6), (-2, -12), (0, 0)$ **27.** $(2, -2)$
29. $(3, 0), (3, -3)$ **31.** 12 inches   **33.** $(5, 10)$ **35.** $(3, 3), (-3, 3), (5, 5), (-5, 5)$

37. $y = 13 + 1.5x$

Bags Collected	Cost in Dollars
5	20.50
7	23.50
9	26.50
11	29.50
13	32.50

39. $y = 7 + 1.1x$

Gallons per Month	Monthly Cost in Dollars
15	23.50
20	29.00
25	34.50
30	40.00
35	45.50

41. -3 **43.** 2   **45.** 0   **47.** $y = -5x + 4$ **49.** $y = \frac{3}{2}x - 3$

Problem Set 3.3

1.

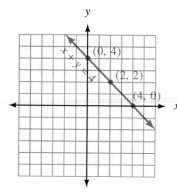

3.

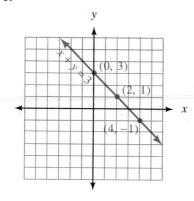

5.

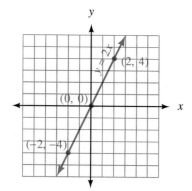

7.

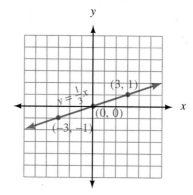

9.

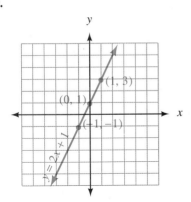

11.

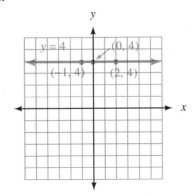

13.

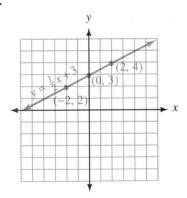

15.

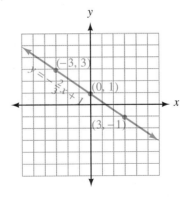

17. $y = -2x + 3$

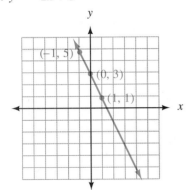

19. $y = \dfrac{-3}{2}x + 3$

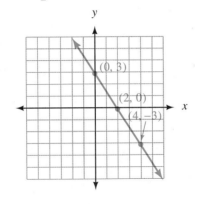

21. $y \frac{1}{2}x + 3$

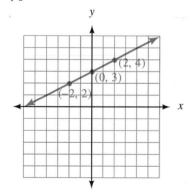

23.

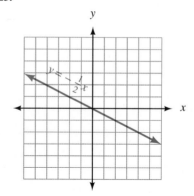

25.

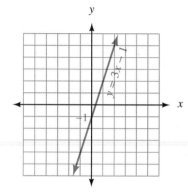

27.

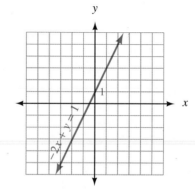

29.

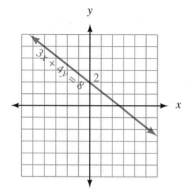

31.

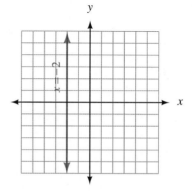

33.

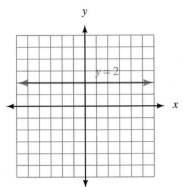

35.

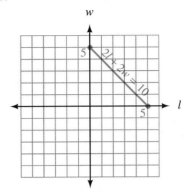

Note: The graph does not go beyond the *x*- and *y*- axes because *l* and *w* are the length and width of a rectangle and therefore cannot be negative.

37.

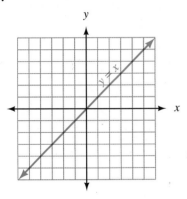

39.

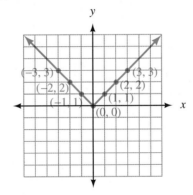

41.

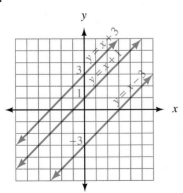

43. 5   **45.** −2   **47.** 3   **49.** −6

Problem Set 3.4

1.

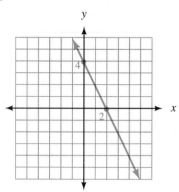

3.

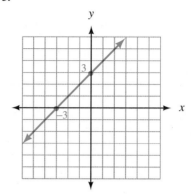

5.

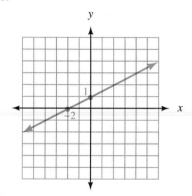

7.

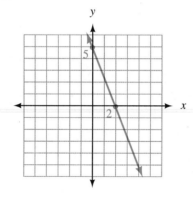

9.

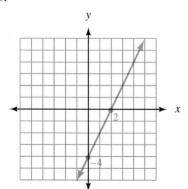

11.

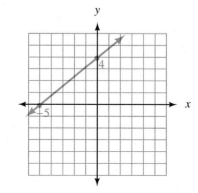

13.

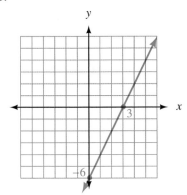

15.

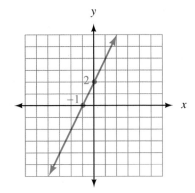

17.

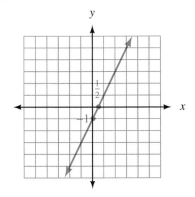

19.

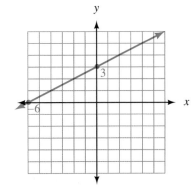

21.

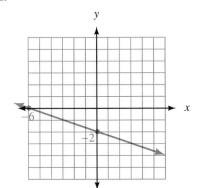

23.

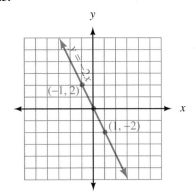

25.

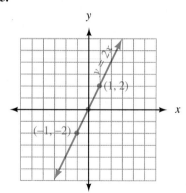

27.

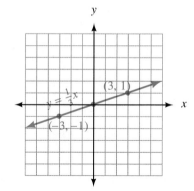

29.

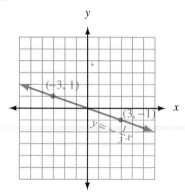

31.

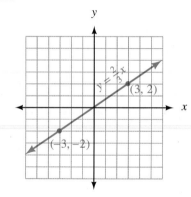

33.

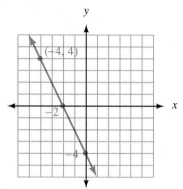

35.

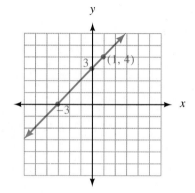

37.

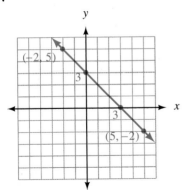

39.

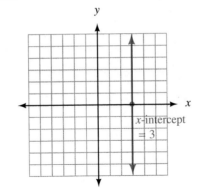

41.

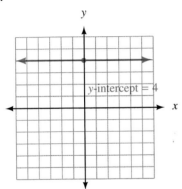

43. 49   **45.** −7   **47.** −12   **49.** −18   **51.** −17

Problem Set 3.5

1.

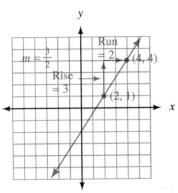

3.

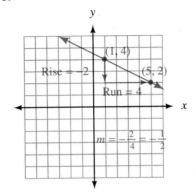

5.

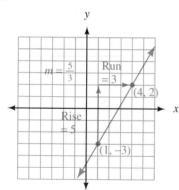

7.

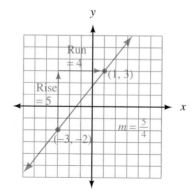

9.

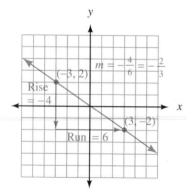

11.

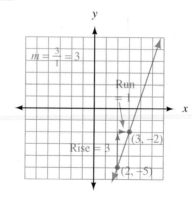

13.

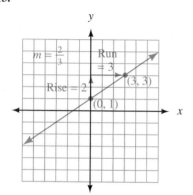

15.

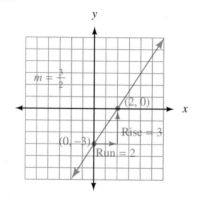

17.

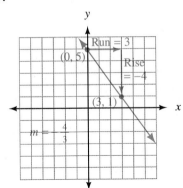

19.

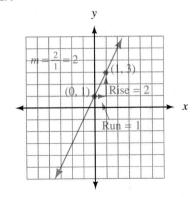

21.

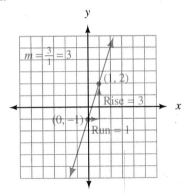

23.

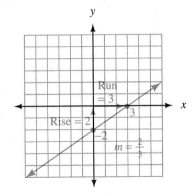

25.

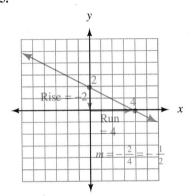

27.

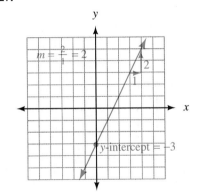

29.

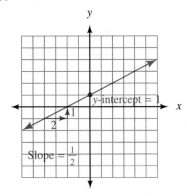

31. 6   **33.** $I = \dfrac{V}{R}$ **35.** $l = \dfrac{P - 2w}{2}$ **37.** $y = 2x - 4$ **39.** $y = -\frac{5}{4}x + 2$

Problem Set 3.6

1. $y = \frac{2}{3}x + 1$ **3.** $y = \frac{3}{2}x - 1$ **5.** $y = -\frac{2}{5}x + 3$ **7.** $y = 2x - 4$ **9.** $y = -3x + 2$

11. $m = 2, b = 4$ **13.** $m = -3, b = 3$

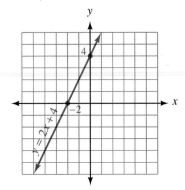

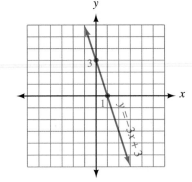

15. $m = \dfrac{-3}{2}, b = 3$ **17.** $m = \frac{4}{5}, b = -4$

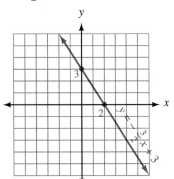

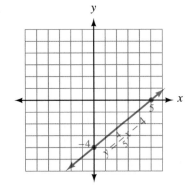

19. $m = \dfrac{-2}{5}, b = -2$

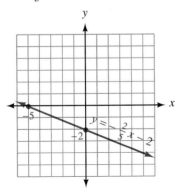

21. $y = 2x - 1$ **23.** $y = -\frac{1}{2}x - 1$ **25.** $y = \frac{3}{2}x - 6$ **27.** $y = -3x + 1$ **29.** $y = x + 2$
31. $y = x - 2$ **33.** $y = 2x - 3$ **35.** $y = \frac{4}{3}x + 2$ **37.** $y = -\frac{2}{3}x - 3$ **39.** $y = -x - 1$
41. $y = -\frac{2}{3}x + 2$ **43.** $y = -\frac{5}{2}x - 5$ **45.** $x = 3$ **47.** $y = 3$ **49.** 75   **51.** 20%   **53.** 400

55.

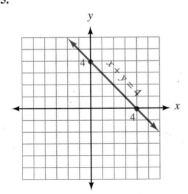

57.

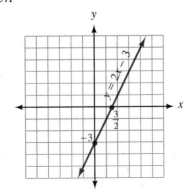

59.

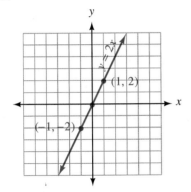

Problem Set 3.7

1. $(2, 1)$

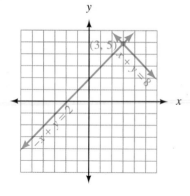

3. $(-1, 2)$

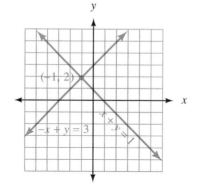

5. $(3, 5)$

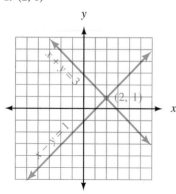

7. $(4, 3)$

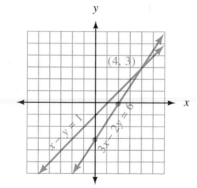

9. $(0, -6)$

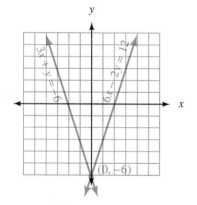

11. $(1, 0)$

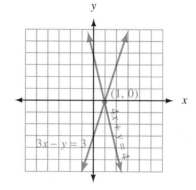

13. $(0, 0)$

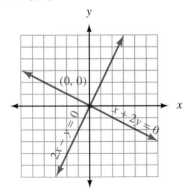

15. $(-5, -6)$

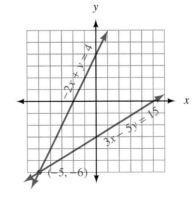

17. $(-1, -1)$

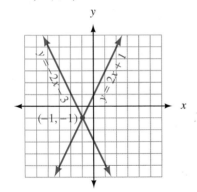

19. $(-3, 2)$

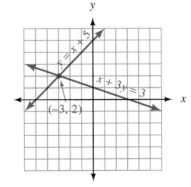

21. $(-3, 5)$

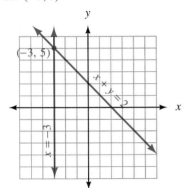

23. $(-4, 6)$

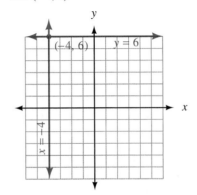

25. Lines are parallel; there is no solution to the system

27. Lines coincide; any solution to one equation is a solution to the other

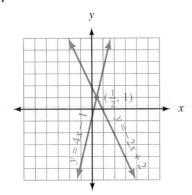

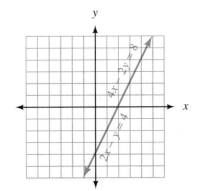

29.

31.

33. -15 **35.** 27

37. -27 **39.** 13

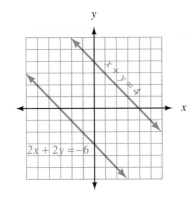

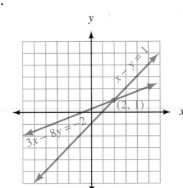

Problem Set 3.8

1. $(2, 1)$ **3.** $(3, 7)$ **5.** $(2, -5)$ **7.** $(-1, 0)$ **9.** Lines coincide   **11.** $(4, 8)$ **13.** $(\frac{1}{5}, 1)$ **15.** $(1, 0)$
17. $(-1, -2)$ **19.** $(-5, \frac{3}{4})$ **21.** $(-4, 5)$ **23.** $(-3, -10)$ **25.** $(3, 2)$ **27.** $(5, \frac{1}{3})$ **29.** $(-2, \frac{2}{3})$
31. $(2, 2)$ **33.** Lines are parallel   **35.** $(1, 1)$ **37.** Lines are parallel   **39.** $(7, 5)$ **41.** $(10, 12)$
43. $x < 5$ **45.** $x \leq -4$ **47.** $x \geq 3$ **49.** $x > -4$

Problem Set 3.9

1. $(4, 7)$ **3.** $(3, 17)$ **5.** $(\frac{3}{2}, 2)$ **7.** $(2, 4)$ **9.** $(0, 4)$ **11.** $(-1, 3)$ **13.** $(1, 1)$ **15.** $(2, -3)$
17. $(-2, \frac{3}{5})$ **19.** $(-3, 5)$ **21.** Lines are parallel   **23.** $(3, 1)$ **25.** $(\frac{1}{2}, \frac{3}{4})$ **27.** $(2, 6)$
29. $(4, 4)$ **31.** $(5, -2)$ **33.** $(18, 10)$ **35.** Lines coincide   **37.** $(10, 12)$
39. 40 gallons   **41.** $l = 9$ meters; $w = 3$ meters   **43.** 12 nickels, 15 dimes   **45.** 480
47. $800 at 8%, $1,600 at 10%

Problem Set 3.10

As you can see, in addition to the answers to the problems we have sometimes included the system of equations used to solve the problems. Remember, you should attempt the problem on your own before looking here to check your answers or equations.

1. $x + y = 25$ The two numbers   **3.** 3, 12   **5.** $x - y = 5$ The two numbers
   $y = x + 5$ are 10 and 15. $x = 2y + 1$ are 9 and 4.

7. 6, 29   **9.** Let $x =$ the amount invested at 6% and $y =$ the amount invested at 8%.
                  $x + y = 20,000$ He has $9,000 at 8%
             $0.06x + 0.08y = 1,380$ and $11,000 at 6%.

11. $8,000 at 5%, $2,000 at 6%   **13.** 6 nickels, 8 quarters

15. Let $x =$ the number of dimes and $y =$ the number of quarters.
                  $x + y = 21$ He has 12 dimes
             $0.10x + 0.25y = 3.45$ and 9 quarters.

17. Let $x =$ the number of liters of 50% solution and $y =$ the number of liters of 20% solution.
                  $x + y = 18$ 6 liters of 50% solution
             $0.50x + 0.20y = 0.30(18)$ 12 liters of 20% solution

19. 10 gallons of 10% solution, 20 gallons of 7% solution   **21.** 20 adults, 50 kids

23. 16 feet wide, 32 feet long   **25.** 33 of the $5 chips, 12 of the $25 chips

27. 50 at $11, 100 at $20   **29.** $x < -4$ **31.** $x \le -3$ **33.** $x \le -1$

35. $x > 1$

Chapter 3 Review

1–6.

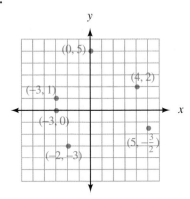

7. $(4, -6), (0, 6), (1, 3), (2, 0)$ **8.** $(5, -2), (0, -4), (15, 2), (10, 0)$ **9.** $(4, 2), (2, -2), (\frac{9}{2}, 3)$
10. $(2, 13), (-\frac{3}{5}, 0), (-\frac{6}{5}, -3)$ **11.** $(2, -3), (-1, -3), (-3, -3)$ **12.** $(6, 5), (6, 0), (6, -1)$
13. $(2, -\frac{3}{2})$ **14.** $(6, -1)$ **15.** $(-\frac{8}{3}, -1), (-3, -2)$
16. $(1, -3)$

17.

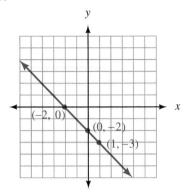

18.

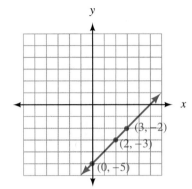

19.

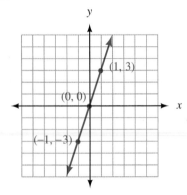

20.

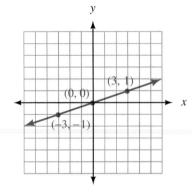

21.

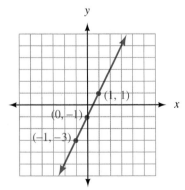

22.

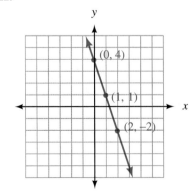

23.

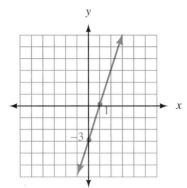

24.

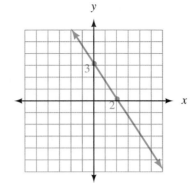

25.

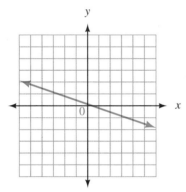

26.

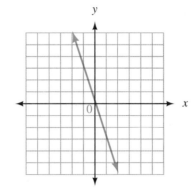

27.

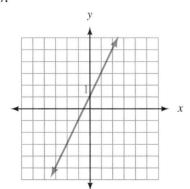

28.

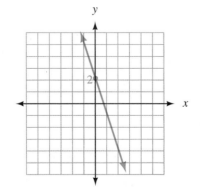

29.

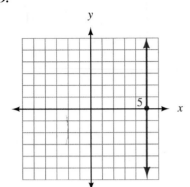

30.

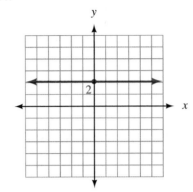

31.

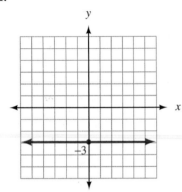

32.

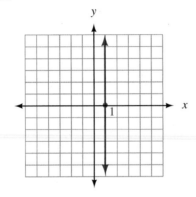

33. x-intercept: 2
 y-intercept: -6

34. x-intercept: 5   **35.** x-intercept: 3   **36.** x-intercept: 7   **37.** x-intercept: 2   **38.** x-intercept: 5
 y-intercept: 2 y-intercept: -3 y-intercept: 7 y-intercept: -6 y-intercept: 10
39. 2   **40.** -5 **41.** -1 **42.** $\frac{7}{9}$ **43.** 2   **44.** -2 **45.** $y = 3x + 2$ **46.** $y = -5x - 1$
47. $y = -x + 6$ **48.** $y = -4x - 3$ **49.** $y = -\frac{1}{3}x + \frac{3}{4}$ **50.** $y = -\frac{2}{5}x - \frac{2}{3}$ **51.** $m = 4, b = -1$
52. $m = -3, b = 2$ **53.** $m = -2, b = -5$ **54.** $m = 3, b = -4$ **55.** $m = -2, b = 3$ **56.** $m = 2, b = 3$
57. $m = -\frac{5}{2}, b = 4$ **58.** $m = \frac{3}{4}, b = 4$ **59.** $y = 4x - 11$ **60.** $y = -2x + 2$ **61.** $y = \frac{1}{2}x + 1$
62. $y = \frac{2}{3}x + 8$ **63.** $y = -\frac{3}{4}x + \frac{1}{4}$ **64.** $y = -\frac{2}{5}x - \frac{9}{5}$ **65.** $y = -2x + 8$ **66.** $y = \frac{3}{2}x + 2$
67. $y = x$ **68.** $y = 3x$ **69.** $y = -\frac{4}{3}x - 3$ **70.** $y = -\frac{1}{3}x + 4$ **71.** $(4, -2)$ **72.** $(-3, 2)$
73. $(3, -2)$ **74.** $(2, 0)$ **75.** $(2, 1)$ **76.** $(-1, -2)$ **77.** $(1, -3)$ **78.** $(-2, 5)$ **79.** Lines coincide
80. $(2, -2)$ **81.** $(1, 1)$ **82.** Lines are parallel   **83.** $(-2, -3)$ **84.** $(5, -1)$ **85.** $(-2, 7)$
86. $(-4, -2)$ **87.** $(-1, 4)$ **88.** $(2, -6)$ **89.** Lines are parallel   **90.** $(-3, -10)$ **91.** $(1, -2)$
92. Lines coincide   **93.** 10, 8   **94.** 24, 8   **95.** \$4,000 at 4%, \$8,000 at 5%
96. \$3,000 at 6%, \$11,000 at 8%   **97.** 10 dimes, 7 nickels   **98.** 9 dimes, 6 quarters
99. 40 liters of 10% solution, 10 liters of 20% solution   **100.** 20 liters of 25% solution, 20 liters of 15% solution

101.

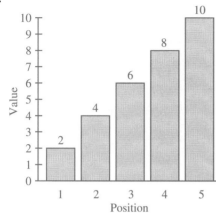

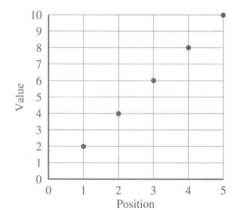

102.

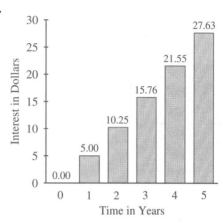

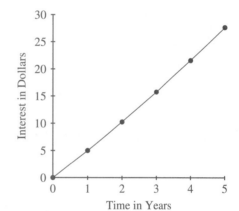

Chapter 3 Test

1. $(0, -2)$, $(5, 0)$, $(10, 2)$, $(-\frac{5}{2}, -3)$ **2.** $(2, 5)$, $(0, -3)$

3.

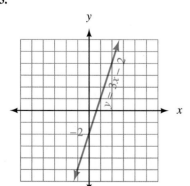

4.

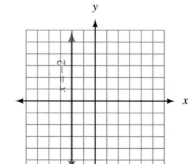

5.

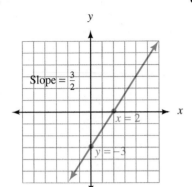

6.

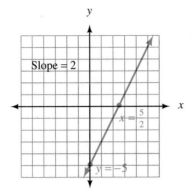

7. $y = 4x + 8$ **8.** $y = 3x + 10$ **9.**

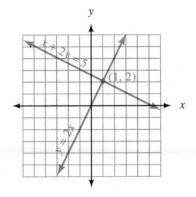

10. $(-3, -4)$ **11.** $(5, -3)$ **12.** $(2, -2)$ **13.** Lines coincide   **14.** $(2, 7)$ **15.** $(3, \frac{5}{2})$
16. $(8, 2)$ **17.** 7, 5   **18.** 3, 12   **19.** $6,000   **20.** 7 nickels, 5 quarters
21.

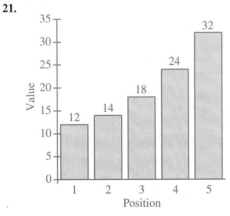

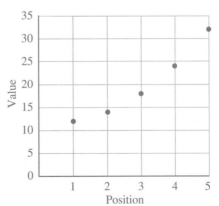

CHAPTER 4

Problem Set 4.1

1. Base 4; exponent 2; 16   **3.** Base 0.3; exponent 2; 0.09   **5.** Base 4; exponent 3; 64
7. Base -5; exponent 2; 25   **9.** Base 2; exponent 3; -8 **11.** Base 3; exponent 4; 81
13. Base $\frac{2}{3}$; exponent 2; $\frac{4}{9}$ **15.** Base $\frac{1}{2}$; exponent 4; $\frac{1}{16}$ **17.** x^9 **19.** 7^7 **21.** 2^{12} **23.** x^{28} **25.** x^{10}
27. 5^{12} **29.** 2^{50} **31.** a^{3x} **33.** b^{xy} **35.** $16x^2$ **37.** $81x^4$ **39.** $0.25a^2b^2$ **41.** $64x^3y^3z^3$
43. $8x^{12}$ **45.** x^{14} **47.** $128x^7$ **49.** $432x^{10}$ **51.** $16x^4y^6$ **53.** $\frac{8}{27}a^{12}b^{15}$ **55.** $30x^9$ **57.** x^8y^5
59. 4.32×10^4 **61.** 5.7×10^2 **63.** 2.38×10^5 **65.** 2,490   **67.** 352   **69.** 28,000   **71.** 27 inches³
73. 15.6 inches³   **75.** 36 inches³   **79.** 6.5×10^8 **81.** \$740,000   **83.** \$180,000
85. 100,000; the exponent and the number of zeros are the same   **87.** $(2^3)^2 = 2^6 = 64; 2^{32} = 2^9 = 512$
89. (b)   **91.** -3 **93.** 11   **95.** -5 **97.** 5

Problem Set 4.2

1. $\frac{1}{9}$ **3.** $\frac{1}{36}$ **5.** $\frac{1}{125}$ **7.** $\frac{2}{x^3}$ **9.** $\frac{1}{8x^3}$ **11.** $\frac{1}{100}$ **13.** 25   **15.** $\frac{1}{25}$ **17.** x^6 **19.** 6^{10}

21. $\frac{1}{6^{10}}$ **23.** $\frac{1}{2^8}$ **25.** 2^8 **27.** $81x^4y^4$ **29.** 1   **31.** $2a^2b$ **33.** $\frac{1}{49y^6}$ **35.** $\frac{1}{x^8}$ **37.** $\frac{1}{y^3}$

39. x^2 **41.** a^6 **43.** y^{40} **45.** $\frac{1}{x}$ **47.** x^9 **49.** a^{16} **51.** $\frac{1}{a^4}$ **53.** x **55.** 3.57×10^{-4}

57. 3.57×10^4 **59.** 4.8×10^{-3} **61.** 9×10^{-6} **63.** 0.00423   **65.** 56,000   **67.** 0.00008
69. 78.9   **71.** 0.002   **73.** 6×10^{-3} **75.** 2.5×10^4 **77.** 2.35×10^5 **79.** 8.2×10^{-4}
81. 100 inches², 400 inches²; 4   **83.** $x^2; 4x^2; 4$ **85.** 216 inches³; 1,728 inches³; 8   **87.** $x^3; 8x^3; 8$
89. $7x$ **91.** $2a$ **93.** $10y$
95.   **97.**

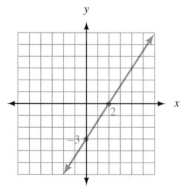

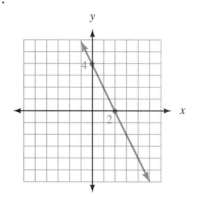

Problem Set 4.3

1. $12x^7$ **3.** $-16y^{11}$ **5.** $32x^2$ **7.** $200a^6$ **9.** $-24a^3b^3$ **11.** $24x^6y^8$ **13.** $3x$ **15.** $\frac{6}{y^3}$

17. $\frac{1}{2a}$ **19.** $-\frac{3a}{b^2}$ **21.** $\frac{x^2}{9z^2}$ **23.** $-\frac{12}{xy^6}$ **25.** 6×10^8 **27.** 1.75×10^{-1} **29.** 1.21×10^{-6}

31. 4.2×10^3 **33.** 3×10^{10} **35.** 5×10^{-3} **37.** $8x^2$ **39.** $-11x^5$ **41.** 0   **43.** $4x^3$ **45.** $31ab^2$

47. $-14abc$ **49.** $4x^3$ **51.** $\dfrac{1}{b^2}$ **53.** $\dfrac{6y^{10}}{x^4}$ **55.** 2×10^6 **57.** 1×10^1 **59.** 4.2×10^{-6}

61. $9x^3$ **63.** $-20a^2$ **65.** $6x^5y^2$ **67.** 2 **69.** 4 **71.** $(4 + 5)^2 = 9^2 = 81; 4^2 + 5^2 = 16 + 25 = 41$

73. Both equal 49   **75.** $P = 6x$ **77.** $V = 8x^2$ inches^3 **79.** $V = 2x^3$
$A = 2x^2$

81. $(-4, 16), (-2, 4), (-1, 1), (0, 0), (1, 1), (2, 4), (4, 16)$

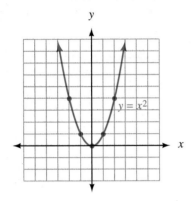

83. $(-3, 18), (-2, 8), (-1, 2), (0, 0), (1, 2), (2, 8), (3, 18)$

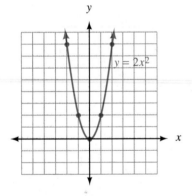

85. $(-4, 8), (-2, 2), (-1, \frac{1}{2}), (0, 0), (1, \frac{1}{2}), (2, 2), (4, 8)$

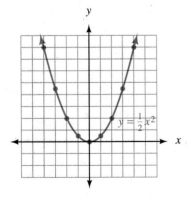

87. $(-4, 4)$, $(-2, 1)$, $(-1, \frac{1}{4})$, $(0, 0)$, $(1, \frac{1}{4})$, $(2, 1)$, $(4, 4)$

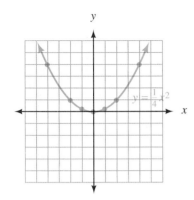

89. -8 **91.** 9 **93.** 0

95.

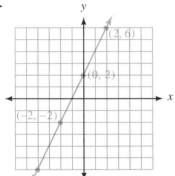

97.

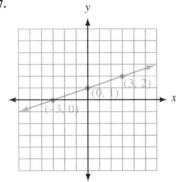

Problem Set 4.4

1. Trinomial, 3   **3.** Trinomial, 3   **5.** Binomial, 1   **7.** Binomial, 2   **9.** Monomial, 2
11. Monomial, 0   **13.** $5x^2 + 5x + 9$ **15.** $5a^2 - 9a + 7$ **17.** $x^2 + 6x + 8$ **19.** $6x^2 - 13x + 5$
21. $x^2 - 9$ **23.** $3y^2 - 11y + 10$ **25.** $6x^3 + 5x^2 - 4x + 3$ **27.** $2x^2 - x + 1$ **29.** $2a^2 - 2a - 2$
31. $-\frac{1}{9}x^3 - \frac{2}{3}x^2 - \frac{5}{2}x + \frac{7}{4}$ **33.** $-4y^2 + 15y - 22$ **35.** $x^2 - 33x + 63$ **37.** $8y^2 + 4y + 26$
39. $75x^2 - 150x - 75$ **41.** $12x + 2$ **43.** 4 **45.** 25 **47.** 16 **49.** $10x^2$ **51.** $-15x^2$
53. $6x^3$ **55.** $6x^4$

Problem Set 4.5

1. $6x^2 + 2x$ **3.** $6x^4 - 4x^3 + 2x^2$ **5.** $2a^3b - 2a^2b^2 + 2ab$ **7.** $3y^4 + 9y^3 + 12y^2$
9. $8x^5y^2 + 12x^4y^3 + 32x^2y^4$ **11.** $x^2 + 7x + 12$ **13.** $x^2 + 7x + 6$ **15.** $x^2 + 2x + \frac{3}{4}$
17. $a^2 + 2a - 15$ **19.** $xy + bx - ay - ab$ **21.** $x^2 - 36$ **23.** $y^2 - \frac{25}{36}$ **25.** $2x^2 - 11x + 12$
27. $2a^2 + 3a - 2$ **29.** $6x^2 - 19x + 10$ **31.** $2ax + 8x + 3a + 12$ **33.** $25x^2 - 16$ **35.** $2x^2 + \frac{5}{2}x - \frac{3}{4}$
37. $3 - 10a + 8a^2$ **39.** $56 - 83x + 30x^2$ **41.** $x^3 + 4x^2 - x - 4$ **43.** $a^3 - 6a^2 + 11a - 6$
45. $x^3 + 8$ **47.** $2x^3 + 17x^2 + 26x + 9$ **49.** $5x^4 - 13x^3 + 20x^2 + 7x + 5$ **51.** $2x^4 + x^2 - 15$
53. $6a^6 + 15a^4 + 4a^2 + 10$ **55.** $x^3 + 12x^2 + 47x + 60$ **57.** $A = x(2x + 5) = 2x^2 + 5x$
59. $A = x(x + 1) = x^2 + x$

61. (a)

Corner Square Side	Dimensions and Volume of the Box			
	Length	Width	Height	Volume
0	17	11	0	0 in.^{3}
1	15	9	1	135 in.^{3}
1.5	14	8	1.5	168 in.^{3}
2	13	7	2	182 in.^{3}
2.5	12	6	2.5	180 in.^{3}
3	11	5	3	165 in.^{3}
3.5	10	4	3.5	140 in.^{3}
4	9	3	4	108 in.^{3}
4.5	8	2	4.5	72 in.^{3}
5	7	1	5	35 in.^{3}
5.5	6	0	5.5	0 in.^{3}

(b)

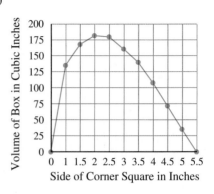

63. $R = (1{,}200 - 100p)p$ **65.** $R = (1{,}700 - 100p)p$ **67.** $y = \frac{1}{2}x + 2$ **69.** $y = 3x$ **71.** $y = x - 2$
$m = \frac{1}{2}, b = 2$

Problem Set 4.6

1. $x^2 - 4x + 4$ **3.** $a^2 + 6a + 9$ **5.** $x^2 - 10x + 25$ **7.** $a^2 - a + \frac{1}{4}$ **9.** $x^2 + 20x + 100$
11. $a^2 + 1.6a + 0.64$ **13.** $4x^2 - 4x + 1$ **15.** $16a^2 + 40a + 25$ **17.** $9x^2 - 12x + 4$
19. $9a^2 + 30ab + 25b^2$ **21.** $16x^2 - 40xy + 25y^2$ **23.** $49m^2 + 28mn + 4n^2$
25. $36x^2 - 120xy + 100y^2$ **27.** $x^4 + 10x^2 + 25$ **29.** $a^4 + 2a^2 + 1$ **31.** $x^6 - 14x^3 + 49$
33. $x^2 - 9$ **35.** $a^2 - 25$ **37.** $y^2 - 1$ **39.** $81 - x^2$ **41.** $4x^2 - 25$ **43.** $16x^2 - \frac{1}{9}$ **45.** $4a^2 - 49$
47. $36 - 49x^2$ **49.** $x^4 - 9$ **51.** $a^4 - 16$ **53.** $25y^8 - 64$ **55.** $2x^2 - 34$ **57.** $-12x^2 + 20x + 8$
59. $a^2 + 4a + 6$ **61.** $8x^3 + 36x^2 + 54x + 27$ **63.** $(50 - 1)(50 + 1) = 2{,}500 - 1 = 2{,}499$
65. Both equal 25   **67.** $x^2 + (x + 1)^2 = 2x^2 + 2x + 1$ **69.** $x^2 + (x + 1)^2 + (x + 2)^2 = 3x^2 + 6x + 5$

71.

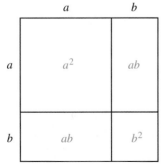

$a^2 + ab + ab + b^2 = a^2 + 2ab + b^2$

73. $2x^2$ **75.** $5x$ **77.** $\dfrac{a^4}{2b^2}$

79. $(3, -1)$

81. $(-1, 1)$

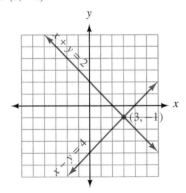

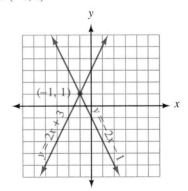

Problem Set 4.7

1. $x - 2$ **3.** $3 - 2x^2$ **5.** $5xy - 2y$ **7.** $7x^4 - 6x^3 + 5x^2$ **9.** $10x^4 - 5x^2 + 1$ **11.** $-4a + 2$

13. $-8a^4 - 12a^3$ **15.** $-4b - 5a$ **17.** $-6a^2b + 3ab^2 - 7b^3$ **19.** $-\dfrac{a}{2} - b - \dfrac{b^2}{2a}$ **21.** $3x + 4y$

23. $-y + 3$ **25.** $5y - 4$ **27.** $xy - x^2y^2$ **29.** $-1 + xy$ **31.** $-a + 1$ **33.** $x^2 - 3xy + y^2$

35. $2 - 3b + 5b^2$ **37.** $-2xy + 1$ **39.** $xy - \frac{1}{2}$ **41.** $\dfrac{1}{4x} - \dfrac{1}{2a} + \dfrac{3}{4}$ **43.** $\dfrac{4x^2}{3} + \dfrac{2}{3x} + \dfrac{1}{x^2}$

45. $3a^{3m} - 9a^m$ **47.** $2x^{4m} - 5x^{2m} + 7$ **49.** $3x^2 - x + 6$ **51.** 4 **53.** $x + 5$ **55.** Both equal 7

57. $\dfrac{3(10) + 8}{2} = \dfrac{38}{2} = 19;\ 3(10) + 4 = 34$ **59.** $(7, -1)$ **61.** $(2, 3)$ **63.** $(1, 1)$ **65.** Lines coincide

Problem Set 4.8

1. $x - 2$ **3.** $a + 4$ **5.** $x - 3$ **7.** $x + 3$ **9.** $a - 5$ **11.** $x + 2 + \dfrac{2}{x + 3}$ **13.** $a - 2 + \dfrac{12}{a + 5}$

15. $x + 4 + \dfrac{9}{x - 2}$ **17.** $x + 4 + \dfrac{-10}{x + 1}$ **19.** $a + 1 + \dfrac{-1}{a + 2}$ **21.** $x - 3 + \dfrac{17}{2x + 4}$

23. $3a - 2 + \dfrac{7}{2a + 3}$ **25.** $2a^2 - a - 3$ **27.** $x^2 - x + 5$ **29.** $x^2 + x + 1$ **31.** $x^2 + 2x + 4$

33. 5, 20   **35.** \$800 at 8%, \$400 at 9%   **37.** Eight \$5 bills, twelve \$10 bills
39. 10 gallons of 20%, 6 gallons of 60%

Chapter 4 Review

1. -1 **2.** -64 **3.** -64 **4.** -16 **5.** $\frac{9}{49}$ **6.** $\frac{1}{27}$ **7.** y^{12} **8.** y^{16} **9.** x^{30} **10.** x^{50}
11. x^{35} **12.** x^{40} **13.** 2^{24} **14.** 3^{40} **15.** $27y^3$ **16.** $16y^4$ **17.** $-8x^3y^3z^3$ **18.** $\frac{4}{9}a^4b^6c^8$

19. $\frac{1}{49}$ **20.** $\frac{1}{81}$ **21.** $\dfrac{4}{x^5}$ **22.** $\dfrac{7}{x^6}$ **23.** $\dfrac{1}{27y^3}$ **24.** $\dfrac{1}{36y^2}$ **25.** a^6 **26.** a^8 **27.** $\dfrac{1}{x^4}$

28. $\dfrac{1}{x^4}$ **29.** x^{15} **30.** x^{12} **31.** $\dfrac{1}{x^5}$ **32.** x^4 **33.** 1 **34.** 1 **35.** -3 **36.** 6 **37.** $9x^6y^4$

38. $-8x^{12}y^9$ **39.** $64a^{22}b^{20}$ **40.** $108a^{18}b^{29}$ **41.** $\dfrac{-1}{27x^3y^6}$ **42.** $\dfrac{1}{16x^{12}y^{20}}$ **43.** b **44.** $\dfrac{1}{b^6}$

45. $\dfrac{1}{x^{35}}$ **46.** 1   **47.** $5x^7$ **48.** $\dfrac{4}{x^2}$ **49.** $\dfrac{6y^7}{x^5}$ **50.** $\dfrac{8y^3}{x^5}$ **51.** $4a^6$ **52.** $7a^9$ **53.** $2x^5y^2$

54. $4x^4y^6$ **55.** 6.4×10^7 **56.** 1.05×10^{-1} **57.** 2.3×10^8 **58.** 5×10^{-5} **59.** 8×10^3

60. 4×10^6 **61.** $8a^2 - 12a - 3$ **62.** $-2a^3 - 10a^2 - 5a + 13$ **63.** $-12x^2 + 5x - 14$

64. $-13x + 10$ **65.** $-4x^2 - 6x$ **66.** $7x - 5$ **67.** 14   **68.** 121   **69.** $12x^2 - 21x$

70. $40x^2 + 15x$ **71.** $24x^5y^2 - 40x^4y^3 + 32x^3y^4$ **72.** $15x^4y^2 + 20x^3y^3 + 50x^2y^4$ **73.** $a^3 + 6a^2 + a - 4$

74. $a^3 - 8a^2 + 19a - 14$ **75.** $x^3 + 125$ **76.** $x^3 - 1$ **77.** $6x^2 - 29x + 35$ **78.** $15x^2 + 14x - 8$

79. $25y^2 - \frac{1}{25}$ **80.** $\frac{1}{4}y^2 - 4$ **81.** $a^4 - 9$ **82.** $a^4 - 49$ **83.** $a^2 - 10a + 25$ **84.** $a^2 + 12a + 36$

85. $9x^2 + 24x + 16$ **86.** $25x^2 - 10x + 1$ **87.** $y^4 + 6y^2 + 9$ **88.** $y^4 - 4y^2 + 4$ **89.** $-2b - 4a$

90. $3a^2 - 2ab - 4b^2$ **91.** $-5x^4y^3 + 4x^2y^2 + 2x$ **92.** $-4x^3y^2 + 2xy - 3$ **93.** $-8y + 5$

94. $-4x^2 - 2x$ **95.** $5 - 4b + 6b^2$ **96.** $3 - \dfrac{2}{x^3} + \dfrac{4}{x^4}$ **97.** $x + 9$ **98.** $x - 7$ **99.** $2x + 5$

100. $2x + 1$ **101.** $x^2 - 4x + 16$ **102.** $x^2 + 3x + 9$ **103.** $x - 3 + \dfrac{16}{3x + 2}$ **104.** $2x + 7 + \dfrac{11}{2x - 3}$

105. $x^2 - 4x + 5 + \dfrac{5}{2x + 1}$ **106.** $x^2 - 2x + 4 + \dfrac{-6}{3x - 1}$ **107.** $V = 3x^3$ **108.** Yes   **109.** Yes

111. 3   **112.** 24

Chapter 4 Test

1. 81   **2.** $\frac{9}{16}$ **3.** $72x^{18}$ **4.** $\frac{1}{9}$ **5.** 1   **6.** a^2 **7.** x **8.** 2.78×10^{-2} **9.** 243,000

10. $\dfrac{y^2}{2x^4}$ **11.** $3a$ **12.** $10x^5$ **13.** 9×10^9 **14.** $8x^2 + 2x + 2$ **15.** $3x^2 + 4x + 6$ **16.** $3x - 4$

17. 10   **18.** $6a^4 - 10a^3 + 8a^2$ **19.** $x^2 + \frac{5}{6}x + \frac{1}{6}$ **20.** $8x^2 + 2x - 15$ **21.** $x^3 - 27$

22. $x^2 + 10x + 25$ **23.** $9a^2 - 12ab + 4b^2$ **24.** $9x^2 - 16y^2$ **25.** $a^4 - 9$ **26.** $2x^2 + 3x - 1$

27. $4x + 3 + \dfrac{4}{2x - 3}$ **28.** $3x^2 + 9x + 25 + \dfrac{76}{x - 3}$ **29.** 15.625 centimeters³   **30.** x^3

CHAPTER 5

Problem Set 5.1

1. $5(3x + 5)$ **3.** $3(2a + 3)$ **5.** $4(x - 2y)$ **7.** $3(x^2 - 2x - 3)$ **9.** $3(a^2 - a - 20)$

11. $4(6y^2 - 13y + 6)$ **13.** $x^2(9 - 8x)$ **15.** $13a^2(1 - 2a)$ **17.** $7xy(3x - 4y)$ **19.** $11ab^2(2a - 1)$

21. $7x(x^2 + 3x - 4)$ **23.** $11(11y^4 - x^4)$ **25.** $25x^2(4x^2 - 2x + 1)$ **27.** $8(a^2 + 2b^2 + 4c^2)$

29. $4ab(a - 4b + 8ab)$ **31.** $11a^2b^2(11a - 2b + 3ab)$ **33.** $12x^2y^3(1 - 6x^3 - 3x^2y)$ **35.** $(x + 3)(y + 5)$

37. $(x + 2)(y + 6)$ **39.** $(a - 3)(b + 7)$ **41.** $(a - b)(x + y)$ **43.** $(2x - 5)(a + 3)$

45. $(b - 2)(3x - 4)$ **47.** $(x + 2)(x + a)$ **49.** $(x - b)(x - a)$ **51.** $(x + y)(a + b + c)$

53. $(3x + 2)(2x + 3)$ **55.** $(10x - 1)(2x + 5)$ **57.** $(4x + 5)(5x + 1)$ **59.** $(x + 2)(x^2 + 3)$

61. $(3x - 2)(2x^2 + 5)$ **63.** 6   **65.** $3(4x^2 + 2x + 1)$

67. $A = 1{,}000(1 + r)$; when $r = 0.12$, $A = 1{,}000(1.12) = \$1{,}120$

69. $x^2 + 7x + 12$ **71.** $x^2 + 5x - 14$ **73.** $x^2 - 5x - 14$ **75.** $x^2 - x - 6$ **77.** $x^3 + 27$

79. $2x^3 + 9x^2 - 2x - 3$

Problem Set 5.2

1. $(x + 3)(x + 4)$ **3.** $(x + 1)(x + 2)$ **5.** $(a + 3)(a + 7)$ **7.** $(x - 2)(x - 5)$ **9.** $(y - 3)(y - 7)$
11. $(x - 4)(x + 3)$ **13.** $(y + 4)(y - 3)$ **15.** $(x + 7)(x - 2)$ **17.** $(r - 9)(r + 1)$
19. $(x - 6)(x + 5)$ **21.** $(a + 7)(a + 8)$ **23.** $(y + 6)(y - 7)$ **25.** $(x + 6)(x + 7)$ **27.** $2(x + 1)(x + 2)$
29. $3(a + 4)(a - 5)$ **31.** $100(x - 2)(x - 3)$ **33.** $100(p - 5)(p - 8)$ **35.** $x^2(x + 3)(x - 4)$
37. $2r(r + 5)(r - 3)$ **39.** $2y^2(y + 1)(y - 4)$ **41.** $x^3(x + 2)(x + 2)$ **43.** $3y^2(y + 1)(y - 5)$
45. $4x^2(x - 4)(x - 9)$ **47.** $(x + 2y)(x + 3y)$ **49.** $(x - 4y)(x - 5y)$ **51.** $(a + 4b)(a - 2b)$
53. $(a - 5b)(a - 5b)$ **55.** $(a + 5b)(a + 5b)$ **57.** $(x - 6a)(x + 8a)$ **59.** $(x + 4b)(x - 9b)$
61. $(x^2 - 3)(x^2 - 2)$ **63.** $(x - 100)(x + 20)$ **65.** $(x - \frac{1}{2})(x - \frac{1}{2})$ **67.** $(x + 0.2)(x + 0.4)$ **69.** $(x + 16)$
71. $4x^2 - x - 3$ **73.** $h = 16(4 - t)(1 + t)$; when $t = 4$, $h = 16(4 - 4)(1 + 4) = 0$ **75.** $6a^2 + 13a + 2$
77. $6a^2 + 7a + 2$ **79.** $6a^2 + 8a + 2$ **81.** $2x^2 + 7x - 11$ **83.** $(7x + 3) - (4x - 5) = 3x + 8$
85. $(5x^2 - 5) - (2x^2 - 4x) = 3x^2 + 4x - 5$

Problem Set 5.3

1. $(2x + 1)(x + 3)$ **3.** $(2a - 3)(a + 1)$ **5.** $(3x + 5)(x - 1)$ **7.** $(3y + 1)(y - 5)$ **9.** $(2x + 3)(3x + 2)$
11. $(2x - 3y)(2x - 3y)$ **13.** $(4y + 1)(y - 3)$ **15.** $(4x - 5)(5x - 4)$ **17.** $(10a - b)(2a + 5b)$
19. $(4x - 5)(5x + 1)$ **21.** $(6m - 1)(2m + 3)$ **23.** $(4x + 5)(5x + 3)$ **25.** $(3a - 4b)(4a - 3b)$
27. $(3x - 7y)(x + 2y)$ **29.** $(2x + 5)(7x - 3)$ **31.** $(3x - 5)(2x - 11)$ **33.** $(5t - 19)(3t - 2)$
35. $2(2x + 3)(x - 1)$ **37.** $2(4a - 3)(3a - 4)$ **39.** $x(5x - 4)(2x - 3)$ **41.** $x^2(3x + 2)(2x - 5)$
43. $2a(5a + 2)(a - 1)$ **45.** $3x(5x + 1)(x - 7)$ **47.** $5y(7y + 2)(y - 2)$ **49.** $a^2(3a - 1)(5a + 1)$
51. $3y(4x + 5)(2x - 3)$ **53.** $2y(2x - y)(3x - 7y)$ **55.** Both equal 25   **57.** $4x^2 - 9$ **59.** $x^4 - 81$
61. $x^2 - 9$ **63.** $36a^2 - 1$ **65.** $x^2 + 8x + 16$ **67.** $4x^2 + 12x + 9$

Problem Set 5.4

1. $(x + 3)(x - 3)$ **3.** $(a + 6)(a - 6)$ **5.** $(x + 7)(x - 7)$ **7.** $4(a + 2)(a - 2)$ **9.** Cannot be factored
11. $(5x + 13)(5x - 13)$ **13.** $(3a + 4b)(3a - 4b)$ **15.** $(3 + m)(3 - m)$ **17.** $(5 + 2x)(5 - 2x)$
19. $2(x + 3)(x - 3)$ **21.** $32(a + 2)(a - 2)$ **23.** $2y(2x + 3)(2x - 3)$ **25.** $(a^2 + b^2)(a + b)(a - b)$
27. $(4m^2 + 9)(2m + 3)(2m - 3)$ **29.** $3xy(x + 5y)(x - 5y)$ **31.** $(x - 1)^2$ **33.** $(x + 1)^2$ **35.** $(a - 5)^2$
37. $(y + 2)^2$ **39.** $(x - 2)^2$ **41.** $(m - 6)^2$ **43.** $(2a + 3)^2$ **45.** $(7x - 1)^2$ **47.** $(3y - 5)^2$
49. $(x + 5y)^2$ **51.** $(3a + b)^2$ **53.** $3(a + 3)^2$ **55.** $2(x + 5y)^2$ **57.** $5x(x + 3y)^2$

59. $(x + 3 + y)(x + 3 - y)$ **61.** $(x + y + 3)(x + y - 3)$ **63.** 14   **65.** 25   **67.** $x - 2 + \dfrac{2}{x - 3}$

69. $3x - 2 + \dfrac{9}{2x + 3}$

Problem Set 5.5

1. $(x + 9)(x - 9)$ **3.** $(x + 5)(x - 3)$ **5.** $(x + 3)^2$ **7.** $(y - 5)^2$ **9.** $2ab(a^2 + 3a + 1)$
11. Cannot be factored   **13.** $3(2a + 5)(2a - 5)$ **15.** $(3x - 2y)^2$ **17.** $4x(x^2 + 4y^2)$ **19.** $2y(y + 5)^2$
21. $a^4(a^2 + 4b^2)$ **23.** $(x + 4)(y + 3)$ **25.** $(x^2 + 4)(x + 2)(x - 2)$ **27.** $(x + 2)(y - 5)$ **29.** $5(a + b)^2$
31. Cannot be factored   **33.** $3(x + 2y)(x + 3y)$ **35.** $(2x + 19)(x - 2)$ **37.** $100(x - 2)(x - 1)$
39. $(x + 8)(x - 8)$ **41.** $(x + a)(x + 3)$ **43.** $a^5(7a + 3)(7a - 3)$ **45.** Cannot be factored
47. $a(5a + 1)(5a + 3)$ **49.** $(x + y)(a - b)$ **51.** $3a^2b(4a + 1)(4a - 1)$ **53.** $5x^2(2x + 3)(2x - 3)$
55. $(3x + 41y)(x - 2y)$ **57.** $2x^3(2x - 3)(4x - 5)$ **59.** $(2x + 3)(x + a)$ **61.** $(y^2 + 1)(y + 1)(y - 1)$
63. $3x^2y^2(2x + 3y)^2$ **65.** 5   **67.** $-\frac{3}{2}$ **69.** $-\frac{3}{4}$ **71.** x^{15} **73.** $72x^{18}$ **75.** 5.76×10^4

Problem Set 5.6

1. $-2, 1$ **3.** $4, 5$ **5.** $0, -1, 3$ **7.** $-\frac{2}{3}, -\frac{3}{2}$ **9.** $0, -\frac{4}{3}, \frac{4}{3}$ **11.** $0, -\frac{1}{3}, -\frac{3}{5}$ **13.** $-1, -2$
15. $4, 5$ **17.** $6, -4$ **19.** $2, 3$ **21.** -3 **23.** $4, -4$ **25.** $\frac{3}{2}, -4$ **27.** $-\frac{2}{3}$ **29.** 5 **31.** $4, -\frac{5}{2}$
33. $\frac{5}{3}, -4$ **35.** $\frac{7}{2}, -\frac{7}{2}$ **37.** $0, -6$ **39.** $0, 3$ **41.** $0, 4$ **43.** $0, 5$ **45.** $2, 5$ **47.** $\frac{1}{2}, -\frac{4}{3}$
49. $4, -\frac{5}{2}$ **51.** $8, -10$ **53.** $5, 8$ **55.** $6, 8$ **57.** -4 **59.** $5, 8$ **61.** $6, -8$ **63.** $0, -\frac{3}{2}, -4$
65. $0, 3, -\frac{5}{2}$ **67.** $0, \frac{1}{2}, -\frac{5}{2}$ **69.** $0, \frac{3}{5}, -\frac{3}{2}$ **71.** $-3, -2, 2$ **73.** $-4, -1, 4$
75. \$75 for the bicycle, \$15 for the suit   **77.** \$2,400 for the house, \$600 for the lot   **79.** $\frac{1}{8}$ **81.** x^8
83. x^{18} **85.** 5.6×10^{-3}

Problem Set 5.7

1. Two consecutive even integers are x and $x + 2$; $x(x + 2) = 80$; 8, 10 and $-10, -8$ **3.** 9, 11 and $-11, -9$
5. $x(x + 2) = 5(x + x + 2) - 10$; 8, 10 and 0, 2   **7.** 8, 6
9. The numbers are x and $5x + 2$; $x(5x + 2) = 24$; 2, 12 and $-\frac{12}{5}, -10$ **11.** 5, 20 and 0, 0
13. Let $x = $ the width; $x(x + 1) = 12$; width 3 inches, length 4 inches
15. Let $x = $ the base; $\frac{1}{2}(x)(2x) = 9$; base 3 inches, height 6 inches
17. $x^2 + (x + 2)^2 = 10^2$; 6 inches and 8 inches
19. 12 meters   **21.** $1,400 = 400 + 700x - 100x^2$; 2 items or 5 items
23. 2 videotapes or 8 videotapes   **25.** $R = xp = (1,200 - 100p)p = 3,200$; \$4 or \$8   **27.** \$7 or \$10
29. $42a^5b^3$ **31.** $2x^2y^2$ **33.** $25x^5$ **35.** 6×10^{-3} **37.** $2a^3$

Chapter 5 Review

1. $10(x - 2)$ **2.** $x^2(4x - 9)$ **3.** $5(x - y)$ **4.** $x(7x^2 + 2)$ **5.** $4(2x + 1)$ **6.** $2(x^2 + 7x + 3)$
7. $8(3y^2 - 5y + 6)$ **8.** $6a^3(3a - 5)$ **9.** $15xy^2(2y - 3x^2)$ **10.** $4y(2y^3 - 5y - 8)$ **11.** $7(7a^3 - 2b^3)$
12. $12x^2(4x^3 - 3x - 5)$ **13.** $6ab(b + 3a^2b^2 - 4a)$ **14.** $3a^4b^3(11b^2 - 16a + 6b)$ **15.** $(x + a)(y + b)$
16. $(x + a)(y + 5)$ **17.** $(x - 5)(y + 4)$ **18.** $(x + 5)(y - 4)$ **19.** $(2x - 3)(y + 5)$
20. $(3x - 2)(y + 5)$ **21.** $(5x - 4a)(x - 2b)$ **22.** $(3x - 2b)(5x - b)$ **23.** $(y + 7)(y + 2)$
24. $(w + 5)(w + 10)$ **25.** $(a - 6)(a - 8)$ **26.** $(r + 5)(r - 4)$ **27.** $(r - 6)(r - 12)$ **28.** $(a + 7)(a - 5)$
29. $(y + 9)(y + 11)$ **30.** $(x - 8)(x + 3)$ **31.** $(y + 6)(y + 2)$ **32.** $(n - 9)(n + 4)$
33. $(2x + 3)(x + 5)$ **34.** $(2y - 5)(2y - 1)$ **35.** $(5y + 6)(y + 1)$ **36.** $(4n - 5m)(2n + m)$
37. $(5a - 3)(4a - 3)$ **38.** $(7x - 2y)(2x + 5y)$ **39.** $(2r + 3t)(3r - 2t)$ **40.** $(10y - 3)(2y - 3)$
41. $(2x - 7)(5x + 3)$ **42.** $(4x + 3y)(3x - 5y)$ **43.** $(n + 9)(n - 9)$ **44.** $(2y + 3)(2y - 3)$
45. Cannot be factored   **46.** $(2r + 3t)(2r - 3t)$ **47.** $(6y + 11x)(6y - 11x)$ **48.** $(5 + y)(5 - y)$
49. $(8a + 11b)(8a - 11b)$ **50.** $(y^2 + 9)(y + 3)(y - 3)$ **51.** $(8 + 3m)(8 - 3m)$
52. $(4 + a^2)(2 + a)(2 - a)$ **53.** $(y + 10)^2$ **54.** $(m - 8)^2$ **55.** $(8t + 1)^2$ **56.** $(7x - 1)^2$
57. $(4n - 3)^2$ **58.** $(7w + 8)^2$ **59.** $(2r - 3t)^2$ **60.** $(4x + 9y)^2$ **61.** $(3m + 5n)^2$ **62.** $(5x - 3y)^2$
63. $2(x + 4)(x + 6)$ **64.** $a(a - 3)(a - 7)$ **65.** $3m(m + 1)(m - 7)$ **66.** $x^3(x - 6)(x + 5)$
67. $5y^2(y - 2)(y + 4)$ **68.** $6y^2(y + 6)(y - 1)$ **69.** $2(2x + 1)(2x + 3)$ **70.** $a(3a + 1)(a - 5)$
71. $2m(2m - 3)(5m - 1)$ **72.** $3x(3x + y)(2x - y)$ **73.** $5y(2x - 3y)(3x - y)$ **74.** $4y(4x - y)(2x - 3y)$
75. $4(x + 5)^2$ **76.** $x(2x + 3)^2$ **77.** $5(x + 3)(x - 3)$ **78.** $5y(3x - y)^2$ **79.** $3x(2x + 3y)(2x - 3y)$
80. $5x(4x + y)(4x - y)$ **81.** $3ab(2a + b)(a + 5b)$ **82.** $x^3(x + 1)(x - 1)$ **83.** $y^4(4y^2 + 9)$
84. $4x^3(x + 2y)(3x - y)$ **85.** $5a^2b(3a - b)(2a + 3b)$ **86.** $3ab^2(2a - b)(3a + 2b)$ **87.** $-2, 5$
88. $0, -\frac{2}{3}, 4$ **89.** $-\frac{5}{2}, \frac{5}{2}$ **90.** $0, 7, \frac{7}{2}$ **91.** $2, -5$ **92.** $-\frac{5}{3}, 4$ **93.** $7, -7$ **94.** $-\frac{9}{4}, \frac{9}{4}$ **95.** $0, 9$
96. $-\frac{3}{7}, \frac{3}{7}$ **97.** $-\frac{3}{2}, -\frac{2}{3}$ **98.** $-3, 0$ **99.** $0, -\frac{5}{3}, \frac{2}{3}$ **100.** $0, 2, \frac{7}{2}$ **101.** 10, 12 and $-12, -10$
102. 10, 11 and $-11, -10$ **103.** 5, 7 and $-1, 1$ **104.** 5, 15   **105.** 6, 11 and $-\frac{11}{2}, -12$
106. 3 inches, 16 inches   **107.** 2 inches   **108.** 8 meters

Chapter 5 Test

1. $5(x - 2)$ **2.** $9xy(2x - 1 - 4y)$ **3.** $(x + 2a)(x - 3b)$ **4.** $(x - 7)(y + 4)$ **5.** $(x - 2)(x - 3)$
6. $(x - 3)(x + 2)$ **7.** $(a - 4)(a + 4)$ **8.** Cannot be factored   **9.** $(x^2 + 9)(x - 3)(x + 3)$
10. $3(3x - 5y)(3x + 5y)$ **11.** $(x + 5)(x + 3)(x - 3)$ **12.** $(x - b)(x + 5)$ **13.** $2(2a + 1)(a + 5)$
14. $3(m - 3)(m + 2)$ **15.** $(2y - 1)(3y + 5)$ **16.** $2x(2x + 1)(3x - 5)$ **17.** $-3, -4$ **18.** 2
19. $-6, 6$ **20.** $5, -4$ **21.** $5, 6$ **22.** $0, 4, -4$ **23.** $3, -\frac{5}{2}$ **24.** $0, 1, -\frac{1}{3}$ **25.** $4, 16$
26. $3, 5$ and $-3, -1$ **27.** 3 feet, 14 feet   **28.** 5 meters and 12 meters   **29.** 200 or 300 items   **30.** \$3 or \$6

CHAPTER 6

Problem Set 6.1

1. $\dfrac{1}{x - 2}, x \neq 2$ **3.** $\dfrac{1}{a + 3}, a \neq -3, 3$ **5.** $\dfrac{1}{x - 5}, x \neq -5, 5$ **7.** $\dfrac{(x + 2)(x - 2)}{2}$ **9.** $\dfrac{2(x - 5)}{3(x - 2)}, x \neq 2$

11. 2 **13.** $\dfrac{5(x - 1)}{4}$ **15.** $\dfrac{1}{x - 3}$ **17.** $\dfrac{1}{x + 3}$ **19.** $\dfrac{1}{a - 5}$ **21.** $\dfrac{1}{3x + 2}$ **23.** $\dfrac{x + 5}{x + 2}$

25. $\dfrac{2m(m + 2)}{m - 2}$ **27.** $\dfrac{x - 1}{x - 4}$ **29.** $\dfrac{2(2x - 3)}{2x + 3}$ **31.** $\dfrac{2x - 3}{2x + 3}$ **33.** $\dfrac{1}{(x^2 + 9)(x - 3)}$ **35.** $\dfrac{3x - 5}{(x^2 + 4)(x - 2)}$

37. $\dfrac{2x(7x + 6)}{2x + 1}$ **39.** $\dfrac{x + 2}{x + 5}$ **41.** $\dfrac{x + a}{x + b}$ **43.** $\dfrac{x + a}{x + 3}$ **45.** $\frac{4}{3}$ **47.** $\frac{4}{5}$ **49.** $\frac{8}{1}$ **51.** $\frac{5}{x}$

53. 40.7 miles/hour   **55.** 39.25 feet/minute   **57.** 12.95 feet/second
59. Level ground: 10 minutes/mile and 0.1 mile/minute; downhill: 6.67 minutes/mile and 0.15 mile/minute
61. 48 miles/gallon   **63.** $5 + 4 = 9$ **69.** 0 **71.** $16a^2b^2$ **73.** $19x^4 + 21x^2 - 42$
75. $7a^4b^4 + 9b^3 - 11a^3$

Problem Set 6.2

1. 2 **3.** $\dfrac{x}{2}$ **5.** $\frac{3}{2}$ **7.** $\dfrac{1}{2(x - 3)}$ **9.** $\dfrac{4a(a + 5)}{7(a + 4)}$ **11.** $\dfrac{y - 2}{4}$ **13.** $\dfrac{2(x + 4)}{x - 2}$ **15.** $\dfrac{x + 3}{(x - 3)(x + 1)}$

17. 1 **19.** $\dfrac{y - 5}{(y + 2)(y - 2)}$ **21.** $\dfrac{x + 5}{x - 5}$ **23.** 1 **25.** $\dfrac{a + 3}{a + 4}$ **27.** $\dfrac{2y - 3}{y - 6}$ **29.** $\dfrac{x - 1}{x - 2}$ **31.** $\frac{3}{2}$

33. $2(x - 3)$ **35.** $2a$ **37.** $(x + 2)(x + 1)$ **39.** $-2x(x - 5)$ **41.** $\dfrac{2(x + 5)}{x(y + 5)}$ **43.** $\dfrac{2x}{x - y}$ **45.** $\dfrac{1}{x - 2}$

47. $\frac{1}{5}$ **49.** $\frac{1}{100}$ **51.** 2.7 miles   **53.** 742 miles/hour   **55.** 0.45 miles/hour   **57.** 8.8 miles/hour
59. Level ground: 6 miles/hour; downhill: 9 miles/hour   **61.** 3 **63.** $\frac{11}{4}$ **65.** $\frac{11}{35}$ **67.** $9x^2$ **69.** $6ab^2$

Problem Set 6.3

1. $\dfrac{7}{x}$ **3.** $\dfrac{4}{a}$ **5.** 1 **7.** $y + 1$ **9.** $x + 2$ **11.** $x - 2$ **13.** $\dfrac{6}{x + 6}$ **15.** $\dfrac{(y + 2)(y - 2)}{2y}$

17. $\dfrac{3 + 2a}{6}$ **19.** $\dfrac{7x + 3}{4(x + 1)}$ **21.** $\frac{1}{5}$ **23.** $\frac{1}{3}$ **25.** $\dfrac{3}{x - 2}$ **27.** $\dfrac{4}{a + 3}$ **29.** $\dfrac{2(x - 10)}{(x + 5)(x - 5)}$

31. $\dfrac{x + 2}{x + 3}$ **33.** $\dfrac{a + 1}{a + 2}$ **35.** $\dfrac{1}{(x + 3)(x + 4)}$ **37.** $\dfrac{y}{(y + 5)(y + 4)}$ **39.** $\dfrac{3(x - 1)}{(x + 4)(x + 1)}$ **41.** $\frac{1}{3}$

43. $\dfrac{x+1}{x}$ **45.** $\dfrac{x}{x+1}$ **47.** $\dfrac{x+3}{x+2}$ **49.** $\dfrac{x+2}{x+3}$ **51.** $x+\dfrac{2}{x}=\dfrac{x^2+2}{x}$ **53.** $\dfrac{1}{x}+\dfrac{1}{2x}=\dfrac{3}{2x}$ **55.** 3

57. -2 **59.** $\frac{1}{5}$ **61.** $-2, -3$ **63.** $3, -2$ **65.** $0, 5$

Problem Set 6.4

1. -3 **3.** 20   **5.** -1 **7.** 5   **9.** -2 **11.** 4   **13.** 3, 5   **15.** $-8, 1$ **17.** 2   **19.** 1
21. 8   **23.** 5   **25.** Possible solution 2, which does not check; $\varnothing$ **27.** 3
29. Possible solution -3, which does not check; $\varnothing$ **31.** 0   **33.** Possible solutions 2 and 3, but only 2 checks; 2
35. -4 **37.** -1 **39.** $-6, -7$ **41.** Possible solutions -3 and -1, but only -1 checks; -1 **43.** 7
45. Length 13 inches, width 4 inches   **47.** 6, 8 or $-8, -6$ **49.** 6 inches, 8 inches

Problem Set 6.5

1. $\dfrac{1}{x}+\dfrac{1}{3x}=\dfrac{16}{3}; \dfrac{1}{4}$ and $\dfrac{3}{4}$ **3.** $x+\dfrac{1}{x}=\dfrac{13}{6}; \dfrac{2}{3}$ and $\dfrac{3}{2}$ **5.** $\dfrac{7+x}{9+x}=\dfrac{5}{7}; -2$ **7.** $\dfrac{1}{x}+\dfrac{1}{x+2}=\dfrac{5}{12};$ 4 and 6

9. Let x = the speed of the boat in still water

	d	r	t
Upstream	26	$x-3$	$\dfrac{26}{x-3}$
Downstream	38	$x+3$	$\dfrac{38}{x+3}$

The equation is $\dfrac{26}{x-3}=\dfrac{38}{x+3}; x = 16$ miles per hour

11. 300 miles per hour   **13.** 170 miles per hour and 190 miles per hour   **15.** 9 miles per hour

17. 8 miles per hour   **19.** $\dfrac{1}{12}-\dfrac{1}{15}=\dfrac{1}{x};$ 60 hours   **21.** $\frac{60}{11}$ minutes   **23.** 12 minutes

25.

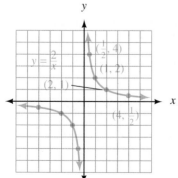

27.

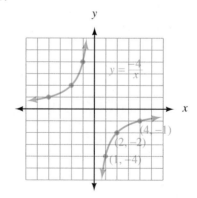

29.

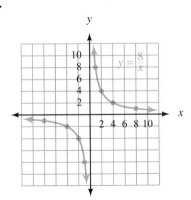

31.

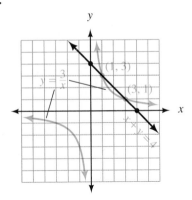

33. $(2, -1)$ **35.** $(4, 3)$ **37.** $(1, 1)$ **39.** $(5, 2)$

Problem Set 6.6

1. 6   **3.** $\frac{1}{6}$ **5.** xy^2 **7.** $\dfrac{xy}{2}$ **9.** $\dfrac{y}{x}$ **11.** $\dfrac{a+1}{a-1}$ **13.** $\dfrac{x+1}{2(x-3)}$ **15.** $a-3$ **17.** $\dfrac{y+3}{y+2}$

19. $x + y$ **21.** $\dfrac{a+1}{a}$ **23.** $\dfrac{1}{x}$ **25.** $\dfrac{2a+3}{3a+4}$ **27.** $\dfrac{x-1}{x+2}$ **29.** $\dfrac{x+4}{x+1}$ **31.** $\frac{7}{3}, \frac{17}{7}, \frac{41}{17}$ **33.** $x < 1$

35. $x \geq -7$ **37.** $x < 6$ **39.** $x \leq 2$

Problem Set 6.7

1. 1   **3.** 10   **5.** 40   **7.** $\frac{5}{4}$ **9.** $\frac{7}{3}$ **11.** $3, -4$ **13.** $4, -4$ **15.** $2, -4$ **17.** $6, -1$
19. 15 hits   **21.** 21 milliliters alcohol   **23.** 45.5 grams of fat   **25.** 14.7 inches   **27.** 343 miles

29. $\dfrac{x+2}{x+3}$ **31.** $\dfrac{2(x+5)}{x-4}$ **33.** $\dfrac{1}{x-4}$

Chapter 6 Review

1. $\dfrac{1}{2(x-2)}, x \neq 2$ **2.** $\dfrac{-1}{x+3}, x \neq -3$ **3.** $\dfrac{1}{a-6}, a \neq -6, 6$ **4.** $\dfrac{1}{a+9}, a \neq -9, 9$

5. $\dfrac{2x-1}{x+3}, x \neq -3$ **6.** $5, x \neq -3$ **7.** $\dfrac{1}{x+4}$ **8.** $x - 7$ **9.** $\dfrac{3x-2}{2x-3}$ **10.** $\dfrac{2x+3}{5x+2}$

11. $\dfrac{1}{(x-2)(x^2+4)}$ **12.** $\dfrac{1}{x^2+9}$ **13.** $x - 2$ **14.** $x - 3$ **15.** $a + 8$ **16.** $a - 6$ **17.** $\dfrac{y+b}{y+5}$

18. $\dfrac{2x+3}{3x+2}$ **19.** $\dfrac{x}{2}$ **20.** $\dfrac{x+1}{4(x+2)}$ **21.** $\dfrac{x+2}{x-4}$ **22.** $\dfrac{x(x-7)}{(x+2)(x+3)}$ **23.** $(a-6)^2$ **24.** $(a+8)^2$

25. $\dfrac{x-1}{x+2}$ **26.** $\dfrac{2x+1}{x+3}$ **27.** 1   **28.** 1   **29.** $x - 9$ **30.** $x - 7$ **31.** $\dfrac{13}{a+8}$ **32.** $\dfrac{-13}{a-7}$

33. $\dfrac{x^2+5x+45}{x(x+9)}$ **34.** $\dfrac{x^2-9x+27}{x(x-3)}$ **35.** $\dfrac{4x+5}{4(x+5)}$ **36.** $\dfrac{5x+3}{5(x+3)}$ **37.** $\dfrac{1}{(x+6)(x+2)}$

38. $\dfrac{1}{(x-2)(x-6)}$ **39.** $\dfrac{a-6}{(a+5)(a+3)}$ **40.** $\dfrac{6(2a+5)}{(a-9)(a+6)}$ **41.** 4   **42.** 6   **43.** 9   **44.** −7

45. 1, 6   **46.** −5, −7   **47.** Possible solution 2, which does not check; $\varnothing$

48. Possible solution 5, which does not check; $\varnothing$ **49.** 5   **50.** −6   **51.** 3 and $\frac{7}{3}$ **52.** 1 and 5

53. 15 miles per hour   **54.** 20 miles per hour   **55.** 84 hours   **56.** 4 hours

57.   **58.**

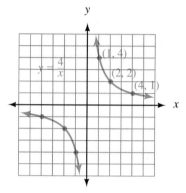
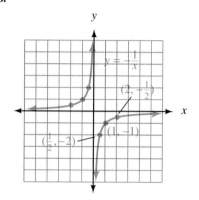

59. $\dfrac{x-2}{4(x+2)}$ **60.** $\dfrac{y+3}{y+7}$ **61.** $\dfrac{y-4}{y-8}$ **62.** $\dfrac{4a-7}{a-1}$ **63.** $\frac{2}{5}$ **64.** $\frac{4}{3}$ **65.** $\frac{2}{9}$ **66.** $\frac{1}{6}$ **67.** 12

68. $\frac{12}{7}$ **69.** −6, 6   **70.** −8, 8   **71.** −6, 8   **72.** −3, 6   **73.** 279 miles   **74.** 2,604 miles

Chapter 6 Test

1. $\dfrac{x+4}{x-4}$ **2.** $\dfrac{2}{a+2}$ **3.** $\dfrac{y+7}{x+a}$ **4.** 3   **5.** $(x-7)(x-1)$ **6.** $\dfrac{x+4}{3x-4}$ **7.** $(x-3)(x+2)$

8. $\dfrac{-3}{x-2}$ **9.** $\dfrac{2x+3}{(x+3)(x-3)}$ **10.** $\dfrac{3x}{(x+1)(x-2)}$ **11.** $\frac{11}{5}$ **12.** 1   **13.** 6   **14.** 15 miles per hour

15. 60 hours   **16.**   **17.** $\frac{1}{2}$ and $\frac{1}{3}$

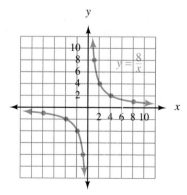

18. 132   **19.** $\dfrac{x+1}{x-1}$ **20.** $\dfrac{x+4}{x+2}$

CHAPTER 7

Problem Set 7.1

1. 3   **3.** −3   **5.** Not a real number   **7.** −12   **9.** 25   **11.** Not a real number   **13.** −8   **15.** −10
17. 35   **19.** 1   **21.** −2   **23.** −5   **25.** −1   **27.** −3   **29.** −2   **31.** x **33.** $3x$ **35.** xy
37. $a + b$ **39.** $7xy$ **41.** x **43.** $2x$ **45.** x^2 **47.** $6a^3$ **49.** $5a^4b^2$ **51.** x^2 **53.** $3a^4$
55. x^2 **57.** $3 + 4 = 7$ **59.** $\sqrt{25} = 5$ **61.** $12 + 5 = 17$ **63.** $\sqrt{169} = 13$ **65.** 6 and −1

67. 3 and −1   **69.** $x + 3$ **71.** 5 and 7   **73.** $r = \dfrac{\sqrt{65} - \sqrt{50}}{\sqrt{50}} \approx 0.14 = 14\%$

75. Approximately 0.095 or 9.5%   **77.** 5   **79.** $\sqrt{125} \approx 11.2$ **81.** 30 feet   **83.** $x - 4$ **85.** $\dfrac{2}{a - 2}$

87. $\dfrac{2x + 1}{x}$ **89.** $\dfrac{x + 2}{x + a}$

Problem Set 7.2

1. $2\sqrt{2}$ **3.** $2\sqrt{3}$ **5.** $2\sqrt[3]{3}$ **7.** $5x\sqrt{2}$ **9.** $3ab\sqrt{5}$ **11.** $3x\sqrt[3]{2}$ **13.** $4x^2\sqrt{2}$ **15.** $20\sqrt{5}$
17. $x\sqrt{7x}$ **19.** $2x^2\sqrt[3]{x}$ **21.** $6a^2\sqrt[3]{a^2}$ **23.** $4a\sqrt{5a}$ **25.** $15y\sqrt{2x}$ **27.** $14x\sqrt{3y}$ **29.** $\frac{4}{5}$ **31.** $\frac{2}{3}$

33. $\frac{2}{3}$ **35.** $\frac{2}{3}$ **37.** $2x$ **39.** $3ab$ **41.** $\dfrac{3x}{2y}$ **43.** $\dfrac{5\sqrt{2}}{3}$ **45.** $\sqrt{3}$ **47.** $\dfrac{8\sqrt{2}}{7}$ **49.** $\dfrac{12\sqrt{2x}}{5}$

51. $\dfrac{3a\sqrt{6}}{5}$ **53.** $\dfrac{15\sqrt{2}}{2}$ **55.** $\dfrac{14y\sqrt{7}}{3}$ **57.** $5ab\sqrt{2}$ **59.** $6x\sqrt{2y}$ **61.** $\dfrac{8xy\sqrt{3y}}{5}$

63. $\sqrt{12} \approx 3.464$; $2\sqrt{3} \approx 2(1.732) = 3.464$

65. $\frac{5}{4}$ seconds   **67.** $\dfrac{2(x + 5)}{x(x + 1)}$ **69.** $\dfrac{x + 1}{x - 1}$ **71.** $(x + 6)(x + 3)$

Problem Set 7.3

1. $\dfrac{\sqrt{2}}{2}$ **3.** $\dfrac{\sqrt{3}}{3}$ **5.** $\dfrac{\sqrt{10}}{5}$ **7.** $\dfrac{\sqrt{6}}{2}$ **9.** $\dfrac{2\sqrt{15}}{3}$ **11.** $\dfrac{\sqrt{30}}{2}$ **13.** 2   **15.** $\sqrt{7}$ **17.** $\sqrt{5}$ **19.** $2\sqrt{5}$

21. $2\sqrt{3}$ **23.** $\dfrac{\sqrt{7}}{2}$ **25.** $xy\sqrt{2}$ **27.** $\dfrac{x\sqrt{15y}}{3}$ **29.** $\dfrac{4a^2\sqrt{5}}{5}$ **31.** $\dfrac{6a^2\sqrt{10a}}{5}$ **33.** $\dfrac{2xy\sqrt{15y}}{3}$

35. $\dfrac{4xy\sqrt{5y}}{3}$ **37.** $\dfrac{18ab\sqrt{6b}}{5}$ **39.** $18x\sqrt{x}$ **41.** $\dfrac{\sqrt[3]{4}}{2}$ **43.** $\dfrac{\sqrt[3]{3}}{3}$ **45.** $\dfrac{\sqrt[3]{12}}{2}$ **47.** $\dfrac{\sqrt[3]{6}}{2}$

49. Both are 0.707   **51.** 6 miles   **53.**

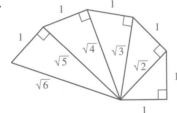

55. $\sqrt{2}, \sqrt{3}, \sqrt{4} = 2$ **57.** $10x$ **59.** $23x$ **61.** $10a$ **63.** $x + 5$ **65.** $\dfrac{5a + 6}{15}$ **67.** $\dfrac{1}{(a + 2)(a + 3)}$

Problem Set 7.4

1. $7\sqrt{2}$ **3.** $2\sqrt{5}$ **5.** $7\sqrt{3}$ **7.** $\frac{11}{56}\sqrt{5}$ **9.** $13\sqrt{13}$ **11.** $6\sqrt{10}$ **13.** $6\sqrt{5}$ **15.** $4\sqrt{2}$ **17.** 0
19. $-30\sqrt{3}$ **21.** $-6\sqrt{3}$ **23.** 0 **25.** $\frac{3}{2}\sqrt{2} + \frac{3}{2}\sqrt{3}$ **27.** 0 **29.** $\frac{23}{4}\sqrt{2}$ **31.** $-3\sqrt{7}$ **33.** $35\sqrt{3}$
35. $-6\sqrt{2} - 6\sqrt{3}$ **37.** $2x\sqrt{x}$ **39.** $4a\sqrt{3}$ **41.** $15x\sqrt{2x}$ **43.** $13x\sqrt{3xy}$ **45.** $-b\sqrt{5a}$
47. $27x\sqrt{2x} - 8x\sqrt{3x}$ **49.** $23x\sqrt{2y}$ **51.** $\dfrac{4 - \sqrt{6}}{3}$ **53.** $3 + \sqrt{2}$ **55.** $\dfrac{-2 + \sqrt{2}}{2}$

57. 3.968 is not equal to the decimal approximation of $\sqrt{8}$, which is 2.828 **59.** $9\sqrt{3}$
61. $9x^2 + 6xy + y^2$ **63.** $9x^2 - 16y^2$ **65.** 9 **67.** $2, 3$ **69.** Possible solutions 2 and 4; only 2 checks; 2

Problem Set 7.5

1. $\sqrt{6}$ **3.** $2\sqrt{3}$ **5.** $10\sqrt{21}$ **7.** $24\sqrt{2}$ **9.** $\sqrt{6} - \sqrt{2}$ **11.** $\sqrt{6} + 2$ **13.** $2\sqrt{6} + 3$ **15.** $6 - \sqrt{15}$
17. $2\sqrt{6} + 2\sqrt{15}$ **19.** $3 + 2\sqrt{2}$ **21.** $x + 6\sqrt{x} + 9$ **23.** $27 - 10\sqrt{2}$ **25.** $a - \sqrt{a} + \frac{1}{4}$
27. $16 + 6\sqrt{7}$ **29.** $11 + 5\sqrt{5}$ **31.** $-28 + \sqrt{2}$ **33.** $\sqrt{6} + \frac{1}{3}\sqrt{3} + \frac{1}{2}\sqrt{2} + \frac{1}{6}$ **35.** $x - 36$
37. $a + \sqrt{a} + \frac{2}{9}$ **39.** 1 **41.** $30 + \sqrt{7}$ **43.** $6x + 16\sqrt{x} + 8$ **45.** $49a - 4b$ **47.** $\dfrac{\sqrt{15} + \sqrt{6}}{3}$
49. $\dfrac{5 - \sqrt{10}}{3}$ **51.** $6 + 2\sqrt{5}$ **53.** $5 + 2\sqrt{6}$ **55.** $\dfrac{5 - \sqrt{21}}{2}$ **57.** $\dfrac{x + 4\sqrt{x} + 4}{x - 4}$
59. $\dfrac{5 - \sqrt{15} - \sqrt{10} + \sqrt{6}}{2}$ **61.** $6\sqrt{5}$ **63.** $52 + 14\sqrt{3}$ **65.** $-6, 1$ **67.** $0, 3$ **69.** $-9, 9$
71. $3, -5$ **73.** 500 miles

Problem Set 7.6

1. 3 **3.** 44 **5.** $\varnothing$ **7.** $\varnothing$ **9.** 8 **11.** 4 **13.** $\varnothing$ **15.** $-\frac{2}{3}$ **17.** 25 **19.** 4 **21.** 7
23. 8 **25.** $\varnothing$ **27.** Possible solutions 1 and 6; only 6 checks; 6   **29.** $-1, -2$ **31.** -4
33. Possible solutions 5 and 8; only 8 checks; 8   **35.** $0, 3$ **37.** $\varnothing$ **39.** 1 **41.** $x + 2 = \sqrt{8x}$; $x = 2$
43. $x - 3 = 2\sqrt{x}$; possible solutions 1 and 9; only 9 checks; 9   **45.** $\frac{392}{121} \approx 3.2$ feet
Note: Decimals in 47–51 have been rounded to the nearest tenth.

47.

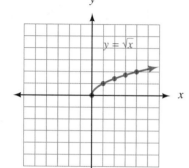

x	y
0	0
1	1
2	1.4
3	1.7
4	2

49.

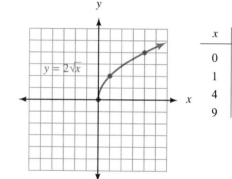

x	y
0	0
1	2
4	4
9	6

51.

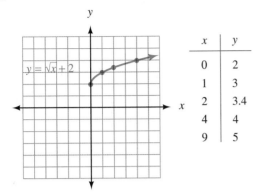

x	y
0	2
1	3
2	3.4
4	4
9	5

53. $\frac{3}{2}$ **55.** $\dfrac{a-1}{a+1}$ **57.** $\frac{1}{3}$ and 3   **59.** Walk 4 miles per hour and jog 8 miles per hour

61. $\frac{12}{7}$ minutes

Chapter 7 Review

1. 5   **2.** 9   **3.** 13   **4.** 20   **5.** −1   **6.** 4   **7.** 5   **8.** −3   **9.** $10xy^2$ **10.** $11x^2y$ **11.** $2a$

12. $-3a$ **13.** $2\sqrt{6}$ **14.** $2\sqrt{11}$ **15.** $2x\sqrt{15}$ **16.** $4x\sqrt{5}$ **17.** $3xy^2\sqrt{10x}$ **18.** $2x^2y\sqrt{30y}$

19. $-4\sqrt{2}$ **20.** $-6\sqrt{2}$ **21.** $6x\sqrt{5xy}$ **22.** $8xy\sqrt{7y}$ **23.** $\dfrac{\sqrt{3}}{7}$ **24.** $\dfrac{\sqrt{5}}{8}$ **25.** $\dfrac{2\sqrt{2}}{9}$ **26.** $\dfrac{3\sqrt{2}}{11}$

27. $\frac{7}{8}$ **28.** $\frac{11}{12}$ **29.** $\dfrac{7ab}{4}$ **30.** $\dfrac{10ab}{9}$ **31.** $\dfrac{4\sqrt{5}}{7}$ **32.** $\dfrac{10\sqrt{2}}{9}$ **33.** $\dfrac{2a\sqrt{10}}{11}$ **34.** $\dfrac{3a\sqrt{10}}{13}$

35. $\dfrac{10\sqrt{21}}{7}$ **36.** $\dfrac{12\sqrt{6}}{5}$ **37.** $\dfrac{6ab\sqrt{30}}{5}$ **38.** $\dfrac{40ab\sqrt{6}}{9}$ **39.** $\dfrac{-5xy\sqrt{5x}}{6}$ **40.** $\dfrac{-9xy\sqrt{5y}}{2}$ **41.** $\dfrac{2\sqrt{7}}{7}$

42. $\dfrac{5\sqrt{3}}{3}$ **43.** $\dfrac{4\sqrt{10}}{5}$ **44.** $\dfrac{2\sqrt{30}}{3}$ **45.** $\dfrac{\sqrt{15}}{12}$ **46.** $\dfrac{\sqrt{14}}{10}$ **47.** $-6\sqrt{3}$ **48.** $\dfrac{-20\sqrt{15}}{3}$

49. $\dfrac{4b\sqrt{6a}}{3}$ **50.** $\dfrac{2a\sqrt{70b}}{7}$ **51.** $\dfrac{\sqrt[3]{6}}{2}$ **52.** $\dfrac{\sqrt[3]{18}}{3}$ **53.** $\dfrac{-3\sqrt{3}-12}{13}$ **54.** $\dfrac{-\sqrt{5}+1}{2}$ **55.** $3-\sqrt{7}$

56. $-12-6\sqrt{5}$ **57.** $\sqrt{5}+\sqrt{2}$ **58.** $\dfrac{5\sqrt{7}-5\sqrt{3}}{4}$ **59.** $\dfrac{-\sqrt{15}-5}{2}$ **60.** $\dfrac{\sqrt{10}-2}{3}$ **61.** $\dfrac{7-2\sqrt{10}}{3}$

62. $\dfrac{5+\sqrt{21}}{2}$ **63.** $\dfrac{x+6\sqrt{x}+9}{x-9}$ **64.** $\dfrac{x-4\sqrt{x}+4}{x-4}$ **65.** $-4\sqrt{5}$ **66.** $4\sqrt{11}$ **67.** $-11\sqrt{3}$

68. $95\sqrt{2}$ **69.** $-22\sqrt{5}$ **70.** $2\sqrt{3}$ **71.** $5x\sqrt{2}$ **72.** $13x\sqrt{7}$ **73.** $-ab\sqrt{10a}$ **74.** $-7ab\sqrt{11}$

75. $3+3\sqrt{3}$ **76.** $5-3\sqrt{5}$ **77.** $4\sqrt{6}+4\sqrt{10}$ **78.** $3\sqrt{35}-3\sqrt{14}$ **79.** $x-49$ **80.** $x-16$

81. $2\sqrt{5}-2$ **82.** $27-3\sqrt{7}$ **83.** $x+10\sqrt{x}+25$ **84.** $x-4\sqrt{x}+4$ **85.** 12   **86.** 21

87. 7   **88.** 11   **89.** 16   **90.** 81   **91.** No solution   **92.** No solution   **93.** 0, 5

94. Possible solutions 3 and 1; only 3 checks; 3   **95.** $x=\sqrt{3}$ **96.** $8\sqrt{2}$

97.

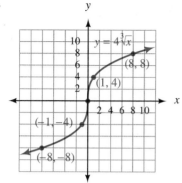

98.

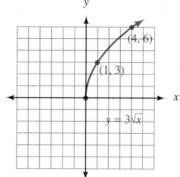

99.

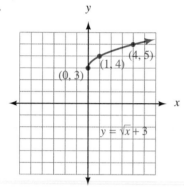

100.

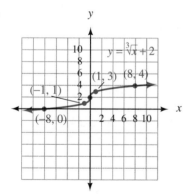

Chapter 7 Test

1. 4   **2.** -6 **3.** 7, -7 **4.** 3   **5.** -2 **6.** -3 **7.** $5\sqrt{3}$ **8.** $4\sqrt{2}$ **9.** $\dfrac{\sqrt{6}}{3}$ **10.** $\dfrac{\sqrt[3]{2}}{2}$

11. $15x\sqrt{2}$ **12.** $\dfrac{2xy\sqrt{15y}}{5}$ **13.** $4\sqrt{3}$ **14.** $11x\sqrt{2}$ **15.** $\sqrt{15} - 2\sqrt{3}$ **16.** $-51 - \sqrt{5}$

17. $x - 36$ **18.** $8 - 2\sqrt{15}$ **19.** $\dfrac{5 - \sqrt{21}}{2}$ **20.** $\dfrac{x - 5\sqrt{x}}{x - 25}$ **21.** 12   **22.** $\varnothing$

23. Possible solutions 2 and 6; only 6 checks; 6   **24.** Possible solutions 1 and 16; only 16 checks; 16

25. $x = \sqrt{6}$ **26.**

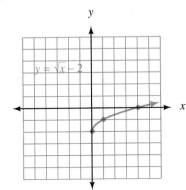

x	y
0	−2
1	−1
4	0
9	1
16	2

CHAPTER 8

Problem Set 8.1

1. ± 3 **3.** ± 5 **5.** $\pm 2\sqrt{2}$ **7.** $\pm 5\sqrt{2}$ **9.** $\pm 3\sqrt{2}$ **11.** $0, -4$ **13.** $4, -6$ **15.** $5 \pm 5\sqrt{3}$

17. $-1 \pm 5\sqrt{2}$ **19.** $2, -3$ **21.** $\frac{11}{4}, -\frac{1}{4}$ **23.** $\dfrac{1 \pm 2\sqrt{3}}{3}$ **25.** $\dfrac{-2 \pm 3\sqrt{3}}{6}$ **27.** $3 \pm \sqrt{3}$

29. $-2 \pm \sqrt{5}$ **31.** $2 \pm \sqrt{2}$ **33.** $\frac{7}{3}, -1$ **35.** $\dfrac{-1 \pm \sqrt{7}}{2}$ **37.** $\dfrac{4 \pm 2\sqrt{3}}{5}$ **39.** $-5 \pm \sqrt{7}$

41. $4, -2$ **43.** $-6 \pm 2\sqrt{2}$ **45.** $(-1 + 5\sqrt{2} + 1)^2 = (5\sqrt{2})^2 = 50$ **47.** $-7, 1$

49. $r = -1 + \dfrac{\sqrt{A}}{10}$ **51.** $5\sqrt{3}$ feet

53. No, the height is $3\sqrt{3}$ feet ≈ 5.20 feet, which is less than 5 feet 8 inches $= 5\frac{8}{12}$ feet $= \approx 5.67$ feet

55. 3 feet   **57.** $x^2 + 6x + 9$ **59.** $x^2 - 10x + 25$ **61.** $(x - 6)^2$ **63.** $(x + 2)^2$ **65.** 2

67. 2

Problem Set 8.2

1. 9   **3.** 1   **5.** 16   **7.** 1   **9.** 64   **11.** $\frac{9}{4}$ **13.** $\frac{49}{4}$ **15.** $\frac{1}{4}$ **17.** $\frac{9}{16}$ **19.** $-6, 2$ **21.** $-2, 8$

23. $-3, 1$ **25.** $0, 10$ **27.** $-5, 3$ **29.** $-2 \pm \sqrt{7}$ **31.** $2 \pm 2\sqrt{2}$ **33.** $8, -1$ **35.** $-1 \pm \sqrt{2}$

37. $-2, 1$ **39.** $\dfrac{-2 \pm \sqrt{3}}{2}$ **41.** $\dfrac{1 \pm \sqrt{3}}{2}$ **43.** $\frac{1}{2}$ **45.** $\dfrac{9 \pm \sqrt{105}}{6}$

47. $-2 + 2.646 = 0.646; -2 - 2.646 = -4.646$ **51.** $(-2 + \sqrt{7}) + (-2 - \sqrt{7}) = -4$

53.   **55.**

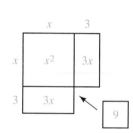   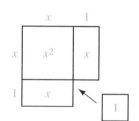

57. 4   **59.** -24 **61.** $2\sqrt{10}$ **63.** $2\sqrt{3}$ **65.** $2xy\sqrt{5y}$ **67.** $\frac{9}{5}$

Problem Set 8.3

1. $-1, -2$ **3.** $-2, -3$ **5.** -3 **7.** $-3 \pm \sqrt{2}$ **9.** $-1, -\frac{3}{2}$ **11.** $\dfrac{-2 \pm \sqrt{3}}{2}$ **13.** 1

15. $\dfrac{5 \pm \sqrt{53}}{2}$ **17.** $\frac{2}{3}, -\frac{1}{2}$ **19.** $\dfrac{1 \pm \sqrt{21}}{2}$ **21.** $\dfrac{-1 \pm \sqrt{57}}{4}$ **23.** $\frac{5}{2}, -1$ **25.** $\dfrac{-3 \pm \sqrt{23}}{2}$

27. $\dfrac{-2 \pm \sqrt{10}}{3}$ **29.** $\dfrac{1 \pm \sqrt{11}}{2}$ **31.** $0, \dfrac{-3 \pm \sqrt{41}}{4}$ **33.** $0, \frac{4}{3}$ **35.** $\dfrac{3 \pm \sqrt{21}}{6}$ **37.** $6\sqrt{15}$

39. $-4 - 3\sqrt{6}$ **41.** 5   **43.** $\dfrac{3 - \sqrt{5}}{2}$

Problem Set 8.4

1. $3 + i$ **3.** $6 - 8i$ **5.** 11   **7.** $9 + i$ **9.** $-1 - i$ **11.** $11 - 2i$ **13.** $4 - 2i$ **15.** $8 - 7i$
17. $-1 + 4i$ **19.** $6 - 3i$ **21.** $14 + 16i$ **23.** $9 + 2i$ **25.** $11 - 7i$ **27.** 34   **29.** 5

31. $\dfrac{6 + 4i}{13}$ **33.** $\dfrac{-9 - 6i}{13}$ **35.** $\dfrac{-3 + 9i}{5}$ **37.** $\dfrac{3 + 4i}{5}$ **39.** $\dfrac{-6 + 13i}{15}$ **41.** $x^2 + 9$

43. $\dfrac{1}{i} \cdot \dfrac{i}{i} = \dfrac{i}{i^2} = \dfrac{i}{-1} = -i$ **45.** $8, -2$ **47.** $5, 1$ **49.** $-3 \pm 2\sqrt{3}$ **51.** $\dfrac{\sqrt{2}}{2}$ **53.** $\dfrac{2xy\sqrt{6y}}{3}$

55. $\dfrac{\sqrt[3]{2}}{2}$

Problem Set 8.5

1. $4i$ **3.** $7i$ **5.** $i\sqrt{6}$ **7.** $i\sqrt{11}$ **9.** $4i\sqrt{2}$ **11.** $5i\sqrt{2}$ **13.** $2i\sqrt{2}$ **15.** $4i\sqrt{3}$ **17.** $1 \pm i$

19. 2   **21.** $\frac{3}{2}, -4$ **23.** $2 \pm 2i$ **25.** $\dfrac{-1 \pm 3i}{2}$ **27.** $\dfrac{1 \pm i\sqrt{3}}{2}$ **29.** $\dfrac{-1 \pm i\sqrt{3}}{2}$ **31.** $2, 3$

33. $\dfrac{-1 \pm i\sqrt{2}}{3}$ **35.** $\frac{1}{2}, -2$ **37.** $\dfrac{1 \pm 3\sqrt{5}}{2}$ **39.** $4 \pm 3i$ **41.** $\dfrac{4 \pm \sqrt{22}}{2}$ **43.** Yes   **45.** $3 - 7i$

47.

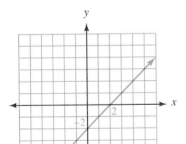

49.

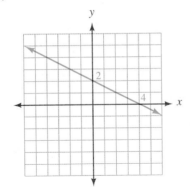

51. $23\sqrt{2}$

53. $-6\sqrt{6}$ **55.** $2x\sqrt{3}$

Problem Set 8.6

1.

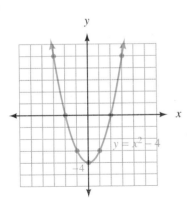

3.

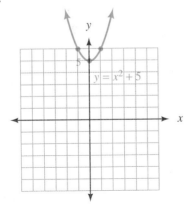

5.

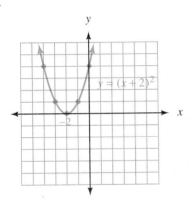

7.

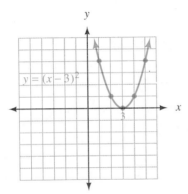

9.

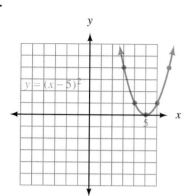

$y = (x - 5)^2$

11.

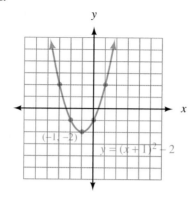

$(-1, -2)$

$y = (x + 1)^2 - 2$

13.

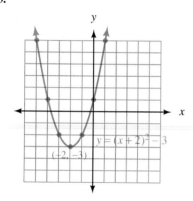

$y = (x + 2)^2 - 3$

$(-2, -3)$

15.

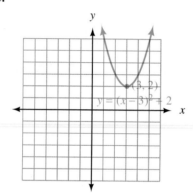

$(3, 2)$

$y = (x - 3)^2 + 2$

17.

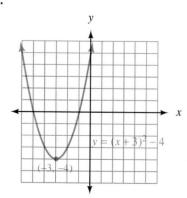

$y = (x + 3)^2 - 4$

$(-3, -4)$

19.

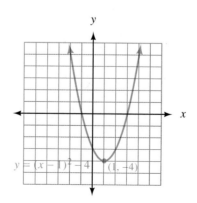

$y = (x - 1)^2 - 4$ $(1, -4)$

21.

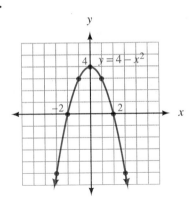

23.

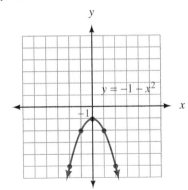

25.

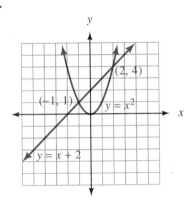

27.

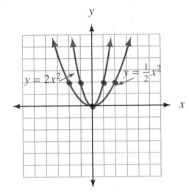

29. 11   **31.** ∅   **33.** $\frac{62}{3}$ **35.** Possible solutions -1 and 6; only 6 checks; 6

Chapter 8 Review

1. $\pm 4\sqrt{2}$ **2.** $\pm 2\sqrt{15}$ **3.** ± 4 **4.** ± 3 **5.** $-9, 3$ **6.** $-7, 11$ **7.** $-2, \frac{2}{3}$ **8.** $-2, 5$

9. $\dfrac{-5 \pm 4\sqrt{2}}{2}$ **10.** $\dfrac{4 \pm 3\sqrt{3}}{3}$ **11.** $\dfrac{2 \pm 5i}{3}$ **12.** $\dfrac{-1 \pm 3i\sqrt{2}}{2}$ **13.** $-4 \pm 2\sqrt{5}$ **14.** $4 \pm 2\sqrt{5}$

15. $2 \pm \sqrt{11}$ **16.** $-3, -1$ **17.** $\dfrac{9 \pm \sqrt{93}}{2}$ **18.** $-1, 6$ **19.** $-3, 1$ **20.** $\dfrac{3 \pm \sqrt{15}}{3}$ **21.** $-4, -3$

22. 4   **23.** $\dfrac{-5 \pm i\sqrt{3}}{2}$ **24.** $-2 \pm \sqrt{6}$ **25.** $\dfrac{-4 \pm \sqrt{26}}{2}$ **26.** $\dfrac{-3 \pm i\sqrt{39}}{6}$ **27.** $-\frac{1}{2}, 3$

28. $\dfrac{1 \pm i\sqrt{17}}{3}$ **29.** $\dfrac{1 \pm i\sqrt{2}}{2}$ **30.** $\dfrac{2 \pm \sqrt{5}}{2}$ **31.** $4 + 2i$ **32.** $2 - 8i$ **33.** $10 + 5i$ **34.** $5 - 2i$

35. $-i$ **36.** $-1 - 5i$ **37.** $-1 - 3i$ **38.** $1 - 6i$ **39.** $2 - 2i$ **40.** $1 + 3i$ **41.** $6 - 2i$
42. $-24 - 12i$ **43.** $20 + 24i$ **44.** $18 - 21i$ **45.** $7 - i$ **46.** $19 - 17i$ **47.** $8 - 6i$

48. $-16 + 30i$ **49.** 17   **50.** 13   **51.** $\dfrac{1 + 3i}{10}$ **52.** $\dfrac{-1 + 2i}{5}$ **53.** $\dfrac{10 - 25i}{29}$ **54.** $\dfrac{9 + 12i}{25}$

55. $\dfrac{-2 + 8i}{17}$ **56.** $\dfrac{6 - 9i}{13}$ **57.** $\dfrac{4 + 3i}{5}$ **58.** $\dfrac{-9 - 40i}{41}$ **59.** $\dfrac{-9 + 19i}{34}$ **60.** $\dfrac{17 - 7i}{26}$ **61.** $6i$

62. $12i$ **63.** $i\sqrt{17}$ **64.** $i\sqrt{31}$ **65.** $2i\sqrt{10}$ **66.** $6i\sqrt{2}$ **67.** $10i\sqrt{2}$ **68.** $11i\sqrt{2}$

69.

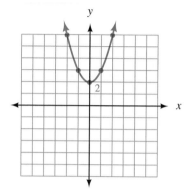

70.

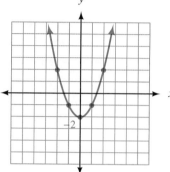

71.

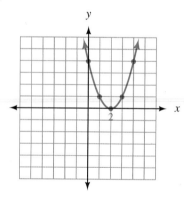

72.

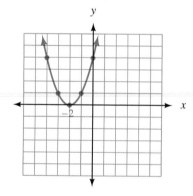

73.

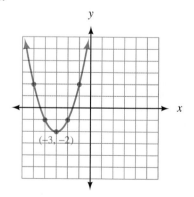

74.

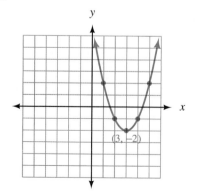

75.

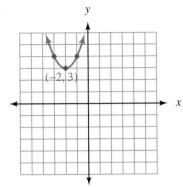

76.

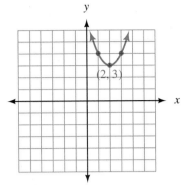

Chapter 8 Test

1. $-1, 8$ **2.** $3 \pm 2\sqrt{3}$ **3.** $\dfrac{5 \pm 5i\sqrt{3}}{2}$ **4.** $\dfrac{3 \pm i\sqrt{31}}{4}$ **5.** $\tfrac{1}{3}, -1$ **6.** $\dfrac{-1 \pm \sqrt{33}}{2}$ **7.** $-\tfrac{2}{3}$

8. $3 \pm \sqrt{15}$ **9.** $3i$ **10.** $11i$ **11.** $6i\sqrt{2}$ **12.** $3i\sqrt{2}$ **13.** $3 + 8i$ **14.** $-1 + 2i$ **15.** 5

16. $1 + 5i$ **17.** $\dfrac{-1 + 3i}{10}$ **18.** $\dfrac{3 + 4i}{5}$

19.

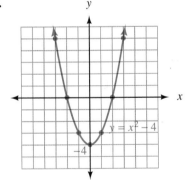

20.

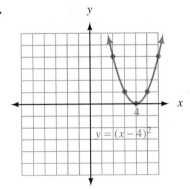

21.

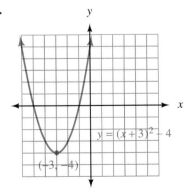

22.

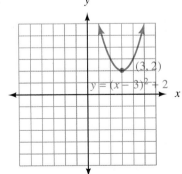

CHAPTER 9

Problem Set 9.1

1.

3.

5.

7.

9.

11.

13.

15.

17.

19.

21.

23.

25.

27.

29.

31.

33.

35.

37.

39.

41. $5 < x < 15$ **43.** $4 < x < 5$ **45.** The width is between 3 inches and $\frac{11}{2}$ inches   **47.** -5 **49.** 5

51. -7 **53.** 9   **55.** 6   **57.** $2x - 3$ **59.** $-3, 0, 2$ **61.** $-6x + 11$ **63.** -8 **65.** -4

67. $w = \dfrac{P - 2l}{2}$ **69.** $x < -1$    **71.** $y \geq \frac{3}{2}x - 6$

73. Width is 2 inches; length is 11 inches

Problem Set 9.2

1.

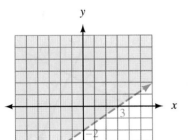

3.

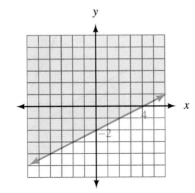

5.

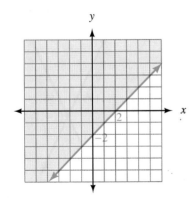

7.

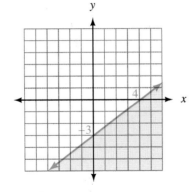

9.

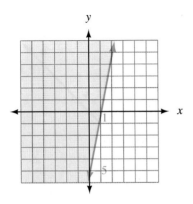

11.

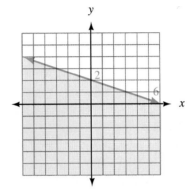

13.

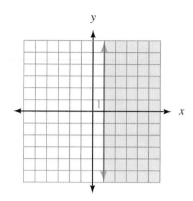

15.

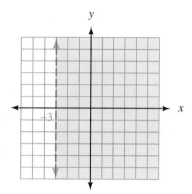

17.

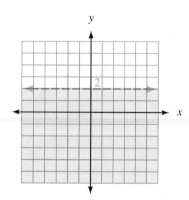

19.

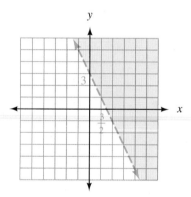

21.

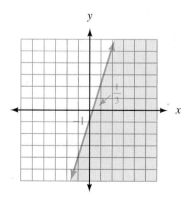

23.

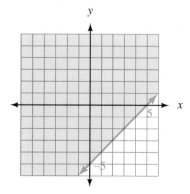

25.

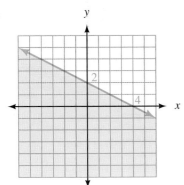

27.

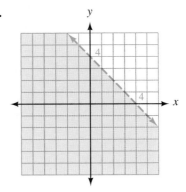

29. $(-2, 2)$ $(0, 3)$ $(2, 4)$ **31.**

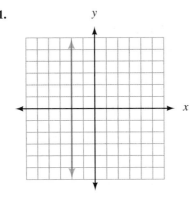

33. -1

35. $y = -3x + 1$ **37.** $(5, -7)$ **39.** $(3, 17)$ **41.** $800 at 8\%, $400 at 10\%

Problem Set 9.3

1. $x^2 - 16$ **3.** $4x^2 - 81$ **5.** $x^3 + 8$ **7.** $x^3 - 1$ **9.** $x^3 + 64$ **11.** $8x^3 - 27$ **13.** $x^3 + y^3$
15. $(x - 2)(x^2 + 2x + 4)$ **17.** $(x + 7)(x^2 - 7x + 49)$ **19.** $(3x - 5y)(9x^2 + 15xy + 25y^2)$
21. $(x + 2)(x^2 - 2x + 4)$ **23.** $(x - 5)(x^2 + 5x + 25)$ **25.** $(x + 1)(x^2 - x + 1)$
27. $(2x + y)(4x^2 - 2xy + y^2)$ **29.** $(3x - 2y)(9x^2 + 6xy + 4y^2)$ **31.** $(5x + 2)(25x^2 - 10x + 4)$
33. $(x + 4y)(x^2 - 4xy + 16y^2)$ **35.** $200x^{24}$ **37.** x^7 **39.** 8×10^1 **41.** $10ab^2$
43.

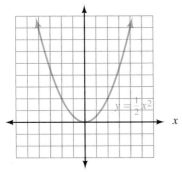

45. $6x^4 + 6x^3 - 2x^2$ **47.** $9y^2 - 30y + 25$

49. $4a^4 - 49$ **51.** $x + 2 + \dfrac{14}{x - 4}$ **53.** $(a + 5)(3x - 2)$ **55.** $(2x - 5y)^2$ **57.** $(2x + 7y)(x - 3y)$

59. $3, 6$ **61.** $\frac{5}{4}, -\frac{3}{2}$ **63.** $4, 6$ or $-2, 0$

Problem Set 9.4

1. 8   **3.** 130   **5.** 5   **7.** -3 **9.** 3   **11.** 243   **13.** 2   **15.** $\frac{1}{2}$ **17.** 1   **19.** 5   **21.** 25

23. 9   **25.** 84 pounds   **27.** $\frac{735}{2}$ or 367.5   **29.** \$235.50   **31.** 96 pounds   **33.** 12 amps   **35.** $\dfrac{x + 2}{x + 3}$

37. $\dfrac{2(x + 5)}{x - 4}$ **39.** $\dfrac{1}{x - 4}$ **41.** $\dfrac{x + 5}{x - 3}$ **43.** -2 **45.** 24 hours

Problem Set 9.5

1. $D = \{1, 3, 5\}, R = \{2, 4, 6\}$ **3.** $D = \{1, 3, 4\}, R = \{a, b, c\}$ **5.** $D = \{1, 2, 3, 4\}, R = \{1, 2, 3, 4\}$
7. $D = \{1, 3, 5, 7, 9\}, R = \{2, 8, 14, 20, 26\}$ **9.** $D = \{0, 2, 4\}, R = \{2, \frac{2}{3}, -\frac{2}{3}\}$
11. $D = \{3, 4, 5, 6, 7\}, R = \{2, 4, 6, 8, 10\}$ **13.** 2   **15.** -3 **17.** $-1, -2$ **19.** $-5, 1$ **21.** $3, -3$
23. $2, 3$ **25.** $-\frac{1}{2}, 3$ **27.** $0, 2, 4$ **29.** $-3, -5, -6, a - 5$ **31.** $-1, 1, -2, 2z - 3$
33. $5, 10, 17, a^2 + 1$ **35.** $-1, -\frac{1}{2}, \frac{1}{3}, 2, \frac{3}{4}$ **37.** $15, 0, a^2 + 2a, a^2 - 2a$ **39.** $0, 7, -28, a^3 - 1$ **41.** 7
43. $5\sqrt{2}$ **45.** $\dfrac{\sqrt{10}}{5}$ **47.** $21\sqrt{3}$ **49.** $-4 - 3\sqrt{6}$ **51.** $4\sqrt{5} + 4\sqrt{3}$ **53.** 7

Problem Set 9.6

1. 2   **3.** 4   **5.** 3   **7.** 5   **9.** 3   **11.** 9   **13.** 4   **15.** 25   **17.** 8   **19.** 64   **21.** 8   **23.** 4
25. -2 **27.** 5   **29.** 17   **31.** 6   **33.** x **35.** x^2 **37.** $a^{2/5}$ **39.** $3y^2$ **41.** $3a^2b$ **43.** x

45. $\frac{1}{5}$ **47.** $\frac{1}{4}$ **49.** $\frac{1}{9}$ **51.** $-3, 2$ **53.** $\dfrac{2 \pm \sqrt{10}}{3}$ **55.** $\dfrac{2 \pm i\sqrt{3}}{2}$ **57.** $2 + 8i$ **59.** $23 - 2i$

Chapter 9 Review

1.

2.

3.

4.

5.

6.

7.

8.

9.

10.

11.

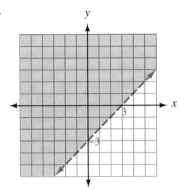

12.

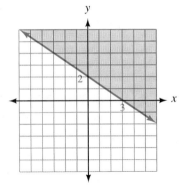

13.

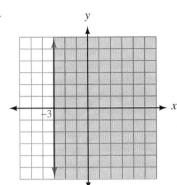

14.

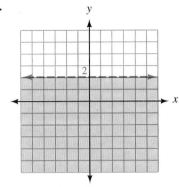

15.

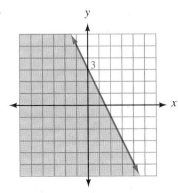

16.

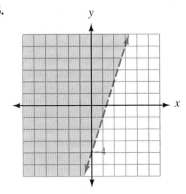

17. $(a - 7)(a^2 + 7a + 49)$ **18.** $(a + 6)(a^2 - 6a + 36)$ **19.** $(x - 2)(x^2 + 2x + 4)$
20. $(x + 2)(x^2 - 2x + 4)$ **21.** $(3x + y)(9x^2 - 3xy + y^2)$ **22.** $(3x - y)(9x^2 + 3xy + y^2)$
23. $(4x + 5)(16x^2 - 20x + 25)$ **24.** $(2x - 3)(4x^2 + 6x + 9)$ **25.** -35 **26.** 125 **27.** $\frac{1}{2}$
28. 4   **29.** 9   **30.** 30 amps   **31.** $D = \{2, -3, 4\}, R = \{1, 5, -2\}$
32. $D = \{-3, 1, -5\}, R = \{-2, -4\}$ **33.** $D = \{2, 4, 6\}, R = \{5, 11, 17\}$
34. $D = \{-2, -1, 0\}, R = \{-1, 1, 3\}$ **35.** $-3, -7$ **36.** $-5, 2$ **37.** $3, 5$ **38.** $-4, 7$
39. $f(-3) = 6, f(0) = 0, f(1) = -2$ **40.** $f(-2) = 3, f(1) = -3, f(3) = -7$
41. $f(-2) = 9, f(2) = 9, f(a) = 2a^2 + 1$ **42.** $f(-1) = -4, f(3) = -20, f(c) = -3c^2 + 2c + 1$
43. $f(-2) = -\frac{1}{2}, f(4) = \frac{1}{4}, f(\frac{1}{2}) = 2, f(\frac{2}{3}) = \frac{3}{2}$ **44.** $f(36) = 6, f(81) = 9, f(b) = \sqrt{b}$

45. $f(-2) = -7, f(2) = 9, f(a) = a^3 + 1$ **46.** $f(-1) = -4, f(3) = 24, f(b) = b^3 - 3$ **47.** 6
48. 10   **49.** 9   **50.** 16   **51.** 4   **52.** 4   **53.** 41   **54.** 2   **55.** $2y^5$ **56.** $3y^5$
57. $\frac{1}{3}$ **58.** $\frac{1}{4}$

Chapter 9 Test

1.
```
◄━━━━⊕━━━━━━━⊕━━►
      2        3
```

2.
```
◄━━━━●━━━━━━━●━━►
     -2        3
```

3.
```
◄━━━━●━━━━━━━●━━►
      2        5
```

4.
```
◄━━━━⊕━━━━━━━⊕━━►
     -3        5
```

5.

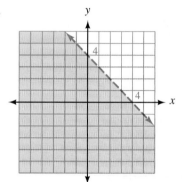

6.

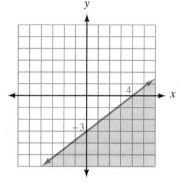

7. $(x - 7)(x^2 + 7x + 49)$ **8.** $(x + 3)(x^2 - 3x + 9)$ **9.** $(2x + 5y)(4x^2 - 10xy + 25y^2)$
10. $(2x - 3y)(4x^2 + 6xy + 9y^2)$ **11.** 100   **12.** 2   **13.** $D = \{1, 3, 5\}, R = \{2, 4, 6\}$
14. $x \neq 5$ and $x \neq -5$ **15.** 26   **16.** 11   **17.** 2   **18.** -23 **19.** 5   **20.** 4   **21.** x **22.** $3x^2y^3$

INDEX